MW01048302

DISCRETE MATHEMATICS

Proof Techniques and Mathematical Structures

DISCRETE MATHEMATICS

Proof Techniques and Mathematical Structures

R C Penner

Departments of Mathematics and Physics
University of Southern California

Published by

World Scientific Publishing Co. Pte. Ltd.
P O Box 128, Farrer Road, Singapore 912805
USA office: Suite 1B, 1060 Main Street, River Edge, NJ 07661
UK office: 57 Shelton Street, Covent Garden, London WC2H 9HE

Library of Congress Cataloging-in-Publication Data
Penner, R. C., 1956–
Discrete mathematics : proof techniques and mathematical structures / R. C. Penner.
p. cm.
Includes bibliographical references and index.
ISBN 9810240880
1. Mathematics. 2. Computer science--Mathematics. I. Title.
QA39.2.P44 1999
510.21--dc21 99-044793
CIP

British Library Cataloguing-in-Publication Data
A catalogue record for this book is available from the British Library.

First published 1999
Reprinted 2001

Printed in Singapore by Uto-Print

Dedicated to my mother,
Beverly Preston Penner,
a teacher

Written for my children,
Kirby and Michael

Contents

PART 2

General Preface

I am neither joking nor exaggerating when I say that there are no real pre-requisites for reading this book: Chapter 1 begins *tabula rasa*, that is, with a clean slate, except for a passing familiarity with basic mathematical results from high school such as the quadratic formula, Pythagoras' theorem, and so on (and these results are actually used only for motivation in the early chapters).

At the same time that the pre-requisites are minimal, the goal is no less than to teach the "beginning" math student how to discover, prove, and check mathematics as well as how to write down proofs; this involves not only the discovery and exposition of proofs but also learning the basic constructions and vocabulary of mathematics. Often at American colleges and universities, this is the purview of the first post-calculus mathematics course often for computer science as well as math majors (and sometimes it is hurriedly treated as a first topic in a junior- or senior-level course on modern algebra). On the other hand, this volume does not depend at all upon differential calculus (the only exception being that certain exercises, which are independent of the main text, sometimes do depend upon calculus). The book could quite well be studied much earlier though it does assume a modicum of that elusive attribute "intellectual maturity" on the part of the reader. It is also a suitable text for an upper-division college course with several different possible emphases.

This text has two parts, and Part 1 is the "core" of a one-semester course upon which this volume is based which covers proofs and the foundations of discrete mathematics. Part 2 consists of several independent chapters which apply the constructions and techniques of Part 1 to a more advanced study of various topics in pure math, applied math, and computer science. Part 2 is certainly more advanced and intrinsically interesting than Part 1, and the reader might rightly regard studying Part 1 as the price to be paid in order that Part 2 be accessible.

At the same time that there have been the two goals of explaining proofs and teaching the basic constructions of mathematics, there have likewise been two guiding virtues: truth and beauty. By this I mean that there have been no pedagogical short-cuts or lies (for how can one teach intellectual rigor without being explicit and honest about the assumptions?), and I have at each stage strived to go beyond the standard material and naively present more advanced and "beautiful" mathematics than is usually covered in a course at this level in both Parts 1 and 2. This deeper and more interesting material is often covered in optional sections, which can be judiciously included to tailor a suitable course for a particular audience. This has actually been another guiding principle: to keep the text as flexible as possible by allowing many different courses based upon different paths through the volume. Thus, this text surely contains more material than any particular syllabus based upon it, I believe this is appropriate and desirable, and the reader is heartily encouraged to "poke around" in this volume beyond any assigned reading.

There are two acknowledgements I want to make here, one from my mathematical childhood (by which I mean my undergraduate days) and one more recent. For the former, I thank my long-ago teachers Anil Nerode and Michael Morley for instilling my abiding love for the foundations of mathematics, and I can only hope that this volume might inspire a student as Professor Nerode's notes on related material once inspired me on to a career in research mathematics.

More recently, I have taught the course *Proof Techniques and Mathematical Structures* at the University of Southern California often over the last decade, and I thank the many students who have helped in the evolution of this manuscript as well as various colleagues at USC, especially the late Dennis Estes, for valuable input. Let me finally acknowledge that work on this volume has been partially supported by the National Science Foundation.

Preface for Instructors

As mentioned above, Part 1 more or less covers the "core" of a one-semester course on the foundations of math. (As an aside, we note that this book has also been used successfully for independent reading both by strong undergraduates and as a remedial for graduate students.) This foundational material includes predicate calculus, set theory, elementary number theory, relations, and functions. At the college level, there has typically been sufficient time in this course for a "final topic" following this core material, which can either be culled from the independent chapters of Part 2 or else treated concurrently as included optional sub-topics from Part 1. These optional sections in Part 1 are marked with an asterisk (*), and they may be skipped entirely without affecting the continuity of core sections. (The only logical dependence is that later optional sections may depend upon earlier ones but only within the same chapter.) There are many sensible paths through this volume, therefore, by judiciously including various consecutive optional sections within a chapter from Part 1, or by omitting any or all such optional sections and instead moving on to more advanced topics in Part 2.

The various chapters in Part 2 are entirely independent, and each is organized so that the sections become increasingly more sophisticated; we have tried to include numerous intellectually satisfying places within each chapter in Part 2 at which a course might end. The final sections of a chapter in Part 2 marked with a dagger (†) are entirely independent from one another, and each treats a major principle or proves a major theorem. Thus, a chapter or various fractions of various chapters from Part 2 could reasonably be included in a syllabus.

Though the topics and general flow of Part 1 can be gleaned by perusing the Table of Contents, we next serially discuss aspects of each chapter with an emphasis on the optional sections.

Chapter 1 contains Parts A and B, and Part A is a short and gentle introduction to induction and to basic concepts in logic. Part B discusses methods of proof giving numerous classical examples. We have typically lectured only selectively and briefly on the material in Part B and assigned the bulk of this material to independent reading in the first week or so of class. The idea is that with just a minimal background, essentially all of the standard methods of proof can be palatably described and presented. At the same time, comprehensively presenting the predicate calculus before discussing methods of proof tends to overwhelm the student, who may lose sight of the forest for the trees. In particular, it is useful and practical to present a first view of induction in this somewhat informal setting. We have included general exercises on proof techniques (though we usually do not assign such problems ourselves for reasons mentioned parenthetically above).

Chapter 2 is dedicated to a complete treatment of the predicate calculus with an emphasis on the nuts and bolts of calculational techniques. The section §2.6 on Aristotelean logic is optional because it is too elementary for many students (and gives practice in "translating" between English and the predicate calculus for whomever it may be necessary) and is a sensible reading assignment. §2.8 on commutativity and associativity is "extremely optional" and should almost certainly be relegated to reading or ignored altogether; it is included only for (the perspicacious student who may demand) completeness.

Chapter 3 treats set theory, and the optional §3.3 on ZFC gives a complete treatment of Zermelo-Fraenkel set theory and the Axiom of Choice; actually, most of the axioms of ZFC are presented in the (non-optional) previous section since they are required for our careful treatment of sets. Furthermore, ZFC gives a nice application of the predicate calculus as well as a nice example of an axiomatic mathematical definition. (In some texts at this level, the real numbers are given as the prototypical example of an axiomatic definition, and we believe this is overwhelming for the student; furthermore, a careful definition of the reals is the purview of a first course in analysis or topology in our opinion. The real numbers are employed in this book just in order to give illustrative examples of various constructions, we require essentially no formal properties of them, and hence the term "discrete mathematics" in our title.) We also allow ourselves a certain license in discussing philosophical aspects of these foundational issues. The other optional section §3.7 gives a complete and careful definition of the natural numbers and includes a survey of recursion and arithmetic (again for that perspicacious student).

In Chapter 4, we break from an exposition of the foundations of mathematics to discuss elementary number theory. The point of this chapter is to convince the reader of the efficacy and power of what has come before in the context of a topic which the reader presumably "mastered" in elementary school. The chapter requires roughly two weeks of lectures, begins with the basics, and ends with a brief discussion of more advanced material, namely, the Euclidean algorithm and continued fractions. In fact, this entire chapter is "optional" (and later chapters are essentially independent of it) but highly recommended to prevent the student from being overwhelmed by the definitions and formalism and losing sight of the goals of the core course.

Chapter 5 returns to the systematic development of foundations with a complete treatment of relations, and it is truly the "heart" of Part 1. Optional sections include treatments of iterations, closure operations, meets and joins of partitions, and the basics of lattice theory. An upper-division college course could begin with Chapters 5 and 6 and then quickly move on to selected topics in Part 2.

We turn finally to a discussion of functions in Chapter 6 covering more material than is customary including restriction and extension of domain and range, algebras of real-valued functions, and the usual "categorical" material of inclusion, projection, quotient maps, and so on. There is an optional section on dynamical systems with applications in the next optional section to the study of the symmetric groups. We believe these are essentially ideal additional topics since the treatment of permutations and dynamics here helps to instill a proper intuition regarding functions.

This completes our serial discussion of Part 1. Highlights from the optional sections of Part 2 include treatments of generating functions and characteristic equations (where we proceed by "guessing" and then proving the main theorem) in Chapter 8. Kleene's Theorem and a treatment of semi-groups and monoids are given in Chapter 9 (and a further discussion of algebraic structures is the purview, we believe, of a first course in modern algebra), and Chapter 10 contains some of the great combinatorics results of the 1920-30's of Hall and Menger as well as a glimpse of Ramsey theory, the chromatic polynomial of a graph, and an explicit description of Poincaré duality for planar graphs.

Exercises follow each section of this book, and there is a dramatic difference between Parts 1 and 2 in this regard. Many of the exercises in Part 1 are of an "easy" or "low-brow" nature in that they ask for verifications of the theorems presented in the text for instance (but to be sure, some of the exercises in Part 1 are of a more interesting sort); this has been done in order to fulfill the needs of several different possible courses based upon this book. In contrast, there are many fewer exercises in Part 2, they are included partly for the convenience of the instructor, are almost all of a substantial nature, and though eclectic, present interesting and/or beautiful aspects of the relevant theory. (It is worth saying that for some optional sections the best exercises are probably just applications and examples of the main results from the text.)

We can have no defense for the choice of topics included as "beautiful" in the optional sections and in Part 2–and for the omission of many other beautiful and important topics as well–other than to ascribe it to our own personal tastes and proclivities and to the constraints of space and time.

As was mentioned above, this book is independent of any knowledge of differential calculus except for certain exercises, and in each of these exercises, we explicitly state that it depends upon calculus. There are furthermore a few exercises in Chapter 10 which depend upon elementary linear algebra, and again this is stated explicitly in each problem.

As to our self-imposed independence from calculus, we might admit that this text has been written in part in reaction to certain trends of calculus reform, which, for better or worse–and simply as a practical matter–have deleted proofs from the traditional syllabus in favor of other algorithmic and electronic aspects. Under the circumstances, we believe it is only sensible that students can profitably learn intellectual rigor and its necessary formalism in studying the foundations of discrete mathematics rather than calculus. (The now-traditional but threatened role of calculus in American college education can apparently be traced to G. H.

Hardy's lovely and influential book *A Course of Pure Mathematics*, Cambridge University Press, 1908).

We believe there will be a growing population of students at all levels for courses on discrete mathematics, and we hope this book with its flexible design may be of some pleasant general utility.

Robert C. Penner

Los Angeles, California
July, 1999

Chapter 1

Proof Techniques

The notion of what constitutes a proof in mathematics can be elusive, especially for the beginner. In this chapter, we shall address the questions: What is a proof? How can one go about discovering a proof? How does one write down a proof? It is worth emphasizing that discovery and exposition of a proof actually require different skills, and both will be discussed here.

We study various examples of proofs in this chapter working through both their discovery and their exposition. The reader will likely find that the discovery process is familiar from the usual logical deductions of everyday life. In contrast, writing down a proof requires learning the foundations and language of mathematics, and much of this volume is dedicated to learning these fundamentals. Of course, as our knowledge of basic mathematics deepens in subsequent chapters, we shall uncover and prove progressively deeper facts. At the same time, the proof techniques which we discuss in this introductory chapter are those used throughout this volume and indeed throughout all mathematics.

This chapter breaks up naturally into Part A, where we give a gentle introduction to important concepts from logic which will be covered in greater detail in the next chapter and describe a basic method of proof called "induction", and Part B, where we survey a wide variety of proof techniques in mathematics and give notable examples of various "famous" proofs. The only exercises in this chapter are on induction in §A.5 and collectively on proof techniques at the end of §B.5. Part B can reasonably be relegated to assigned reading.

Part A *ELEMENTS OF LOGIC AND INDUCTION*

Various fundamental concepts from mathematical logic are introduced and discussed in this first part of the chapter, and we turn to a systematic study of this material only in the next chapter. We describe in §A.5 the first and simplest of several incarnations of the method of proof by "induction". Our summary discussion of logic here is sufficient for the general discussion of methods of proof covered in the remainder of this chapter.

A.1 DEFINITIONS

A definition in mathematics is an explicit interpretation of a word or symbol in terms of other words or symbols. There is never anything to prove about a definition: It is an assertion which must simply be taken for granted. In fact, a definition or a collection of definitions is often the starting point for an assertion or a collection of assertions to be proved. The usual syntax (that is, the notation) for presenting a definition is to print the word being defined in italic typeface. Thus, we might have said above that

> A *mathematical definition* is an explicit interpretation of a word or symbol in terms of other words or symbols.

The reader should infer from the simple notational fact that the phrase "mathematical definition" is in italic typeface that we are here defining that phrase. Subsequent use of the unitalicized term refers implicitly to the definition. Sometimes for purposes of exposition, we enclose a term in quotation marks in order to refer to the term before actually giving its formal definition; for instance, we shall rigorously discuss the "predicate calculus" in the next chapter, and we gently introduce aspects of this predicate calculus here in order to then discuss proof techniques. Some other examples of mathematical definitions are:

> Three points are said to be *collinear* if they lie on a common line. If three points are not collinear, then we say that they are *non-collinear.*
>
> A *triangle* is a collection of three non-collinear points together with the three line segments connecting them in pairs. The points are called the *vertices* of the triangle, and the line segments are called the *sides* of the triangle.
>
> A *circle* is the collection of all points in the plane at a fixed distance from a given point.
>
> A *dog* is a mammal of the canine species.
>
> A *duck* is a mammal of the canine species.

A *natural number* is simply a positive integer $1, 2, 3, \ldots$.

In the first of these definitions, it is assumed that the reader knows what a point and a line are, and one is defining the words "collinear" and "non-collinear" in terms of this *a priori* (i.e., already given) knowledge. The second statement defines the words "triangle", "vertex", and "side" in terms of "line segments" and "collinearity". The third statement defines "circle" in terms of the the known words "point" and "distance", and so on. The fifth definition is pretty silly since a duck would not satisfy the definition of being a duck, but we cannot object since a definition must simply be accepted. On the other hand to be reasonable, we hereby retract the dumb duck definition once and for all. The sixth definition expresses "natural numbers" in terms of "integers" and "positivity".

We can infer from the silly definition of duck above that some care should be exercised in making definitions. Indeed, there are examples in mathematics where highly developed and logically correct theories about certain objects based on shaky definitions are subsequently found to be vapid in the sense that no such object could actually exist. Thus, a correct theory based on spurious definitions can come crashing to the ground!

A.2 PROPOSITIONS AND PREDICATES

A *proposition* is a statement which is either true or false but not both true and false. The *truth value* of a proposition is 0 if the proposition is false and is 1 if the proposition is true. Some examples of propositions and their truth values are:

Proposition	Truth Value
2+2=4	1
2=3	0
A dog is a mammal.	1
The earth is a tostada.	0

A *propositional variable* is a symbol, for instance the symbol P or Q, which represents an arbitrary proposition. In particular, the propositional variable P might represent any of the propositions above.

To clarify this concept, consider the definition

> An integer n is *even* if there is another integer m so that $n = 2m$, and n is *odd* if there is another integer m so that $n = 2m+1$.

Here we define the notion of "even" and "odd" integers in terms of addition and multiplication of integers. In this definition, the variables m and n represent integer variables, that is, arbitrary integers. In the perfectly analogous sense, a propositional variable P represents an arbitrary proposition. Furthermore, just as we might specialize the value of an integer variable n, for instance $n = 2$, we may also sometimes specialize the value of a propositional variable P, and we write, for instance,

$$P = [\text{The earth is a tostada}]$$

to signify that P represents the proposition "The earth is a tostada". As a point of terminology, given a proposition P, we may sometimes say simply "Suppose $P\ldots$" or "Suppose that $P\ldots$" with the explicit meaning "Suppose that P is true ...", i.e., "Suppose that P has truth value 1...".

In contrast, the statement $n + 2 = 3$ is not a proposition, for it has no particular associated truth value (that is, it is neither true nor false) until the value of n is specified. Of course, it has truth value 1 if $n = 1$ and truth value 0 otherwise. Statements of this form are called "predicates". The precise definition of a predicate will be given presently though we shall only study them carefully later.

First, let us remark that given the predicate $P(n) = [n + 2 = 3]$, we may specify a value n_0 for n to get a proposition $P(n_0) = [n_0 + 2 = 3]$, which itself has an associated truth value. For instance, if $n_0 = 1$, then the associated proposition is $P(1) = [1 + 2 = 3]$ with truth value 1, while if $n_0 = 0$, then the proposition is $P(0) = [0 + 2 = 3]$ with truth value 0. Thus, the predicate $P(n) = [n + 2 = 3]$ is really an assignment of one proposition to each possible value of $n = n_0$. Implicit in this discussion is the specification of a collection, called the "universe of discourse", of possible values that $n = n_0$ might take, and a sensible universe of discourse for the predicate $n + 2 = 3$ is the collection of all integers.

A *predicate* $P(x)$ is the specification of a collection $\mathcal{U}$ of objects together with a family of propositions, one proposition $P(x_0)$ for each element x_0 of $\mathcal{U}$. The collection $\mathcal{U}$ is called the *universe of discourse* of the predicate, and we require that $\mathcal{U}$ must be non-empty, that is, $\mathcal{U}$ has at least one element. Just as for propositions, we define a *predicate variable* to be a symbol such as $P(n)$ which may represent an arbitrary predicate. For instance, $P(n)$ might represent the predicate $n + 2 = 3$ over the universe of all integers.

For some further examples, take the predicate $P(x) = [x \text{ has two legs}]$ on the universe of all mammals or the predicate $Q(x, y) = [x \text{ has } y \text{ legs}]$ on the universe of all pairs x, y, where x is a mammal and y is a natural number. The predicate $R(n) = [n \geq 1]$ on the universe of all integers has truth value 0 exactly when n is not a natural number. Let us emphasize immediately that the universe of discourse of a predicate is really an essential part of its specification. For instance, the predicate $R(n)$ on the universe of natural numbers always has truth value 1.

Thus, we may derive from a predicate $P(x)$ together with an element x_0 of the associated universe $\mathcal{U}$ a proposition $P(x_0)$, which has a truth value 0 or 1. This process of passing from the predicate $P(x)$ to a proposition $P(x_0)$ for some (that is, any) particular x_0 in the universe is called *specialization* of the predicate.

There are two other propositions, denoted $\exists x P(x)$ and $\forall x P(x)$, which we can associate to the predicate, by defining:

$\exists x P(x)$ has truth value 1 if and only if $P(x_0)$ is true for some x_0 in $\mathcal{U}$

$\forall x P(x)$ has truth value 1 if and only if $P(x_0)$ is true for each x_0 in $\mathcal{U}$

The symbol $\exists$ is called the *existential quantifier* and should be read in English as "there exists"; the symbol $\forall$ is called the *universal quantifier* and should be read in English as "for all". These definitions of $\exists$ and $\forall$ evidently faithfully correspond to their usual meanings in English in the sense that the truth values of $\exists x P(x)$ and $\forall x P(x)$ correspond to their English translations read from left to right. The passage from the predicate $P(x)$ together with its universe $\mathcal{U}$ to the proposition $\forall x P(x)$ or $\exists x P(x)$ is called *quantification* of the predicate $P(x)$, and each quantified predicate $\exists x P(x)$ or $\forall x P(x)$ together with the specification of its universe of discourse itself has an associated truth value 0 or 1 and is therefore itself a proposition.

There are methods of proof associated with the very definition of the truth value associated to a quantified predicate: To prove that $\exists x P(x)$ has truth value 1, we must show that there is indeed some x_0 in the universe of discourse so that $P(x_0)$ has truth value 1; on the other hand, to prove that $\forall x P(x)$ has truth value 1, we must show that for an arbitrary x_0 in the universe of discourse, $P(x_0)$ has truth value 1. The former method of proof is called *existential generalization*, and the latter is called *universal generalization.* Let us here formalize these notions:

EXISTENTIAL GENERALIZATION To prove that $\exists x P(x)$ has truth value 1, we show that there is some x_0 in the universe of discourse so that $P(x_0)$ has truth value 1.

UNIVERSAL GENERALIZATION To prove that $\forall x P(x)$ has truth value 1, we choose a fixed but arbitrary x_0 in the universe of discourse and prove that $P(x_0)$ has truth value 1.

For some examples, let $E(n) = [n \text{ is even}]$ and $O(n) = [n \text{ is odd}]$ on the universe of integers. The predicate $\exists n E(n)$ has truth value 1 since $E(0)$ has truth value 1 (i.e., 0 is even), so we have successfully exhibited $x_0 = 0$ in the universe so that $E(x_0)$ is true, and the method of existential generalization applies. In a similar way, $\exists n O(n)$ also has truth value 1, since $O(1)$ is true for instance.

On the other hand, on the universe of all even integers, $\forall n E(n)$ has truth value 1 since an arbitrary element of the universe in this case is an even integer n_0, so $E(n_0)$ is automatically true, and the method of universal generalization applies. Of course, $\exists n E(n)$ is true on the universe of even integers by existential generalization as before.

For a final example, on the universe consisting of the single element 2, the truth values are exactly as in the previous paragraph.

The flip side of universal or existential generalization is called universal or existential "instantiation", and we begin by discussing the former.

UNIVERSAL INSTANTIATION Suppose that $\forall x P(x)$ has truth value 1, and let x_0 be a particular member of the universe of discourse of $P(x)$. Then $P(x_0)$ has truth value 1.

The reader should observe that universal generalization is, in effect, the "reverse" of universal instantiation in the following sense: The former allows us to conclude that the general case $\forall x P(x)$ is true provided that each special case $P(x_0)$ is true, whereas the latter allows us to conclude that any special case $P(x_0)$ is true from the assumed truth of the general case $\forall x P(x)$, where x_0 is any element of the universe of discourse of $P(x)$. In other words, universal instantiation amounts to "if $\forall x P(x)$ is true, then for any particular x_0 in the universe, $P(x_0)$ is also true". Students sometimes find universal instantiation silly or vapid (presumably because it is so "obvious" in English), but it is nevertheless not inconsequential.

In the same way, existential generalization allows us to conclude that $\exists x P(x)$ is true if there is some x_0 in the universe of discourse of $P(x)$ so that $P(x_0)$ is true. On the other hand, if we assume $\exists x P(x)$, then we may conclude that $P(x_0)$ is true for some particular x_0 in the universe. Formalizing this, we have

EXISTENTIAL INSTANTIATION If $\exists x P(x)$ has truth value 1, then there is some x_0 in the universe so that $P(x_0)$ has truth value 1.

Just as before, the reader should observe that existential generalization is effectively the "reverse" of existential instantiation. In other words, existential instantiation amounts to "if $\exists x P(x)$ is true, then $P(x_0)$ is also true for some particular x_0 in the universe". Again, students sometimes find existential instantiation silly or vapid, but it is not.

Let us finally give some examples of the utility of quantified predicates in definitions. Indeed, our definitions above of even and odd integers themselves actually tacitly involved quantified predicates: For instance, let $T(n,m)$ denote the predicate $n = 2m$ on the universe of all pairs n, m of integers, and specialize $n = n_0$ to get a predicate $T(n_0, m)$ defined as m varies over the universe of all integers. With this notation, n_0 was defined above to be even if and only if $\exists m T(n_0, m)$ has truth value 1.

For another important example of such a definition, consider the predicate $D(n,k,m)$ given by $[n = km]$ defined on the universe of all triples n, k, m of integers. As before, we may specialize $n = n_0$ and $m = m_0$ to get a predicate $D(n_0, k, m_0)$ on the universe of all integers k, and we say that m_0 *divides* n_0 if $\exists k D(n_0, k, m_0)$ has truth value 1. If m_0 divides n_0, then we write $m_0 | n_0$ (so $2|6$ and $5|25$ for instance), and if m_0 does not divide n_0, then we write $m_0 \nmid n_0$ (so $2 \nmid 5$ and $8 \nmid 2$ for instance). Notice that given two integers n_0 and m_0, exactly one of $m_0 | n_0$ or $m_0 \nmid n_0$ has truth value 1, and for each integer n, we have $n|0$ (since $0 = 0 \cdot n$) and $1|n$ (since $n = n \cdot 1$). An integer n was defined above to be even if and only if $2|n$,

We shall return to a detailed discussion of predicates and quantifiers later. It is incumbent upon us to point out here the logical gap above in that we have been imprecise by not specifying exactly what a "universe of discourse" actually is. This gap will be filled only much later, when we discuss axiomatic set theory. For now, the reader should have a firm grasp of propositions and a reasonable intuitive idea about predicates, their universes of discourse, their specialization, and their quantification. This intuition should suffice for the informal discussion of proof techniques which we undertake later in this chapter.

A.3 CONJUNCTION AND NEGATION

Let P and Q be two propositional variables. There are numerous interesting ways to construct a new propositional variable R from P and Q, where the truth value of R depends upon the truth values of P and Q. As a simple example, we might attempt to formalize the English word "and", defining R to be true exactly when both P and Q are true. Precisely, we introduce the symbol $\wedge$ to be read in English as the word "and", set $R = P \wedge Q$ called the *conjunction* of P and Q, and define $R = P \wedge Q$ to have truth value 1 exactly when both P and Q have truth value 1. This obviously faithfully formalizes the English word "and" in our current context. For instance, if $P =$ [A dog is a mammal] and $Q = [2 \cdot 2 = 4]$, then

$$R \;=\; P \wedge Q \;=\; [\text{A dog is a mammal}] \wedge [2 \cdot 2 = 4]$$

has the same truth value as

$$[\text{A dog is a mammal, and } 2 \cdot 2 = 4],$$

and this common truth value is 1 since a dog is indeed a mammal and $2 \cdot 2$ is indeed 4. If either of P or Q has truth value 0, then $R = P \wedge Q$ has truth value 0 by definition; for instance, if P is as before, and $Q = [6|7]$ (that is, "6 divides 7"), then $R = P \wedge Q$ has truth value 0 since in fact $6 \nmid 7$ (that is, "6 does not divide 7").

A systematic way to record this definition of the symbol $\wedge$ is to separately enumerate all of the possible truth values of the propositional variables P and Q and to list the corresponding truth values of $P \wedge Q$ in a table, as in:

P	Q	$P \wedge Q$
0	0	0
0	1	0
1	0	0
1	1	1

Thus, the various rows of the table correspond to the various possible assignments of truth values to P and Q, and the final column of the table expresses the associated truth value of $P \wedge Q$. In this way, we completely determine the truth value of the propositional variable $P \wedge Q$ in terms of the truth values of the propositional variables P and Q. As a point of terminology, we shall actually refer to an expression such as $P \wedge Q$ (together with its associated truth values in terms of the truth values of P and Q) as a "propositional form". We shall systematically study propositional forms and their relationships later.

Such a presentation of truth values is called a *truth table.* Truth tables are an efficient way of defining new propositions (i.e., of defining propositional forms) as we have already seen in the case of $\wedge$. Furthermore, truth tables lead to a systematic if somewhat boring and inefficient method of proof, as we shall see later.

We turn next to the operation of "negating" a proposition P, that is, the construction of a new proposition which is true if and only if P is false and is false if and only if P is true. To be explicit, we define the propositional form $\neg P$ with a truth table as before:

P	$\neg P$
0	1
1	0

The propositional form $\neg P$ is called the *negation* of P, and the symbol $\neg$ should be read in English as "not". This obviously reflects the English meaning of negation in our current context though the English sentence often has to be modified for grammatical reasons. It is worth emphasizing immediately that P and $\neg(\neg P)$ have exactly the same truth values, as the reader can check mentally, or the reader might write down the truth table for P, $\neg P$, $\neg(\neg P)$ for a complete proof of this fact.

Here are some examples of negation:

P	$\neg P$
I am not bored.	I am bored.
Elvis lives.	Elvis is dead.
$2 = 3$	$2 \neq 3$

These examples are self-evident and require no further comment.

For an example involving predicates, let $E(n)$ denote the predicate $[n$ is even$]$ on the universe of all integers as before, so the negation of $E(n)$ is $\neg E(n) =$ $[n$ is not even$]$. Consider the proposition $P = \exists n E(n)$, which is true since $E(2)$ is true for instance. The English translation of P is [There is some even integer], and the negation of this in English is [There is no even integer] or in other words, [For each integer n, n is not even], that is, the negation $\neg P$ of $P = \exists n E(n)$ is $\forall n \neg E(n)$, where the universe of discourse is again the collection of integers. For an example of negating a universally quantified predicate, consider the proposition $Q = \forall n E(n)$, where $E(n)$ and its universe are as before. The English translation of Q is [Every integer is even], which is false since $E(1)$ is false for instance. The negation of this in English is [Not every integer is even], or in other words, [There is some integer which is not even]. These are examples of general phenomena of negating a quantified predicate which we next formalize.

NEGATION OF QUANTIFIED PREDICATES Given a predicate $P(x)$ defined over some universe $\mathcal{U}$, the negation of $\exists x P(x)$ has the same truth values as $\forall x \neg P(x)$, and the negation of $\forall x P(x)$ has the same truth values as $\exists x \neg P(x)$, where the universe of discourse is $\mathcal{U}$ in each case.

Thus, in negating a quantified predicate, one changes the quantifier from existential to universal or from universal to existential and then negates the predicate itself.

To prove the first assertion above, suppose that $\forall x \neg P(x)$ is true, so $\neg P(x)$ is true for every x in the universe by universal instantiation. In other words, $P(x)$ is false for any x in the universe, so $\exists x P(x)$ is false by definition of the truth value associated with existential quantification. Thus, when $\forall x \neg P(x)$ is true, we have $\exists x P(x)$ is false, so $\neg(\exists x P(x))$ is also true. On the other hand, if $\forall x \neg P(x)$ is false, then there is some x in the universe so that $P(x)$ is true by definition of the truth values associated with universal quantification, so $\exists x P(x)$ is true by existential generalization. The first assertion above therefore follows.

Rather than give a similar argument to prove the second assertion, we might simply take $P(x) = \neg Q(x)$ in the first assertion to conclude that $\forall x Q(x)$ and $\neg(\exists x \neg Q(x))$ have the same truth values, and negating each of these expressions, we find that $\neg(\forall x Q(x))$ and $\exists x \neg Q(x)$ also have the same truth values, as desired.

This completes our first of several proofs about predicates in this volume. Whereas the proof of the first assertion involved simply manipulating the definitions of the truth values associated with quantified predicates, the proof of the second assertion involved symbolic manipulations. The latter is perhaps more typical of the proofs we shall describe in the next chapter, and indeed the term "predicate calculus" accurately indicates that we shall systematically perform "calculations" with predicates. It is important to emphasize that proofs in mathematics and science are more than simply such calculations with predicates: In fact, we shall find that certain formulas involving predicates are actually "blueprints" for the proofs which arise in practice.

A.4 IMPLICATION

Given propositional variables P and Q, we introduce another propositional form $P \Rightarrow Q$, to be read in English as "P implies Q" or as "if P, then Q", where $P \Rightarrow Q$ is defined by the truth table

P	Q	$P \Rightarrow Q$
0	0	1
0	1	1
1	0	0
1	1	1

P is called the *hypothesis* and Q the *conclusion* of the implication $P \Rightarrow Q$. This is intended to formalize the notion of implication in English, where "P implies Q" is true if Q is true whenever P is true. Insofar as this is a definition of $\Rightarrow$, no further comment is actually necessary, but we briefly discuss the relationship of $\Rightarrow$ to the English word "implies". In case both P and Q are true propositions, then Q is true, so, in particular, if P is true then Q is true; thus, the last line of the previous truth table accurately reflects the English. Similarly, in the next to last line, P is true and Q is false, so it is not the case that P true implies Q true, so $P \Rightarrow Q$ is false, again accurately reflecting the English. Notice that by the first two lines in the truth table, $P \Rightarrow Q$ is true whenever P is false. The rationale is that since P is false, it is the case that whenever P is true (namely, never!), Q is true. This aspect of implication is not usually made explicit in its English usage, but the demands of precision in mathematics require us to assign truth values to $P \Rightarrow Q$ in this case in which P is false. Students sometimes find this nuance of the definition of implication unsettling and should take on faith for now that this is the sensible definition.

Given propositional variables P_1, P_2, P_3, define corresponding propositional variables $P_{ij} = (P_i \Rightarrow P_j)$ for any $i, j = 1, 2, 3$, and let

$$R = [P_{12} \wedge P_{23}] = [(P_1 \Rightarrow P_2) \wedge (P_2 \Rightarrow P_3)].$$

We claim that $R \Rightarrow P_{13}$ has truth value 1 for any possible assignment of truth values to P_1, P_2, P_3. Enumerating all possible truth values for P_1, P_2, P_3 in a table as before and computing (from the previous truth tables defining $\wedge$ and $\Rightarrow$) the truth values of P_{ij}, R, and $R \Rightarrow P_{13}$, we find

P_1	P_2	P_3	P_{12}	P_{23}	R	P_{13}	$R \Rightarrow P_{13}$
0	0	0	1	1	1	1	1
0	0	1	1	1	1	1	1
0	1	0	1	0	0	1	1
0	1	1	1	1	1	1	1
1	0	0	0	1	0	0	1
1	0	1	0	1	0	1	1
1	1	0	1	0	0	0	1
1	1	1	1	1	1	1	1

It is incumbent upon the reader to check the computations of entries in the table above using the definitions of $\wedge$ and $\Rightarrow$. A table reflecting the truth values of a propositional form in terms of its component propositional variables, as above, is called a *truth table* for the propositional form.

The table above shows that $R \Rightarrow P_{13}$ is always true, i.e., the last column in the table above consists of all ones, or in other words

$$(*) \qquad \big[(P_1 \Rightarrow P_2) \wedge (P_2 \Rightarrow P_3)\big] \Rightarrow \big(P_1 \Rightarrow P_3\big)$$

is always true. A propositional form which always has truth value 1 (independent of the specification of particular propositions or truth values) is called a *tautology*, and we have proved with the previous truth table that the inset equation above is a tautology. The symbol "$(*)$" next to this equation above indicates that we have given a name, the name "$(*)$", to the equation, so we could say here simply that "$(*)$ is a tautology".

As we have just seen, one can use truth tables to prove that propositional forms are tautologies. Actually, we shall essentially never resort to such proofs by truth table as they tend to be mindless manipulations leading to inefficient proofs which are prone to errors. We shall return to a detailed calculational study of tautologies in the next chapter.

There is much more to our investigations in this volume than just the verification of tautologies. Indeed, we shall see that certain tautologies form the basis for techniques of proof which one can successfully apply in various contexts; thus, certain tautologies are "blue-prints" for whole families of proofs. For instance, suppose we wish to prove that

$$(\dagger) \qquad [\text{I am a duck}] \Rightarrow [\text{I waddle}],$$

that is, we wish to prove that this implication has truth value 1. We might separately prove each of the implications [I am a duck]$\Rightarrow$[I have webbed feet] and [I have webbed feet]$\Rightarrow$[I waddle] to conclude using the tautology $(*)$ above that the original implication $(\dagger)$ is indeed true. This is a typical sort of application of the predicate calculus to proving a specific assertion.

Associated with a given implication $P \Rightarrow Q$, there are other logically related implications which we next discuss. Indeed, we define the following implications derived from $P \Rightarrow Q$.

Propositional Form	Name
$P \Rightarrow Q$	*Implication*
$Q \Rightarrow P$	*Converse*
$\neg P \Rightarrow \neg Q$	*Inverse*
$\neg Q \Rightarrow \neg P$	*Contrapositive*

For instance, given the implication $P \Rightarrow Q$, the related implication $\neg Q \Rightarrow \neg P$ is called the contrapositive of $P \Rightarrow Q$. There is a method of proof associated to the contrapositive of an implication as we shall later. For now, we simply give some examples and analyze truth values.

Of course, the implication [It is $1:45$ p.m.] $\Rightarrow$ [It is afternoon] has truth value 1 in English. The converse of this implication is

$$[\text{It is afternoon}] \Rightarrow [\text{It is } 1:45 \text{ p.m.}],$$

which has truth value 0 in English since if it is 2:00 p.m., then it is indeed afternoon, while it is not 1:45 p.m.. The inverse of the original implication is

$$[\text{It is not } 1:45 \text{ p.m.}] \Rightarrow [\text{It is not afternoon}],$$

which similarly has truth value 0. The contrapositive of the original implication is

$$[\text{It is not afternoon}] \Rightarrow [\text{It is not } 1:45 \text{ p.m.}],$$

which has truth value 1.

For an example with predicates over the universe of all integers, consider the implication $[n \text{ is even}] \Rightarrow [n \text{ is not odd}]$. The converse is

$$[n \text{ is not odd}] \Rightarrow [n \text{ is even}],$$

the inverse is

$$[n \text{ is not even}] \Rightarrow [n \text{ is odd}],$$

and the contrapositve is

$$[n \text{ is odd}] \Rightarrow [n \text{ is not even}].$$

In fact, all of the implications of this paragraph are well-known to be true for each integer n, so each universally quantified implication is actually true. We shall carefully prove these facts in Example B.2.2 below.

Let us next consider the truth values of these various implications.

P	Q	$P \Rightarrow Q$	$Q \Rightarrow P$	$\neg P \Rightarrow \neg Q$	$\neg Q \Rightarrow \neg P$
0	0	1	1	1	1
0	1	1	0	0	1
1	0	0	1	1	0
1	1	1	1	1	1

It is again incumbent on the reader to verify the entries of this truth table using the definitions of $\Rightarrow$ and $\neg$. We observe immediately that $P \Rightarrow Q$ and its contrapositive $\neg Q \Rightarrow \neg P$ have exactly the same truth values in each case of truth values of P and Q. In the same way, the converse $Q \Rightarrow P$ and the inverse $\neg P \Rightarrow \neg Q$ also have exactly the same truth values. Indeed, this follows immediately from the previous observation insofar as the inverse is the contrapositive of the converse, as the reader can check. It is also worth pointing out that the contrapositive of the contrapositive is none other than the original implication with a similar remark for the converse and inverse.

A.5 PROOFS BY INDUCTION FOR NATURAL NUMBERS

We present here an especially powerful and useful method of proof for an assertion of the form $\forall n P(n)$, where n varies over the universe of all natural numbers. This method is called "proof by induction", and it is actually of greater utility than simply for natural numbers, as we shall see later. We present this proof technique here in the context of natural numbers since the method is especially intuitive and intellectually accessible in this context, and we shall require it in the next chapter. There are also extensions and reformulations of this method of proof as we shall discuss in subsequent chapters.

The method depends upon a basic fact, which we state here and will prove below when we have a precise definition of the natural numbers, as follows:

BASIC FACT *Any non-empty collection of natural numbers has a least element.*

For example, the collection of all even natural numbers has least element two. Presumably the Basic Fact is self-evident to the reader from everyday experience with natural numbers. A complete proof of this fact is actually rather involved and is relegated to the optional §3.7, where we first give a careful definition of the the set of natural numbers.

There are actually two different methods of proof by induction, and we next discuss the simplest such method.

PRINCIPLE OF MATHEMATICAL INDUCTION In order to prove $\forall n P(n)$, where n ranges over the universe of natural numbers, we separately prove that the following two statements are true.

Basis Step: Prove that $P(1)$ is true.

Induction Step: Prove that $\forall n[P(n) \Rightarrow P(n+1)]$ is true, where n varies over the universe of natural numbers.

To understand the method, the reader might imagine a "proof machine" which begins by checking the basis step, namely that $P(1)$ is true. The "machine" then considers the statement $P(2)$. Insofar as one has proved the induction step to be true, one concludes by universal instantiation that $P(1) \Rightarrow P(2)$. Since we have checked $P(1)$ in the basis step, the "machine" concludes $P(2)$ is true and moves on to consider $P(3)$. In the same way, $P(2) \Rightarrow P(3)$, so $P(3)$ is true, and so on. Thus, the "machine" serially checks that $P(n)$ holds for each natural number $n = 1, 2, 3, \ldots$, and so $\forall n P(n)$ indeed has truth value 1.

We next show that the validity of the principle of mathematical induction follows easily from the Basic Fact above. Indeed, let us suppose that the Basic Fact is true and consider the quantified predicate $\forall n P(n)$, where the universe of discourse is the collection of natural numbers. Suppose that $\forall n P(n)$ admits a proof by induction and yet is false, so there is some integer m so that $P(m)$ is false. Thus, the collection of all natural numbers m so that $P(m)$ is false is non-empty, and therefore there is a least such natural number, say the natural number m_0, by the Basic Fact. According to the basis step $P(1)$ of the inductive argument, $m_0 > 1$, and we consider the natural number $m_0 - 1$. Since $m = m_0$ is the least natural number so that $P(m)$ is false and $m_0 - 1 < m_0$, we conclude that $P(m_0 - 1)$ is true. Finally, according to the induction step, we have $P(m_0 - 1) \Rightarrow P(m_0)$, so $P(m_0)$ must also be true. This contradiction, i.e., that $P(m_0)$ is both true and false, shows that our supposition that $\forall n P(n)$ is false and yet admits a proof by induction is not tenable and therefore establishes the claim that a proof by induction is a legitimate proof, where we have assumed the Basic Fact in our argument.

The first several examples below illustrate prototypical applications of the principle of mathematical induction in the setting of evaluating finite sums, but we must first introduce some standard notation which the reader has probably seen in previous math courses, as follows: If $x_1, x_2, \ldots, x_n$ is a collection of real numbers, then we use the "sigma notation"

$$\sum_{i=1}^{n} x_i = x_1 + x_2 + \cdots + x_n$$

for the sum of these numbers. More generally, given integers $m \leq n$ and a collection $x_m, x_{m+1}, \ldots, x_n$ of real numbers, we write

$$\sum_{i=m}^{n} x_i = x_m + x_{m+1} + \cdots + x_n$$

for the sum. Furthermore, it sometimes happens that we consider an expression $\sum_{i=m}^{n} x_i$ where $m > n$, and in this trivial case, we refer to the sum as an "empty sum" and set $\sum_{i=m}^{n} x_i = 0$ by convention. Similarly, given a collection $x_m, x_{m+1}, \ldots, x_n$ of real numbers, we write

$$\prod_{i=m}^{n} x_i = x_m \cdot x_{m+1} \cdot \cdots \cdot x_n$$

for the product with the convention that the "empty product", i.e., the case $m > n$, has value 1.

■ **Example A.5.1** We prove $\forall n P(n)$, where n varies over the universe of natural numbers and $P(n)$ is the equality $\sum_{j=0}^{n} 2^j = 2^{n+1} - 1$. According to the statement of proof by induction, we must check two assertions, the basis step $P(1)$ and the induction step that if the equation holds for the natural number n, then it also holds for $n+1$. In the basis step, $n = 1$, and $P(1)$ asserts that

$$2^{1+1} - 1 = 4 - 1 = 3 = 1 + 2 = 2^0 + 2^1 = \sum_{j=0}^{1} 2^j,$$

so the basis step of our inductive proof is complete.

To check the inductive step, we suppose that $P(n)$ holds, so indeed $\sum_{j=0}^{n} 2^j = 2^{n+1} - 1$. To check that $P(n+1)$ holds, we write

$$\sum_{j=0}^{n+1} 2^j = \sum_{j=0}^{n} 2^j + 2^{n+1},$$

so the sum occurring in $P(n+1)$ can be expressed in terms of the sum occurring in $P(n)$. Applying the hypothesis $P(n)$, we conclude $\sum_{j=0}^{n} 2^j = 2^{n+1} - 1$, and substituting this into the previous inset equation, we find

$$\sum_{j=0}^{n+1} 2^j = (2^{n+1} - 1) + 2^{n+1} = 2 \cdot 2^{n+1} - 1 = 2^{n+2} - 1 = 2^{(n+1)+1} - 1,$$

so we have completed the induction step.

According to the principle of induction, this is a valid proof of the original universally quantified assertion, and we next give a formal presentation of this argument.

A.5.1 CLAIM *For any natural number n, we have $\sum_{j=0}^{n} 2^j = 2^{n+1} - 1$.*

Proof Proceed by induction, and for the basis step, compute

$$2^{1+1} - 1 = 4 - 1 = 3 = 1 + 2 = 2^0 + 2^1 = \sum_{j=0}^{1} 2^j.$$

For the induction step, compute

$$\begin{aligned}\sum_{j=0}^{n+1} 2^j &= \sum_{j=0}^{n} 2^j + 2^{n+1} = (2^{n+1} - 1) + 2^{n+1} \\ &= 2 \cdot 2^{n+1} - 1 = 2^{n+2} - 1 = 2^{(n+1)+1} - 1,\end{aligned}$$

where the second equality follows from the induction hypothesis. *q.e.d.*

The symbols "*q.e.d.*" abbreviate "*quod erat demonstrandum*" which is Latin for "as was to be shown". Mathematicians use many other symbols, such as the box " ■ " or the symbol " ⊠ " to indicate the end of a proof, and the latter symbol is typical when presenting mathematics at a blackboard. The particular symbol used is not important, but it is useful to have a special symbol which delineates the end of a proof. We shall always use "*q.e.d.*" for this purpose in this volume.

As in the previous example, it is often useful to explicitly state that one is using the method of induction, and the terminology of "basis step" and "induction step" or "inductive step" are universally recognized and may be used without further comment when presenting a proof by induction. When proving the induction step $\forall n[P(n) \Rightarrow P(n+1)]$, the assertion $P(n)$ is called the "induction hypothesis", as above, and one may also use this term freely in general with no further comment. Thus, the formal exposition of a proof by induction is quite efficient owing to this standardized terminology. It is worth emphasizing immediately that there is nothing sacred about using the symbol n for an arbitrary natural number in the induction step: For instance, rather than assume that $P(n)$ is true for $n \geq 1$

and prove that $P(n+1)$ is true, we might equivalently assume that $P(n-1)$ is true for $n \geq 2$ and prove that $P(n)$ is true. This is just a point of notation.

For another similar example, we prove

■ **Example A.5.2** *For each natural number n, we have $\sum_{k=1}^{n} k = n(n+1)/2$.*

Proof We proceed by induction, and the basis step follows easily

$$\sum_{k=1}^{1} k = 1 = \frac{1(1+1)}{2}.$$

For the induction we must check

$$\begin{aligned}\sum_{k=1}^{n+1} k &= \sum_{k=1}^{n} k + (n+1) = \frac{n(n+1)}{2} + (n+1) \\ &= \frac{n(n+1) + 2(n+1)}{2} = \frac{n^2 + 3n + 2}{2} = \frac{(n+1)(n+2)}{2} \\ &= \frac{(n+1)(n+1+1)}{2},\end{aligned}$$

as desired, where the second equality follow from the induction hypothesis. *q.e.d.*

A famous anecdote about Example A.5.2 is the story that Carl Friedrich Gauss as a boy was given the tedious assignment of adding all the numbers from 1 to 100. His teacher expected this to be a grueling exercise in addition, but Gauss quickly derived the equality above and gave the correct answer.

As the reader can surmise, the two examples above are typical of a whole class of important computational problems involving finite sums, where the basis step is easily checked to be an equality of numbers, and the induction step amounts to some arithmetic manipulation. This routine proof technique is fundamental and must be mastered by the reader. For another example of a proof by induction in this mold, we have

■ **Example A.5.3** *For any natural number n and any number $q \neq 1$, we have $\sum_{k=1}^{n} q^{k-1} = (q^n - 1)/(q-1)$.*

A sum of the form $\sum_{k=1}^{n} q^{k-1}$ is called a *geometric series*, and this putative (i.e., alleged or supposedly true) formula for a geometric series is absolutely basic.

Proof For the basis step, we find

$$\sum_{k=1}^{1} q^{k-1} = q^0 = 1 = \frac{q-1}{q-1} = \frac{q^1 - 1}{q-1},$$

and for the induction, compute

$$\begin{aligned}\sum_{k=1}^{n} q^{k-1} &= \sum_{k=1}^{n-1} q^{k-1} + q^{n-1}\\ &= \frac{q^{n-1}-1}{q-1} + q^{n-1}\\ &= \frac{q^{n-1}-1+q^n-q^{n-1}}{q-1}\\ &= \frac{q^n-1}{q-1},\end{aligned}$$

where the second equality follows from the inductive hypothesis. *q.e.d.*

Observe that in the examples of induction above, we proved a stated numerical equality while in practice, one often does not know the "answer" in advance in this way. In fact, given a finite sum, one can enumerate (that is, list) the first several values of the sum and on the basis of this data simply guess an expression for the general term. Of course, such inspired guessing is problematic, but having thereby guessed the "answer", one can then prove that the guess is correct or discover that it is not correct using a proof by induction.

■ **Example A.5.4** Consider the sum $s_n = \sum_{i=1}^{n}(2i-1)$ for $n \geq 1$. Listing the first several values, we find

$$s_1 = 1,\ s_2 = 4,\ s_3 = 9,\ s_4 = 16$$

and are led to guess that in fact $s_n = n^2$. We then verify our guess by induction.

A.5.2 CLAIM *For any natural number n, we have $\sum_{i=1}^{n}(2i-1) = n^2$.*

Proof The basis step $n = 1$ was computed above, and for the induction, we have

$$\begin{aligned}\sum_{i=1}^{n+1}(2i-1) &= \sum_{i}^{n}(2i-1) + (2n+1)\\ &= n^2 + (2n+1) = (n+1)^2,\end{aligned}$$

where the second equality follows from the inductive hypothesis. *q.e.d.*

There is nothing systematic that we can say here about this process of guessing except that success clearly depends upon the repertoire of formulas with which one is familiar. Notice that having successfully guessed, induction provides a completely rigorous proof, and the process by which one has guessed is then completely irrelevant. More generally, it is actually not so uncommon in mathematics that one first makes a well-informed guess and then verifies or refutes and modifies the guess.

We have seen the utility of the method of induction for evaluating finite sums. For an example which is not of this type, we have

■ **Example A.5.5** *For each natural number n, we have $6|(n^3 - n)$.*

Proof The proof is by induction, and the basis step is trivial since for $n = 1$, we have $n^3 - n = 0$, and $6|0$ since $0 = 0 \cdot 6$. For the induction, suppose that 6 indeed divides $n^3 - n$, so $n^3 - n = 6k$ for some integer k and compute

$$\begin{aligned}(n+1)^3 - (n+1) &= n(n+1)(n+2)\\ &= 3n(n+1) + (n^3 - n)\\ &= 3n(n+1) + 6k.\end{aligned}$$

Turning to the summand $3n(n+1)$, we observe that $n(n+1)$ must be even (since it is the product of consecutive integers one of which is necessarily even), so $n(n+1) = 2\ell$ for some integer ℓ. It follows that

$$(n+1)^3 - (n+1) = 6\ell + 6k = 6(\ell + k),$$

completing the induction step. *q.e.d.*

One can easily extend the method of induction to prove an assertion of the form $\forall n P(n)$, where n varies over the universe of all natural numbers greater than or equal to some fixed natural number n_0; that is, the method can be used to prove $\forall n[(n \geq n_0) \Rightarrow P(n)]$ for any fixed natural number n_0. Namely, one simply replaces the basis step in the induction proof by $P(n_0)$ and proves the induction step, as before, assuming that n (and hence also $n + 1$) is greater than or equal to n_0. For an example of this type of extension of the method, we have

■ **Example A.5.6** *For each natural number $n \geq 2$, we have $n^2 > n + 1$.*

Proof The proof is by induction and the basis step $n = 2$ is trivial since $2^2 = 4 > 3 = 2 + 1$. For the induction, suppose that $n \geq 2$ and $n^2 \geq n + 1$ and compute

$$(n+1)^2 = n^2 + 2n + 1 \geq (n+1) + 2n + 1 = 3n + 2 \geq n + 2,$$

where the first inequality follows from the induction hypothesis, and the last follows from the fact that $n \geq 2$. *q.e.d.*

To present another such example, define the *factorial function*

$$n! = n \cdot (n-1) \cdot (n-2) \cdot \cdots \cdot 2 \cdot 1 \quad \text{for } n \geq 1$$

which the reader has undoubtedly seen before, where the standard convention is to set $0! = 1$ as well.

■ **Example A.5.7** *For each integer $|n| \geq 4$, we have $|n|! > n^2$.*

Proof Begin by considering the case that n is positive. The proof is by induction, and the basis step is trivial because $4! = 24 > 16 = 4^2$. For the induction, suppose $n \geq 4$ and $n! > n^2$ and compute

$$(n+1)! = (n+1)n! > (n+1)n^2 \geq (n+1)^2,$$

where the first inequality follows from the induction hypothesis and the second from the previous example. In case n is negative, we may perform a similar induction on $|n|$ to obtain the desired conclusion. *q.e.d.*

Observe that we are here employing the obvious extension of the method of induction to the collection of all non-negative integers and then to the collection of all non-positive integers in the natural way. Similarly, for any integer n_0, one can employ the method of induction to prove any quantified assertion of the form $\forall n(n \geq n_0 \Rightarrow P(n))$ or $\forall n(n \leq n_0 \Rightarrow P(n))$, where n varies over the universe of integers.

In all of the examples so far, the basis step has amounted to a trivial verification, while the induction step was more complicated or difficult. It is not always so; in fact, the whole spectrum of possibilities actually occurs in practice. In particular and in contrast to the previous examples, there are cases in which the basis step is most difficult and the induction step is essentially trivial!

As was mentioned above, there is actually a more general method of induction, which we next state.

> **PRINCIPLE OF STRONG INDUCTION** To prove $\forall nP(n)$, where n ranges over the universe of natural numbers, we separately prove that the following two statements are true.
>
> Basis Step: Prove that $P(1)$ holds.
>
> Induction Step: Prove that $\forall n[\forall k[k \leq n \Rightarrow P(k)] \Rightarrow P(n+1)]$, where n and k each vary over the universe of natural numbers.

Let us begin by comparing this method of induction with the first one. Of course, the two methods have exactly the same basis step. The conclusions in the induction step are also identical, but the induction hypotheses differ: Indeed, in the first method we suppose simply $P(n)$, while in strong induction, we suppose $\forall k[k \leq n \Rightarrow P(k)]$. Thus, in strong induction, the induction hypothesis is that $P(k)$ holds for each k less than $n+1$, while in the first method, the induction hypothesis $P(n)$ is a special case of this assumption, namely the case $k = n$. In particular by universal instantiation, the induction hypothesis for strong induction implies the induction hypothesis for induction.

Regarding the principle of strong induction as a "proof machine" as before, we see that the "machine" first checks $P(1)$, then assumes $P(1)$ and checks $P(2)$, then assumes $P(1) \wedge P(2)$ and checks $P(3)$, then assumes $P(1) \wedge P(2) \wedge P(3)$ and checks $P(4)$, and so on.

We proceed as before to show that a proof by strong induction is a valid proof assuming the Basic Fact and suppose that the basis and induction steps have been verified for the predicate $P(n)$. If there is some natural number m so that $P(m)$ is false, then there is a least such natural number m_0 again by the Basic Fact. By the basis step, $m_0 > 1$, and since m_0 is least, $P(k)$ is true for each $1 \leq k < m_0$. Since the induction step has been verified, we conclude that $P(m_0)$ is true which is contrary to hypothesis, and this completes the proof of the validity of a proof by strong induction.

Strong induction seems to be a more powerful method of proof since we are allowed more assumptions and yet try to derive the same conclusion in the induction step, but it turns out that for the natural numbers, any statement which can be proved using strong induction could actually be proved using the first method.

Before giving some examples of strong induction, we must give some definitions. Given a natural number n, we say that n is *composite* if there is some natural number k, where $1 < k < n$, so that $k|n$. For instance, 6 is composite since $2|6$, and 1 is not composite. We say a number $n > 1$ is *prime* if it is not composite; in particular, our definition explicitly makes the convention that $n = 1$ is not prime. Thus, every natural number other than 1 is either composite or prime, and 1 is neither prime nor composite.

To recognize a prime number, we observe that in general if $m|n$ for natural numbers m, n, then $m \leq n$. To see this, just notice that if $n = km$ and m, n are natural, i.e., $n, m \geq 1$, then so too is k (i.e., k must be positive), so $k \geq 1$, whence $n = mk \geq m$ since $m \geq 1$. Thus, to check if a natural number n is prime we need simply list the finite collection of numbers $m > 1$ less than n and check that $m \nmid n$. For each fixed $1 < m \leq n$, we verify that $m \nmid n$ by listing the numbers k less than n and checking that $n \neq km$ for k on our finite list. Actually, we need only list the primes less than or equal to $n/2$ and check that none of them divides n, as the reader should verify. We shall study prime numbers in detail later and for now just uncover some basic facts about them which rely on strong induction.

■ **Example A.5.8** *Any natural number $n \geq 2$ is a product of primes.*

Proof The proof is by induction, and for the basis step, simply observe that 2 is written as the product with one factor of the prime 2. For the induction, we suppose that every number less than $n+1$ may be written as a product of primes and show that $n+1$ may be so written.

If $n+1$ is prime, then $n+1$ is itself a product of primes as before, and otherwise $n+1$ is composite, say $n+1 = ab$ with $1 < a, b < n+1$. By the inductive hypothesis, each of a and b can be expressed as a product of primes, so as a product of such, $n+1=ab$ is itself such a product. *q.e.d.*

We hope that the reader recognizes the simplicity and elegance of this inductive argument.

In preparation for the next examples, we let p_n denote the nth prime in the usual order on the natural numbers. For instance, $p_1 = 2$, $p_2 = 3$, $p_3 = 5$, and so on. Before giving our next example of an inductive argument, we give the following famous argument due to Euclid which depends on the previous example.

■ **Example A.5.9** *There is no largest prime number.*

Proof Suppose to the contrary that there is a largest prime number p_n, so $p_n > p_{n-1} > \cdots > p_1$ is a complete list of all the primes. Define the natural number

$$q = p_1 p_2 \cdots p_n + 1,$$

so $p_i|(q-1)$ for each $i = 1, \ldots, n$.

We claim that no prime p_i can divide q. Indeed, if $p_i|q$, say $q = s \cdot p_i$, then since also $p_i|(q-1)$, say $q - 1 = t \cdot p_i$, where $t = p_1 p_2 \cdots p_{i-1} p_{i+1} \cdots p_n$, we conclude that $1 = q - (q-1) = s \cdot p_i - t \cdot p_i = (s-t) \cdot p_i$. Thus, p_i must divide 1, so then $p_i = 1$, which contradicts that p_i is prime.

On the other hand, by Example A.5.8, q may be written as a product of primes, and we let p denote a prime dividing q. Of course, it may be that $p = q$ if q is itself prime. Since $p|q$, we have $p \leq q$ as was proved above, so either q is prime and $q > p_n > \cdots p_1$, or p is a prime dividing q, where p is not among $p_1, p_2, \ldots, p_n$. Thus, our premise that there are only finitely many primes is not tenable, and the proof is complete. *q.e.d.*

■ **Example A.5.10** *For each natural number n, the nth prime p_n satisfies the inequality $p_n < 2^{2^n}$.*

Proof The proof is by induction, and for the basis step, simply observe that $p_1 = 2 < 4 = 2^2 = 2^{2^1}$. For the induction, we suppose that $p_k < 2^{2^k}$ for each $k \leq n$, where $n \geq 2$ and show that $p_{n+1} < 2^{2^{n+1}}$. To this end, consider the primes $p_1 < p_2 < \cdots < p_n$, and define $q = p_1 p_2 \cdots p_n + 1$. As in the previous argument, either q is prime or there is some prime p not among $p_1, p_2, \ldots, p_n$. In either case, the next prime p_{n+1} satisfies the inequality $p_{n+1} \leq q$.

To finish the argument, we must show that $q < 2^{2^{n+1}}$, and to this end, we first notice that

$$q = p_1 p_2 \cdots p_n + 1 = 2(p_2 \cdots p_n) + 1 < 2^2(p_2 \cdots p_n) + 1,$$

where we have used that $p_1 = 2$ in the first equality, so $q \leq 2^2(p_2 \cdots p_n)$ using the fact that if $k < m+1$ then $k \leq m$ for any integers k, m. Thus,

$$q \leq 2^{2^1}(2^{2^2} \cdots 2^{2^n}) = 2^{2^1+2^2+\cdots+2^n} = 2^{2^{n+1}-1} < 2^{2^{n+1}},$$

where we have used the inductive hypothesis, namely, $p_k < 2^{2^k}$ for each $k \leq n$, in the first inequality, and we have used Claim A.5.1 above in the second equality. *q.e.d.*

To recapitulate, there are actually two methods of proof by induction, the principle of induction and the principle of strong induction. The validity of either method of proof has been demonstrated here assuming the Basic Fact which began this section (and which we shall prove later) that any non-empty collection of natural numbers has a least element. These powerful proof techniques are actually applicable in a much more general setting than simply for assertions about the natural numbers, as we shall see, and in the general setting, the principle of strong induction is actually more powerful than the principle of induction. Each method of proof by induction is a valuable tool in many diverse situations and must be mastered by the reader.

EXERCISES

A.5 PROOFS BY INDUCTION FOR NATURAL NUMBERS

1. Prove the following identities.

(a) $\sum_{k=1}^{n} (2k) = n(n+1)$

(a) $\sum_{k=1}^{n} (4k-3) = n(2n-1)$

(c) $\sum_{i=0}^{n} i^2 = \frac{n(n+1)(2n+1)}{6}$

(d) $\sum_{i=0}^{n} i^3 = \left[\frac{n(n+1)}{2}\right]^2$

(e) $\sum_{i=2}^{n} \frac{1}{i^2-1} = \frac{3}{4} - \frac{2n+1}{2n(n+1)}$

(f) $\sum_{i=1}^{n}(2i-1)^2 = \frac{n(4n^2-1)}{3}$

(g) $\prod_{i=2}^{n} (1 + \frac{(-1)^i}{i}) = 1$ for any odd natural number $n \geq 3$

2. Compute the value of the sum $\sum_{i=0}^{n} i(i!)$.

3. Compute the value of the sum $\sum_{i=1}^{n}(2i-1)^3$.

4. Compute the value of the product $\prod_{i=2}^{n} (1 - \frac{(-1)^i}{i})$ for each even natural number n.

5. Prove that for each natural number n, $7^n - 2^n$ is divisible by 5.

6. Prove that for each natural number n, $7^n - 1$ is divisible by 6.

7. Prove that $3^n - 1$ is divisible by 2 for each natural number n.

8. Prove that $2^{2n-1} + 3^{2n-1}$ is divisible by 5 for each natural number n.

9. Prove that for each natural number n, we have $1^3 + 2^3 + \cdots + n^3 = (1 + 2 + \cdots + n)^2$.

10. Prove that for each integer $n \geq 0$ and each pair $a, b \geq 0$ of integers, we have the equality $a^n + b^n \geq [(a+b)/2]^n$.

11. Fix real numbers a and d. A sequence of real numbers of the form

$$a, \ a+d, \ a+2d, \ a+3d, \ldots$$

is called an *arithmetic progression.* Prove that the sum of the terms in a finite arithmetic progression is given by the formula $\sum_{i=0}^{n} (a+id) = \frac{(n+1)(2a+nd)}{2}$.

12. Prove that for each natural number $n \geq 4$, we have the inequality $n^2 \leq 2^n$.

13. Prove that $\sum_{i=1}^{n} \frac{1}{i^2} \leq \frac{2n}{n+1}$ for each natural number n.

14. Prove that every natural number $n \geq 4$ can be written a sum of numbers each of which is a 2 or a 5.

15. Prove that every natural number $n \geq 14$ can be written as a sum of numbers each of which is a 3 or an 8.

16. Prove that every natural number $n \geq 12$ can be written as a sum of numbers each of which is a 4 or a 5.

17. Here is another proof of Example A.5.2:

$$\begin{aligned} (q-1)\sum_{k=1}^{n} q^{k-1} &= q\sum_{k=1}^{n} q^{k-1} - \sum_{k=1}^{n} q^{k-1} \\ &= \sum_{k=2}^{n+1} q^{k-1} - \sum_{k=1}^{n} q^{k-1}. \\ &= q^n - 1 \end{aligned}$$

A sum such as this, where certain terms cancel in pairs, is called a *telescoping sum.* Fill in the gaps of the proof above with careful inductive arguments.

18. Give a proof of Example A.5.8 by induction (not strong induction).

19. Use induction and the product rule $d/dx(fg) = g\ df/dx + f\ dg/dx$ of differential calculus to derive the formula $d/dx(x^n) = nx^{n-1}$ for the derivative of x^n for each $n \geq 0$.

(∗) Part B *METHODS OF PROOF*

It is not a vast oversimplification to assert that much of mathematics boils down to proving that propositions of the form $P \Rightarrow Q$ are true, where P and Q are particular propositions. From the very definition, $P \Rightarrow Q$ is proved to be true if whenever P is true, Q is also true, for $P \Rightarrow Q$ is automatically true whenever P is false. Thus, to prove that $P \Rightarrow Q$ in practice, we may assume that P is true and try to conclude that Q is true.

This second part of Chapter 1 is dedicated to various methods of proving implications. Each subsection contains a discussion of one or several such methods, and each method is associated with a certain tautology just as our proof above that [I am a duck]⇒[I waddle] relied on tautology (∗) in §A.4. Each subsection therefore begins with a short discussion of the relevant tautological propositional forms and then presents numerous examples of the associated proof techniques. In these various examples, we shall prove that $P \Rightarrow Q$ is true for particular propositions P and Q. In our examples, P and Q will often involve mathematical objects such as triangles, natural numbers, and so on, but the reader should keep in mind that our intent is to exhibit logical techniques many of which are already familiar from everyday life. Exercises on the various methods of proof taken together are given at the end of the chapter.

It is probably best in a normal course to just survey in lectures and discussions the material presented in this second half of Chapter 1 and to relegate its more comprehensive study to a serious reading assignment. There is no real loss of continuity in skipping directly to Chapter 2, but the reader who is untiring and carefully studies the examples here will surely profit in insight there. As mentioned in the preface, we do not usually assign problems on this material at this point, but we include a comprehensive section of final exercises for completeness. For another overview of proof techniques which emphasizes discovery rather than formal logic, we refer the reader to the lovely book by Polya, *How to Solve It*, Doubleday Anchor, 1957.

B.1 *CHAINS OF IMPLICATIONS*

Perhaps the most basic method of proof for an implication $P \Rightarrow Q$ relies on tautology (∗) in §A.4. The idea of the method is to find a collection $P_1, P_2, \ldots, P_{n+1}$ of propositions together with a "chain"

$$P_1 \Rightarrow P_2, P_2 \Rightarrow P_3, \ldots, P_n \Rightarrow P_{n+1}$$

of implications beginning with $P = P_1$ and ending with $Q = P_{n+1}$. We claim that if each implication $P_i \Rightarrow P_{i+1}$ is true for each $i = 1, \ldots, n$, then the implication $P \Rightarrow Q$ is true as well.

Indeed, we proved this assertion for $n = 2$ using truth tables in the last subsection, that is, tautology (∗) of §A.4 is exactly our current assertion when $n = 2$ and forms the basis step for an inductive argument. For the inductive step, if $n \geq 3$, then consider the first pair $P_1 \Rightarrow P_2$, $P_2 \Rightarrow P_3$ of implications, so again $P_1 \Rightarrow P_3$, and we have $P_1 \Rightarrow P_3$, $P_3 \Rightarrow P_4$, $\cdots$, $P_n \Rightarrow P_{n+1}$. The inductive hypothesis therefore implies that indeed $P_1 \Rightarrow P_{n+1}$, as required.

To be explicit, we formally present the method of proof just discussed:

> **CHAINS OF IMPLICATIONS** To prove that $P \Rightarrow Q$, find a chain $P_1 \Rightarrow P_2, \ldots, P_n \Rightarrow P_{n+1}$ of implications beginning with $P_1 = P$ and ending with $P_{n+1} = Q$, and separately prove that each implication $P_i \Rightarrow P_{i+1}$ is true, for $i = 1, \ldots, n$.

It is not misleading to think of a path which begins at $P = P_1$, ends at $Q = P_{n+1}$, and passes successively through the "points" $P_1, P_2, \ldots, P_{n+1}$, where there is a directed line segment running from P_i to P_{i+1} and representing the implication $P_i \Rightarrow P_{i+1}$ for each $i = 1, 2, \ldots n$, as in

$$P = P_1 \Rightarrow P_2 \Rightarrow P_3 \Rightarrow \cdots \Rightarrow P_n \Rightarrow P_{n+1} = Q.$$

To apply this method to the proof of an implication given only its hypothesis and conclusion, we must first find some chain of implications as above and then separately prove each implication. It is often not so obvious how to find a suitable chain of implications, and indeed there can be many different such chains. Thus, there can be many different proofs. This is an important point to emphasize: when proving an implication (indeed for *any* proof), there is no single right answer. There are often many different correct proofs of a given implication, and among the correct proofs, each is equally legitimate though some correct proofs may be simpler or more interesting than others. There are likewise many ways to err in discovering or presenting a proof!

Suppose we wish to prove $P \Rightarrow Q$ by finding a chain of implications, and we therefore seek a chain $P_1 \Rightarrow P_2, \ldots, P_n \Rightarrow P_{n+1}$ of provable implications, where $P_1 = P$ and $P_{n+1} = Q$. There are two essentially different approaches with which one might begin: One could start from P with Q in mind and try to find a true implication $P_1 \Rightarrow P_2$ where P_2 is logically more similar to Q than is P, or one could start from Q and try to find a true implication $P_n \Rightarrow P_{n+1}$ where P_n is logically more similar to P than Q is. We refer to the former procedure as a *forward step* and to the latter procedure as a *backward step*.

$P_1 \Rightarrow P_2 \ \cdots \ P_{n+1}$	$P_1 \ \cdots \ P_n \Rightarrow P_{n+1}$
forward step	backward step

Having applied a forward step, for instance, it remains to find a chain of true implications beginning with P_2 and ending with Q, and to this end, we might either apply a forward step from P_2 or a backward step from Q. Taking successive forward and backward steps as one's imagination and intellect suggest, one hopes to discover the required chain of implications beginning with P and ending with Q. Thus, in discovering a suitable chain of implications, one takes small backward and forward steps which hopefully result in the desired chain of implications. It is a kind of "divide-and-conquer" logical strategy. The following examples should illuminate this discussion.

■ **Example B.1.1** Define the propositions

$$M = [\text{I am a mammal}],\ C = [\text{I am a cat}],$$
$$H = [\text{I have hair}],\ N = [\text{I am not a fish}],$$

and assume the implications

$$M \Rightarrow H,\ C \Rightarrow M,\ H \Rightarrow N$$

which are well-known to be true in English and whose truth we assume. There is evidently the chain $C \Rightarrow M,\ M \Rightarrow H,\ H \Rightarrow N$ of implications, and we conclude that $C \Rightarrow N$ is true with two forward steps from C or with two backward steps from N. Thus, we have here discovered a proof that $C \Rightarrow N$. One possible exposition of this result and its proof is:

B.1.1 CLAIM *A cat is not a fish.*

Proof Since a mammal has hair, a cat is a mammal, and an animal with hair is not a fish, it follows that a cat is not a fish. *q.e.d.*

We explicitly state the claim which we wish to prove, and the statement itself will be written in slanted typeface for emphasis as above. There then follows the proof, whose beginning is explicitly indicated with bold typeface and whose end is indicated by the letters "*q.e.d.*", as before. Finally, notice that we do not explicitly say that this is a proof relying on chains of implications since we assume that the reader is sufficiently sophisticated to recognize the underlying logic of the proof.

Before proceeding with further examples, we observe that the previous claim should naturally be interpreted as a quantified predicate. Namely, on the universe of all animals, we might have defined the predicates

$$M(x) = [x \text{ is a mammal}],\ C(x) = [x \text{ is a cat}]$$
$$H(x) = [x \text{ has hair}],\ N(x) = [x \text{ is not a fish}]$$

and assumed that each of the quantified predicates

$$\forall x(M(x) \Rightarrow H(x)),\ \forall x(C(x) \Rightarrow M(x)),\ \forall x(H(x) \Rightarrow N(x)$$

has truth value 1. For each and every $x = x_0$ in the universe of all animals, the implications $M(x_0) \Rightarrow H(x_0)$, $C(x_0) \Rightarrow M(x_0)$, $H(x_0) \Rightarrow N(x_0)$ therefore hold (i.e., have truth value 1 by universal instantiation) since we assume that the universally quantified predicates above are true. We may therefore prove $C(x_0) \Rightarrow N(x_0)$ with a chain of implications as before, and since $x = x_0$ was arbitrary, the implication $C(x_0) \Rightarrow N(x_0)$ therefore holds for each x_0 in the universe. By universal generalization, it follows that $\forall x(C(x) \Rightarrow N(x))$ has truth value 1.

In our examples below, we shall typically prove quantified implications of the form $\forall x(P(x) \Rightarrow Q(x))$ using universal instantiation and generalization, where $P(x)$ and $Q(x)$ have the same universe of discourse. Just as above, we shall introduce various techniques to prove the implication of propositions $P(x_0) \Rightarrow Q(x_0)$ for an arbitrary but fixed x_0, and then conclude that $\forall x(P(x) \Rightarrow Q(x))$ is true by universal generalization. For instance, in this subsection we shall use chains of implications to prove that $P(x_0) \Rightarrow Q(x_0)$ as x_0 ranges over the universe of discourse to conclude that $\forall x(P(x) \Rightarrow Q(x))$. The reasons for presenting the proof techniques below in this context (rather than simply in the context of implications of particular propositions) are purely pedagogical: we can present more interesting examples of proofs at this level of generality.

■ **Example B.1.2** Prove that over the universe of integers n, if $n^2+2n+1=0$, then $n=-1$, that is, prove $\forall n[(n^2+2n+1=0) \Rightarrow (n=-1)]$.

Fixing a particular natural number $n = n_0$, we might take the forward step

$$[n_0^2 + 2n_0 + 1 = 0] \Rightarrow \left[n_0 = \frac{-2 \pm \sqrt{4-4}}{2} = -1 \pm 0\right]$$

using the well-known quadratic formula $[ax^2+bx+c=0] \Rightarrow [x = \frac{-b\pm\sqrt{b^2-4ac}}{2a}]$, which we assume here. Taking finally another forward step $[n_0 = -1 \pm 0] \Rightarrow [n_0 = -1]$ provides a suitable chain of implications and hence a proof. One possible formal presentation is as follows.

B.1.2 CLAIM *For each integer n, if $n^2+2n+1=0$, then $n=-1$.*

Proof If an integer n satisfies the quadratic equation $n^2+2n+1=0$, then $n = \frac{-2\pm\sqrt{4-4}}{2} = -1$ by the quadratic formula. *q.e.d.*

Observe that in our proof, we did not introduce notation for the arbitrary but fixed $n = n_0$ as we did before. This is just a point of notation insofar as n is as reasonable a symbol as n_0 for an arbitrary fixed integer. Also, notice that we do not explicitly say that we are using universal generalization or instantiation to prove the quantified implication; we assume that the reader is sufficiently sophisticated to recognize this fact.

For a proof of Claim B.1.2 using backward steps, we might first observe that $[n+1=0] \Rightarrow [n=-1]$, so then of course, we have $[(n+1)^2=0] \Rightarrow [n+1=0]$ since an integer is zero if its square is zero. Taking the final forward (or backward) step $[n^2+2n+1=0] \Rightarrow [(n+1)^2=0]$, which is an algebraic fact, we find the chain

$$[n^2 + 2n + 1 = 0] \Rightarrow [(n + 1)^2 = 0],$$
$$[(n + 1)^2 = 0] \Rightarrow [n + 1 = 0],$$
$$[n + 1 = 0] \Rightarrow [n = -1]$$

of implications, which give another legitimate proof of Claim B.1.2. A formal presentation of this second proof follows.

Proof Given the integer n, if $0 = n^2 + 2n + 1 = (n + 1)^2$, then also $n + 1 = 0$ since if the square of an integer is zero, then so too is the integer itself. Thus, $n = -1$, as desired. *q.e.d.*

It is interesting to notice that the facts are presented in this proof in the reverse order from that in which they were discovered above. Of course, in part this is a consequence of the fact that we took backward steps in our discovery of the chain of implications. It is not uncommon in mathematics that the order in which a proof is discovered is very nearly the reverse of the order in which it is written down.

It is important to emphasize that there is a distinction between the discovery of a proof and its exposition, as we have already seen in the examples above, and this distinction is fundamental. Whereas some analogue of discovery is surely familiar to the reader from everyday life, proper mathematical exposition is presumably a new skill which must be mastered in bits and pieces as we proceed through this volume.

We next turn to an extended example of a proof using chains of implications and prove a result of Pythagoras. The discovery process will involve a rather long chain of implications, sometimes taking forward and sometimes taking backward steps, and it is therefore a more typical application of chains of implications than those above. At the same time, the distinction between discovery and exposition in our example will be quite pronounced, and the exposition of Pythagoras' result will lead us to define some standard mathematical terms.

■ **Example B.1.3** A *polygon* p is a collection $x_1, x_2, \ldots, x_n$, for $n \geq 3$, of points in the plane together with the line segments connecting x_i and x_{i+1}, for $i = 0, 2 \ldots n$, where we set $x_0 = x_n$ and $x_{n+1} = x_1$ for convenience; we furthermore require that no three consecutive points x_{i-1}, x_i, x_{i+1} are collinear for $i = 1, \ldots, n$. The points x_i are called the *vertices* of p and the line segments are called the *sides* of p. Thus, a polygon with three sides (i.e., $n = 3$) is a triangle. A *chord* of a circle c is a line segment with its endpoints lying in c, and we say that a polygon p *inscribes* in c if each side of p is a chord of c.

We shall prove here that every triangle in the plane inscribes in some circle. More formally, we might introduce the predicate $I(p, c) = [p$ inscribes in $c]$ defined on the universe of all pairs p, c, where p is a polygon and c is a circle in the plane. Thus, a given polygon p inscribes in some circle if $\exists c I(p, c)$ has truth value 1, where c varies over the universe of all circles in the plane. We shall prove here that

$$\forall p[p \text{ is a triangle } \Rightarrow \exists c I(p,c)],$$

where p varies over the universe of all polygons and c over the universe of all circles in the plane.

To prove this, we may take an arbitrary but fixed polygon and show that it inscribes in some circle if it is a triangle. In other words, we may suppose that t is an arbitrary triangle and prove that $\exists c I(t,c)$, that is, we must exhibit a circle c so that $I(t,c)$ has truth value 1. Thus, we consider the implication

"if t is a triangle, then t inscribes in some circle".

In a forward step from the hypothesis, we observe that if t is a triangle, then t determines three non-collinear points in the plane, namely, its vertices; taking a backward step from the conclusion, we observe that if there is a circle c containing the vertices of t, then t inscribes in c. Thus, if we could prove that given any three non-collinear points in the plane, there is some point ζ (ζ is the Greek letter "zeta") equidistant from the vertices, then we can find a chain of implications and complete our proof. In fact, the circle with center at ζ passing through the vertices is the circle in which t inscribes.

This first chain of implications has served simply to better identify the problem at hand: We must show that given three non-collinear points x, y, z in the plane, namely the vertices of t, there is some point ζ equidistant from them.

To attack this problem, consider first just the pair of points x, y, which are necessarily distinct since x, y, z are assumed to be non-collinear. Consider the line segment s with endpoints x, y, and let ℓ denote the line perpendicular to s and passing through the midpoint of s.

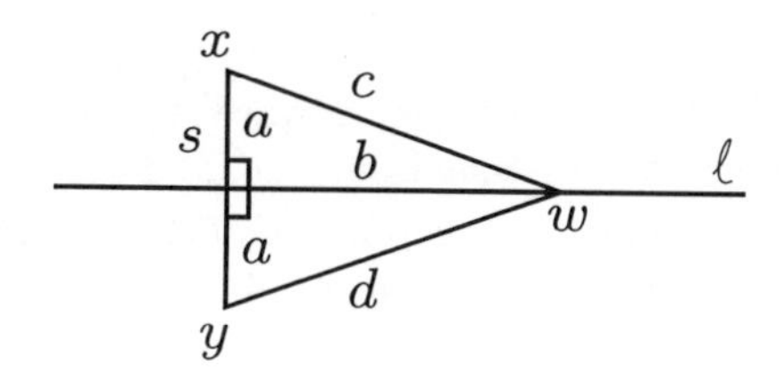

The line ℓ is called the *perpendicular bisector* of x and y. Our interest in this line ℓ stems from the fact that each point on ℓ is equidistant to x and y, that is, we claim that

"if w is a point of ℓ, then w is equidistant from x and y".

To prove this implication, introduce the notation indicated in the previous figure: Let a denote the distance from x to ℓ (so a is also the distance from y to ℓ by definition), let w be an arbitrary point on ℓ, let b denote the distance along ℓ from w to s, let c denote the distance from x to w, and let d denote the distance from y to w.

Taking a backward step from the conclusion (that is, from the proposition that w is equidistant from x and y), we see that if $c = d$, then w is equidistant from x and y. Taking another backward step, we see that if $c^2 = a^2+b^2$ and $d^2 = a^2+b^2$, then $c = d$. Thus, to complete the chain of implications, we must show that if w is a point of ℓ, then $c^2 = a^2 + b^2$ and $d^2 = a^2 + b^2$, and this follows directly from two applications of Pythagoras' famous result on right triangles, which we assume here, since ℓ makes a right angle with s by construction. We have thus proved that points on ℓ are indeed equidistant from x and y.

Of course, we seek a point ζ equidistant from all of x, y, z, and it must evidently be that that ζ lies on the line ℓ. In fact, we shall let $\ell_{x,y}$ denote the perpendicular bisector of x, y and let $\ell_{y,z}$ denote the perpendicular bisector of y, z. In light of what we proved above, each point of $\ell_{y,z}$ is equidistant from y and z as well. The lines $\ell_{x,y}$ and $\ell_{y,z}$ must intersect since x, y, z are non-collinear, and we let ζ denote this point of intersection.

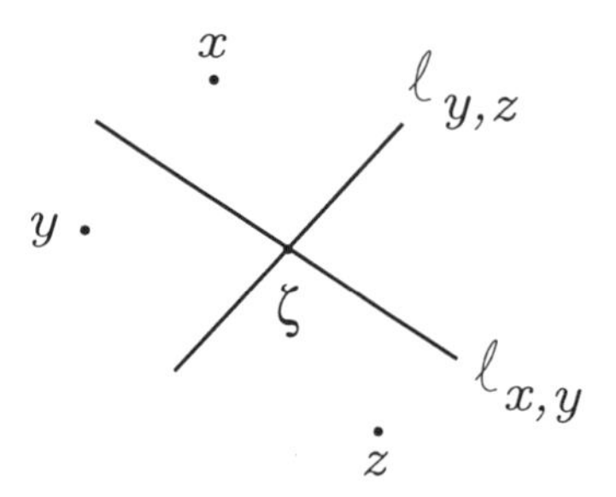

We claim finally that ζ is equidistant from x, y and z. To see this, we take a forward step from "ζ lies on $\ell_{x,y}$ and on $\ell_{y,z}$" to conclude in light of our claims above that ζ is equidistant from x, y and ζ is equidistant from y, z, or in other words $d(\zeta, x) = d(\zeta, y)$ and $d(\zeta, y) = d(\zeta, z)$, where $d(u, v)$ denotes the distance between points u, v in the plane. Taking a final forward step, we conclude $d(\zeta, x) = d(\zeta, y) = d(\zeta, z)$, so ζ is indeed equidistant from x, y, and z.

At this point, we have completed the process of discovering the proof and are logically satisfied that the assertion is correct. Notice that our chain of implications is rather complicated in this example, and that certain inspirations were necessary at various points in order to see exactly how to proceed. These aspects are typical of more complicated proofs using chains of implications, as we shall see.

One possible mathematical presentation of the argument we have just concluded is as follows.

> **LEMMA** *Given two distinct points in the plane, there is a line of points equidistant from them.*
>
> **Proof** Let ℓ be the line perpendicular to and bisecting the segment between the given points. Two applications of Pythago-

ras' result on right triangles shows that points on ℓ are equidistant. *q.e.d.*

PROPOSITION *Given three non-collinear points in the plane, there is a point equidistant from them.*

Proof Let x, y, z denote the given points. The previous lemma asserts the existence of lines $\ell_{x,y}$ and $\ell_{y,z}$ of points equidistant from x, y and y, z, respectively. Since x, y, z are non-collinear, the lines $\ell_{x,y}$ and $\ell_{y,z}$ intersect in a point which is equidistant from x, y, z by construction. *q.e.d.*

THEOREM *Any triangle inscribes in some circle.*

Proof The vertices of a triangle t determine three non-collinear points. Applying the previous proposition, there is some point ζ equidistant from the vertices of t, and the circle centered at ζ and passing through the vertices is a circle in which t inscribes. *q.e.d.*

Again the reader will observe that the order in which the proof is presented is essentially the reverse of that in which it was discovered.

Let us explain the new terms that have arisen in this exposition. A *lemma* is a subsidiary result, sometimes technical, which is used in the proof of another result. A main result is called either a *proposition* or a *theorem*, the key distinction being that a theorem is often a more important result or a more difficult result to prove than a proposition. A *claim* is a result less important than a proposition or even a self-contained fact which one wants to state explicitly; we have seen several examples of claims already.

Thus, the distinction between claim, proposition, and theorem is largely a matter of taste. The appearance of the word "proposition" in its two guises here (as a result less important than a theorem and as a statement which is either true or false as in the first part of this chapter) is unfortunate but usually causes no confusion in practice.

Another related term, a *corollary*, is a result which follows from a particular claim, proposition, or theorem. For instance, we have

B.1.3 COROLLARY *There is a point equidistant from Hong Kong, Moscow, and Los Angeles.*

Proof The three cities are non-collinear, and we simply apply the proposition above to the points determined by these cities in the plane in space containing them. *q.e.d.*

In fact, the equidistant point to these three cities in the plane containing them constructed above actually lies inside the earth. We leave it to the reader to prove as an exercise using further applications of Pythagoras' theorem that there are also two points on the surface of the earth equidistant from these three cities. As a hint for this exercise, we include the following figure.

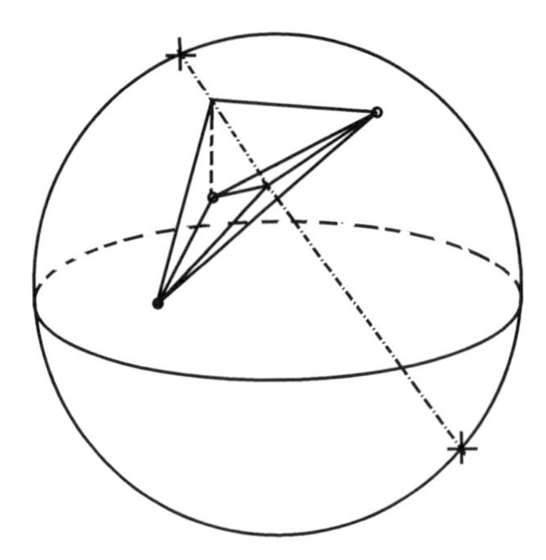

At the risk of waxing too philosophical, suppose for a moment that Pythagoras himself were to magically appear at the door. Believe it or not, to the Pythagoreans, the proof that a triangle inscribes in a circle would have involved a ritualistic sign and whisper, so a Pythagorean would "prove" a theorem with some mystical incantation which was understood by his brethren as a "proof". This cabala would presumably be meaningless to the reader (let alone a convincing proof), and we can only imagine how Pythagoras himself would apprehend our proof given above (let alone the suggestion that the Earth is round). An amazing fact is that despite this lack of intellectual common ground, the reader and Pythagoras *would* agree on the fact that every triangle inscribes in a circle or on the verity of Pythagoras' Theorem itself. This perpetuity of mathematics is truly remarkable.

B.2 PROOF BY CONTRADICTION

This subsection is dedicated to a discussion of the following method of proof.

> **PROOF BY CONTRADICTION** To prove the implication $P \Rightarrow Q$, assume that P is true, assume that $\neg Q$ is true, and prove that something *a priori* known to be true is false. That is, assume that we already know that some proposition R is true, and prove the implication $(P \wedge \neg Q) \Rightarrow \neg R$.

The logic of this method requires recalling from the definition of $\Rightarrow$ that $P \Rightarrow Q$ is false only when P is true and Q is false. Thus, $P \Rightarrow Q$ is true unless these truth values are correct, so we must show that these truth values cannot be correct, that is, we must show that $P \wedge \neg Q$ cannot have truth value 1. To recognize this, we assume that $P \wedge \neg Q$ does actually have truth value 1 and derive from this assumption that some proposition R is false, where R is already known to be true for some reason. The proof that R is false might, for instance, involve a chain of implications starting with $P \wedge \neg Q$ and ending with $\neg R$. The assumption $P \wedge \neg Q$ therefore leads to the contradiction that R is true and $\neg R$ is true, which is impossible according to our definition of proposition. Thus, our assumption cannot have been correct, so it cannot be that $P \wedge \neg Q$ has truth value 1, that is, $P \Rightarrow Q$ cannot be false, so $P \Rightarrow Q$ is true.

The tricky point about a proof by contradiction is that we must find the proposition R. In some examples, the proposition R might be $\neg P$ (or it might be Q), and since we assume P (and we assume $\neg Q$), R is indeed *a priori* false by assumption. In other examples, R amounts to a convention which arises in the course of proving that $(P \wedge \neg Q) \Rightarrow \neg R$. In still other examples, R might be some known fact, for instance, an already-proved result. We shall give instances of each of these possibilities in the examples below and shall let R denote the proposition playing this role in each case. The symbol R is thus reserved in this subsection for this special proposition in each of our proofs by contradiction.

■ **Example B.2.1** We wish to prove the implication $P \Rightarrow Q$, where

$$P = [\text{I am shiftless}], Q = [\text{I am bored}],$$

and we assume that P is true and Q is false, that is, we assume

$$[\text{I am shiftless}] \wedge [\text{I am not bored}].$$

We also introduce the proposition S = [I am idle] and suppose the implications $P \Rightarrow S$, $S \Rightarrow Q$ are true in English.

Now, we assume $P \wedge \neg Q$ is true, so, in particular, P is true. Taking a forward step using $P \Rightarrow S$, we conclude that S is therefore also true. A final forward step from S using $S \Rightarrow Q$, shows that Q is true, but we assume in this proof by contradiction that Q is false. Thus, the proposition denoted R above is actually the proposition $\neg Q$ (that is, the negation of the conclusion of the original implication) in this example.

It is important to emphasize here the different status of the assumptions $P \wedge \neg Q$ and $(P \Rightarrow S) \wedge (S \Rightarrow Q)$ above. The latter assumption is regarded as a known fact in English, and we prove here $P \Rightarrow Q$ in English by showing that the assumption $P \wedge \neg Q$ is not tenable in the sense that it leads to a contradiction. The former assumption is therefore an artifice of our proof, while the latter assumption is one about the English language.

■ **Example B.2.2** In this extended example, we shall prove that each integer is either even or odd but not both even and odd, as was promised above. In other words, letting $E(n) = [n \text{ is even}]$ and $O(n) = [n \text{ is odd}]$ on the universe of integers as before, we shall prove that for each integer n,

$$E(n) \Rightarrow \neg O(n), \tag{1}$$

$$\neg E(n) \Rightarrow O(n). \tag{2}$$

Of course, implication (2) is the inverse of implication (1), so it follows from remarks in §A.4 above that the converse and the contrapositive of implication (1) are also true for each integer n.

A preliminary result which is easily proved using generalization, instantiation, and chains of implications is

B.2.1 LEMMA *For each integer n, if n is even, then $n+1$, $n-1$ are odd, while if n is odd, then $n+1$, $n-1$ are even.*

Proof Fix an arbitrary integer n. If n is even, then $n = 2k$ for some integer k by definition, so $n+1 = 2k+1$ and $n-1 = 2(k-1)+1$ are indeed odd. Similarly, if n is odd, then $n = 2k+1$ for some integer k, so $n+1 = 2(k+1)$ and $n-1 = 2k$ are indeed even. *q.e.d.*

We do not work through the discovery process here since we have already seen several proofs using chains of implications; we have simply presented the result and its formal proof. We shall often abbreviate or even omit a discussion of discovery from now on.

Another preliminary result we shall need is

B.2.2 LEMMA *The integer 0 is even and is not odd.*

Proof To see that 0 is even, we must prove that $\exists k(0 = 2k)$, and this follows from the equality $0 = 2 \cdot 0$. To see that 0 is not odd, we proceed by contradiction and suppose that 0 is odd so $0 = 2k+1$ for some integer k. Solving the previous equation for k, we find $k = -1/2$, which contradicts that k is an integer. *q.e.d.*

It is worth pointing out that this exposition does explicitly indicate that this is a proof by contradiction. One does not have to make such an explicit statment, but it is often a valuable clue to the reader about the proof. In this proof by contradiction, the proposition denoted R above is the statement $[k \text{ is an integer}]$, which is implicit in the specification of the universe over which k varies. As in this example, one typically omits the explicit reference to existential generalization as usual assuming a sufficient level of sophistication on the part of the reader.

Turning finally to the main assertion in this extended example, we have

B.2.3 THEOREM *No integer is both even and odd, and each integer is either even or odd.*

Proof For the first assertion, suppose to get a contradiction that an integer n is both even and odd. Thus, $n \neq 0$ by Lemma B.2.2 above, and we suppose first that n is positive. In this case, $n - 1$ is also both even and odd by Lemma B.2.1, and by induction, we conclude that 0 must likewise be both even and odd, which is in contradiction to Lemma B.2.2 above. In the same way, if n is negative, even, and odd, then $n + 1$ is also even and odd by Lemma B.2.2, and again we conclude by induction that 0 must also be both even and odd, which is again in contradiction to Lemma B.2.2.

For the second assertion, we prove that each integer must be either even or odd and suppose to the contrary that the integer n is neither even nor odd. Thus, $n \neq 0$ by Lemma B.2.2, and we suppose first that n is positive. In this case, $n - 1$ must be neither even nor odd, since if $n - 1$ were even, for instance, then n would be odd by Lemma B.2.1. Thus, $n - 1$ is neither even nor odd, and by induction, we conclude that 0 is neither even nor odd, which is in contradiction to Lemma B.2.2. Similarly, if n is negative and neither even nor odd, then $n + 1$ is neither even nor odd by Lemma B.2.1, and by induction, we again conclude that 0 is neither even nor odd in contradiction to Lemma B.2.2. *q.e.d.*

In this example of proof by contradiction, the proposition playing the role of R above is the previously proved fact Lemma B.2.2, namely, that 0 is even and is not odd.

■ **Example B.2.3** We prove that for any real number x, if $1 \leq x \leq 2$, then $x^2 - 3x + 2 \leq 0$. In this case, in order to derive a contradiction, we assume that $1 \leq x \leq 2$ and $x^2 - 3x + 2 > 0$. Taking a forward step from the second assumption, we find $(x^2 - 3x + 2 > 0) \Rightarrow ((x - 2)(x - 1) = x^2 - 3x + 2 > 0)$. Thus, the product of the numbers $x - 2$ and $x - 1$ is positive, and this can happen in only one of two ways: Either $x - 2$ and $x - 1$ are both positive, or they are both negative. Thus, either $(x > 2) \wedge (x > 1)$ or $(x < 2) \wedge (x < 1)$, i.e., $x > 2$ or $x < 1$, and this contradicts that $1 \leq x \leq 2$.

Thus, the proposition denoted R above is in this case the negation of the hypothesis of the original implication. Having discovered this proof, a formal exposition might be

B.2.4 CLAIM *For any real number x, if $1 \leq x \leq 2$, then $x^2 - 3x + 2 \leq 0$.*

Proof Suppose to the contrary that $1 \leq x \leq 2$ and $(x-2)(x-1) = x^2 - 3x + 2 > 0$. Thus, either $x > 2$ or $x < 1$, and this contradicts that $1 \leq x \leq 2$. *q.e.d.*

It is worth remarking that this exposition again explicitly indicates that this is a proof by contradiction; the words "suppose to the contrary" are intended to communicate this fact to the reader. Before we were even more explicit and began such a proof with a statement such as "We proceed by contradiction and suppose..." for instance.

Notice that in the previous example, we assumed the negation of the conclusion and proved the negation of the hypothesis. Thus, we simply proved the contrapositive of the original assertion. Since we have seen that an implication and its contrapositive have the same truth values, this proves the original implication. We next formalize this method of proof.

PROOF BY CONTRAPOSITIVE In order to prove the implication $P \Rightarrow Q$, we prove instead the contrapositive $\neg Q \Rightarrow \neg P$.

Thus, in this method of proof, one simply replaces the implication by its contrapositive, which one attempts to prove by using one of the various other methods of proof we discuss. This may not seem particularly advantageous, but this reformulation can often be useful especially in that it often sheds light on the meaning of the original implication.

It is worth emphasizing that this is just a special case of proof by contradiction: Rather than assuming $P \wedge \neg Q$ and deriving a contradiction, we assume only $\neg Q$ and derive $\neg P$ from this assumption. One should therefore think of a proof by contrapositive as a special case of a proof by contradiction, where the proposition denoted R above is known *a priori* to be the proposition $\neg P$, i.e., the negation of the hypothesis.

We next give two examples of proofs using the contrapositive.

■ **Example B.2.4** We prove that for any two non-negative integers x, y, if $\sqrt{xy} \neq (x+y)/2$, then $x \neq y$, or more formally as a quantified predicate $\forall x \forall y((\sqrt{xy} \neq (x+y)/2) \Rightarrow (x \neq y))$, where x, y vary over the universe of non-negative integers. As usual in proving a universally quantified predicate, we suppose that x, y are arbitrary non-negative integers and prove the implication $(\sqrt{xy} \neq (x+y)/2) \Rightarrow (x \neq y)$, whose contrapositive is $(x = y) \Rightarrow (\sqrt{xy} = (x+y)/2)$. In this form, the implication is essentially trivial since if $x = y$, then $\sqrt{xy} = \sqrt{x^2} = x = 2x/2 = (x+y)/2$. As in this example, the contrapositive is sometimes an easier implication to understand and therefore to prove than the original implication itself.

A formal presentation of this result might be

B.2.5 CLAIM *For any non-negative integers x, y, if $\sqrt{xy} \neq (x+y)/2$, then $x \neq y$.*

Proof Suppose x, y are arbitrary non-negative integers, and consider the contrapositive. We must show that if $x = y$, then $\sqrt{xy} = (x+y)/2$, and to this end, we observe that if $x = y$, then $\sqrt{xy} = \sqrt{x^2} = x = 2x/2 = (x+y)/2$. *q.e.d.*

Just as in this example, it is sometimes useful to explicitly tell the reader that the argument will rely on the contrapositive; as before, this can be a valuable clue to the reader about the structure of the upcoming proof. It is not, however, necessary to explicitly state the contrapositive since we assume the reader is sufficiently sophisticated to do this herself or himself.

■ **Example B.2.5** We prove that if the square of an integer is even, then so too is the integer itself. That is, we prove that $\forall n(E(n^2) \Rightarrow E(n))$, where $E(n)$ is the predicate $[n \text{ is even}]$ over the universe of integers as before. The natural proof relies on the contrapositive and Theorem B.2.3. Indeed, the contrapositive of the implication $E(n^2) \Rightarrow E(n)$ is $\neg E(n) \Rightarrow \neg E(n^2)$. Fixing an arbitrary integer n, we assume $\neg E(n)$, so n is not even. It follows from Theorem B.2.3 that n is therefore odd, so $n = 2k + 1$ for some integer k. Thus,

$$n^2 = (2k+1)^2 = 4k^2 + 4k + 1 = 2(2k^2 + 2) + 1,$$

so n^2 is odd, whence n^2 is not even again by Theorem B.2.3.

A formal presentation of this argument might be

B.2.6 CLAIM *If the square of an integer is even, then the integer itself is also even.*

Proof Consider the contrapositive, and suppose that n is an not an even integer, so $n = 2k+1$ for some integer k by Theorem B.2.3. Thus, $n^2 = (2k+1)^2 = 2(2k^2 + 2) + 1$ is also odd. *q.e.d.*

We close with an interesting and famous example of a proof by contradiction due to Pythagoras (or perhaps his follower Hippasus) that $\sqrt{2}$ is irrational. To explain this, we recall that a number x is said to be *rational* if x can be written in the form $x = p/q$, where p is an integer and q is a non-zero integer. A real number which is not rational is said to be *irrational.*

More formally, we might introduce the predicate $[x = p/q]$ defined on the universe of all triples x, p, q, where x is a real number, p is an integer, and q is a non-zero integer. Thus, a real number x is rational if and only if $\exists p \exists q(x = p/q)$, and the implication to be proved here is

$$\forall x\big[(x = \sqrt{2}) \Rightarrow \neg(\exists p \exists q[x = \frac{p}{q}])\big].$$

The reader might take a moment to prove that the negation of $\exists p \exists q(x = p/q)$ is $\forall p \forall q(x \neq p/q)$.

■ **Example B.2.6** We prove that $\sqrt{2}$ is irrational. To proceed by contradiction, we suppose that $x = \sqrt{2}$ and x is rational, so

$$x = \sqrt{2} = \frac{p}{q}$$

for some integers p, q, where q is non-zero. We may assume that this fraction p/q is in reduced form, so p and q have no common factors. That is, we assume that $\forall n[(n|p \Rightarrow n \nmid q) \wedge (n|q \Rightarrow n \nmid p)]$.

Now, since $x = \sqrt{2} = p/q$,

$$x^2 = 2 = \frac{p^2}{q^2},$$

so $p^2 = 2q^2$, so p^2 is even. It follows from Claim B.2.6 that p must also be even, so there is some integer k so that $p = 2k$, but then $p^2 = (2k)^2 = 4k^2$, so $2q^2 = p^2 = 4k^2$. Dividing this last equation by two, we find $q^2 = 2k^2$, so q^2 is also even. Again applying Claim B.2.6, we conclude that q itself is even.

We find that both p and q are even, and this contradicts the assumption that p and q have no common factors, for $2|p$ and $2|q$.

Thus, the proposition we contradict, which is denoted R above, is the technical condition which arose during the proof that p and q have no common factors. This completes our discovery of the proof of the Pythagorean result, and a formal presentation of this result follows.

B.2.7 CLAIM $\sqrt{2}$ *is irrational.*

Proof Suppose to get a contradiction that $\sqrt{2} = p/q$, where $q \neq 0$ and p and q are integers with no common factors. Thus, $2 = p^2/q^2$, so $p^2 = 2q^2$, whence p^2 is even. This implies that p itself is even by Claim B.2.6, so $p = 2k$ for some integer k. Thus, $4k^2 = p^2 = 2q^2$ so $q^2 = 2k^2$ is even, whence q is also even as before. We have shown that both p and q are even, which is contrary to our assumption that they have no common factors. *q.e.d.*

To close, we recapitulate the remark made above that the tricky point about a proof by contradiction is that one does not know in advance what contradiction one should aim to derive (except in the case of a direct proof of the contrapositive), that is, one does not know the proposition R in advance.

B.3 INSTANTIATION

We have several times used generalization and instantiation to prove quantified assertions in the previous examples. In this subsection, we discuss how one might use instantiation to prove implications when the hypothesis is a quantified predicate. In contrast to parts of the previous discussion, we shall here be quite explicit about applications of instantiation and generalization.

■ **Example B.3.1** We claim that the only integer dividing 1 is $+1$ or -1. That is, we claim that $\forall n(n|1 \Rightarrow n = \pm 1)$, where n varies over the universe of all integers. We begin by choosing some fixed but arbitrary integer n (intending to later apply universal generalization), and we assume that $n|1$, or, in other words, $\exists k(1 = kn)$, so in particular, $n \neq 0$. By existential instantiation, there is thus some integer k so that $1 = kn$, whence $k = 1/n$, which is an integer if and only if $n = \pm 1$. Finally, since n was arbitrary, the claim follows from universal generalization.

A formal presentation of this proof might be

B.3.1 CLAIM *If n is an integer which divides 1, then $n = \pm 1$.*

Proof Suppose that $n|1$, so there is some integer k so that $1 = kn$. Solving for k, we have $k = 1/n$, which is an integer if and only if $n = \pm 1$. *q.e.d.*

Our next two examples illustrate the use of universal instantiation when the hypothesis is a universally quantified assertion.

■ **Example B.3.2** We claim that $\forall m([\forall n(m|n)] \Rightarrow [m = \pm 1])$, where m and n vary over the universe of integers, that is, we claim that the only integers dividing each and every integer are $+1$ and -1. To see this, choose (in order to later apply universal generalization) an integer m and prove $[\forall n(m|n)] \Rightarrow [m = \pm 1]$. We proceed directly and assume that $\forall n(m|n)$, so by universal instantiation, $m|n$ for the integer $n = 1$, that is, $m|1$. It follows from Claim B.3.1 that $m = \pm 1$, as was asserted.

A formal presentation of this argument is

B.3.2 CLAIM *The only integers which divide every integer are ± 1.*

Proof Suppose that n is an integer which divides every integer. In particular, n divides 1, so $n = \pm 1$ by Claim B.3.1. *q.e.d.*

■ **Example B.3.3** We prove that $\forall n([\forall m(m|n)] \Rightarrow [n = 0])$, where both m and n vary over the universe of all integers. We begin (intending to use universal generalization) by picking some fixed but arbitrary integer n. In order to prove the implication $[\forall m(m|n)] \Rightarrow [n = 0]$, we argue directly and assume that $\forall m(m|n)$ is indeed true. In particular (applying universal instantiation), $m|n$ for $m = 0$, that is, $0|n$, or, in other words, $\exists k(n = k \cdot 0)$. By existential instantiation, there is some particular k so that $n = k \cdot 0 = 0$, so $n = 0$, as was asserted.

A formal presentation of this might be

B.3.3 CLAIM *If n is an integer which is divided by every other integer, then $n = 0$.*

Proof Suppose that n is an integer which is divided by every other integer. In particular, 0 divides n, so $n = k \cdot 0 = 0$ for some integer k. *q.e.d.*

Our next example illustrates a hybrid proof technique involving existential instantiation and the contrapositive.

■ **Example B.3.4** We claim that $\forall m \forall n[(n \neq \pm 1) \Rightarrow (n|(m-1) \Rightarrow n \nmid m)]$, where m, n vary over the universe of integers. That is, we claim that no integer other than ± 1 can divide both a number and its successor. To prove this, let us fix arbitrary integers n, m and consider the contrapositive, so we assume that $n|(m-1) \Rightarrow n \nmid m$ is false, that is, we assume that $n|m$ and $n|(m-1)$. By two applications of existential instantiation, we find integers k, ℓ satisfying $m = \ell n$ and $m - 1 = kn$. Taking the difference of these two equations, we find $1 = m - (m-1) = \ell n - kn = (\ell - k)n$, so by existential generalization $n|1$. By Claim B.3.1, we conclude that $n = \pm 1$, thus establishing the contrapositive. Since m and n were arbitrary, we conclude using universal generalization that the original quantified implication indeed has truth value 1.

A possible formal presentation of this argument is

B.3.4 CLAIM *No integer other than* ± 1 *can divide both an integer and its successor.*

Proof Suppose to the contrary that $n \neq \pm 1$ and m are integers where $n|m$ and $n|(m-1)$. There are therefore integers k, ℓ satisfying $m = \ell n$ and $m-1 = kn$, so taking the difference, we find $1 = m - (m-1) = \ell n - kn = (\ell - k)n$. Thus, $n|1$, so $n = \pm 1$ by Claim B.3.1, which is contrary to hypothesis. *q.e.d.*

As we have seen in these examples, universal and existential instantiation can be useful techniques for proving an implication whose hypothesis is a quantified predicate.

B.4 CONSTRUCTIVE AND NON-CONSTRUCTIVE PROOFS

This subsection is dedicated to a discussion of methods of proof for existentially quantified predicates. According to existential generalization (indeed, from the very definition of the truth values of $\exists x P(x)$), in order to prove $\exists x P(x)$, we might exhibit some x_0 in the universe of discourse of $P(x)$ so that $P(x_0)$ has truth value 1. Though we have seen this method before, we begin with some examples.

■ **Example B.4.1** We prove that $\forall a \forall b \forall c([(a|b) \wedge (b|c)] \Rightarrow [a|c])$, where a, b, c vary over the universe of integers. To see this, we may choose arbitrary a, b, c, and suppose that $a|b$ and $b|c$, so by existential instantiation, there are integers k, ℓ satisfying $b = ka$, $c = \ell b$. We wish to prove that $a|c$, or, in other words, that $\exists m(c = ma)$, where m varies over the universe of integers. We prove this by exhibiting such an integer m, namely, since $c = \ell b = \ell(ka) = (\ell k)a$, we may take $m = \ell k$. Having exhibited the integer m, we conclude $a|c$, and since a, b, c were arbitrary, the original quantified assertion follows from universal generalization. A formal exposition of this argument follows.

B.4.1 CLAIM *Given integers a, b, c, if $a|b$ and $b|c$, then $a|c$.*

Proof Assuming $a|b$ and $b|c$, there are integers k, ℓ so that $b = ka$ and $c = \ell b$. Thus, $c = \ell b = \ell(ka) = (k\ell)a$, as desired. *q.e.d.*

In this example, we have proved the existentially quantified assertion $a|c$ by exhibiting the required integer $k\ell$. It is worth emphasizing that no explanation of how we found this integer is actually necessary here; all we must do to give a legitimate proof is exhibit this integer, but a more complete proof, as above, is certainly preferable. For a more extreme example of this phenomenon, we have

■ **Example B.4.2** *Given integers a, b, c, d so that $ad - bc \neq 0$ and given integers u, v, there are rational numbers x, y so that*

$$ax + by = u \text{ and } cx + dy = v.$$

Proof Take

$$x = \frac{du - bv}{ad - bc} \quad \text{and} \quad y = \frac{av - cu}{ad - bc},$$

and check that the required equations hold. *q.e.d.*

Again this proof relies on existential instantiation and universal generalization, but here we give no indication of how we discovered the solution. We simply exhibit the solution and tell the reader to check that the required equations hold. This is simply a matter of arithmetic, which the reader should actually check. Though the reader may be unsatisfied to have not gained any real mathematical insight into solving such equations (and there is a whole theory here called "linear algebra"), this is a perfectly legitimate proof of our claim.

Each of the two examples above involve existentially quantified predicates, and in each case we in fact construct the required object, namely, the integer m in Example B.4.1 and the rational numbers x, y in Example B.4.2. There is actually another method of proving an existentially quantified predicate $\exists x P(x)$, as we next formalize

CONSTRUCTIVE PROOF To prove $\exists x P(x)$, exhibit or construct some x_0 in the universe so that $P(x_0)$ is true.

NON-CONSTRUCTIVE PROOF To prove $\exists x P(x)$, prove that there must be some x_0 in the universe so that $P(x_0)$ is true without actually exhibiting it.

The two proofs above are indeed constructive, and we next give some examples of non-constructive proofs.

■ **Example B.4.3** *For any predicate $P(x)$ over a universe $\mathcal{U}$, we have the implication $[\forall x P(x)] \Rightarrow [\exists x P(x)]$.*

Proof Suppose $\forall x P(x)$ and choose some x_0 in the universe $\mathcal{U}$, which we may do since $\mathcal{U}$ is assumed to be non-empty. By universal instantiation $P(x_0)$ holds, so $\exists x P(x)$ holds by existential generalization. *q.e.d.*

Of course, this proof gives no clue as to how to find x_0. However it assures us of the existence of x_0 just the same and is therefore a legitimate proof. For another example, consider

■ **Example B.4.4** *For each integer n, we have $3|(n^3 - n)$.*

Proof Choose an integer n and observe that

$$n^3 - n = n(n^2 - 1) = n[(n-1)(n+1)] = (n-1)\ n\ (n+1),$$

so $n^3 - n$ is the product of three consecutive integers. Since every third integer is a multiple of 3, i.e., 3 divides every third integer, exactly one of $n-1, n, n+1$ is actually a multiple of 3. In other words, for one of these three integers, say the integer m, we have $3|m$. On the other hand, $m|(n^3 - n)$ by definition, so it follows from Claim B.4.1 that $3|(n^3 - n)$. *q.e.d.*

Again the proof is non-constructive in the sense that we do not explicitly construct the integer k so that $n^3 - n = 3k$. We could actually easily pursue the ideas above to give another constructive proof of Example B.4.4, as follows: Setting $m = 3\ell$, if $m = n$, then $k = (n^2-1)\ell$, while if $m = n\pm 1$, then $k = (n^2 \mp n)\ell$. The reader should not infer that one can always find a constructive proof if one has a non-constructive proof as in this example.

In fact, there are statements which admit no known constructive proof yet can be proved non-constructively. For instance, Example B.4.3 above is of this type, and there are many other more mathematically profound examples as well. As a historical and philosophical point, we mention that there have been schools of philosophers and mathematicians called "intuitionists" who have rejected a non-constructive proof as invalid, i.e., not legitimate. Nowadays, essentially all mathematicians are perfectly willing to accept a non-constructive proof as valid though a constructive proof is usually regarded as "better" than a non-constructive one for the obvious reason: For instance, a constructive proof is often more enlightening.

An interesting consequence of the previous comments is that the notion of what constitutes a legitimate proof is apparently culturally dependent. Our imaginary conversation with Pythagoras at the end of §B.1 already illustrated this intriguing phenomenon: There is no absolute standard of proof from a historical perspective, and the notion of "rigor" in mathematics has actually changed over the ages. There are several aspects to this: There can be philosophical objections to a proof (for instance, intuitionism), and there are also cultural agreements on what constitutes a valid proof and especially its exposition. The perpetuity of mathematical results which was discussed above in the context of our imaginary conversation with Pythagoras is all the more remarkable in light of this mutability of what constitutes a legitimate proof. Students sometimes find this mutability disturbing, but one can only hope to reflect the standards of rigor of her or his own era. We hasten to add that the methods of proof we present in this chapter are all well within the bounds of what is considered valid proof these days.

B.5 DISPROOFS

Having spent most of this chapter illustrating how one goes about proving an assertion is true, we turn finally to the question of how one might prove that an assertion is false. For the simplest example, we may show a particular proposition P is false by showing that its negation $\neg P$ is true. To prove that $P \Rightarrow Q$ is false for particular propositions P and Q, we must show that P is true and Q is false; this follows immediately from the definition of implication. To show that $\exists x P(x)$ is false, we must show that $P(x_0)$ is false for each member x_0 of the universe (that is, we must show $\forall x \neg P(x)$ is true), whereas to show that $\forall x P(x)$ is false, we must simply exhibit some x_0 in the universe so that $P(x_0)$ is false (that is, we must show $\exists x \neg P(x)$ is true).

Implicit in this discussion is the understanding which follows from our definitions of proposition and negation that either a proposition or its negation must be true, and no proposition can be both true and false. We make this remark primarily to mention that there have been schools of philosophers who have rejected these tenets, which are called the "law of the excluded middle" (since there is no "middle ground" between true and false) and the "law of contradiction" respectively. In particular, the intuitionists (who were briefly discussed in §B.4 above) rejected these so-called "rules of thought" and in particular did not regard as valid a proof by contradiction. Furthermore, rejecting only the law of excluded middle but maintaining most of the other features of the mathematical logic which we describe here leads to so-called "fuzzy" logic", which actually has important programming applications. We hasten to add that the law of excluded middle and the law of contradiction are entirely accepted by current standards, and the disproofs above would therefore be considered valid by essentially every contemporary mathematician.

EXERCISES

B METHODS OF PROOF

1. Prove or disprove.

(a) If n is prime, then $2^n + 1$ is prime.

(b) The converse of part (a).

2. Consider the equation $a^3 + b^3 = c^3$ for natural numbers a, b, c.

(a) Prove that this equation has no solution where a, b, c are all prime.

(b) Prove or disprove that this equation has no solution where a, b, c are consecutive natural numbers.

(c) Prove or disprove that this equation has no solution where a, b, c are consecutive integers.

3. Prove that if p is a prime and n is an integer so that p divides n^2, then in fact, p^2 divides n^2.

4. Prove that at a party of $n \geq 2$ people, there must be at least two people at the party who have the same number of friends at the party

5. Prove that any natural number n is a prime, a square, or divides $(n-1)!$.

6. A *perfect number* is a natural number which is equal to the sum of all its divisors smaller than itself; for instance 6=3+2+1 is perfect. Prove that a perfect number cannot be prime.

7. Prove that if a, b, c are odd integers, then the equation $ax^2 + bx + c = 0$ does not have a rational solution x.

8. Prove that $n! + 1$ is odd if n is a natural number other than 1.

9. How many pairs p, q of primes are there so that $p - q = 3$? Prove your assertion.

10. How many triplets of primes are there of the form $p, p+2, p+4$? Prove your assertion.

11. Suppose that x, y are positive real numbers. Their *arithmetic mean* is $A = (x+y)/2$, their *geometric mean* is $G = \sqrt{xy}$, and their *harmonic mean* is $H = 2xy/(x+y)$. Prove that $A \geq G \geq H$ with equality if and only if $x = y$.

12. Prove that $\sqrt{8}$ is irrational.

13. Prove that $\sqrt{6}$ is irrational.

14. Prove that $x^2 + y^2 + z^2 \geq xy + yz + xz$ for each triple x, y, z of real numbers.

15. Given positive real numbers a, b, c, prove that $a < b+c$, $b < a+c$, and $c < a+b$ simultaneously hold if and only if $2a^2b^2 + 2b^2c^2 + 2a^2c^2 - a^4 - b^4 - c^4 > 0$.

16. Prove or disprove that every polygon inscribes in a circle.

17. Prove or disprove that for every real number x satisfying $-1 < x < 1$, there is a real number y satisfying $-1 < y < 1$ and so that $x^2 + y^2 \leq 1$.

18. Prove that $t + 1/t \geq 2$ for any positive real number t.

19. Prove that if x and y are real numbers, then $|x+y| \leq |x| + |y|$.

Chapter 2

Predicate Calculus

Having discussed various methods of proof in the first chapter, we now turn to a systematic study of propositions, predicates, and their logical relationships. We admit here that we were somewhat cavalier in the first chapter with regard to explicitly specifying our assumptions about English, the integers, and so on; we can only hope that the reader was not dismayed by this, and the various methods of proof were satisfactorily communicated. There will be no more such funny business. We shall be most precise about any assumptions made from now on. Indeed, it is the purpose of this chapter and the next to discuss the basic assumptions of all mathematics!

This chapter is dedicated to studying the "predicate calculus" which gives precise rules for manipulating propositions and predicates. We assume here that the reader has studied Part A of Chapter 1 and adopt the definitions given there. The word "calculus" is used in this context to mean a method of calculation, and the calculations we perform involve propositions and predicates. As we have seen in the first chapter, certain abstract logical relationships among propositional and predicate variables can be used as paradigms (that is, models or "blue-prints") for methods of proof which are applicable in many diverse particular situations, mathematical and otherwise. Thus, we study here a calculational tool for analyzing logical relationships between propositions and predicates, and these logical relationships find applications throughout mathematics. Furthermore, as we shall see, the foundations of mathematics rest squarely on the predicate calculus itself.

2.1 LOGICAL OPERATORS

Suppose that P and Q are particular propositions. We may construct new propositions from P and Q, where the truth value of the new proposition depends in some specified way upon the truth values of P and Q. For instance, the conjunction $P \wedge Q$ considered before is such an example, where the truth value of $P \wedge Q$ was defined by specifying a truth table. A symbol such as "$\wedge$" which allows the construction of new propositions from old is called a *logical operator*, and "$\wedge$" is just one among five standard logical operators which we consider here. We begin by simply listing the various symbols designating the logical operators, their names, and their English translations.

- The symbol $\neg$ is called the *negation operator.* $\neg P$ is called the *negation* of P, and $\neg P$ is read in English as "not P".

- The symbol $\wedge$ is called the *conjunction operator.* $P \wedge Q$ is the *conjunction* of P and Q, and $P \wedge Q$ is read in English as "P and Q".

- The symbol $\vee$ is called the *disjunction operator.* $P \vee Q$ is the *disjunction* of P and Q, and $P \vee Q$ is read in English as "P or Q".

- The symbol $\Rightarrow$ is called the *implication operator.* $P \Rightarrow Q$ is the *implication* with *hypothesis* P and *conclusion* Q, and $P \Rightarrow Q$ is read in English as "P implies Q" or as "if P, then Q".

- The symbol $\Leftrightarrow$ is called the *bi-implication operator.* $P \Leftrightarrow Q$ is the *logical equivalence* of P and Q, and $P \Leftrightarrow Q$ is read in English as "P if and only if Q".

Notice that the negation operator is applied to a single proposition, whereas the others are all applied to pairs of propositions. For this reason, $\neg$ is called a *unary logical operator*, whereas the other logical operators $\wedge$, $\vee$, $\Rightarrow$, and $\Leftrightarrow$ are called *binary logical operators.*

Just as before for $\wedge$ and $\neg$, we specify truth values in a truth table to formally define the various logical operators:

P	Q	$\neg P$	$P \wedge Q$	$P \vee Q$	$P \Rightarrow Q$	$P \Leftrightarrow Q$
0	0	1	0	0	1	1
0	1	1	0	1	1	0
1	0	0	0	1	0	0
1	1	0	1	1	1	1

Of course, this truth table serves to define the logical operators (and no further comment is really necessary), but we include here various remarks on the definitions. We have already observed in Part A of Chapter 1 that these definitions of negation, conjunction, and implication faithfully correspond to their English equivalents. Notice that $P \vee Q$ is true if either P or Q is true or if both P and Q are true, and this is the usual meaning of the word "or" in English. Also, $P \Leftrightarrow Q$ is true only when P and Q have the same truth values, and again this faithfully captures the English expression "if and only if". As a point of notation, we mention that one sometimes uses the symbol "iff" abbreviating the phrase "if and only if" in mathematics at the blackboard in place of the bi-implication operator.

EXERCISES

2.1 LOGICAL OPERATORS

1. Construct truth tables for the following expressions.

(a) $(Q \wedge \neg P) \Rightarrow P$

(b) $\neg(P \wedge Q) \Rightarrow (\neg P \wedge \neg Q)$

(c) $(\neg P \vee Q) \wedge (P \wedge \neg Q)$

(d) $[P \wedge (Q \vee R)] \Leftrightarrow [(P \wedge Q) \vee (P \wedge R)]$

(e) $[(P \wedge Q) \Rightarrow R] \Rightarrow [P \Rightarrow (Q \Rightarrow R)]$

2. Define another binary logical operator $\oplus$ called the *exclusive or* operator with the following truth table:

P	Q	$P \oplus Q$
0	0	0
0	1	1
1	0	1
1	1	0

(a) Construct truth tables for the expressions $P \oplus P$ and $P \oplus \neg P$.

(b) Construct truth tables for the expressions $(P \wedge Q) \oplus (P \wedge R)$ and $P \wedge (Q \oplus R)$.

(c) Find an expression in the propositional variables P and Q using $\vee, \wedge, \neg$ which has the same truth values as $P \oplus Q$.

3. Define another binary logical operator $\uparrow$ called the *Sheffer stroke* or *nand* (which is short for "not-and") operator with the following truth table:

P	Q	$P \uparrow Q$
0	0	1
0	1	1
1	0	1
1	1	0

(a) Construct truth tables for the expressions $P \uparrow P$ and $P \uparrow \neg P$.

(b) Construct truth tables for the expressions $P \wedge (Q \uparrow R)$ and $(P \wedge Q) \uparrow (P \wedge R)$.

(c) Find an expression in the propositional variables P and Q using $\vee, \wedge, \neg$ which has the same truth values as $P \uparrow Q$.

4. Define another binary logical operator $\downarrow$ called the *Pierce arrow* or *nor* operator with the following truth table:

P	Q	$P \downarrow Q$
0	0	1
0	1	0
1	0	0
1	1	0

(a) Construct truth tables for the expressions $P \downarrow P$ and $P \downarrow \neg P$.

(b) Construct truth tables for the expressions $P \wedge (Q \downarrow R)$ and $(P \wedge Q) \downarrow (P \wedge R)$.

(c) Find an expression in the propositional variables P and Q using $\vee, \wedge, \neg$ which has the same truth values as $P \downarrow Q$.

2.2 PROPOSITIONAL FORMS

Now suppose that $P, Q, R, \ldots$ are a list of propositions and propositional variables. We may produce expressions, by which we mean simply strings of symbols, involving these propositional variables by repeatedly applying the logical operators above, and as a point of notation, we shall always enclose the result of a logical operation in parentheses for now. For instance, we write $(P \wedge (Q \wedge R))$ to indicate that we first construct the expression $(Q \wedge R)$ and then finally construct the expression $(P \wedge (Q \wedge R))$. We shall also sometimes use square brackets rather than parentheses in order to improve the readability of an expressions; thus, we might write $[P \wedge (Q \vee R)]$ instead of $(P \wedge (Q \vee R))$ for instance.

We introduce here the *logical constants* 0 and 1 (which we have used before to represent the truth values of propositions), where 0 is the "constant proposition" which is always false, and 1 is the "constant proposition" which is always true. The logical constants may be used in place of propositions or propositional variables in logical operators in the natural way. We may write, for instance, $(P \wedge 1)$ for the conjunction of P and 1, and the associated truth value is determined from the truth table above for $(P \wedge Q)$ by specializing to the rows where Q has truth value 1; thus, $(P \wedge 1)$ has the same truth values as P itself.

With these rules, various strings of symbols involving logical operators, logical constants, propositions, and propositional variables are "legal" (that is, legitimate) expressions. For instance, $[P \vee (Q \wedge R)]$ and $(0 \vee 1)$ are legitimate expressions. On the other hand, $((P$ and $(P \neg Q)$ are not legitimate expressions since we are not combining symbols in accordance with their rules of usage; in the former case, we do not close the parentheses which we have opened, and in the latter case, we seem to be pretending that $\neg$ is a binary operator rather than a unary one. A "legal" string of symbols is called a *propositional form* or simply a *form*, and these are the basic objects of interest for us now. As a point of notation, if P and Q are propositional forms, then we write "$P = Q$" if P and Q are identical as string of symbols. A more formal treatment of such "legal" expressions will be given in the next chapter. Also, we wish to emphasize that our discussion so far does not involve predicates; we shall extend our discussion to include predicates only later in this chapter.

Actually, certain of the logical operators above can be defined as propositional forms in terms of the others. For instance, the reader can check using truth tables that $(P \Leftrightarrow Q)$ has exactly the same truth values as $([P \Rightarrow Q] \wedge [Q \Rightarrow P])$ and that $(P \vee Q)$ has exactly the same truth values as $(\neg[(\neg P) \wedge (\neg Q)])$. For a more interesting and important example, the reader should check again using truth tables that $(P \Rightarrow Q)$ has the same truth values as $[(\neg P) \vee Q]$. In particular, it follows that the various logical operators above can actually be defined as propositional forms using conjunction and negation alone.

More generally, we define two propositional forms to be *logically equivalent* if they have exactly the same truth values for any assignment of truth values to their component propositional variables. Thus, we have seen in the previous paragraph that $(P \Rightarrow Q)$ is logically equivalent to $[(\neg P) \vee Q]$, for instance. One of the basic problems in the predicate calculus is recognizing when two propositional forms are logically equivalent. Of course, one method of recognizing this involves comparing truth values in a truth table, but we shall develop more interesting and powerful techniques for this problem here.

There is a basic trichotomy for propositional forms, as follows: We say a propositional form P is a *tautology* if it is true for any assignment of truth values to its component propositional variables, that is, if $P \Leftrightarrow 1$; at the other extreme, P is an *absurdity* if it is false for any assignment of truth values to its component propositional variables, that is, if $P \Leftrightarrow 0$; finally, in the remaining case that P is false for some assignment of truth values and true for some other assignment of truth values to its component propositional variables, we say that the propositional form P is a *contingency*.

For some examples, $([P \vee Q] \Leftrightarrow [\neg((\neg P) \wedge (\neg Q))])$ and $(P \Rightarrow [P \vee Q])$ are tautologies for instance. Numerous other examples of tautologies will be given in the next sections. The reader can easily check that $(P \wedge (\neg P))$ and $([(P \Rightarrow Q) \wedge P] \wedge (\neg Q))$, for instance, are absurdities, and that P and $(P \Rightarrow Q)$, for instance, are contingencies.

Observe that if two propositional forms P and Q are logically equivalent, then $(P \Leftrightarrow Q)$ is a tautology by definition, so in this sense a logical equivalence is just a special case of a tautology. As a point of terminology, a tautology of the form $(P \Leftrightarrow Q)$ is called a *logical identity*.

As another point of terminology, when we speak of "proving" a propositional form P, we mean proving that P is a tautology, whereas by "disproving" P, we mean showing that it is a contingency or an absurdity, that is, showing it is false for some assignment of truth values to its component propositional variables.

EXERCISES

2.2 PROPOSITIONAL FORMS

1. Establish with a truth table whether each of the following propositional forms is a tautology, contingency or absurdity.

(a) $P \vee \neg P$

(b) $P \Rightarrow [\neg(\neg P)]$

(c) $\neg(P \wedge Q) \Leftrightarrow [(\neg P) \vee (\neg Q)]$

(d) $(P \Rightarrow Q) \Leftrightarrow [(\neg Q) \Rightarrow (\neg P)]$

(e) $(P \Rightarrow Q) \wedge (Q \Rightarrow P)$

(f) $[P \vee (\neg Q)] \Rightarrow Q$

(g) $P \Rightarrow (P \vee Q)$

(h) $[(P \wedge Q) \Leftrightarrow Q] \Leftrightarrow [P \Leftrightarrow Q]$

(i) $[(\neg P) \vee Q] \wedge [P \wedge (\neg Q)]$

2. (a) Establish the tautology $[P \Rightarrow Q] \Leftrightarrow [(\neg P) \vee Q]$ with a truth table.

(b) Establish the tautology $[P \vee Q] \Leftrightarrow [\neg((\neg P) \wedge (\neg Q))]$ with a truth table.

(c) For each of the propositional forms $P \vee Q$, $P \Rightarrow Q$, and $P \Leftrightarrow Q$ find another propositional form involving only $\wedge$ and $\neg$ with the same truth values.

3. For each of the propositional forms $P \oplus Q$, $P \uparrow Q$, and $P \downarrow Q$, where $\oplus$, $\uparrow$, and $\downarrow$ denote exclusive or, nand, and nor respectively, find propositional forms involving only $\wedge$ and $\neg$ with the same truth values.

4. For each possible choice of two logical operators from among $\vee$, $\wedge$, $\neg$, $\Rightarrow$, and $\Leftrightarrow$, determine whether the truth values of the other specified logical operators can be expressed as propositional forms in terms of the two chosen logical operators. You may also freely use the logical constants 0 and 1.

2.3 PARENTHESES

We introduce conventions here to hereafter avoid the debauch of parentheses and brackets inherent in our definitions, and the first convention is quite simple:

2.3.1 CONVENTION *Always drop the outermost parentheses in a propositional form.*

For instance, we could write $P \vee Q$ rather than $(P \vee Q)$ for the disjunction of P and Q using this convention. Our next convention involves the negation operator:

2.3.2 CONVENTION *The negation operator applies to the smallest possible sub-expression consistent with parentheses.*

Armed with this convention, we can often drop the parentheses surrounding the negation of a propositional form. For instance, rather than writing $(\neg P) \wedge Q$, we could write simply $\neg P \wedge Q$. If, in fact, we intended to negate the conjunction $P \wedge Q$, then we would write $\neg(P \wedge Q)$ to indicate that the entire expression inside parentheses is to be negated.

Our final convention, about which we shall speak further below, involves conjunctions and disjunctions:

2.3.3 CONVENTION *Suppose that $P_1, \ldots, P_n$ are a collection of propositions, where $n \geq 1$. We let $P_1 \wedge P_2 \wedge \cdots \wedge P_n$ denote the particular conjunction $(\cdots(P_1 \wedge P_2) \wedge \cdots \wedge P_n)$, and we let $P_1 \vee P_2 \vee \cdots \vee P_n$ denote the particular disjunction $(\cdots(P_1 \vee P_2) \vee \cdots \vee P_n)$*

The first point of this convention is that given the ordered collection $P_1, \ldots, P_n$, there are many possible ways to insert parentheses and $\wedge$'s to get a legal propositional form, and we are choosing one of these to be written entirely without parentheses with a similar comment for $\vee$'s. For instance for $n = 3$, there are exactly two such propositional forms for $\wedge$, namely $(P_1 \wedge P_2) \wedge P_3$ and $P_1 \wedge (P_2 \wedge P_3)$, and Convention 2.3.3 specifies that the former propositional form is denoted simply $P_1 \wedge P_2 \wedge P_3$. Furthermore, one proves these two propositional forms are logically equivalent (as is clear in English) by considering the truth table, and we therefore say that $\wedge$ is "associative". Of course, $\vee$ is also "associative" in the sense that $(P_1 \vee P_2) \vee P_3$ and $P_1 \vee (P_2 \vee P_3)$ are logically equivalent. When $n = 4$, there are three such propositional forms for $\wedge$, namely, $(P_1 \wedge P_2) \wedge (P_3 \wedge P_4)$, $P_1 \wedge (P_2 \wedge (P_3 \wedge P_4))$, and $((P_1 \wedge P_2) \wedge P_3) \wedge P_4$; the convention allows us to write the last such form as $P_1 \wedge P_2 \wedge P_3 \wedge P_4$, and the three forms are again found to be logically equivalent.

A good analogy to imagine is the sum $x_1 + x_2 + \cdots + x_n$ of an ordered collection $x_1, x_2, \ldots, x_n$ of numbers where $n \geq 3$. As the reader well knows from elementary school, "addition is associative", i.e., $(x_1 + x_2) + x_3 = x_1 + (x_2 + x_3)$ for any numbers x_1, x_2, x_3. Thus, the order of additions does not affect the value of the resulting sum, and we may write simply $x_1 + x_2 + \cdots + x_n$ with no parentheses to unambiguously designate this number. In fact, $\wedge$ and $\vee$ are also "associative" in the sense discussed above, and we are writing the conjunction or disjunction without parentheses to signify a particular propositional form among these logically equivalent ones in analogy to addition.

Of course, the reader also learned in elementary school that "addition is commutative", i.e., $x_1 + x_2 = x_2 + x_1$ for any numbers x_1, x_2, so we might even re-order the numbers $x_{i_1}, x_{i_2}, \ldots, x_{i_n}$ and get the same sum

$$x_{i_1} + x_{i_2} + \cdots + x_{i_n} = x_1 + x_2 + \cdots + x_n,$$

where each $i = 1, 2, \ldots, n$ occurs once and only once in the list $i_1, i_2, \ldots, i_n$. The analogy between addition and $\wedge$ or $\vee$ actually goes deeper in the sense that $\wedge$ and $\vee$ are also "commutative" in the sense that

$$[P_1 \wedge P_2] \Leftrightarrow [P_2 \wedge P_1] \text{ and } [P_1 \vee P_2] \Leftrightarrow [P_2 \vee P_1],$$

as the reader may again prove by truth table. Thus, Convention 2.3.3 has the further point that $P_1 \wedge P_2 \wedge \cdots \wedge P_n$ is a form which is logically equivalent to any one derived from $P_1, P_2, \ldots, P_n$ by listing them in some order and then inserting $\wedge$'s to produce a propositional form and likewise for $\vee$.

The reader can certainly imagine an attempted proof by truth table of these consequences of "commutativity" and "associativity" of $\wedge$ and $\vee$; of course, one could produce and check the truth tables for any fixed $n \geq 3$, but this fails to be a rigorous proof of the general result. We shall formally prove these assertions by induction below and warn the reader in advance that the proofs are tedious especially since the results are obvious by analogy with addition. Though these arguments provide further insights into proofs by induction, they are actually included primarily for completeness and are postponed until the optional §2.8 for pedagogical reasons. We have discussed this material here only to indicate the significance of Convention 2.3.3.

Recapitulating, we have introduced conventions which allow us to judiciously omit parentheses in propositional forms. It is worth emphasizing that it is never incorrect to include all or some of the parentheses in a propositional form thereby not taking full advantage of the conventions above, but the omission of parentheses can often improve the readability of a propositional form. At the other extreme, one even allows irrelevant but "legal" parentheses surrounding propositional forms, as in $((((P)) \wedge (Q)))$, i.e., if P is a propositional form, then so too is (P).

EXERCISES

2.3 PARENTHESES

1. For each of the following expressions written in keeping with our conventions, insert the implicit parentheses to get a fully parenthesized propositional form.

   (a) $\neg\neg\neg P \wedge \neg P$

   (b) $\neg(P \Rightarrow Q) \wedge (P \vee Q \vee \neg P)$

   (c) $(P_1 \wedge P_2 \wedge P_3) \vee (Q_1 \wedge Q_2 \wedge Q_3)$

2. (a) For each $n = 1, 2, 3$, give all possible fully parenthesized propositional forms arising from the list of propositional variables $P_1, P_2, \ldots, P_n$ by inserting parentheses and $\wedge$'s. Check in each case by truth table that the truth values of the resulting propositional form depend only on n.

(b) For each $n = 1, 2, 3$, give all possible lists $P_{i_1}, P_{i_2}, \ldots, P_{i_n}$, where each of $P_1, P_2, \ldots, P_n$ occurs exactly once on the list. Check in each case by truth table that the truth values of $P_{i_1} \wedge P_{i_2} \wedge \cdots \wedge P_{i_n}$ depend only on n.

(c) Perform analyses analogous to parts (a) and (b) for $n = 4$.

3. Further common conventions involving parentheses (which we shall not adopt in the text and only examine in this exercise) are as follows: By convention, we shall first apply $\neg$, then $\vee$ and $\wedge$, then $\Rightarrow$, and finally $\Leftrightarrow$ and may therefore omit redundant parentheses. As a point of terminology, one says, for instance, that $\neg$ "binds more tightly" than $\vee$, $\wedge$, $\Rightarrow$, or $\Leftrightarrow$ in this case. For an example, we may write simply $\neg P \vee Q \Rightarrow P \Leftrightarrow Q$ for the parenthesized expression

$$[[((\neg P) \vee Q) \Rightarrow P] \Leftrightarrow Q].$$

Similarly insert parentheses in each of the following expressions written in keeping with the conventions here on the "binding" of the logical operators.

(a) $P \wedge Q \Rightarrow R \wedge S$

(b) $P \wedge Q \Rightarrow \neg Q \vee R$

(c) $P \Rightarrow Q \Leftrightarrow R \vee Q$

(d) $(P \Rightarrow Q) \vee (S \wedge \neg Q) \Rightarrow R$

2.4 STANDARD LOGICAL IDENTITIES

There are certain logical identities which play a central role in our subsequent discussion, and the purpose of this section is to explicitly state and prove these logical identities. The reader should not infer that the list of identities here is exhaustive: There are many other interesting logical identities, and we are simply bringing attention to various simple but important examples here.

We begin by simply listing the logical identities of current interest together with the name in English associated with each one:

Name	*Logical Identity*
(1) Idempotence of $\wedge$	$P \Leftrightarrow (P \wedge P)$
(2) Idempotence of $\vee$	$P \Leftrightarrow (P \vee P)$
(3) Commutativity of $\wedge$	$(P \wedge Q) \Leftrightarrow (Q \wedge P)$
(4) Commutativity of $\vee$	$(P \vee Q) \Leftrightarrow (Q \vee P)$
(5) Associativity of $\wedge$	$[P \wedge (Q \wedge R)] \Leftrightarrow [(P \wedge Q) \wedge R]$
(6) Associativity of $\vee$	$[P \vee (Q \vee R)] \Leftrightarrow [(P \vee Q) \vee R)]$
(7) Distributivity of $\wedge$ over $\vee$	$[P \wedge (Q \vee R)] \Leftrightarrow [(P \wedge Q) \vee (P \wedge R)]$
(8) Distributivity of $\vee$ over $\wedge$	$[P \vee (Q \wedge R)] \Leftrightarrow [(P \vee Q) \wedge (P \vee R)]$
(9) Laws of 0	$[(P \vee 0) \Leftrightarrow P]$ and $[(P \wedge 0) \Leftrightarrow 0]$

(10) Laws of 1 $[(P \wedge 1) \Leftrightarrow P]$ and $[(P \vee 1) \Leftrightarrow 1]$

(11) Rules of Thought:
- Law of Excluded Middle $[(P \vee \neg P) \Leftrightarrow 1]$
- Law of Contradiction $[(P \wedge \neg P) \Leftrightarrow 0]$
- Law of Identity $P \Leftrightarrow P$

(12) DeMorgan's Laws: $[\neg(P \wedge Q)] \Leftrightarrow [\neg P \vee \neg Q]$
$[\neg(P \vee Q)] \Leftrightarrow [\neg P \wedge \neg Q]$

(13) Law of Implication $[P \Rightarrow Q] \Leftrightarrow [\neg P \vee Q]$

(14) Law of Equivalence $[P \Leftrightarrow Q] \Leftrightarrow [(P \Rightarrow Q) \wedge (Q \Rightarrow P)]$

(15) Law of Exportation $[(P \wedge Q) \Rightarrow R] \Leftrightarrow [P \Rightarrow (Q \Rightarrow R)]$

(16) Law of Contrapositive $[P \Rightarrow Q] \Leftrightarrow [\neg Q \Rightarrow \neg P]$

(17) Law of Absurdity $[\neg P] \Leftrightarrow [(P \Rightarrow Q) \wedge (P \Rightarrow \neg Q)]$

Each of these logical identities is sufficiently important in practice that the reader should make a point of memorizing or absorbing them once and for all! The English names are not so important to memorize, but the reader should become facile with the various symbolic statements.

Before discussing the proofs, let us make some simple observations about these identities. All of (1)-(6) and (9)-(11) are really quite obvious. We have already discussed certain of the rules of thought in § 1.B.5. DeMorgan's laws (12) are quite useful in practice; the first law states that the negation of a conjunction is the disjunction of the negations, and the second law states that the negation of a disjunction is the conjunction of the negations. (Thus, $\wedge$ is replaced by $\vee$ in De Morgan's laws, and vice versa.) We have seen the laws of implication (13) and equivalence (14) in §2.2. The law (16) of the contrapositive has been proved in a previous truth table.

As to the proofs of these logical equivalences, we leave it to the reader to check them all directly by truth tables now once and for all, and most of the truth tables can be easily done just mentally. Actually, we shall prove (15)-(17) later assuming (1)-(14) as sample calculations in the predicate calculus.

EXERCISES

2.4 STANDARD LOGICAL IDENTITIES

1. Refering here to standard logical identity (n) in the text simply as "2.4.n" for each $n = 1, 2, \ldots, 17$, prove the following logical identities by truth table.

(a) 2.4.7 (b) 2.4.8

(c) 2.4.12 (d) 2.4.15

(e) 2.4.16 (f) 2.4.17

2. Find a propositional form not involving $\Rightarrow$ which is logically equivalent to the contrapositive, converse, inverse, and negation of each of the following implications.

(a) $P \Rightarrow (Q \vee R)$ (b) $(P \wedge Q) \Rightarrow R$

(c) $P \Rightarrow (Q \wedge R)$ (d) $(P \vee Q) \Rightarrow R$

3. Prove each of the following logical equivalences.

(a) $(P \vee Q) \Leftrightarrow (\neg P \Rightarrow Q)$ (b) $[(P \wedge Q) \Rightarrow P] \Leftrightarrow 1$

(c) $\neg[\neg(P \vee Q) \Rightarrow \neg P] \Leftrightarrow 0$ (d) $[(Q \Rightarrow P) \wedge (\neg P \Rightarrow Q)] \Leftrightarrow P$

2.5 STANDARD RULES OF INFERENCE

The last section enumerated a useful list of logical equivalences, and this section is dedicated to a similar list of tautologies of the form $P \Rightarrow Q$, where P and Q are particular propositional forms. Such a tautology, namely, an implication which is a tautology, is called a *rule of inference*, and, in light of our discussion of proof techniques before, it should come as no surprise that rules of inference are essential tools: Again, they are paradigms for methods of proof. As before, we begin by simply presenting various important rules of inference giving their English names, and the proofs are left as exercises with truth tables. Let us again emphasize that this list is not exhaustive, rather these are just some of the more commonly used rules of inference.

Name	*Rule of Inference*
(1) Addition	$P \Rightarrow (P \vee Q)$
(2) Simplification	$(P \wedge Q) \Rightarrow P$
(3) *Modus Ponens*	$[P \wedge (P \Rightarrow Q)] \Rightarrow Q$
(4) *Modus Tolens*	$[\neg Q \wedge (P \Rightarrow Q)] \Rightarrow \neg P$
(5) Vacuous Proof	$(P \Leftrightarrow 0) \Rightarrow (P \Rightarrow Q)$
(6) Trivial Proof	$(Q \Leftrightarrow 1) \Rightarrow (P \Rightarrow Q)$
(7) Disjunctive Syllogism	$[(P \vee Q) \wedge \neg P] \Rightarrow Q$
(8) Hypothetical Syllogism	$[(P \Rightarrow Q) \wedge (Q \Rightarrow R)] \Rightarrow (P \Rightarrow R)$
(9) Constructive Dilemma	$[(P \Rightarrow Q) \wedge (R \Rightarrow S) \wedge (P \vee R)] \Rightarrow [Q \vee S]$
(10) Destructive Dilemma	$[(P \Rightarrow Q) \wedge (R \Rightarrow S) \wedge (\neg Q \vee \neg S)]$ $\Rightarrow [\neg P \vee \neg R]$
(11) Transitivity of $\Leftrightarrow$	$[(P \Leftrightarrow Q) \wedge (Q \Leftrightarrow R)] \Rightarrow (P \Leftrightarrow R)$

(12) Replacement:

$$[(P \Leftrightarrow Q) \wedge (R \Leftrightarrow S)] \Rightarrow [(P \wedge R) \Leftrightarrow (Q \wedge S)]$$
$$[(P \Leftrightarrow Q) \wedge (R \Leftrightarrow S)] \Rightarrow [(P \vee R) \Leftrightarrow (Q \vee S)]$$
$$[(P \Leftrightarrow Q) \wedge (R \Leftrightarrow S)] \Rightarrow [(P \Rightarrow R) \Leftrightarrow (Q \Rightarrow S)]$$
$$[(P \Leftrightarrow Q) \wedge (R \Leftrightarrow S)] \Rightarrow [(P \Leftrightarrow R) \Leftrightarrow (Q \Leftrightarrow S)]$$

Again, we serially discuss the tautologies above, each of which should be memorized or absorbed by the reader and proved by truth table now once and for all, and again several of the truth tables can be done just mentally. Rules (1) and (2) are obvious and require no further comment. Similarly, rules (3)-(6) evidently follow directly from the definition of $\Rightarrow$. The phrase "*modus ponens*" means "method of putting forward" in Latin, and rule (3) "puts forward" (that is, has as conclusion) Q, hence the terminology; similarly, "*modus tolens*" means "method of taking away", and in rule (4) one "takes away" (that is, concludes the negation of) P, a nice classical imagery of truth. Also, from (5) and (6), we might prove the implication $P \Rightarrow Q$ by proving that P is false or by proving that Q is true; the former is called a *vacuous proof* and the latter a *trivial proof* of the implication $P \Rightarrow Q$. The rule (7) is again evident from the definitions. We proved (8) in §1.A.4 (it was our equation $(*)$), and the reader should observe that this rule of inference was therefore the basis of our proofs by chains of implications in §1.B.1. Notice also the similarity between rules (8) and (11); indeed, one simply replaces all of the logical operators $\Leftrightarrow$ in (11) by $\Rightarrow$ to obtain (9). As a point of terminology, we could just as well have called (9) "transitivity of $\Rightarrow$", but we prefer the more classical terminology of "hypothetical syllogism". The reader should probably prove rules (9)-(12) by actually writing down truth tables as they are perhaps a tad too complicated to prove just mentally.

To close, we remark that universal and existential generalization and instantiation (which we discussed in §1.A.2) should also naturally be regarded as standard rules of inference, as each is a tautological implication. For instance, one might think of universal instantiation as the tautology $(\forall x P(x)) \Rightarrow P(x_0)$, where x_0 is any element of the universe of discourse of $P(x)$.

EXERCISES

2.5 STANDARD RULES OF INFERENCE

1. Refering here to standard rule of inference (n) in the text simply as "2.5.n" for each $n = 1, 2, \ldots, 12$, prove the following logical identities by truth table.

(a) 2.5.9 (b) 2.5.10

(c) 2.5.11 (d) 2.5.12

2. Prove the following rules of inference.

(a) $(P \Leftrightarrow Q) \Rightarrow (\neg P \Leftrightarrow \neg Q)$ (b) $[(P \Rightarrow \neg Q) \wedge (S \vee Q)] \Rightarrow (P \Rightarrow S)$

(c) $[(P \vee Q) \Rightarrow R] \Rightarrow [P \Rightarrow R]$ (d) $[(P \Rightarrow Q) \wedge [(P \wedge Q) \Rightarrow R]] \Rightarrow (P \Rightarrow R)$

(*) 2.6 ARISTOTELEAN LOGIC

We have learned some of the fundamentals of the predicate calculus in the preceding pages. In this optional section we digress from this our main theme to briefly discuss Aristotelean logic, and our interest in this is largely historical and terminological. Furthermore, the formalism of Aristotelean logic allows us to give some examples here of applying rules of inference with carefully articulated hypotheses in contrast to our discussion of proof techniques in the first chapter, where we were somewhat cavalier about the underlying assumptions about English, the natural numbers, and so on. Thus, the current section illustrates complete English examples in contrast to some of those in the previous chapter. The examples we consider here give some practice which may well be unnecessary for the reader of translating from English into predicate calculus explaining why this section is optional.

We are interested in proving implications of the form $(P_1 \wedge P_2 \wedge \ldots \wedge P_n) \Rightarrow Q$, and each proposition P_i is called a *hypothesis*, while Q is called the conclusion as before. The formal presentation of hypotheses and conclusion is called a *syllogism*. The syntax of presenting a syllogism involves simply listing the collection of all hypotheses in a column underneath which one draws a horizontal line segment; below this line segment, one repeats a statement of the conclusion preceded by the symbol "$\therefore$ ", which is to be read in English as "therefore". For instance the syllogism corresponding to *modus ponens* is just

$$\begin{array}{l} P \\ P \Rightarrow Q \\ \hline \therefore\ Q \end{array}$$

and the syllogism corresponding to the implication $(P_1 \wedge \cdots \wedge P_n) \Rightarrow Q$ is

$$\begin{array}{l} P_1 \\ P_2 \\ \vdots \\ P_n \\ \hline \therefore\ Q \end{array}$$

In practice, a syllogism usually involves explicit propositions (as opposed to propositional variables) both for hypotheses P_i and conclusion Q, and one attempts to "prove" the syllogism, that is, prove that the conjunction of hypotheses actually implies the conclusion. The syntax of presenting such a proof in this setting is extremely formal. One proves separately the rules of inference such as *modus ponens* once and for all (say using truth tables) and assumes them to be known and given names as before. One then formally lists a collection $Q_1, \ldots, Q_m$

of true statements in a column underneath which one draws a line segment, and the last entry in this column (below the line segment) is $\therefore\ Q$ just as before. At the same time, next to statement Q_j for $j = 1, \ldots, m$, (and next to statement Q as well), one gives a proof of Q_j (and of Q itself) using the assumed rules of inference, the original hypotheses $P_1, \ldots, P_n$, and the previously proved statements Q_i, for $1 \leq i < j$. An illustration should fully explain this.

■ **Example 2.6.1** Consider the following explicit statement of hypotheses and conclusion in English:

> If today is Wednesday and the sky is blue, then I am learning mathematics. If I am learning mathematics, then I am happy and stimulated. It is Wednesday. The sky is blue. Therefore, I am happy and stimulated.

In order to explicitly state the corresponding syllogism, we introduce the following notation $W =$ [It is Wednesday], $B =$ [The sky is blue], $M =$ [I am learning mathematics], $H =$ [I am happy], $S =$ [I am stimulated] for the various propositions in the English paragraph above. Thus, the transcription into a syllogism for this example is:

$$\begin{array}{l} (W \wedge B) \Rightarrow M \\ M \Rightarrow (H \wedge S) \\ W \\ B \\ \hline \therefore\ H \wedge S \end{array}$$

A proof of this syllogism is given by:

W	1) hypothesis 3
B	2) hypothesis 4
$W \wedge B$	3) addition
M	4) modus ponens, hypothesis 1, and line 3
$\therefore\ H \wedge S$	5) modus ponens, hypothesis 2, and line 4

The reader may have seen such formal proofs of syllogisms in earlier mathematics classes. The notion of a syllogism and its proof goes back to Aristotle. In fact, one is essentially never this formal when doing mathematics except for practice. There was actually a school of mathematicians, called the "formalist Hilbert school", who propounded such an explicit treatment of all mathematics (using notation different from syllogisms however) in the early part of this century. Indeed, the famous tome *Principia Mathematica* by Bertrand Russell and Alfred North Whitehead was intended to be the first volume establishing all of mathematics from this very formal point of view. The subsequent volumes of *Principia Mathematica* were never written, in part because the philosophy of this formalist school was dealt a death blow by profound theorems of Gödel in mathematical logic! We shall comment further on Gödel's theorems in §3.3 below.

■ **Example 2.6.2** For an example of a logically more complicated syllogism than the previous one, consider the following formal expression of hypotheses and conclusion in English.

> If dogs are mammals, then cats are reptiles. If I am a mammal, then either I am a dog or I am a cat. Either dogs are mammals, or I am a mammal. Cats are not reptiles. I am not a dog. Therefore, I am a cat.

We introduce the propositions $P =$ [Dogs are mammals], $Q =$ [Cats are reptiles], $A =$ [I am a dog], $B =$ [I am a cat], and $R =$ [I am a mammal] to find the syllogism

$$\begin{array}{l} P \Rightarrow Q \\ R \Rightarrow (A \vee B) \\ P \vee R \\ \neg Q \\ \neg A \\ \hline \therefore\ B \end{array}$$

A proof of the syllogism is

$P \Rightarrow Q$	1) hypothesis 1
$R \Rightarrow (A \vee B)$	2) hypothesis 2
$P \vee R$	3) hypothesis 3
$Q \vee (A \vee B)$	4) constructive dilemma and lines 1-3
$(Q \vee A) \vee B$	5) associativity of $\vee$
$\neg Q \wedge \neg A$	6) hypotheses 4,5, and conjunction
$\neg(Q \vee A)$	7) DeMorgan
$\therefore\ B$	8) disjunctive syllogism and lines 5,7

Part of the point of this example is that the particular propositions involved in a syllogism need not actually be true in English: The syllogism and its proof simply depend upon stated logical relationships between given propositions.

On a more pessimistic note, we close with an example of a syllogism involving propositions which are quantified predicates.

■ **Example 2.6.3** Consider the following enumeration of hypotheses and conclusion in English:

> There is a human being who is kind or honest. If a human being has green eyes, then she or he is not honest. If a human being has black hair, then she or he is not kind. A human being is either not kind or not honest. Therefore, there is a human being who does not have green eyes and black hair.

We introduce the following predicates $K(x) = [x \text{ is kind}]$, $H(x) = [x \text{ is honest}]$, $G(x) = [x \text{ has green eyes}]$, and $B(x) = [x \text{ has black hair}]$ on the universe of all human beings x, so the syllogism is

$$\begin{array}{l} \exists x(K(x) \vee H(x)) \\ \forall x(G(x) \Rightarrow \neg H(x)) \\ \forall x(B(x) \Rightarrow \neg K(x)) \\ \forall x(\neg H(x) \vee \neg K(x)) \\ \hline \therefore \ \exists x \neg(G(x) \wedge B(x)) \end{array}$$

A proof of this syllogism is

$K(c) \vee H(c)$	1) hypothesis 1 and existential generalization
$G(c) \Rightarrow \neg H(c)$	2) hypothesis 2 and universal instantiation
$B(c) \Rightarrow \neg K(c)$	3) hypothesis 3 and universal instantiation
$\neg G(c) \vee \neg B(c)$	4) destructive dilemma and lines 1,2,3
$\neg(G(c) \wedge B(c))$	5) DeMorgan and line 4
$\therefore \ \exists x \neg(G(x) \wedge B(x))$	6) existential generalization and line 5

Recapitulating, we have seen in this section that syllogisms are simply an alternative notation for certain manipulations involving logical operators and rules of inference. Though syllogisms and their proofs play an important historical role in the development of logic, we shall always employ the formalism of the predicate calculus in the sequel (rather than the formalism of syllogisms).

EXERCISES

2.6 ARISTOTELEAN LOGIC

1. There are logical "mistakes", i.e., incorrect manipulations in the predicate calculus, which are called *fallacies*, and this exercise is dedicated to two famous fallacies.

(a) Consider the following hypotheses and conclusion in English:

> If the butler did it, then he will nervous when interrogated. The butler was quite nervous when interrogated. The butler therefore did it.

Exhibit the syllogism associated with this errant logical deduction, and give the truth table for the associated propositional form. This common fallacy is called the *fallacy of affirming the consequent.*

(b) Consider the following hypotheses and conclusion in English:

> If the butler is covered with the blood of the victim, then the butler did it. The butler is impeccably clean. The butler therefore did not do it.

Exhibit the syllogism associated with this errant logical deduction, and give the truth table for the associated propositional form. This common fallacy is called the *fallacy of denying the antecedent.*

2. Determine which of the following syllogisms are valid; supply a proof of the valid syllogisms, and give an example of the syllogism in English for the invalid ones.

(a)
$$\begin{array}{l} P \vee Q \\ P \Rightarrow R \\ \hline \therefore Q \vee R \end{array}$$

(b)
$$\begin{array}{l} P \wedge Q \\ P \Rightarrow R \\ \hline \therefore Q \wedge R \end{array}$$

(c)
$$\begin{array}{l} (P \wedge R) \Rightarrow Q \\ P \vee R \\ R \wedge Q \\ \hline \therefore Q \vee R \end{array}$$

(d)
$$\begin{array}{l} P \Rightarrow (Q \vee R) \\ S \Rightarrow \neg R \\ Q \Rightarrow \neg P \\ P \\ S \\ \hline \therefore P \wedge Q \end{array}$$

3. For each of the following English paragraphs, list the relevant conclusions and prove the corresponding syllogism.

(a) I am either green with envy or red with rage. I am not red with rage.

(b) On a good day, I am courageous, but on a bad day I am sleepy. Today is either good or bad.

(c) All cows are quadrupeds, and some quadrupeds have udders.

(d) Whatever is good for the American auto industry is good for the American economy. Whatever is good for the American economy is good for you. It would be good for the American auto industry if you bought an American car.

(e) No prime natural number other than 2 is even. 6 is even, but 7 is not even.

2.7 LOGICAL EQUIVALENCE

This section develops the basic tools for proving logical equivalence and gives numerous examples. Rules of inference 2.5.11 and 2.5.12 play special roles in this discussion, as we shall explain.

We begin with rule 2.5.11 and suppose that we are trying to prove that two propositional forms P and R are logically equivalent. According to rule 2.5.11, if we can find yet another propositional form Q so that $P \Leftrightarrow Q$ and $Q \Leftrightarrow R$, then we may conclude that indeed $P \Leftrightarrow R$. Expanding on this observation, we have

2.7.1 PROPOSITION *Suppose that $P_1, \ldots, P_{n+1}$ is a collection of propositional forms, where $n \geq 2$, and suppose that $P_1 \Leftrightarrow P_2$, $P_2 \Leftrightarrow P_3$, $\ldots, P_n \Leftrightarrow P_{n+1}$. Then $P_1 \Leftrightarrow P_{n+1}$.*

Proof The proof is by induction on n, and the basis step $n = 2$ is simply a recapitulation of rule 2.5.11 above. For the induction, suppose that $P_1 \Leftrightarrow P_2$, $\ldots, P_n \Leftrightarrow P_{n+1}$. In particular, the first $n - 1$ of these bi-implications imply that $P_1 \Leftrightarrow P_n$ by the inductive hypothesis. Thus, we have $P_1 \Leftrightarrow P_n$ and also $P_n \Leftrightarrow P_{n+1}$, so a final application of rule 2.5.11 completes the induction step. *q.e.d.*

This proposition typifies a general situation where one first proves a specialized predicate $Q(2)$ and then argues by induction using this result that in fact $Q(n)$ holds for each $n \geq 2$. For instance, the predicate playing the role of $Q(n)$ in Proposition 2.7.1 is

$$[(P_1 \Leftrightarrow P_2) \wedge (P_2 \Leftrightarrow P_3) \wedge \cdots \wedge (P_n \Leftrightarrow P_{n+1})] \Rightarrow [P_1 \Leftrightarrow P_{n+1}],$$

so $Q(2)$ is simply 2.5.11. We shall see numerous examples of this sort of "extension by induction" in the sequel. For now, we just observe that this proof also applies verbatim to give a complete proof of the validity of our method of proof by chains of implications in §1.B.1: Simply replace the appearances of $\Leftrightarrow$ in the proof above with $\Rightarrow$. Just as with chains of implications, to prove that $P \Leftrightarrow Q$, it suffices to find a chain of bi-implications as in the previous result beginning with $P_1 = P$ and ending with $P_{n+1} = Q$. This is a valuable observation regarding logical equivalence just as chains of implications are valuable tools for proving implications (as in §1.B.1).

Turning to the significance of rule 2.5.12, consider a propositional form P and suppose that we can find a sub-propositional form of P, say the sub-propositional form is R_1, and we have proved that R_1 is logically equivalent to some other propositional form R_2; for instance, we gave an extensive list of such logical equivalences in §2.4. Let us formally replace the string of symbols R_1 by the string of symbols R_2 in P to get another propositional form Q. For an example, we might have

$$P = [\neg(R_1 \wedge T) \vee S] \quad \text{and} \quad Q = [\neg(R_2 \wedge T) \vee S].$$

2.7.2 LEMMA *With the notation above, P is logically equivalent to Q.*

Proof We proceed by induction on the number n of logical operations required to construct P from R_1 or in other words, the number of logical operations required to construct Q from R_2. For the basis step $n = 0$, we have that $P = R_1$ and $Q = R_2$ are logically equivalent by hypothesis. The induction step follows easily from the induction hypothesis, rule 2.5.12, and the fact that $(S \Leftrightarrow T) \Leftrightarrow (\neg S \Leftrightarrow \neg T)$ for any propositional forms S and T. *q.e.d.*

The combination of the previous two results suggests the following method of proving two propositional forms P and Q are logically equivalent: Start with P, for instance, and use known logical equivalences, e.g., the ones the reader was asked to memorize, to replace sub-expressions of P by logically equivalent expressions. Suppose that we can start with P and obtain Q by a finite sequence of such replacements, that is, suppose that we have a collection $P = P_1, P_2, \ldots, P_{n+1} = Q$, where P_{i+1} arises from P_i by a single such replacement for each $i = 1, \ldots, n$.

2.7.3 THEOREM *With the notation above, P and Q are logically equivalent.*

Proof According to Lemma 2.7.2, we have the chain $P_1 \Leftrightarrow P_2, \ldots, P_n \Leftrightarrow P_{n+1}$ of logical equivalences, and the result thus follows immediately from Proposition 2.7.1. *q.e.d.*

Using this result, we finally illustrate typical manipulations of propositional forms in the predicate calculus. Indeed, the rest of this section is dedicated to sample calculations based on Theorem 2.7.3, but we introduce first one small bit of notation, as follows: We shall allow ourselves to write a chain of bi-implications as one long equation

$$P_1 \Longleftrightarrow P_2 \Longleftrightarrow \cdots \Longleftrightarrow P_{n+1}$$

rather than in the notation used before. This allows us to avoid inserting various parentheses as well as allowing us to write one long equation which proves $P_1 \Leftrightarrow P_{n+1}$. For definiteness in this volume, we shall reserve the "long" bi-implication symbol "$\Longleftrightarrow$" for such expressions rather than using the usual symbol "$\Leftrightarrow$" for the bi-implications in the chain. More generally (for instance at a blackboard), one can judiciously indent to visually communicate which are the predicate forms $P_1, P_2, \ldots, P_{n+1}$ in the chain of bi-implications. For instance, an extreme such presentation would be

$$\begin{aligned} P_1 &\Longleftrightarrow P_2 \\ &\Longleftrightarrow P_3 \\ &\quad\vdots \\ &\Longleftrightarrow P_n \\ &\Longleftrightarrow P_{n+1} \end{aligned}$$

The examples below illustrate these notations.

In the first three examples, we supply the promised proofs of the laws of exportation, contrapositive, and absurdity respectively (namely, 2.4.15-17) assuming the logical identities 2.4.1-14.

■ **Example 2.7.1** We prove here the law of exportation

$$[(P \wedge Q) \Rightarrow R] \Leftrightarrow [P \Rightarrow (Q \Rightarrow R)],$$

and the computation runs as follows:

$$\begin{aligned} P \Rightarrow (Q \Rightarrow R) &\Longleftrightarrow P \Rightarrow (\neg Q \vee R) \text{ by the law of implication} \\ &\Longleftrightarrow \neg P \vee (\neg Q \vee R) \text{ by the law of implication} \\ &\Longleftrightarrow (\neg P \vee \neg Q) \vee R \text{ by associativity of } \vee \\ &\Longleftrightarrow \neg(P \wedge Q) \vee R \text{ by DeMorgan} \\ &\Longleftrightarrow [(P \wedge Q) \Rightarrow R] \text{ by the law of implication.} \end{aligned}$$

Since we have found a chain of bi-implications beginning with $[P \Rightarrow (Q \Rightarrow R)]$ and ending with $[(P \wedge Q) \Rightarrow R]$, the law of exportation has been proved to be a tautology.

A typical aspect of this example which is worth emphasizing immediately is that one essentially always uses the law of implication to replace the implication operator $\Rightarrow$. The reasons for this are that we have at our disposal many standard logical identities involving $\wedge$, $\vee$, and $\neg$, and these standard identities allow a richer manipulation of forms expressed in terms of these operators.

■ **Example 2.7.2** We next prove the law of contrapositive

$$(P \Rightarrow Q) \Leftrightarrow (\neg Q \Rightarrow \neg P),$$

and the computation is

$$P \Rightarrow Q \Longleftrightarrow \neg P \vee Q \Longleftrightarrow \neg\neg Q \vee \neg P \Longleftrightarrow \neg Q \Rightarrow \neg P,$$

where the first and last equivalence follow from the law of implication, and the second equivalence follows from commutativity of disjunction and the fact that $\neg\neg Q \Leftrightarrow Q$.

A typical aspect of this example is that we allow ourselves to perform several replacements in a single equivalence, for instance, in the second equivalence above. Of course, some restraint must be exercised so as to keep the steps intelligible to the reader. To be on the safe side, the student is advised stick to one or two replacements in each step.

■ **Example 2.7.3** We next prove the law of absurdity

$$\neg P \Leftrightarrow [(P \Rightarrow Q) \wedge (P \Rightarrow \neg Q)]$$

as follows:

$$(P \Rightarrow Q) \wedge (P \Rightarrow \neg Q) \Longleftrightarrow (\neg P \vee Q) \wedge (\neg P \vee \neg Q)$$
$$\Longleftrightarrow \neg P \vee (Q \wedge \neg Q) \Longleftrightarrow \neg P \vee 0 \Longleftrightarrow \neg P.$$

The first equivalence follows from two applications of the law of implication, the second from distributivity of $\vee$ over $\wedge$, the third from the law of contradiction, and the fourth from the laws of 0.

■ **Example 2.7.4** We prove the logical identity

$$[(Q \Rightarrow P) \wedge (\neg P \Rightarrow Q)] \Leftrightarrow P$$

as follows:

$$\begin{aligned}(Q \Rightarrow P) \wedge (\neg P \Rightarrow Q) &\Longleftrightarrow (\neg Q \vee P) \wedge (\neg\neg P \vee Q) \text{ by the law of implication}\\ &\Longleftrightarrow (\neg Q \vee P) \wedge (P \vee Q) \text{ since } \neg\neg P \Leftrightarrow P\\ &\Longleftrightarrow (P \vee \neg Q) \wedge (P \vee Q) \text{ commutativity of } \vee\\ &\Longleftrightarrow P \vee (\neg Q \wedge Q) \text{ by distributivity of } \vee \text{ over } \wedge\\ &\Longleftrightarrow P \vee 0 \text{ by the law of contradiction}\\ &\Longleftrightarrow P \text{ by laws of 0.}\end{aligned}$$

This example illustrates another typical feature of such manipulations: One often compares the two propositional forms in question and chooses the more complicated one to manipulate with replacements. In particular, if a propositional variable occurs in one of the forms and not in the other (for instance, Q occurs only on the lefthand side of the equivalence in the previous example), then one must manipulate the form containing it in such a way that it disappears (for instance, using a law of 0 in the previous example).

■ **Example 2.7.5** We prove the logical identity

$$P \Leftrightarrow [P \wedge (P \vee R)]$$

as follows:

$$P \wedge (P \vee R) \Longleftrightarrow (P \vee 0) \wedge (P \vee R) \Longleftrightarrow [P \vee (R \wedge \neg R)] \wedge (P \vee R)$$
$$\Longleftrightarrow (P \vee R) \wedge (P \vee \neg R) \wedge (P \vee R) \Longleftrightarrow (P \vee R) \wedge (P \vee \neg R)$$
$$\Longleftrightarrow P \vee (R \wedge \neg R) \Longleftrightarrow P \vee 0 \Longleftrightarrow P.$$

The first and second equivalences rely on laws of 0, the third on distributivity of $\vee$ over $\wedge$, the fourth on associativity, commutativity and idempotence of $\wedge$, the fifth on distributivity of $\vee$ over $\wedge$, the sixth on the law of contradiction, and the seventh on laws of 0.

Notice that we used laws of 0 in this case to make intermediary expressions "more complicated" (i.e., the intermediary expressions in the first and second steps have more symbols) rather than just simplifying the form as in the previous examples. The judicious use of the laws of 0 and 1 in this way may be necessary in certain cases.

We have the following results, whose proofs are left as truly elementary exercises in proofs by induction, but which will be discussed briefly below.

2.7.4 THEOREM *Suppose that $P_1, \ldots, P_n$ is a collection of propositional forms, where $n \geq 1$. Then we have*

$$\neg(P_1 \wedge \cdots \wedge P_n) \Leftrightarrow (\neg P_1 \vee \cdots \vee \neg P_n)$$
$$\neg(P_1 \vee \cdots \vee P_n) \Leftrightarrow (\neg P_1 \wedge \cdots \wedge \neg P_n)$$

2.7.5 THEOREM *Suppose that $Q, P_1, \ldots, P_n$ are propositional forms. Then we have*

$$[(P_1 \wedge \cdots \wedge P_n) \wedge Q] \Leftrightarrow [(P_1 \wedge \cdots \wedge P_n \wedge Q)]$$
$$[(P_1 \vee \cdots \vee P_n) \vee Q] \Leftrightarrow [(P_1 \vee \cdots \vee P_n \vee Q)]$$

Furthermore, we have

$$[Q \wedge (P_1 \vee \cdots \vee P_n)] \Leftrightarrow [(Q \wedge P_1) \vee \cdots \vee (Q \wedge P_n)]$$
$$[Q \vee (P_1 \wedge \cdots \wedge P_n)] \Leftrightarrow [(Q \vee P_1) \wedge \cdots \wedge (Q \vee P_n)]$$

Of course, the first result gives the analogue of DeMorgan's laws for arbitrary conjunctions and disjunctions, and the second result similarly gives the analogues of associativity and distributivity for arbitrary conjunctions and disjunctions. These are typical of whole families of arguments in mathematics as was mentioned before, where a simple result such as DeMorgan's laws gives rise to a more general result such as Theorem 2.7.4 above using a proof by induction. Owing to the importance of this technique in general, the reader should really sit down and give a rigorous proof by induction of the previous two theorems, and, in each case, one inducts on the number $n \geq 1$.

■ **Example 2.7.6** We prove here

$$[(P_1 \vee \cdots \vee P_n) \Rightarrow Q] \Leftrightarrow [(P_1 \Rightarrow Q) \wedge \cdots \wedge (P_n \Rightarrow Q)]$$

as follows:

$$(P_1 \vee \cdots \vee P_n) \Rightarrow Q \Longleftrightarrow \neg(P_1 \vee \cdots \vee P_n) \vee Q \Longleftrightarrow (\neg P_1 \wedge \cdots \wedge \neg P_n) \vee Q$$
$$\Longleftrightarrow (\neg P_1 \vee Q) \wedge \cdots \wedge (\neg P_n \vee Q) \Longleftrightarrow (P_1 \Rightarrow Q) \wedge \cdots \wedge (P_n \Rightarrow Q).$$

The first and last equivalences follow from the law of implication, the second follows from DeMorgan's law, and the third from distributivity of $\vee$ over $\wedge$.

The previous result of course provides a method of proof in the spirit of the first chapter: To prove that $(P_1 \vee \cdots \vee P_n) \Rightarrow Q$, we may separately prove each implication $P_j \Rightarrow Q$ for each $j = 1, \ldots, n$.

Recapitulating, we have presented a method of proving a logical equivalence with a suitable chain of bi-implications and have given several examples; we continue these investigations in §2.9. Using this method, we turn in the following §2.8 to the consequences of Convention 2.3.3 which depend upon commutativity and associativity of $\wedge$ and $\vee$ as was discussed in §2.3. In fact, §2.8 is technical and included mainly for completeness.

EXERCISES

2.7 LOGICAL EQUIVALENCE

1. Prove the following results from the text.

(a) Theorem 2.7.4 (b) Theorem 2.7.5

2. Prove the following logical equivalences for each $n \geq 1$.

$$[(P_1 \vee P_2 \vee \cdots \vee P_n) \wedge Q] \Leftrightarrow [(P_1 \wedge Q) \vee (P_2 \wedge Q) \vee \cdots \vee (P_n \wedge Q)]$$
$$[(P_1 \wedge P_2 \wedge \cdots \wedge P_n) \vee Q] \Leftrightarrow [(P_1 \vee Q) \wedge (P_2 \vee Q) \wedge \cdots \wedge (P_n \vee Q)]$$

3. Prove the following logical equivalences.

(a) $[P \Rightarrow (Q \vee R)] \Leftrightarrow [(P \wedge \neg Q) \Rightarrow R]$

(b) $[(P \vee Q) \Rightarrow Q] \Leftrightarrow [\neg P \vee Q]$

(c) $[(P \vee Q) \Rightarrow R] \Leftrightarrow [(P \Rightarrow R) \wedge (Q \Rightarrow R)]$

(d) $[(R \Rightarrow P) \wedge (R \Rightarrow Q)] \Leftrightarrow [R \Rightarrow (P \wedge Q)]$

(e) $[(P \wedge \neg Q) \Rightarrow (R \wedge \neg R)] \Leftrightarrow [P \Rightarrow Q]$

(f) $[(R \wedge P) \Rightarrow Q] \Leftrightarrow [R \Rightarrow (P \Rightarrow Q)]$

(g) $[P \Rightarrow (Q \wedge R)] \Leftrightarrow [(P \Rightarrow Q) \wedge (P \Rightarrow R)]$

(h) $[(P \wedge Q) \Rightarrow R] \Leftrightarrow [(P \Rightarrow R) \vee (Q \Rightarrow R)]$

(i) $[P \Rightarrow (Q \vee R)] \Leftrightarrow [(P \wedge \neg Q) \Rightarrow R]$

(j) $[(P \vee Q) \Rightarrow R] \Leftrightarrow [(P \Rightarrow R) \wedge (Q \Rightarrow R)]$

4. Give a shorter proof in Example 2.7.5.

5. Recall the nand operator $(P \uparrow Q) \Leftrightarrow [\neg(P \wedge Q)]$ and prove the following logical equivalences.

(a) $(P \uparrow P) \Leftrightarrow (\neg P)$

(b) $[(P \uparrow P) \uparrow (Q \uparrow Q)] \Leftrightarrow [P \vee Q]$

(c) $[(P \uparrow Q) \uparrow (P \uparrow Q)] \Leftrightarrow [P \wedge Q]$

6. Recall the nor operator $(P \downarrow Q) \Leftrightarrow [\neg P \wedge \neg Q]$. Find a propositional form using only $\vee$ and $\neg$ which is logically equivalent to each of the following.

(a) $P \downarrow P$ (b) $(P \downarrow P) \downarrow (Q \downarrow Q)$ (c) $(P \downarrow Q) \downarrow (P \downarrow Q)$

(*) *2.8 COMMUTATIVITY AND ASSOCIATIVITY*

This optional section is technical, is included primarily for logical completeness, and should probably be relegated to assigned reading or skipped altogether. It is dedicated to a proof of

2.8.1 THEOREM *Suppose that $P_1, \ldots, P_n$ are propositional forms. List these forms in some order $P_{i_1} \cdots P_{i_n}$ and then insert parentheses and $\wedge$'s so as to obtain a propositional form P. Then $P \Leftrightarrow (P_1 \wedge \cdots \wedge P_n)$ for any choice of order and any choice of insertions. The analogous result holds for $\vee$.*

As was discussed in §2.3 above, this is the analogue of a result about addition which follows from commutativity and associativity and is presumably well-known to the reader. In fact, the proof we present is of much greater generality than simply for $\wedge$ and $\vee$ (it actually applies to any "commutative and associative binary operation"), and we shall have occasion later to recall it in other contexts.

We begin with a result which relies only on associativity of $\wedge$ or $\vee$.

2.8.2 LEMMA *Suppose that $P_1, P_2, \ldots, P_n$ is an ordered list of propositional forms, where $n \geq 3$. If P is a propositional form derived from $P_1 P_2 \cdots P_n$ by inserting parentheses and $\wedge$'s, then P is logically equivalent to $P_1 \wedge P_2 \wedge \ldots \wedge P_n$. In the same way, if P is a propositional form derived from $P_1 P_2 \cdots P_n$ by inserting parentheses and $\vee$'s, then P is logically equivalent to $P_1 \vee P_2 \vee \ldots \vee P_n$.*

Proof We concentrate on proving the assertion for $\wedge$, the proof for $\vee$ being entirely analogous. The proof is by induction on n, and the basis step $n = 3$ follows from associativity of $\wedge$ as discussed before. For the induction, consider the propositional form P constructed from the ordered list $P_1, \ldots, P_{n+1}$ by inserting parentheses and $\wedge$'s, and suppose by Convention 2.3.1 that the entire propositonal form P is not enclosed in parentheses, that is, do not write the final parentheses. There is then exactly one $\wedge$ in P which is not enclosed in some pair of parentheses, namely, the "last" binary operation $\wedge$ applied, and we may write $P = Q \wedge R$, where Q arises from the ordered list $P_1, \ldots, P_m$ and R arises from the ordered list $P_{m+1}, \ldots, P_{n+1}$ by inserting parentheses and $\wedge$'s, where $m \leq n$.

By the induction hypothesis, Q is logically equivalent to $P_1 \wedge \ldots \wedge P_m$, and R is logically equivalent to $P_{m+1} \wedge \ldots \wedge P_{n+1}$. Furthermore, according to the induction hypothesis, the propositional form $(P_{m+1} \wedge \cdots \wedge (P_n \wedge P_{n+1}) \cdots)$ is also logically equivalent to $P_{m+1} \wedge \cdots \wedge P_{n+1}$, so it follows from transitivity of $\Leftrightarrow$, that is, from 2.5.11, that R is moreover logically equivalent to $(P_{m+1} \wedge \cdots \wedge (P_n \wedge P_{n+1}) \cdots)$. By the first law of replacement 2.5.12, we conclude that $P = Q \wedge R$ is logically equivalent to

$$P' = (\cdots (P_1 \wedge P_2) \wedge \cdots \wedge P_m) \wedge (P_{m+1} \wedge \cdots \wedge (P_n \wedge P_{n+1}) \cdots).$$

Finally, we claim that P' is logically equivalent to $P_1 \wedge \cdots \wedge P_{n+1}$, and the proof is by a second induction on $n - m \geq 1$. For the basis step $n - m = 1$, there is nothing to prove, for then $P' = P_1 \wedge \cdots \wedge P_{n+1}$. For the induction step, associativity of $\wedge$ assures us of the logical equivalence

$$\begin{aligned}&(\cdots (P_1 \wedge P_2) \wedge \cdots \wedge P_m) \wedge (P_{m+1} \wedge \cdots \wedge (P_n \wedge P_{n+1}) \cdots)\\ &\quad \Leftrightarrow ((\cdots (P_1 \wedge P_2) \wedge \cdots \wedge P_m) \wedge P_{m+1}) \wedge (P_{m+2} \wedge \cdots \wedge (P_n \wedge P_{n+1}) \cdots).\end{aligned}$$

By the inductive hypothesis of our second induction, the latter form is logically equivalent to $P_1 \wedge \cdots \wedge P_{n+1}$, so the assertion follows from a final application of 2.5.11.

To prove the lemma for $\vee$, simply replace each occurrence of $\wedge$ in the argument above by $\vee$. *q.e.d.*

Thus, any propositional form derived from $P_1, \ldots, P_n$ by inserting parentheses and $\wedge$'s (or $\vee$'s) is indeed logically equivalent to $P_1 \wedge \cdots \wedge P_n$ (or to $P_1 \vee \cdots \vee P_n$). Regarding the previous proof, we mention that the general scheme of using a separate proof by induction within another proof by induction, as above, is a fairly common proof technique.

We next expand on the previous result, using both commutativity and associativity of $\wedge$ or $\vee$. To state the result, consider an ordered collection $P_1, \ldots, P_n$ of propositional forms, and list these propositional forms in some order P_{i_1}, P_{i_2}, ..., P_{i_n}, where each subscript $i = 1, 2, \ldots, n$ occurs exactly once among $i_1, i_2, \ldots, i_n$. For instance, if $n = 3$, then P_2, P_3, P_1 is one of the $6 = 3! = 3 \cdot 2 \cdot 1$ possible orders in which we might list the three propositions. For each n, there are in fact $n! = n \cdot (n-1) \cdots 2 \cdot 1$ such lists as the reader could prove as a simple exercise in induction.

2.8.3 LEMMA *Provided $n \geq 2$, the conjunction $(\cdots ((P_{i_1} \wedge P_{i_2}) \wedge P_{i_3}) \wedge \cdots \wedge P_{i_n})$ is logically equivalent to the conjunction $(\cdots ((P_1 \wedge P_2) \wedge P_3) \wedge \cdots \wedge P_n)$, that is*

$$(P_{i_1} \wedge \cdots \wedge P_{i_n}) \Leftrightarrow (P_1 \wedge \cdots \wedge P_n).$$

For disjunction, there is similarly the logical equivalence

$$(P_{i_1} \vee \cdots \vee P_{i_n}) \Leftrightarrow (P_1 \vee \cdots \vee P_n).$$

As was discussed before, the lemma means that the truth value of a multiple conjunction or disjunction of propositional forms is independent of the order in which the conjunction or disjunction is taken. Indeed, given two lists $P_{i_1}, \ldots, P_{i_n}$ and $P_{j_1}, \ldots, P_{j_n}$, each conjunction $P_{i_1} \wedge P_{i_2} \wedge \cdots \wedge P_{i_n}$ and $P_{j_1} \wedge P_{j_2} \wedge \cdots \wedge P_{j_n}$ is logically equivalent to $P_1 \wedge P_2 \wedge \cdots \wedge P_n$ by the theorem and hence to one another by transitivity of $\Leftrightarrow$.

Proof We concentrate on the first assertion about conjunction, and the proof is by induction. For the basis step $n = 2$, there are two possible lists, namely P_1, P_2 and P_2, P_1, and the assertion is trivial for the first list, namely, $(Q \Leftrightarrow Q)$ for $Q = (P_1 \wedge P_2)$. For the second list, the assertion is $(P_1 \wedge P_2) \Leftrightarrow (P_2 \wedge P_1)$, which is just a restatement of commutativity of conjunction. Thus, we have checked the basis step.

For the induction step, we set

$$Q = (\cdots((P_{i_1} \wedge P_{i_2}) \wedge P_{i_3}) \wedge \cdots \wedge P_{i_n}),$$
$$R = (\cdots((P_1 \wedge P_2) \wedge P_3) \wedge \cdots \wedge P_n),$$

so we must prove $(Q \wedge P_{i_{n+1}}) \Leftrightarrow (R \wedge P_{n+1})$. If it happens that $n + 1 = i_{n+1}$, then $Q \Leftrightarrow R$ by the induction hypothesis, and the induction step follows directly from the first rule of replacement 2.5.12.

On the other hand, if $n + 1 \neq i_{n+1}$, then $n + 1 = i_j$ for some $j \leq n$ since P_{n+1} must occur somewhere in the list $P_{i_1}, \ldots, P_{i_{n+1}}$. Applying the induction hypothesis to the collection $P_{i_1}, \ldots, P_{i_n}$ (and the induction hypothesis applies since this collection has only n members), we conclude $Q \Leftrightarrow (S \wedge P_{i_j})$, where

$$S = P_{i_1} \wedge \cdots \wedge P_{i_{j-1}} \wedge P_{i_{j+1}} \wedge \cdots \wedge P_{i_n} \wedge P_{i_j};$$

thus, we have simply moved $i_j = n + 1$ to the end. Finally,

$$\begin{aligned}
[Q \wedge P_{i_{n+1}}] &\Longleftrightarrow [(S \wedge P_{i_j}) \wedge P_{i_{n+1}}] \\
&\Longleftrightarrow [S \wedge (P_{i_j} \wedge P_{i_{n+1}})] \\
&\Longleftrightarrow [S \wedge (P_{i_{n+1}} \wedge P_{i_j})] \\
&\Longleftrightarrow [(S \wedge P_{i_{n+1}}) \wedge P_{i_j}] \\
&\Longleftrightarrow [(S \wedge P_{i_{n+1}}) \wedge P_{n+1}],
\end{aligned}$$

where the various bi-implications follow from commutativity and associativity of $\wedge$. Thus, $Q \wedge P_{i_{n+1}}$ is logically equivalent to $(S \wedge P_{i_{n+1}}) \wedge P_{n+1}$, which, in turn, is logically equivalent to $R \wedge P_{n+1}$ as we observed before, so $Q \wedge P_{i_{n+1}}$ is indeed logically equivalent to $R \wedge P_{n+1}$, as desired.

An entirely analogous argument handles the case of disjunction, and the proof is therefore complete. *q.e.d.*

Theorem 2.8.1 follows directly from the two previous lemmas, and we close with

■ **Example 2.8.1** We prove that for each $j = 1, \dots, n$, there is an equivalence

$$[Q \Rightarrow (P_1 \vee \cdots \vee P_n)] \Leftrightarrow$$
$$[(Q \wedge \neg P_1 \wedge \cdots \wedge \neg P_{j-1} \wedge \neg P_{j+1} \wedge \cdots \wedge \neg P_n) \Rightarrow P_j]$$

as follows:

$$\begin{aligned} Q \Rightarrow (P_1 \vee \cdots \vee P_n) &\Longleftrightarrow \neg Q \vee (P_1 \vee \cdots \vee P_n) \\ &\Longleftrightarrow (\neg Q \vee P_1 \vee \cdots \vee P_{j-1} \vee P_{j+1} \vee \cdots \vee P_n) \vee P_j \\ &\Longleftrightarrow \neg(Q \wedge \neg P_1 \wedge \cdots \wedge \neg P_{j-1} \wedge \neg P_{j+1} \wedge \cdots \wedge \neg P_n) \vee P_j \\ &\Longleftrightarrow (Q \wedge \neg P_1 \wedge \cdots \wedge \neg P_{j-1} \wedge \neg P_{j+1} \wedge \cdots \wedge \neg P_n) \Rightarrow P_j. \end{aligned}$$

The first and last equivalences follow from the law of implication, the second from associativity and commutativity of $\vee$, and the third from DeMorgan's law.

This example depends on Theorem 2.8.1 and gives yet another proof technique in the spirit of the first chapter.

EXERCISES

2.8 COMMUTATIVITY AND ASSOCIATIVITY

1. (a) Determine which of the binary operations $\Rightarrow$, $\Leftrightarrow$, $\oplus$, $\uparrow$, $\downarrow$ is commutative. Prove your assertions.

(b) Determine which of the binary operations $\Rightarrow$, $\Leftrightarrow$, $\oplus$, $\uparrow$, $\downarrow$ is associative. Prove your assertions.

2. Prove that $P \Rightarrow (Q_1 \wedge Q_2 \wedge \cdots \wedge Q_n)$ is logically equivalent to

$$[(P \Rightarrow Q_1) \wedge (P \Rightarrow Q_2) \wedge \cdots \wedge (P \Rightarrow Q_n)]$$

for each $n \geq 1$.

3. Prove that $(P_1 \wedge P_2 \wedge \cdots \wedge P_n) \Rightarrow Q$ is logically equivalent to

$$(P_1 \Rightarrow Q) \vee (P_2 \Rightarrow Q) \vee \cdots \vee (P_n \Rightarrow Q)$$

for each $n \geq 1$.

4. Prove the logical equivalences

$$[Q \wedge (P_1 \wedge \cdots \wedge P_n)] \Leftrightarrow [Q \wedge P_1 \wedge \cdots \wedge P_n]$$
$$[Q \vee (P_1 \vee \cdots \vee P_n)] \Leftrightarrow [Q \vee P_1 \vee \cdots \vee P_n]$$

for each $n \geq 1$.

2.9 PROVING PROPOSITIONAL FORMS

Of course, proving a logical equivalence as in the previous sections is simply a special case of proving that a general propositional form is a tautology, and we turn our attentions now to arbitrary propositional forms. Specifically, given a propositional form P, we prove P by replacing sub-expressions as before, where we begin with P and end with the logical constant 1, so by Theorem 2.7.3, P is therefore a tautology. Thus, the techniques are much as before, and this section is largely dedicated to examples of these manipulations.

It is worth pointing out immediately that in the previous sections where we proved $P \Leftrightarrow Q$ by replacement starting with P and ending with Q, here we describe a proof technique which would manipulate the entire expression $P \Leftrightarrow Q$ itself directly to the logical constant 1. This is a perfectly legitimate proof technique for logical equivalences, but is rather uncommon in practice owing largely to the fact that one must manipulate many more symbols than with our previous techniques for proving logical equivalence.

■ **Example 2.9.1** As was promised before, we prove here the law of hypothetical syllogism as follows:

$[(P \Rightarrow Q) \wedge (Q \Rightarrow R))] \Rightarrow (P \Rightarrow R)$
$\Longleftrightarrow \neg[(\neg P \vee Q) \wedge (\neg Q \vee R)] \vee (\neg P \vee R)$ by implication
$\Longleftrightarrow \neg(\neg P \vee Q) \vee \neg(\neg Q \vee R) \vee (\neg P \vee R)$ by DeMorgan
$\Longleftrightarrow (P \wedge \neg Q) \vee (Q \wedge \neg R) \vee (\neg P \vee R)$ by DeMorgan
$\Longleftrightarrow [\neg P \vee (P \wedge \neg Q)] \vee [R \vee (Q \wedge \neg R)]$ by associativity/commutativity of $\vee$
$\Longleftrightarrow [(\neg P \vee P) \wedge (\neg P \vee \neg Q)] \vee [(R \vee Q) \wedge (R \vee \neg R)]$ by distributivity
$\Longleftrightarrow [1 \wedge (\neg P \vee \neg Q)] \vee [(R \vee Q) \wedge 1]$ by the law of excluded middle
$\Longleftrightarrow (\neg P \vee \neg Q) \vee (R \vee Q)$ by laws of 1
$\Longleftrightarrow (\neg P \vee R) \vee (Q \vee \neg Q)$ by commutativity and associativity of $\vee$
$\Longleftrightarrow (\neg P \vee R) \vee 1$ by the law of excluded middle
$\Longleftrightarrow 1$ by laws of 1

Just as before, one again essentially always uses the law of implication to get rid of the $\Rightarrow$'s. Furthermore, one often distributes $\vee$ over $\wedge$ and vice versa using commutativity and associativity in order to "cancel" pairs $P \vee \neg P$ or $P \wedge \neg P$ using laws of 0 or 1.

Actually, we shall assume a certain level of sophistication on the part of the reader henceforth and shall not explicitly justify each step in future examples as we did previously. This is fairly common and further emphasizes the practical need to "restrain" oneself in exposition and perform only a few replacements in each step of the computation. The reader should strive to exercise such restraint!

■ **Example 2.9.2** We prove here the constructive dilemma as follows:

$$\begin{aligned}
&[(P \Rightarrow Q) \wedge (R \Rightarrow S) \wedge (P \vee R)] \Rightarrow (Q \vee S)\\
&\quad\Longleftrightarrow \neg[(\neg P \vee Q) \wedge (\neg R \vee S) \wedge (P \vee R)] \vee (Q \vee S)\\
&\quad\Longleftrightarrow [(P \wedge \neg Q) \vee (R \wedge \neg S) \vee (\neg P \wedge \neg R)] \vee (Q \vee S)\\
&\quad\Longleftrightarrow (\neg P \wedge \neg R) \vee [Q \vee (P \wedge \neg Q)] \vee [S \vee (R \wedge \neg S)]\\
&\quad\Longleftrightarrow (\neg P \wedge \neg R) \vee [(Q \vee P) \wedge (Q \vee \neg Q)] \vee [(S \vee R) \wedge (S \vee \neg S)]\\
&\quad\Longleftrightarrow (\neg P \wedge \neg R) \vee [(Q \vee P) \wedge 1] \vee [(S \vee R) \wedge 1]\\
&\quad\Longleftrightarrow (\neg P \wedge \neg R) \vee (Q \vee P) \vee (S \vee R)\\
&\quad\Longleftrightarrow (\neg P \wedge \neg R) \vee (P \vee R) \vee (Q \vee S)\\
&\quad\Longleftrightarrow [\neg(P \vee R) \vee (P \vee R)] \vee (Q \vee S)\\
&\quad\Longleftrightarrow 1 \vee (Q \vee S)\\
&\quad\Longleftrightarrow 1
\end{aligned}$$

■ **Example 2.9.3** We prove here the *law of factorization*

$$(P \Rightarrow Q) \Rightarrow [(R \wedge P) \Rightarrow (R \wedge Q)]$$

as follows:

$$\begin{aligned}
&(P \Rightarrow Q) \Rightarrow [(R \wedge P) \Rightarrow (R \wedge Q)]\\
&\quad\Longleftrightarrow \neg(\neg P \vee Q) \vee [\neg(R \wedge P) \vee (R \wedge Q)]\\
&\quad\Longleftrightarrow (P \wedge \neg Q) \vee [(\neg R \vee \neg P) \vee (R \wedge Q)]\\
&\quad\Longleftrightarrow [\neg P \vee (P \wedge \neg Q)] \vee [\neg R \vee (R \wedge Q)]\\
&\quad\Longleftrightarrow [(\neg P \vee P) \wedge (\neg P \vee \neg Q)] \vee [(\neg R \vee R) \wedge (\neg R \vee Q)]\\
&\quad\Longleftrightarrow [1 \wedge (\neg P \vee \neg Q)] \vee [1 \wedge (\neg R \vee Q)]\\
&\quad\Longleftrightarrow (\neg P \vee \neg Q) \vee (\neg R \vee Q)\\
&\quad\Longleftrightarrow (Q \vee \neg Q) \vee (\neg P \vee R)\\
&\quad\Longleftrightarrow 1 \vee (\neg P \vee \neg R)\\
&\quad\Longleftrightarrow 1
\end{aligned}$$

■ **Example 2.9.4** We prove here

$$\begin{aligned}
&[(P \vee Q) \wedge \neg R] \vee [(P \vee Q) \wedge \neg S] \vee (\neg P \wedge \neg Q) \vee (R \wedge S)\\
&\quad\Longleftrightarrow [(P \vee Q) \wedge (\neg R \vee \neg S)] \vee (\neg P \wedge \neg Q) \vee (R \wedge S)\\
&\quad\Longleftrightarrow [(P \vee Q) \wedge (\neg R \vee \neg S)] \vee \neg[(P \vee Q) \wedge (\neg R \vee \neg S)]\\
&\quad\Longleftrightarrow 1
\end{aligned}$$

An important moral of the previous example is that one should not be so quick to fully expand a propositional form using distributivity; rather, one might look for "large" sub-expressions to "cancel" using laws of 0 and 1. For another example of this type, we have

■ **Example 2.9.5** We prove the rule of inference $[((P \Rightarrow Q) \wedge S) \Rightarrow R] \Rightarrow [(P \Rightarrow Q) \Rightarrow (S \Rightarrow R)]$ as follows:

$$\begin{aligned}
&[((P \Rightarrow Q) \wedge S) \Rightarrow R] \Rightarrow [(P \Rightarrow Q) \Rightarrow (S \Rightarrow R)] \\
&\quad \Longleftrightarrow \neg[[(P \Rightarrow Q) \wedge S] \Rightarrow R] \vee [(P \Rightarrow Q) \Rightarrow (S \Rightarrow R)] \\
&\quad \Longleftrightarrow \neg[\neg[(\neg P \vee Q) \wedge S] \vee R] \vee [\neg(\neg P \vee Q) \vee (\neg S \vee R)] \\
&\quad \Longleftrightarrow [(\neg P \vee Q) \wedge S \wedge \neg R] \vee [\neg(\neg P \vee Q) \vee (\neg S \vee R)] \\
&\quad \Longleftrightarrow [(\neg P \vee Q) \wedge \neg(\neg S \vee R)] \vee \neg[(\neg P \vee Q) \vee (\neg S \vee R)] \\
&\quad \Longleftrightarrow 1
\end{aligned}$$

EXERCISES

2.9 PROVING PROPOSITIONAL FORMS

1. Prove the converse of Example 2.9.5.
2. Prove that $[(P \vee R) \wedge (P \Rightarrow S) \wedge (R \Rightarrow S)] \Rightarrow S$.
3. Prove that $[P \wedge (\neg(P \wedge Q))] \Rightarrow \neg Q$.
4. Prove that $[Q \wedge (Q \Rightarrow P) \wedge [(P \wedge Q) \Rightarrow R]] \Rightarrow R$.
5. Prove that $[(P \Rightarrow \neg N) \wedge (\neg P \Rightarrow \neg S) \wedge S \wedge (R \Rightarrow N)] \Rightarrow \neg R$.
6. Prove that $[(P \Rightarrow Q) \wedge [(\neg P \wedge R) \Rightarrow S] \wedge \neg Q] \Rightarrow [(R \vee P) \Rightarrow S]$.
7. Prove that $[(P \Rightarrow M) \wedge (M \Rightarrow T) \wedge M] \Rightarrow (T \vee P)$.
8. Prove that $[(P \Rightarrow N) \wedge (\neg P \Rightarrow \neg S) \wedge (R \Rightarrow \neg N)] \Rightarrow (R \Rightarrow \neg S)$.
9. Prove that $[(P \Rightarrow Q) \wedge [(P \wedge Q) \Rightarrow R]] \Rightarrow (P \Rightarrow R)$.
10. Prove that $[(P \wedge Q) \Leftrightarrow P] \Leftrightarrow (P \Leftrightarrow Q)$.

2.10 PREDICATE FORMS AND QUANTIFICATION

We finally expand our attentions to include predicates and must return to the very formal setting where parentheses are inserted to include the result of each logical operation. In this setting, we again consider "legal" strings of symbols, but here we allow predicate variables and predicates as well as propositional variables, propositions, and the logical constants. Furthermore, another "operation" we can perform on such a legal string is the quantification of a component variable, i.e., we can precede a legal expression by $\forall x$ or $\exists x$ where x is the argument of some component predicate variable $P(x)$. Again to be precise, we enclose each such quantified predicate form by parentheses. For instance, $\forall xP(x)$ must be written $\forall x(P(x))$ for now, but we shall shortly introduce another convention to alleviate the current need for parentheses.

A "legal" string of symbols built from the logical constants, propositions, propositional variables, predicates, and predicate variables using logical operators and quantification inserting parentheses as discussed above is called a *predicate form*, and if P and Q are predicate forms, then we write $P = Q$ if P and Q agree exactly as strings of symbols.

We say that a predicate form P is *valid* if it is true for any possible assignment of propositions and predicates to the component propositional and predicate variables, and we say that P is *logically equivalent* to another predicate form Q if $P \Leftrightarrow Q$ is valid. Furthermore, P is said to be *satisfiable* if it is true for some assignment of propositions and predicates to its component propositional and predicate variables; finally, P is *unsatisfiable* if it is not satisfiable, so P is never true for any assignment of component propositional and predicate variables.

Thus, whereas we had the trichotomy tautology/contingency/absurdity for propositional forms, now we have the analogous valid/satisfiable/unsatisfiable. When we speak of "proving" a predicate form, we mean proving that it is valid, and when we speak of "disproving" a predicate form, we mean proving that it is not valid, so we prove that its negation is satisfiable. In other words, to disprove a predicate form, we must exhibit truth values for the propositions and choices of predicates for the component predicate variables so that the resulting expression is false. We shall see examples of disproofs later in §2.13.

Observe that our notion of "truth", i.e., validity, of a predicate form is most demanding, and it is perhaps surprising that interesting valid predicate forms can be found! Actually, we have already seen the valid predicate forms

$$[[\neg(\exists x(Q(x)))] \Leftrightarrow [\forall x(\neg(Q(x)))]],$$
$$[[\neg(\forall x(Q(x)))] \Leftrightarrow [\exists x(\neg(Q(x)))]],$$

which are easily proved by universal and existential generalization and instantiation. Other such examples will be given in the next section.

Notice that our standard logical equivalences of propositions in §2.4 lead immediately by universal generalization to logical equivalences of predicate forms. For instance,

$$[[\forall x[P(x) \wedge (Q(x) \vee R(y))]] \Leftrightarrow [\forall x[(P(x) \wedge Q(x)) \vee (P(x) \wedge R(y))]]]$$

is valid by the usual distributivity of $\wedge$ over $\vee$. These manipulations of predicate forms together with some other standard identities to be discussed in the next section constitute the full-blown predicate calculus.

In order to relieve the current need for so many parentheses, we shall add to our list yet another convention, and we must first develop some background. Suppose that $P(x, y, z, \ldots)$ is a predicate form which includes the symbols $x, y, z, \ldots$

as expressions for the variables (ranging over their universes) of the component predicate variables. We say a variable, say the variable x, is *bound* if one of the symbols $\exists x$ or $\forall x$ occurs in $P(x, y, z, \ldots)$ so as to quantify the variable x, and otherwise, the variable x is said to be *free*. For instance, in the predicate form $\forall x[\exists z[(Q(x, z) \wedge R(x, y, z)]]$, all occurrences of x and z are bound, while y is free. On the other hand, given a sub-expression $\exists x$ or $\forall x$ of $P(x, y, z, \ldots)$, according to our rules of inserting parentheses, there is a associated sub-expression of $P(x, y, z, \ldots)$ which is enclosed by corresponding parentheses, and this sub-expression is called the *scope* of the corresponding quantifier. For instance, in the example above:

$$[\underbrace{\forall x\,(\exists z\,\underbrace{(Q(x,z) \wedge R(x,y,z))}_{\text{scope of } \exists z})}_{\text{scope of } \forall x})] \,.$$

Of course, from the definitions, each appearance of a variable x is bound, if at all, in the scope of its corresponding quantifier, and the variables which remain unquantified are the free ones. Our final convention for parentheses in the predicate calculus is:

2.10.1 CONVENTION *The scope of a quantifier is the smallest possible sub-expression consistent with parentheses.*

For some examples:

$$\forall x(\underbrace{\exists z \;\; \underbrace{Q(x,z)}_{\text{scope of } \exists z} \;\; \wedge \;\; R(x,y,z)}_{\text{scope of } \forall x}),$$

and here y is free, each appearance of x is bound, and only the appearance of z in $R(x, y, z)$ is free. In contrast, we have

$$\forall x\, \underbrace{\exists z \;\; \underbrace{Q(x,z)}_{\text{scope of } \exists z}}_{\text{scope of } \forall x} \;\; \wedge \;\; R(x,y,z),$$

and here the appearances of x, y, z in $R(x, y, z)$ are all free.

It is worth emphasizing that in the previous example, the same symbols x, z are used in different contexts in the predicate form: The argument x of Q has nothing whatsoever to do with the argument x of R. Although such a double use of a symbol is not specifically ruled out, it can be confusing, and the reader is advised to use different symbols for different arguments of predicate forms when appropriate. For instance, one could write $\forall x \exists z Q(x, z) \wedge R(u, v, w)$ in the previous example. In general, we can evidently replace the symbol for an argument of a predicate by some other symbol provided that all the other related appearances of that symbol are likewise replaced. More precisely, if a variable is bound by a particular quantifier, then we may replace it by another symbol provided we also replace with the same symbol each appearance of the variable within the scope of that quantifier.

We next refine our methods of quantification in several directions. First of all, we shall write $\exists! x P(x)$ in case $P(x)$ is true for exactly one x in the universe of discourse of $P(x)$, and one should read the symbol $\exists!$ in English as "there exists a unique". More precisely, we define $\exists! x P(x)$ to denote the conjunction

$$[\exists x P(x)] \wedge [\forall x \forall y (P(x) \wedge P(y)) \Rightarrow (x = y)],$$

so in particular, we have $\exists! x P(x) \Rightarrow \exists x P(x)$. For instance, consider the predicates $E(n) = [n \text{ is even}]$ and $O(n) = [n \text{ is odd}]$ on the universe consisting of the integers 0,1 and 2. The predicate form $\exists! n O(n)$ is valid, while $\exists! n E(n)$ is not.

Our second refinement of quantification involves a pair $P(x), Q(x)$ of predicates defined over the same universe, and we write

$$\forall Q(x) P(x) \text{ to mean } \forall x (Q(x) \Rightarrow P(x)),$$
$$\exists Q(x) P(x) \text{ to mean } \exists x (Q(x) \wedge P(x)).$$

The English translations, respectively, correspond to "for all x so that $Q(x), P(x)$" and "there exists $Q(x)$ so that $P(x)$". In fact, mathematicians sometimes use the symbol "s.t." to designate the phrase "so that", and one might write instead $\forall x$s.t.Q(x), P(x) and $\exists x Q(x)$s.t.P(x) respectively. We mention parenthetically that the symbol "$\ni$" is sometimes used as shorthand for the phrase "such that", but this notation is only used at the blackboard. Actually, we shall only rarely use the notation $\forall Q(x) P(x)$ and $\exists Q(x) P(x)$ in this volume, but it is extremely common in practice.

A typical example occurs when the universe of discourse is the collection of all real numbers or integers with $Q(x)$ an inequality, say $Q(x) = [x \geq 0]$, for instance. Letting $P(x)$ denote the predicate $[x + 1 \geq 0]$, we find

$$\forall Q(x) P(x) = \forall x \geq 0 (x + 1 \geq 0) \Leftrightarrow \forall x [(x \geq 0) \Rightarrow (x + 1 \geq 0)],$$
$$\exists Q(x) P(x) = \exists x \geq 0 (x + 1 \geq 0) \Leftrightarrow \exists x [(x \geq 0) \wedge (x + 1 \geq 0)].$$

We turn finally to an analysis of the effects of the order of quantification of variables in a predicate form. In fact, the order of quantification can affect the truth value of a predicate form, as is easily seen by example: Consider the predicate $x+y = 1$ on the universe of all pairs x, y of real numbers, so $\forall x \exists y (x+y = 1)$ is true since for each x there is indeed such a y (namely $y = 1 - x$). On the other hand, $\exists y \forall x (x + y = 1)$ is false since we cannot choose a y which works for all x. More precisely, suppose to the contrary that such a y exists. By existential and universal instantiation, y must satisfy both equations $1+y = 1$ and $0+y = 1$, which is absurd.

For another example, consider the predicate $xy = 0$ on the universe of all pairs of real numbers, so $\exists! x \forall y (xy = 0)$ is true since there is indeed the unique choice $x = 0$ so that for all y, we have $xy = 0$. On the other hand, $\forall y \exists! x (xy = 0)$ is false since for $y = 0$ and for any x, we have $xy = 0$. Indeed, for each $y \neq 0$ there is a unique x, while for $y = 0$, there exists such an x, but there is not a unique such x. Notice also that $\forall y \exists x(xy = 0)$, $\exists x \forall y(xy = 0)$, and $\forall y \neq 0 \exists! x(xy = 0)$ are all evidently valid.

EXERCISES

2.10 PREDICATE FORMS AND QUANTIFICATION

1. Let $E(x, y)$ denote the predicate $x = y$ and $L(x, y)$ denote the predicate $x < y$, where each predicate is defined over the universe of all integers. Let $P(x, y, z)$ denote the predicate $x \cdot y = z$ defined over the universe of all ordered triples x, y, z of integers. Using this notation, write the following sentences as quantified predicates.

(a) If $x = 1$, then $x \cdot y = y$ for any y.

(b) If $x \cdot y$ is non-zero, then x and y are non-zero.

(c) If $x \cdot y = 0$, then either $x = 0$ or $y = 0$.

(d) $2x = 4$ if and only if $x = 2$.

(e) The equation $x^4 = y$ has no solution unless $y \geq 0$.

(f) $x < z$ if there is some y between x and z.

(f) If $x < y$, then for some $z < 0$, we have $xz > yz$.

(g) There is some x so that for each y and z we have $xz = xy$.

(h) For each x satisfying $x \geq 1$, there is exactly one $y \leq 1$ satisfying $x \cdot y = 1$.

2. For each part of the previous exercise, express the negation of the quantified predicate after moving the negation sign as far to the right as possible in the predicate form.

3. Let the universe of discourse be the collection of all pairs of natural numbers. For each of the following quantified predicates, specify a predicate which makes the implication true and specify a different predicate which makes the implication false. Prove your assertions.

(a) $\forall x \exists! y P(x, y) \Rightarrow \exists! y \forall x P(x, y)$

(b) $\exists! y \forall x P(x, y) \Rightarrow \forall x \exists! y P(x, y)$

4. Let Q be an arbitrary predicate over the universe consisting of the natural numbers 1,2, and 3, and recall the exclusive or operator $\oplus$, so $(P \oplus Q) \Leftrightarrow [(P \vee Q) \wedge \neg(P \wedge Q)]$. Are the two propositions $\exists! x P(x)$ and $P(1) \oplus P(2) \oplus P(3)$ logically equivalent? Prove your assertion.

5. Let $P(x, y)$ be a predicate defined on the universe of all ordered pairs x, y of natural numbers where $1 \leq x, y \leq 3$. Exhibit finite disjunctions and conjunctions of propositions (without using quantifiers) which are logically equivalent to each of the following.

(a) $\forall x P(x,0) \vee \exists x P(x,0)$

(b) $\forall x \exists ! y P(x,y)$

(c) $\forall x \forall y P(x,y)$

(d) $\exists ! x \exists y P(x,y)$

(e) $\exists x \forall y P(x,y)$

6. Recall from differential calculus that a real-valued function $f(x)$ of a real variable x is said to be *continuous at a point* x_0 if

$$\forall \epsilon > 0 \; \exists \delta > 0 \; \forall x \; (|x - x_0| < \delta \Rightarrow |f(x) - f(x_0)| < \epsilon),$$

where ϵ, δ and x each vary over the universe of all real numbers. Owing to the order of quantification in this definition, to show that a function f is continuous at a point x_0, we are given an arbitrary $\epsilon > 0$ and must exhibit some $\delta > 0$ so that the implication above holds for any x; the quantified predicate then holds by universal generalization. Furthermore, the function f itself is said to be simply *continuous* if it is continuous at x_0 for each real number x_0, i.e.,

$$\forall x_0 \; \forall \epsilon > 0 \; \exists \delta > 0 \; \forall x \; (|x - x_0| < \delta \Rightarrow |f(x) - f(x_0)| < \epsilon),$$

where x_0 also varies over the universe of all real numbers. Owing to the order of quantification, for any specified x_0 and any specified $\epsilon > 0$, we must again exhibit some $\delta > 0$ so that the implication above holds for each x; the quantified predicate again holds by universal generalization.

(a) Show that the function $f(x) = 2x$ is continuous.

(b) Show that the function $f(x) = x^2$ is continuous.

(c) Give the predicate form expressing that a function f is not continuous.

(d) Give an example of a discontinuous function and prove your assertion.

7. Recall from differential calculus that a real-valued function $f(x)$ of a real variable x is said to be *uniformly continuous* if

$$\forall \epsilon > 0 \; \exists \delta > 0 \; \forall x \; \forall x_0 \; (|x - x_0| < \delta \Rightarrow |f(x) - f(x_0)| < \epsilon),$$

where the universe of discourse is in each case again the collection of real numbers. Owing to the order of quantification, we are here given only $\epsilon > 0$ and must exhibit a $\delta > 0$ so that the implication above holds for each x and x_0.

(a) Prove that if a function f is uniformly continuous, then it is also continuous.

(b) Show that the function $f(x) = x$ is uniformly continuous.

(c) Show that $f(x) = 2x + 8$ is uniformly continuous.

(d) Give the predicate form expressing that a function f is not uniformly continuous.

(e) Show that the function $f(x) = x^2$ defined over the whole real line is not uniformly continuous.

(e) Show that the function $g(x) = x^2$ defined on any closed bounded interval in the real line is uniformly continuous.

2.11 STANDARD VALID PREDICATE FORMS

The notations $\forall Q(x)P(x)$ and $\exists Q(x)P(x)$ have a particularly pleasant property which generalizes our result in §1.A.3 on negating quantified predicates, namely

2.11.1 PROPOSITION *We have the following logical equivalences.*

$$\neg\forall Q(x)P(x) \Leftrightarrow \exists Q(x)\neg P(x),$$
$$\neg\exists Q(x)P(x) \Leftrightarrow \forall Q(x)\neg P(x),$$

Proof For the first equivalence, compute

$$\begin{aligned} \neg\forall Q(x)P(x) &\Longleftrightarrow \neg\forall x[Q(x) \Rightarrow P(x)] \\ &\Longleftrightarrow \exists x\neg[Q(x) \Rightarrow P(x)] \\ &\Longleftrightarrow \exists x\neg[\neg Q(x) \vee P(x)] \\ &\Longleftrightarrow \exists x[Q(x) \wedge \neg P(x)] \\ &\Longleftrightarrow \exists Q(x)\neg P(x), \end{aligned}$$

and for the second, compute

$$\begin{aligned} \neg\exists Q(x)P(x) &\Longleftrightarrow \neg\exists x[Q(x) \wedge P(x)] \\ &\Longleftrightarrow \forall x\neg[Q(x) \wedge P(x)] \\ &\Longleftrightarrow \forall x[\neg Q(x) \vee \neg P(x)] \\ &\Longleftrightarrow \forall x[Q(x) \Rightarrow P(x)] \\ &\Longleftrightarrow \forall Q(x)\neg P(x). \end{aligned}$$

q.e.d.

These computations represent our first in the full predicate calculus, where we have again replaced logically equivalent sub-expressions in order to prove that a predicate form is valid, and in this case the valid form is itself actually a logical equivalence.

In contrast to the examples in the previous section where we unsuccessfully attempted to interchange the order of universal and existential quantification, we have

2.11.2 LEMMA *Suppose that $R(x, y)$ is a predicate form with arguments x, y, where R may have further arguments as well. We have the logical equivalences*

$$\forall x \forall y R(x, y) \Leftrightarrow \forall y \forall x R(x, y),$$
$$\exists x \exists y R(x, y) \Leftrightarrow \exists y \exists x R(x, y).$$

Proof We shall concentrate on the first equation and suppose that the lefthand side is true. Choose arbitrary y, x, so by universal instantiation, $R(x, y)$ is true. Since y, x were arbitrary, the righthand side is true by universal generalization. Similar arguments handle the converse as well as the case of existential quantification. *q.e.d.*

The analogous (but slightly more complicated) proofs below by instantiation and generalization will be left as exercises for the reader. As usual, an inductive proof akin to that in §2.8 extends the previous lemma to

2.11.3 PROPOSITION *Fix some predicate $R(x_1, \ldots, x_n, y_1, \ldots, y_m)$ with $m + n$ variables, where perhaps $mn = 0$. For each specification of $y_1, \cdots, y_m$, each $n \geq 2$, and each ordering $i_1, \ldots, i_n$ of $1, \ldots, n$, we have the following logical equivalences*

$$\forall x_1 \cdots \forall x_n R(x_1, \ldots, x_n, y_1, \ldots, y_m) \Leftrightarrow \forall x_{i_1} \cdots \forall x_{i_n} R(x_1, \ldots, x_n, y_1, \ldots, y_m),$$
$$\exists x_1 \cdots \exists x_n R(x_1, \ldots, x_n, y_1, \ldots, y_m) \Leftrightarrow \exists x_{i_1} \cdots \exists x_{i_n} R(x_1, \ldots, x_n, y_1, \ldots, y_m).$$

Another result whose proof is an exercise in generalization and instantiation, we have

2.11.4 PROPOSITION *If a propositional form occurs in a disjunction or conjunction within the scope of a quantifier, then it may be removed from the scope of that quantifier, in the sense that*

$$\forall x(P(x) \vee Q) \Leftrightarrow [(\forall x P(x)) \vee Q] \text{ and } \forall x(P(x) \wedge Q) \Leftrightarrow [(\forall x P(x)) \wedge Q],$$
$$\exists x(P(x) \vee Q) \Leftrightarrow [(\exists x P(x)) \vee Q] \text{ and } \exists x(P(x) \wedge Q) \Leftrightarrow [(\exists x P(x)) \wedge Q].$$

In the same way, if a predicate form occurs in a disjunction or conjunction and its variables are unbound by a quantifier in the scope of that quantifier, then it may be removed from the scope of that quantifier, in the sense that

$$\forall x(P(x) \vee Q(y)) \Leftrightarrow [(\forall x P(x)) \vee Q(y)],$$
$$\forall x(P(x) \wedge Q(y)) \Leftrightarrow [(\forall x P(x)) \wedge Q(y)],$$
$$\exists x(P(x) \vee Q(y)) \Leftrightarrow [(\exists x P(x)) \vee Q(y)],$$
$$\exists x(P(x) \wedge Q(y)) \Leftrightarrow [(\exists x P(x)) \wedge Q(y)].$$

We turn next to two useful equivalences, which say, in effect, that "$\forall$ distributes over $\wedge$" and "$\exists$ distributes over $\vee$". Precisely, we have

2.11.5 THEOREM *We have the following logical equivalences*

$$\forall x[P(x) \wedge Q(x)] \Leftrightarrow [\forall x P(x) \wedge \forall x Q(x)],$$
$$\exists x[P(x) \vee Q(x)] \Leftrightarrow [\exists x P(x) \vee \exists x Q(x)].$$

The proof is again an exercise in instantiation and generalization which is left to the hopefully untiring reader.

We turn finally to two "rules of inference", that is, valid implications, as follows

2.11.6 THEOREM *We have following valid implications*

$$\exists x[P(x) \wedge Q(x)] \Rightarrow [\exists x P(x) \wedge \exists x Q(x)],$$
$$[\forall x P(x) \vee \forall x Q(x)] \Rightarrow \forall x[P(x) \vee Q(x)].$$

Again, the proof is an exercise in instantiation and generalization. It is worth remarking explicitly that the converses of the implications in the previous result are not actually valid, as we shall see.

EXERCISES

2.11 STANDARD VALID PREDICATE FORMS

1. Prove the following results from the text.

(a) Proposition 2.11.3 (b) Proposition 2.11.4

(c) Theorem 2.11.5 (d) Theorem 2.11.6

2. Find an expression logically equivalent to $\forall x P(x)$ using only the existential quantifier $\exists$ and the unary logical operator $\neg$. Similarly, find an expression using only $\forall$ and $\neg$ which is logically equivalent to $\exists x P(x)$.

3. Find an expression logically equivalent to $\exists! x P(x)$ using the quantifiers $\exists$ and $\forall$, the standard logical operators, and the predicate $E(x,y)$ of equality given by $x = y$.

4. (a) Let $P(x)$ and $Q(x)$ be predicates defined on the universe consisting of the numbers 0 and 1. Write each of the predicate forms

$$\exists x[P(x) \wedge Q(x)] \quad \text{and} \quad [\exists x P(x) \wedge \exists x Q(x)]$$

as conjunctions and disjunctions (with no quantifiers) involving the propositions $P(0), P(1), Q(0), Q(1)$. Rearrange the terms in this expression to prove that

$$\exists x[P(x) \wedge Q(x)] \Rightarrow [\exists x P(x) \wedge \exists x Q(x)].$$

(b) Show that the converse of the implication in part (a) is not valid.

(c) Arguing as in part (a), show that

$$\forall x[P(x) \Leftrightarrow Q(x)] \Rightarrow [\forall x P(x) \Leftrightarrow \forall x Q(x)].$$

(d) Show that the converse of the implication in part (c) is not valid.

2.12 PROVING PREDICATE FORMS

This section is primarily dedicated to typical computations in the full predicate calculus where we prove predicate forms, that is, we prove that various predicate forms are valid. Before undertaking these calculations, let us first address the role of truth tables in this setting.

For an example, consider the predicate form

$$R = [\forall x(P(x) \vee Q(x)) \vee \exists y \neg Q(y)],$$

and construct a table of the possible truth values of the underlying quantified predicates and thereby prove R as follows:

$\forall x P(x)$	$\forall x Q(x)$	$\forall x(P(x) \vee Q(x))$	$\exists y \neg Q(y)$	R
0	0	?	1	1
0	1	1	0	1
1	0	1	1	1
1	1	1	0	1

On the basis of the supposed truth values in the first two columns, we can compute (using instantiation, generalization, and our rules for negating quantified predicates) all but one of the remaining entries in the truth table, as indicated. Specifically, we cannot compute a truth value for $\forall x(P(x) \vee Q(x))$ from the supposed truth values when $\forall x P(x)$ and $\forall x Q(x)$ are false. On the other hand, R is the disjunction of this form and $\exists y \neg Q(y)$, which is true under these circumstances. Thus, even though we cannot completely fill in the truth table in this case, the truth table does furnish us with a legitimate proof that R is valid.

This difficulty of being unable to completely fill in the truth table is endemic to proofs of predicate forms by truth tables, and one is not always able to successfully complete a proof with this method. For such an example, consider the predicate form

$$R = [[\exists x P(x) \wedge \exists x Q(x)] \Rightarrow [\exists x(P(x) \wedge Q(x))]]$$

and its corresponding truth table, where we again use instantiation and generalization to fill in much of the table:

$\exists x P(x)$	$\exists x Q(x)$	$\exists x P(x) \wedge \exists x Q(x)$	$\exists x[P(x) \wedge Q(x)]$	R
0	0	0	0	1
0	1	0	0	1
1	0	0	0	1
1	1	1	?	?

In this example, we are unable to completely fill in the truth table, and the missing entry this time is crucial: We are not able in this case to conclude that the form R is valid, and actually it is not. Thus, the (admittedly naive) method of truth tables fails in this example!

In light of the examples above as well as our previous negative comments about even successful proofs by truth table, we are led to the conclusion that the predicate calculus is the important tool for manipulating predicate forms, and we turn our attentions for the rest of this section to sample calculations.

■ **Example 2.12.1** Let us give here another proof of the first example above, namely, let us show that $\forall x[P(x) \vee Q(x)] \vee \exists y \neg Q(y)$ is valid:

$$
\begin{aligned}
&\forall x[P(x) \vee Q(x)] \vee \exists y \neg Q(y)\\
&\quad \Longleftrightarrow \forall x[P(x) \vee Q(x)] \vee \exists y[\neg Q(y) \wedge 1]\\
&\quad \Longleftrightarrow \forall x[P(x) \vee Q(x)] \vee \exists y[\neg Q(y) \wedge (\neg P(y) \vee P(y))]\\
&\quad \Longleftrightarrow \forall x[P(x) \vee Q(x)] \vee \exists y[(\neg Q(y) \wedge \neg P(y)) \vee (\neg Q(y) \wedge P(y))]\\
&\quad \Longleftrightarrow \forall x[P(x) \vee Q(x)] \vee \exists y[\neg Q(y) \wedge \neg P(y)] \vee \exists y[\neg Q(y) \wedge P(y)]\\
&\quad \Longleftrightarrow \forall x[P(x) \vee Q(x)] \vee \neg \forall x[(P(x) \vee Q(x)] \vee \exists y[\neg Q(y) \wedge P(y)]\\
&\quad \Longleftrightarrow 1 \vee \exists y[\neg Q(y) \wedge P(y)]\\
&\quad \Longleftrightarrow 1
\end{aligned}
$$

Of course, we are here proving the form to be valid by replacing logical equivalents to derive the logical constant 1. One moral of this example is that one may have to complicate the form by taking a conjunction or disjunction with a logical constant, and the resulting form is then expanded in some convenient way in order to pursue a calculation; we have made this point before in Example 2.7.5 in the context of propositional forms.

■ **Example 2.12.2** Here we prove the logical equivalence of two predicate forms which were each shown to be valid in the previous section, namely, we prove the logical equivalence of the forms

$$\forall x[P(x) \wedge Q(x)] \Leftrightarrow [\forall x P(x) \wedge \forall x Q(x)]$$

and

$$\exists x[P(x) \vee Q(x)] \Leftrightarrow [\exists x P(x) \vee \exists x Q(x)].$$

To this end, we compute

$$\begin{aligned}
&\forall x[P(x) \wedge Q(x)] \Leftrightarrow [\forall x P(x) \wedge \forall x Q(x)] \\
&\Longleftrightarrow \neg\forall x[P(x) \wedge Q(x)] \Leftrightarrow \neg[\forall x P(x) \wedge \forall x Q(x)] \\
&\Longleftrightarrow \exists x[\neg P(x) \vee \neg Q(x)] \Leftrightarrow [\exists x \neg P(x) \vee \exists x \neg Q(x)] \\
&\Longleftrightarrow \exists x[R(x) \vee S(x)] \Leftrightarrow [\exists x R(x) \vee \exists x S(x)],
\end{aligned}$$

where we define $R(x) = \neg P(x)$ and $S(x) = \neg Q(x)$.

Of course, we are proving the logical equivalence of two forms by manipulating one form to the other using replacement of logical equivalents just as for propositional forms. Furthermore, notice the notational utility of introducing the forms R and S above.

■ **Example 2.12.3** We prove here the following rule of inference

$$\begin{aligned}
&\forall x[P(x) \Rightarrow Q(x)] \Rightarrow [\forall x P(x) \Rightarrow \forall x Q(x)] \\
&\Longleftrightarrow \neg\forall x[\neg P(x) \vee Q(x)] \vee [\neg\forall x P(x) \vee \forall x Q(x)] \\
&\Longleftrightarrow \exists x[P(x) \wedge \neg Q(x)] \vee \exists x \neg P(x) \vee \forall x Q(x) \\
&\Longleftrightarrow \exists x[(P(x) \wedge \neg Q(x)) \vee \neg P(x)] \vee \forall x Q(x) \\
&\Longleftrightarrow \exists x[(P(x) \vee \neg P(x)) \wedge (\neg Q(x) \vee \neg P(x))] \vee \forall x Q(x) \\
&\Longleftrightarrow \exists x[1 \wedge (\neg Q(x) \vee \neg P(x))] \vee \forall x Q(x) \\
&\Longleftrightarrow \exists x[\neg Q(x) \vee \neg P(x)] \vee \forall x Q(x) \\
&\Longleftrightarrow \exists x \neg Q(x) \vee \exists x \neg P(x) \vee \forall x Q(x) \\
&\Longleftrightarrow \exists x \neg P(x) \vee [\neg\forall x Q(x) \vee \forall x Q(x)] \\
&\Longleftrightarrow \exists x \neg P(x) \vee 1 \\
&\Longleftrightarrow 1
\end{aligned}$$

As before, we are proving the validity of the implication by manipulating the entire expression to the logical constant 1, and it is worth emphasizing again that one essentially always uses the law of implication, as above, to get rid of the symbol $\Rightarrow$, and the rationale for this is just as before. For a final similar example of a computation in the predicate calculus, we have

■ **Example 2.12.4** We prove here the following rule of inference

$$\begin{aligned}
&[\exists x P(x) \Rightarrow \forall x Q(x)] \Rightarrow \forall x[P(x) \Rightarrow Q(x)] \\
&\Longleftrightarrow \neg[\neg\exists x P(x) \vee \forall x Q(x)] \vee \forall x[\neg P(x) \vee Q(x)] \\
&\Longleftrightarrow [\neg\neg\exists x P(x) \wedge \neg\forall x Q(x)] \vee \forall x[\neg P(x) \vee Q(x)] \\
&\Longleftrightarrow [\exists x P(x) \wedge \exists x \neg Q(x)] \vee \forall x[\neg P(x) \vee Q(x)] \\
&\Longleftrightarrow [\exists z P(z) \wedge \exists y \neg Q(y)] \vee \forall x[\neg P(x) \vee Q(x)] \\
&\Longleftrightarrow [\forall x(\neg P(x) \vee Q(x)) \vee \exists z P(z)] \wedge [\forall x(\neg P(x) \vee Q(x)) \vee \exists y \neg Q(y)],
\end{aligned}$$

but each predicate form in this last conjunction is logically equivalent to 1 by Example 2.12.1 above, so the original implication is actually logically equivalent to $(1 \wedge 1) \Leftrightarrow 1$, as desired.

It is worth pointing out the notational utility of introducing different symbols (for instance, we replaced x by y, z in the example above) for the various quantified variables in order to avoid confusion in subsequent manipulations.

EXERCISES

2.12 PROVING PREDICATE FORMS

1. Prove the following predicate forms.

(a) $\exists x \exists y[P(x) \wedge Q(y)] \;\Rightarrow\; \exists x P(x)$

(b) $\exists y \forall x P(x,y) \;\Rightarrow \forall x \exists y P(x,y)$

2. Prove the following predicate forms.

(a) $\forall x Q(x) \;\Rightarrow\; \left[\forall x \neg[P(x) \vee \neg Q(x)] \;\vee\; \exists x[P(x) \vee Q(x)]\right]$

(b) $\exists x \forall y[P(x) \wedge Q(y)] \;\vee\; \forall x \exists y\left[[P(x) \Rightarrow Q(y)] \vee P(x)\right]$

3. Prove the following logical equivalences.

(a) $\forall x \forall y[P(x) \wedge Q(y)] \;\Leftrightarrow\; [\forall x P(x) \wedge \forall x Q(x)]$

(b) $\forall x \forall y[P(x) \vee Q(y)] \;\Leftrightarrow\; [\forall x P(x) \vee \forall x Q(x)]$

(c) $\exists x \exists y[P(x) \Rightarrow P(y)] \;\Leftrightarrow\; [\forall x P(x) \Rightarrow \exists x P(x)]$

(d) $\forall x \forall y[P(x) \Rightarrow Q(y)] \;\Leftrightarrow\; [\exists x P(x) \Rightarrow \forall y Q(y)]$

2.13 DISPROOFS

This final section is dedicated to discussing techniques for disproving propositional and predicate forms. We have discussed methods of disproof before in §1.B.5 and next collect previous remarks into formal statements.

> **DISPROVING PROPOSITIONAL FORMS** Suppose that P is a propositional form with component propositional variables $P_1, \ldots, P_m$. To disprove P, that is, to prove that P is not a tautology, we must exhibit truth values for each P_i, for $i = 1, \ldots, m$, which render P false.

More generally for predicate forms, we have

DISPROVING PREDICATE FORMS Suppose that P is a predicate form with component propositional variables $P_1, \ldots, P_m$, if any, and component predicate variables $Q_1, \ldots, Q_n$, if any. To disprove P, that is, to prove that P is not valid, we must exhibit truth values for the P_i, where $i = 1, \ldots, m$, as well as specify particular predicates Q_j, where $j = 1, \ldots, n$, which render P false.

That these are complete disproofs in each case follows immediately from the definitions of "valid" and "tautological". Notice also that the method for propositional forms is simply a special case of the method for predicate forms. In either case, the specification of truth values for the propositional variables, if any, and the specification of particular predicates for the predicate variables, if any, so as to render the predicate form false is called a *counter-example.* Thus, to disprove a predicate or propositional form, we must simply exhibit a counter-example. In a sense, then, it is "easier" to disprove something than to prove it, but such is the human condition.

One can sometimes "guess" a suitable counter-example by pondering the form itself. On the other hand, often one cannot guess so easily, and the natural approach which is usually quite effective is to manipulate the form using replacement by logical equivalents as usual to find another logically equivalent form for which one can more easily "guess" a counter-example. Thus, the manipulations of the predicate calculus can be a useful tool in disproving forms as well as in proving them!

Recapitulating this discussion, one disproves a form by exhibiting a counter-example, and one can use the predicate calculus as a tool to make an effective guess of counter-example. The rest of this section is dedicated to examples which illustrate these remarks as well as some common techniques of disproof.

■ **Example 2.13.1** Consider the converse $(P \vee Q) \Rightarrow P$ of the law of addition. By inspection, if $P = 0$ and $Q = 1$, then $(P \vee Q) = (0 \vee 1) \Leftrightarrow 1$, so the implication in question becomes $1 \Rightarrow 0$, which is false. This disproves the implication.

"By inspection" in the previous paragraph actually means "by guessing", and in examples where one cannot guess so handily, it is sensible to manipulate the form using, as always, replacements by logical equivalents to a "simpler" one where one might more easily guess appropriate truth values. For instance, returning to the form above, we find

$$(P \vee Q) \Rightarrow P \Longleftrightarrow \neg(P \vee Q) \vee P \Longleftrightarrow (\neg P \wedge \neg Q) \vee P$$
$$\Longleftrightarrow 1 \wedge (\neg Q \vee P) \Longleftrightarrow \neg Q \vee P,$$

and this is evidently false only for the truth values before.

Of course, having derived a "simpler" logically equivalent form and "guessed" appropriate truth values, one may plug the computed truth values back into the original form verifying that the form has truth value 0 as a check against mistakes in the computation. Notice that there can even be mistakes in the manipulations leading to a guess of truth values which accidentally render the original form false: Simply plugging in these truth values and checking that the original form is actually false is a complete disproof and is quite independent of any possible mistakes in the computations.

■ **Example 2.13.2** Consider the converse $(P \Rightarrow R) \Rightarrow [(P \Rightarrow Q) \wedge (Q \Rightarrow R)]$ of the hypothetical syllogism. Inspection and a bit of thought might lead the reader to "guess" the truth values $P = 0, Q = 1, R = 0$ for which the implication is false. Short of this, one might compute

$$\begin{aligned}
&(P \Rightarrow R) \Rightarrow [(P \Rightarrow Q) \wedge (Q \Rightarrow R)] \\
&\quad \Longleftrightarrow \neg(\neg P \vee R) \vee [(\neg P \vee Q) \wedge (\neg Q \vee R)] \\
&\quad \Longleftrightarrow (P \wedge \neg R) \vee [(\neg P \vee Q) \wedge (\neg Q \vee R)] \\
&\quad \Longleftrightarrow [(P \wedge \neg R) \vee \neg P \vee Q] \wedge [(P \wedge \neg R) \vee \neg Q \vee R].
\end{aligned}$$

Since this last expression is a conjunction, it is false unless each term is true, so to give a counter-example, it suffices to make false either term in the conjunction. One might easily guess the truth values $P = 0, Q = 1, R = 0$ or $P = 1, Q = 1, R = 0$ by inspection of the second term in this conjunction, and these truth values make the original implication false and either one provides our counter-example.

■ **Example 2.13.3** As was promised before, we now disprove the implication $\forall x[P(x) \vee Q(x)] \Rightarrow [\forall x P(x) \vee \forall x Q(x)]$. For the implication to be false, its hypothesis must be true and its conclusion false. Using the falsity of the conclusion, there must be some x_0 in the universe so that $P(x_0)$ is false, and there must be some x_1 in the universe so that $Q(x_1)$ is false. On the other hand using the truth of the hypothesis, for each x in the universe, $P(x) \vee Q(x)$ must be true. We are led to define the predicates

$$P(x) = \begin{cases} 1, & \text{if } x = x_1; \\ 0, & \text{if } x = x_0; \end{cases} \qquad Q(x) = \begin{cases} 0, & \text{if } x = x_1; \\ 1, & \text{if } x = x_0; \end{cases}$$

on the universe consisting of x_0, x_1. The original implication is evidently false for this choice of predicates, so this counter-example gives a legitimate disproof.

This approach of assuming the truth of the hypothesis and the falsity of the conclusion in order to "guess" truth values is obviously a basic technique for disproving implications.

There are, of course, many other possible counter-examples in the previous example. For instance, we might have taken $P(x)$ (and $Q(x)$ respectively) to be the predicate $x \leq 1$ (and $x \geq 1$) on the universe consisting of the integers $-1, 0, +1$, or we might have taken $P(n)$ (and $Q(n)$ respectively) to be the predicate $[n \text{ is even}]$ (and $[n \text{ is odd}]$) on the universe of all integers. Each of these is a perfectly legitimate counter-example, and the reader may safely infer that there is much flexibility in constructing counter-examples to predicate forms which are not valid.

■ **Example 2.13.4** Here we disprove the implication $[\exists x P(x) \wedge \exists x Q(x)] \Rightarrow \exists x[P(x) \wedge Q(x)]$, as was promised before. On the universe consisting of the integers 0 and 1, simply take $P(x) = [x = 0]$ and $Q(x) = [x = 1]$ for a counter-example.

This is actually a more standard exposition of a disproof, the point being that it is really the responsibility of the reader to check that the asserted form is false for the specified choice of predicates and propositions, and it is the writer's responsibility to make sure that the specified values are in fact a counter-example!

■ **Example 2.13.5** We disprove here the implication $[(P \Rightarrow Q) \vee (R \Rightarrow S)] \Rightarrow [(P \vee R) \Rightarrow (Q \vee S)]$, and it is perhaps a bit much to guess a counter-example directly. We therefore compute

$$\begin{aligned}
&[(P \Rightarrow Q) \vee (R \Rightarrow S)] \Rightarrow [(P \vee R) \Rightarrow (Q \vee S)] \\
&\quad \Longleftrightarrow \neg(\neg P \vee Q \vee \neg R \vee S) \vee [\neg(P \vee R) \vee (Q \vee S)] \\
&\quad \Longleftrightarrow (P \wedge \neg Q \wedge R \wedge \neg S) \vee [\neg(P \vee R) \vee Q \vee S].
\end{aligned}$$

In order that the previous disjunction be false, each term of it must be false. In particular, inspection shows that $P = 1, Q = 0, R = 0, S = 0$ renders both terms false, and we have therefore found a counter-example to the original implication.

Actually, a more typical problem in the predicate calculus is to address the question of what implications, if any, hold between two forms. This is the character of our next example, which says, in effect, that "$\Rightarrow$ is not associative".

■ **Example 2.13.6** We discover which implications hold between the forms $A = [(P \Rightarrow Q) \Rightarrow R]$ and $B = [P \Rightarrow (Q \Rightarrow R)]$. To simplify the situation, we compute

$$\begin{aligned}
A &\Longleftrightarrow \neg(\neg P \vee Q) \vee R \Longleftrightarrow (P \wedge \neg Q) \vee R \\
B &\Longleftrightarrow \neg P \vee \neg Q \vee R.
\end{aligned}$$

Inspection of these two equivalences shows that if $P = 0, Q = 1, R = 0$, then B is true while A is false, so these truth values give a counter-example to the implication $B \Rightarrow A$. After trying to guess a counter-example to $A \Rightarrow B$, one is led to look for a proof of this implication and computes

$$\begin{aligned} A \Rightarrow B &\iff [(P \wedge \neg Q) \vee R] \Rightarrow [\neg P \vee \neg Q \vee R] \\ &\iff [(\neg P \vee Q) \wedge \neg R] \vee \neg P \vee \neg Q \vee R \\ &\iff (\neg P \wedge \neg R) \vee (Q \wedge \neg R) \vee \neg P \vee \neg Q \vee R \\ &\iff [\neg Q \vee (Q \wedge \neg R)] \vee [R \vee (\neg P \wedge \neg R)] \vee \neg P \\ &\iff [(\neg Q \vee Q) \wedge (\neg Q \vee \neg R)] \vee [(R \vee \neg P) \wedge (R \vee \neg R)] \vee \neg P \\ &\iff \neg Q \vee \neg R \vee R \vee \neg P \vee \neg P \\ &\iff (R \vee \neg R) \vee \neg P \vee \neg Q \\ &\iff 1 \vee \neg P \vee \neg Q \\ &\iff 1, \end{aligned}$$

so the implication $A \Rightarrow B$ is actually valid.

■ **Example 2.13.7** Here we disprove the implication $[\forall x P(x) \Rightarrow \forall x Q(x)] \Rightarrow \forall x[P(x) \Rightarrow Q(x)]$, and to begin we compute

$$\begin{aligned} &[\forall x P(x) \Rightarrow \forall x Q(x)] \Rightarrow \forall x[P(x) \Rightarrow Q(x)] \\ &\iff \neg[\neg \forall x P(x) \vee \forall x Q(x)] \vee \forall x[\neg P(x) \vee Q(x)] \\ &\iff [\forall z P(z) \wedge \exists y \neg Q(y)] \vee \forall x[\neg P(x) \vee Q(x)] \\ &\iff [\forall x[\neg P(x) \vee Q(x)] \vee \forall z P(z)] \quad \wedge \quad [\forall x[\neg P(x) \vee Q(x)] \vee \exists y \neg Q(y)]. \end{aligned}$$

The second term in the previous conjunction is logically equivalent to 1 by Example 2.12.1 of the previous section, so by laws of 1, the original implication is logically equivalent to

$$\forall x[\neg P(x) \vee Q(x)] \vee \forall z P(z).$$

Inspection of this last expression leads us to define the predicates $P(x) = [x = 0]$ and $Q(x) = [x = 1]$ on the universe consisting of 0 and 1, which gives a counter-example to the implication and completes our disproof.

■ **Example 2.13.8** We disprove the implication

$$\forall x[P(x) \Rightarrow Q(x)] \Rightarrow [\exists x P(x) \Rightarrow \forall x Q(x)]$$

and begin with the calculation

$$\begin{aligned} &\forall x[P(x) \Rightarrow Q(x)] \Rightarrow [\exists x P(x) \Rightarrow \forall x Q(x)] \\ &\iff \neg \forall x[\neg P(x) \vee Q(x)] \vee [\neg \exists x P(x) \vee \forall x Q(x)] \\ &\iff \exists x[P(x) \wedge \neg Q(x)] \vee \forall x \neg P(x) \vee \forall x Q(x). \end{aligned}$$

It is easy to guess a counter-example to this last expression since it is a three-fold disjunction, and we define, for instance, $P(x) = Q(x) = [x = 0]$ on the universe consisting of 0 and 1 to exhibit the desired counter-example.

Our final example illustrates a predicate form for which we must exhibit both truth values for the propositional variables as well as specify particular predicates.

■ **Example 2.13.9** We address here which implications hold between the two predicate forms

$$A = [\forall x[(P(x) \vee \neg Q) \Rightarrow R(x)]] \text{ and } B = [\forall x(P(x) \Rightarrow R(x))],$$

and, to this end, we compute

$$\begin{aligned} A &= [\forall x[(P(x) \vee \neg Q) \Rightarrow R(x)]] \\ &\iff \forall x[(\neg P(x) \wedge Q) \vee R(x)] \\ &\iff \forall x[(\neg P(x) \vee R(x)) \wedge (Q \vee R(x))] \\ &\iff \forall x(\neg P(x) \vee R(x)) \wedge \forall x(Q \vee R(x)) \\ &\iff \forall x(P(x) \Rightarrow R(x)) \wedge (\forall x R(x) \vee Q) \\ &\iff B \wedge [\forall x R(x) \vee Q]. \end{aligned}$$

Thus, taking $Q = 0$, $P(x) = R(x) = [x = 1]$ on the universe consisting of 0 and 1, we find a counter-example to the implication $B \Rightarrow A$. On the other hand, the implication $A \Rightarrow B$ follows directly from the computation above and the law of simplification.

EXERCISES

2.13 DISPROOFS

1. Prove or disprove each of the following.

(a) $(P \vee \neg Q) \ \vee \ (Q \Rightarrow P)$

(b) $[P \Rightarrow Q] \ \Rightarrow \ [(P \wedge R) \Rightarrow (Q \wedge R)]$

(c) $[(P \wedge Q) \vee R] \ \Rightarrow \ [(P \vee Q) \wedge R]$

(d) $[(P \Rightarrow Q) \wedge \neg Q] \ \Rightarrow \ \neg P$

2. What implications, if any, hold between the following pairs of propositional forms?

(a) $P \Rightarrow Q$ and $Q \Rightarrow P$

(b) $\neg P \Leftrightarrow Q$ and $P \Leftrightarrow \neg Q$

(c) $\neg(P \Leftrightarrow Q)$ and $\neg P \Leftrightarrow \neg Q$

(d) $(P \Rightarrow Q) \Rightarrow R$ and $[(P \vee R) \wedge (Q \Rightarrow R)]$

(e) $\neg P \Rightarrow Q$ and $\neg P \wedge \neg Q$

(f) $(P \Rightarrow Q) \Rightarrow (P \Rightarrow R)$ and $P \Rightarrow (Q \Rightarrow R)$

3. Prove or disprove the following predicate forms.

(a) $[\forall x P(x) \Rightarrow Q] \ \vee \ [Q \Rightarrow \exists x P(x)]$

(b) $\forall x[P \Rightarrow Q(x)] \Leftrightarrow [P \Rightarrow \forall x Q(x)]$

(c) $[\exists x P(x) \Rightarrow Q] \ \vee \ [Q \Rightarrow \forall x P(x)]$

(d) $[\forall x P(x) \Rightarrow Q] \Leftrightarrow [\exists x [P(x) \Rightarrow Q]]$

(e) $[\forall x P(x) \Rightarrow] \ \Rightarrow [\exists x P(x) \Rightarrow Q]$

4. What implications, if any, hold between the forms $\forall x P(x) \Rightarrow \exists x Q(x)$ and $\exists x P(x) \Rightarrow \forall x Q(x)$?

5. What implications, if any, hold between the forms $\forall x[P(x) \vee [P(x) \wedge Q(x)]]$ and $\forall x[P(x) \vee Q(x)]$?

6. What implications, if any, hold between the forms $\exists x P(x) \Rightarrow \exists x Q(x)$ and $\exists x[P(x) \Rightarrow Q(x)]$?

Chapter 3

Set Theory

As defined and discussed here, "sets" are *the* basic objects from which essentially all known mathematics is rigorously constructed, and roughly a set is "a collection of elements". For instance, a school of fish, a deck of cards, or a bunch of grapes are all examples of sets. The precise definition is actually quite involved, as we shall see, but the need for rigor in mathematics demands a careful treatment of this foundational material.

Indeed, the careful definition of sets depends upon a system of ten "axioms", which are statements about sets which serve to define them and, like definitions, must simply be taken for granted. We begin with a discussion of these axioms which should thus be regarded as the underlying assumptions of traditional mathematics. Actually, some of these axioms are immediately useful and are discussed in §3.2 whereas we relegate a discussion of the more technical and esoteric axioms to the optional §3.3. In fact, these axioms amount to a family of particular quantified predicates and therefore represent a basic application of the predicate calculus as well as a complete and interesting example of a system of axioms. There are calculations with sets which are akin to calculations in the predicate calculus, to which we turn our attention and develop the basic computational aspects of set theory. In the optional §3.7, we carefully define the set of natural numbers and prove various basic facts. The next section treats an "inductive" method of constructing sets, and we return to a discussion of proof by induction in this setting. A final section simply introduces the standard notation for the various sets of numbers.

3.1 AXIOMS AND THE PRIMITIVES OF SET THEORY

It is pedagogically convenient to take for granted the notion of a "finite set" for a moment, by which we mean an unordered collection of objects $a_1, \ldots, a_n$ for some natural number n. We write $A = \{a_1, \ldots, a_n\}$ for the set consisting of $a_1, \ldots, a_n$ and refer to each a_i as an *element* or a *member* of A. We write $a \in A$ if a is a member of A and remark that the symbol "$\in$" always designates membership in a set throughout mathematics. In fact, two finite sets A and B are defined to be identical if $\forall u(u \in A \Leftrightarrow u \in B)$, so $A = B$ as sets exactly when A and B have the same elements. As a point of terminology, a set with exactly one element is called a *singleton*, and a set with exactly two elements is called a *doubleton*. For instance, the following sets are all identical

$$\{a_1, a_2, a_3\} = \{a_2, a_3, a_1\} = \{a_1, a_1, a_2, a_3\} = \{a_3, a_3, a_3, a_2, a_2, a_1\}$$

since each has the same elements. As an aside, we mention that mathematicians sometimes study "multisets", which are determined not only by their elements, but also by the numbers of appearances of elements in the multiset; for instance, in the equalities of sets above, there are actually three distinct multisets; we shall not discuss multisets further in this book and mention them here just for completeness.

There may seem to be an absolute distinction between finite sets and their members, but if so, then this is illusory. For instance,

$$A = \{a_1, \{a_1, a_2\}, \{\{a_3\}, \{a_2\}\}\}$$

is a set with three members $a_1 \in A, \{a_1, a_2\} \in A$, and $\{\{a_3\}, \{a_2\}\} \in A$; thus, $\{a_1, a_2\}$ is itself a set, while $\{\{a_3\}, \{a_2\}\}$ is a set of sets.

We informally introduce finite sets here in order that the reader have some intuition about sets at least in this simple setting. Our subsequent definition of general sets in particular guarantees the existence of finite sets and legitimizes the notation and remarks above. To illuminate our discussions, we shall also sometimes give examples in this setting of finite sets.

Turning to the general setting of arbitrary sets, it may be unsettling to the reader to learn that the relationship $a \in A$, that is, a is an element of A, is a "primitive" of the theory; this means that the symbol $\in$ remains undefined, and a suitable interpretation of it is certainly evident to the reader from the previous discussion of finite sets. We also write $a \notin A$ if $a \in A$ is false, that is, if a is not an element of A.

Armed with the primitive $\in$ and given two sets A and B, we say that A is *included in* B if $\forall u(u \in A \Rightarrow u \in B)$. We write $A \subseteq B$ and $B \supseteq A$ in this case and call A a *subset* of B and B a *superset* of A. As a notational point, we mention that some authors write "$A \subset B$" or "$B \supset A$" to indicate that A is a subset of B. Of course, if A is not a subset of B, then

$$\neg\forall u(u \in A \Rightarrow u \in B) \Leftrightarrow \exists u \neg(u \notin A \vee u \in B) \Leftrightarrow \exists u(u \in A \wedge u \notin B),$$

so A has an element which is not an element of B; we write $A \not\subseteq B$ or $B \not\supseteq A$ if A is not a subset of B. For instance, given the finite sets $A = \{a_1, a_2, a_2\}$ and $B = \{a_1, a_2, a_3, A\}$, we have $A \subseteq B$, $B \supseteq A$, $B \not\subseteq A$, and $A \in B$.

3.1.1 THEOREM *Suppose that A, B, and C are sets, $A \subseteq B$, and $B \subseteq C$. Then $A \subseteq C$.*

Proof This follows directly from the definitions and hypothetical syllogism:

$$\begin{aligned}[A \subseteq B \wedge B \subseteq C] &\Leftrightarrow [\forall u(u \in A \Rightarrow u \in B) \wedge \forall u(u \in B \Rightarrow u \in C)] \\ &\Leftrightarrow \forall u[(u \in A \Rightarrow u \in B) \wedge (u \in B \Rightarrow u \in C)] \\ &\Rightarrow \forall u(u \in A \Rightarrow u \in C) \\ &\Leftrightarrow A \subseteq C\end{aligned}$$

q.e.d.

3.1.2 THEOREM *Suppose that A and B are sets, $x \in A$, and $A \subseteq B$. Then $x \in B$.*

Proof Since $A \subseteq B$, we have $\forall u(u \in A \Rightarrow u \in B)$, so by universal instantiation, we find $x \in A \Rightarrow x \in B$. Since $x \in A$ by hypothesis, we conclude that $x \in B$ by modus ponens, as desired. *q.e.d.*

It is worth saying explicitly that the previous computations are typical applications of the predicate calculus to set theory: We "translate" the definition of $\subseteq$ into a particular predicate, perform manipulations in the predicate calculus as before, and then "translate" back using the definition of $\subseteq$.

As a point of terminology, we remark that if $B \subseteq A$, then we may say that "A contains B"; in the same way, if $x \in A$, then we may also say simply "A contains x". Thus, the English usage of the word "contains" does not necessarily distinguish between inclusion of subsets and membership of elements. This imprecision in the English terminology does not lead to confusion in practice. We also remark that one sometimes uses the terms "collection" or "family" as synonyms for "set" to avoid terminological monotony.

To close, we discuss certain generalities about "axiomatic definitions" in mathematics. Namely, one specifies a collection of assertions (for instance, a collection of predicate forms) called *axioms* concerning the object or objects to be defined, and these axioms taken together then serve as a complete definition. Just as with a definition, each axiom must be simply accepted. Sometimes the axioms are expressed, in turn, in terms of some *primitive* which is an object that remains undefined like $\in$ above. In a sense, each axiom describes an attribute, and the attributes together uniquely determine the object or objects. More functionally, the axioms are simply a collection of assertions which must be taken for granted and from which one derives theorems as logical consequences.

Axiomatic definitions in general are more problematic than they may sound, for there are several things that can go wrong with such a definition. For instance, there may be no object with all the required properties or there may be too many. One important notion is that of "consistency", and we say a system of axioms is *inconsistent* if it is possible to derive from the axioms both some proposition R and its negation $\neg R$; otherwise, the axioms are said to be *consistent.* Such a logical argument proving R and $\neg R$ is called a *paradox*, and it shows that there is something deeply wrong with the axioms, i.e., they provide a bad definition and must be discarded.

Another basic notion for axioms, we say a collection of axioms is *independent* if it not possible to logically derive one axiom from the others. One of course seeks independent axioms in order to be as efficient as possible.

In fact, the contemporary definition of "set", which is truly the fundamental object in all of mathematics, is axiomatic, and we shall present the axioms of what is called "Zermelo-Fraenkel" set theory in the optional §3.3, which begins with a description of the colorful intellectual history and development of these axioms. One should not think of these axioms as commandments from upon high, but rather, they evolved over time as mathematicians and philosophers attempted to formulate rigorous foundations for mathematics and understand the infinite.

On the other hand, the next section discusses certain of the less technical axioms that are required for our more naive treatment of set theory in the subsequent sections.

EXERCISES

3.1 AXIOMS AND THE PRIMITIVES OF SET THEORY

1. Describe each of the following finite sets by listing its elements.

(a) The set of natural numbers less than 8.

(b) The set of letters in the English word "mathematics".

(c) The doubleton whose elements are the singletons $\{yin\}$, $\{yang\}$.

(d) The set of natural numbers which divide 42.

(e) The set of positive multiples of 42 which are less than 250.

2. Find sets A and B satisfying the following conditions.

(a) $A \not\subseteq B$ and $B \not\subseteq A$ (b) $A \subseteq B$ but $B \not\subseteq A$

(c) $A \subseteq B$ but $A \notin B$ (d) $A \in B$ but $A \not\subseteq B$

3. Let X, Y, Z be sets and suppose that $X \in Y$ and $Y \in Z$. Can it be that also $X \in Z$? Is it necessarily the case that $X \in Z$? Prove your assertions.

4. Is it possible that $A \subseteq B$ as well as $A \in B$ for sets A and B? Prove your assertions.

5. Enumerate all of the subsets of the following sets.

(a) $\{a,b\}$ (b) $\{\{a,b\},c\}$

(c) $\{\{a,\{b,c\}\}\}$ (d) $\{a,\{a\}\}$

(e) $\{a,b,c\}$ (f) $\{a,\{a\},\{a,\{a\}\}\}$

6. Prove or disprove that if A,B,C are sets so that $A \not\subseteq B$ and $B \subseteq C$, then $A \not\subseteq C$.

7. Let X,Y,Z be sets. Disprove each of the following.

(a) $[X \notin Y \wedge Y \notin Z] \Rightarrow [X \notin Z]$ (b) $[X \subseteq Y] \Rightarrow [\{X\} \subseteq \{Y\}]$

(c) $[X \in Y \wedge Y \notin Z] \Rightarrow [X \notin Z]$ (d) $[X \in Y] \Rightarrow [\{X\} \subseteq \{Y\}]$

8. Suppose that $A_1, A_2, \ldots, A_n$ are sets for some $n \geq 1$ where $A_1 \subseteq A_2$, $A_2 \subseteq A_3$, ..., $A_{n-1} \subseteq A_n$.

(a) Prove that $A_1 \subseteq A_n$.

(b) Prove that if furthermore $A_n \subseteq A_1$, then $A_1 = A_2 = \cdots = A_n$.

9. Give an example of an inconsistent system of axioms. Give an example of a system of axioms which is not independent.

10. [Paradox of Epimenides] Epimenides the Cretan said that all Cretans were liars, and all statements made by Cretans were definitely lies. Was this a lie? A simpler form of this paradox is given by utterance "I am lying", for if I am lying here, then I speak the truth and vice versa. Discuss the resolution of this paradox.

11. [Paradox attributed in *Principia Mathematica* to G. G. Berry of the Bodleian Library] The number of syllables in English names for natural numbers tends to increase as the numbers grow large since only a finite number of names can be made with a fixed finite number of syllables. For instance, there are exactly 11 one-syllable names corresponding to $1,2,\ldots,10,12$, and "one" is the name of the least one-syllable natural number. In the same way, the number 111,777 requires 19 syllables since its name is "one hundred and eleven thousand seven hundred and seventy seven", and it turns out that 111,777 is the least 19-syllable number. The paradox arises when one observes that "the least integer not nameable in fewer than nineteen syllables" is a name in English for 111,777 of 18 syllables, and this contradicts that 111,777 requires 19 syllables. Discuss the resolution of this paradox.

12. [Richard's Paradox] Let E be the set of all decimal numbers $.x_1x_2x_3\cdots$, where each $x_i \in \{0,1,2,\ldots,9\}$, that can be defined by a finite number of English words. It is evident that elements of E can be numbered as the $1^{\text{st}}, 2^{\text{nd}}, 3^{\text{rd}}, \ldots$. Define a number N as follows: If the nth decimal x_n in the nth number of E is p, define the nth decimal of N to be $p+1$ (or 0 if p is 9). The number N so constructed cannot be an element of E, for whatever

$n \in E$ might be, N differs from n in its nth digit. Nevertheless, we have given above a definition of N in finitely many English words, so N must actually be a member of E. Thus, N both is and is not a member of E. Discuss the resolution of this paradox.

13. A historically important and famous axiomatic definition of the set $\mathbb{N}$ of natural numbers is given by the "Peano Postualtes". The primitive for the Peano Postulates is the notion $n' = n + 1$ for the successor of a number n, and the axioms are

PEANO POSTULATES

$1 \in \mathbb{N}$.

For any $n \in \mathbb{N}$, *we have also* $n' \in \mathbb{N}$.

If $X \subseteq \mathbb{N}$, $1 \in X$, *and* $n' \in X$ *whenever* $n \in X$, *then we have* $X = \mathbb{N}$.

For any $n \in \mathbb{N}$, *we have* $1 \neq n'$.

If $m, n \in \mathbb{N}$ *and* $m' = n'$, *then* $m = n$.

These axioms do not actually suffice for a contemporary definition of the natural numbers since the notion of successor can itself be defined in terms of the set theory we shall discuss here, so there is no need to take it as a primitive. We mention the Peano Postulates here primarily to have a specific example of an axiomatic definition in mind and because of their historical importance; we will study them carefully only later in the optional §3.7. Prove the following assertions from the Peano Postulates.

(a) If n is a natural number, then $n \neq n'$.

(b) If $n \neq 1$, then $n = m'$ for some natural number m.

3.2 THE BASICS OF SET THEORY

As mentioned in the previous section, here we serially discuss certain axioms of set theory in order to establish notation and set the stage for later calculations with sets. We shall only present here those axioms that are directly useful in our development of set theory relegating the discussion of the remaining more technical axioms to the next optional section.

Each axiom is expressed as a predicate form, and as a notational point, we shall often let capital letters $A, B, \ldots$ denote sets, namely, the objects defined by the collection of axioms, and let lower-case letters $a, b, \ldots$ denote elements of sets. Of course, one set may be an element of another as we have seen, so this distinction between upper- and lower-case letters is not absolute. Using these conventions, we can often avoid specifying the universes associated with the various predicates below.

The first axiom involves equality of sets:

I. AXIOM OF EXTENSIONALITY

$$\forall A \forall B[A = B \Leftrightarrow \forall u(u \in A \Leftrightarrow u \in B)]$$

This axiom specifies that for any two sets A, B, we have $A = B$ as sets if and only if A and B have exactly the same elements. Thus, a set is determined by its elements. In particular, our notation $A = \{a_1, a_2, \ldots, a_n\}$ for a finite set A with members $a_1, a_2, \ldots, a_n$ depends implicitly upon the Axiom of Extensionality.

3.2.1 THEOREM *Suppose that A, B are sets. Then $A = B$ if and only if $A \subseteq B$ and $B \subseteq A$.*

Proof According to the axiom of extensionality and the definition of inclusion, we have

$$\begin{aligned} A = B &\Leftrightarrow \forall u(u \in A \Leftrightarrow u \in B) \\ &\Leftrightarrow \forall u[(u \in A \Rightarrow u \in B) \wedge (u \in B \Rightarrow u \in A)] \\ &\Leftrightarrow \forall u(u \in A \Rightarrow u \in B) \wedge \forall u(u \in B \Rightarrow u \in A) \\ &\Leftrightarrow A \subseteq B \wedge B \subseteq A, \end{aligned}$$

as was asserted. *q.e.d.*

This result allows us to prove that two sets A and B are equal by separately proving $A \subseteq B$ and $B \subseteq A$, and this elementary observation is actually of basic utility. Let us also remark that the previous result illustrates the typical logical dependence of a mathematical theory upon its axioms: The Axiom of Extensionality is assumed to be true, and we use it together with the predicate calculus to prove a theorem about sets.

The next axiom guarantees the existence of an "empty set" with no elements:

II. AXIOM OF THE NULL SET $\exists \emptyset \forall x(x \notin \emptyset)$

The standard notation for the "empty set" is the symbol $\emptyset$, which is the Norwegian letter "erh", and the set $\emptyset$ is called the *empty set* or the *null set.*

3.2.2. THEOREM *For any set A, we have $\emptyset \subseteq A$ and $A \subseteq A$.*

Proof According to the definition of inclusion of sets, $\emptyset \subseteq A$ if and only if $\forall x(x \in \emptyset \Rightarrow x \in A)$. The implication holds for each x since it is vacuously true, that is, the hypothesis is false, because $x \notin \emptyset$ for each x since $\emptyset$ is empty, proving the first part. The second part similarly follows directly from the valid predicate form $\forall x(x \in A \Rightarrow x \in A)$. *q.e.d.*

Thus, for any set A, we have $\emptyset \subseteq A$ and $A \subseteq A$. A subset $B \subseteq A$ is said to be a *proper subset* if $B \neq \emptyset$ and $B \neq A$, and in this case, we also refer to A as a *proper superset* of B. As a notational point, we mention that some authors reserve the symbol "$\subset$" to indicate a proper subset writing $B \subseteq A$ if B is a subset of A and writing $B \subset A$ if B is a proper subset of A. Still other authors write $B \subsetneq A$ if B is a proper subset of A. For simplicity of notation in this volume, we shall always just write $B \subseteq A$ to indicate that B is a subset of A which may or may not be proper, and when we wish to specify that B is a proper subset of A, we shall explicitly state this.

3.2.3 COROLLARY *The empty set is unique. That is, if $\emptyset$ and $\emptyset'$ each satisfy the Axiom of the Null Set, then $\emptyset = \emptyset'$.*

Proof Applying the first part of the previous theorem to the pair $\emptyset, \emptyset'$, we conclude $\emptyset \subseteq \emptyset'$, and in the same way, we find that $\emptyset' \subseteq \emptyset$. Thus, $\emptyset = \emptyset'$ by Theorem 3.2.1.
q.e.d.

This proof is a typical one in set theory: We use rules of inference together with the previously proved Theorem 3.2.1 to establish a new result. Thus, the theory is built up one step at a time starting from the axioms. We could prove Corollary 3.2.3. directly as a statement in the predicate calculus, but it is more efficient to rely on the already proved Theorem 3.2.1.

The next several axioms allow us to "build" new sets from old in specified ways.

III. SCHEMA OF SEPARATION

$$\forall A \forall P \exists B \forall u [u \in B \Leftrightarrow (u \in A \wedge P(u))],$$

where P varies over all predicates whose universe contains A.

In English, this says that for any predicate P defined on a universe containing a set A, the collection of elements u of A for which $P(u)$ is true is itself a set B. The Schema of Separation therefore allows us to construct the set B from the set A and the predicate P defined on A. There is a standard notation for this situation, where we write

$$B = \{u \in A : P(u)\},$$

and the righthand side of this equation is read in English as "the set of all u in A so that $P(u)$". For instance, if $P(u)$ is the predicate [u is even] on the universe $A = \{0, 1, 2\}$, then we have $B = \{u \in A : P(u)\} = \{0, 2\}$. This axiom has an especially interesting history, as we shall discuss later. As a notational point, we remark that some authors use a vertical line segment "$|$" instead of the colon ":" to represent the English phrase "so that" in this context. This notation $\{u \in A : P(u)\}$ or $\{u \in A | P(u)\}$ is sometimes called the "set-builder notation", and the Schema of Separation allows us to use this notation to define sets.

The next axiom allows us to take two objects a and b and build a doubleton $C = \{a, b\}$ whose elements are exactly a and b:

IV. AXIOM OF PAIRS $\forall a \forall b \exists C[u \in C \Leftrightarrow [(u = a) \vee (u = b)]]$

V. AXIOM OF UNION $\forall A \exists B \forall u[u \in B \Leftrightarrow \exists C(C \in A \wedge u \in C)]$

This axiom allows us to build sets in the following situation: Suppose that A is a set whose elements are themselves sets; we may then construct from A a set B whose members are the elements of the elements of A. A standard notation for the set B in this situation is $B = \cup A$, and one reads the symbol $\cup$ in English as "union", which we are here regarding as a "unary operation on sets of sets". For instance, if $A = \{\{a_1, a_2\}, \{\{a_3\}\}\}$, then $\cup A = \{a_1, a_2, \{a_3\}\}$. To dispel certain possible apprehensions, we explicitly point out that any element of A which is not a set or does not have any members (i.e., the null set $\emptyset$) does not affect $\cup A$, so for instance, if $A = \{\emptyset, x\}$ where x is not a set, then $\cup A = \emptyset$.

Actually, there is another notation for unions which is used when A consists of a pair of sets $A = \{X, Y\}$, and in this case we write $\cup A = X \cup Y$; thus, we are here regarding union as a "binary operation on sets". For instance, in the example above where $X = \{a_1, a_2\}$, $Y = \{\{a_3\}\}$, and $A = \{X, Y\}$, we might write $X \cup Y = \{a_1, a_2\} \cup \{\{a_3\}\} = \{a_1, a_2, \{a_3\}\}$ instead of $\cup A$.

There is another "binary operation on sets" defined as follows: Given sets X, Y, we let $Z = X \cup Y$ and define the *intersection*

$$X \cap Y = \{z \in Z | z \in X \wedge z \in Y\}$$

of X and Y. Notice that we are using the Axiom of Pairs to guarantee that $\{X, Y\}$ is a set, the Axiom of Union to guarantee that $Z = \cup\{X, Y\}$ is a set, and the Schema of Separation to guarantee that $X \cap Y$ is a set. According to the definition, the intersection $X \cap Y$ consists of those elements of X which are also elements of Y, and we have

$$X \cap Y = \{z \in X : z \in Y\} = \{z \in Y : z \in X\}.$$

If A and B are sets and $A \cap B = \emptyset$ so A and B have no common elements, then we say that A and B are *disjoint*. For instance, if $X = \{a_1, a_2, \{a_3\}\}$ and $Y = \{a_1, \{a_2\}, a_3\}$, then X and Y are not disjoint since $X \cap Y = \{a_1\} \neq \emptyset$, while if $X = \{a_1, a_2, a_3\}$ and $Y = \{b_1, b_2, b_3\}$, then X and Y are disjoint provided $a_i \neq b_j$ for any $i, j = 1, 2, 3$.

If $\mathcal{A}$ is a set of sets so that $\forall X \in \mathcal{A}\ \forall Y \in \mathcal{A}\ [X \neq Y \Rightarrow X \cap Y = \emptyset]$, i.e., any two distinct sets in $\mathcal{A}$ are disjoint, then we say that $\mathcal{A}$ is a collection of *pairwise disjoint* sets. For instance, if $\mathcal{A} = \{\{a_0\}, \{a_1\}, \{a_2\}, \{a_3\}\}$, then $\mathcal{A}$ is a collection of pairwise disjoint sets provided $a_i \neq a_j$ for all $i, j = 1, 2, 3$.

It often happens in practice that there is some fixed set $\mathcal{U}$ which contains all the sets under consideration in a given situation, and in this case, we can use the set-builder notation to describe the union and intersection:

$$X \cup Y = \{z \in \mathcal{U} : z \in X \vee z \in Y\},$$
$$X \cap Y = \{z \in \mathcal{U} : z \in X \wedge z \in Y\},$$

indicating the deep analogy between $\wedge, \vee$ and $\cap, \cup$. Notice that we must specify that z is an element of a set $\mathcal{U}$ here in order to apply the Schema of Separation and use the set-builder notation in these expressions. If $\mathcal{A}$ is a set of subsets of $\mathcal{U}$, then we may write the unary union

$$\cup\mathcal{A} = \{z \in \mathcal{U} : \exists X(X \in \mathcal{A} \wedge z \in X)\}$$

using the set-builder notation. Under these circumstances, we may also regard intersection as a "unary operation on sets of sets" by defining

$$\cap\mathcal{A} = \{z \in \mathcal{U} : \forall X(X \in \mathcal{A} \Rightarrow z \in X)\}.$$

It may seem confusing to have both unary and binary interpretations of intersections and unions, but each interpretation is useful in certain contexts. Furthermore, once mastered, this notation does not lead to confusion in practice.

This is an appropriate point to mention "Venn diagrams", which were presumably studied by the reader in elementary school. The idea is to represent sets as collections of points in the plane so as to have a visual image of various intersections and unions. For instance, taking the following sets A and B in the plane

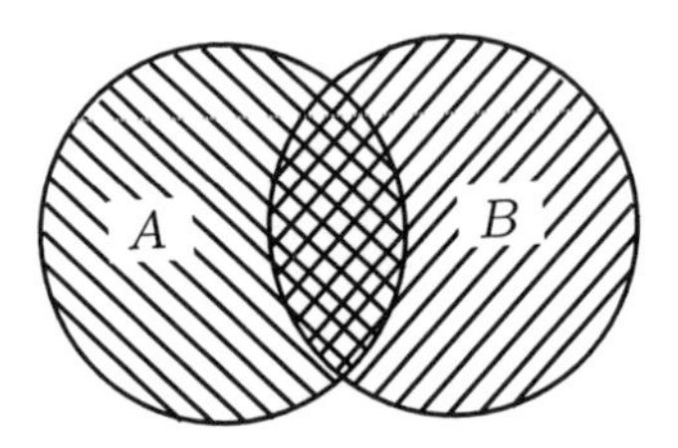

we can similarly represent the union

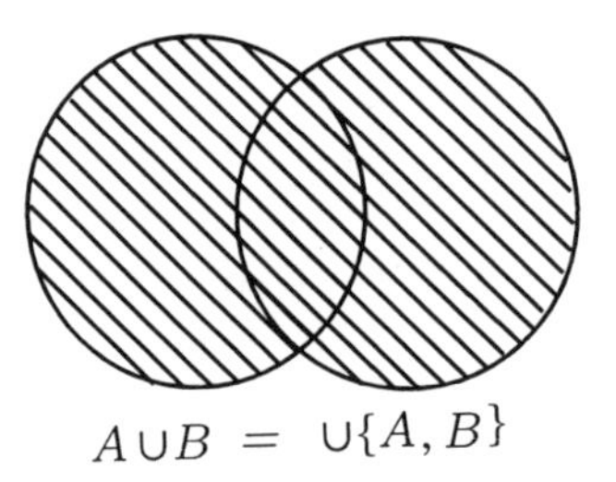

$A \cup B \;=\; \cup\{A, B\}$

and the intersection

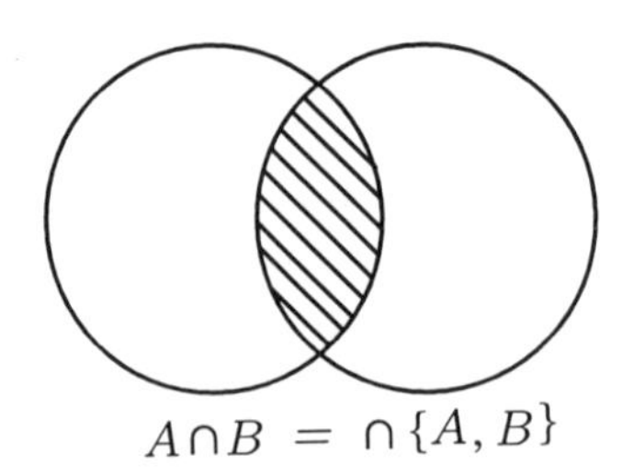

$A \cap B \;=\; \cap\,\{A, B\}$

Venn diagrams thus provide a useful representation in the mind's eye of an assertion about sets and can therefore be valuable when discovering a proof or a counter-example in practice. On the other hand, they are essentially useless for writing down rigorous proofs, and we shall never again mention them in the sequel.

The next axiom allows us to build from a set A another set B whose elements are the subsets of A:

VI. AXIOM OF POWER SET $\forall A \exists B \forall u (u \in B \Leftrightarrow u \subseteq A)$

The set B is called the *power set* of A, and one writes $B = \mathcal{P}(A)$ in this case. For instance, if $A = \emptyset$, then $\mathcal{P}(A) = \{\emptyset\}$, while if $A = \{1, 2\}$, then $\mathcal{P}(A) = \{\emptyset, \{1\}, \{2\}, \{1, 2\}\}$. In particular, notice that we have $\emptyset \in \mathcal{P}(A)$ and $A \in \mathcal{P}(A)$ for any set A by Theorem 3.2.2.

Define an *ordered pair* to be a collection of symbols of the form "(a, b)", where we regard two ordered pairs (a, b) and (a', b') as identical if $a = a'$ and $b = b'$. While intuitively appealing, this casual treatment must be made precise and given in terms of the axioms. To this end, we *carefully* define the ordered pair (a, b) to be the doubleton $\{\{a\}, \{a, b\}\}$ and shall prove in the next section (or leave as an exercise now) that this careful definition guarantees the defining property given above. Once this careful definition is given in terms of sets and we have proved that two ordered pairs indeed have the property described above, we shall never again need the careful definition. It is not uncommon in mathematics when arguing from axioms that one must "pay some dues" to make precise an intuitively clear notion in this way.

One likewise pays some further dues (as detailed in the next section or left as an exercise now) to prove that if A and B are sets, then there is another set denoted $A \times B$ whose elements are exactly the ordered pairs (a, b), where $a \in A$ and $b \in B$. This set $A \times B$ of such ordered pairs is called the *Cartesian product* or simply the *product* of A and B, and "$A \times B$" is often read in English as "A times B".

For an example, define the sets $A = \{a_1, a_2, a_3\}$ and $B = \{b_1, b_2\}$, so the Cartesian products are

$$A \times B = \{(a_1, b_1), (a_1, b_2), (a_2, b_1), (a_2, b_2), (a_3, b_1), (a_3, b_2)\}$$

and

$$B \times A = \{(b_1, a_1), (b_1, a_2), (b_1, a_3), (b_2, a_1), (b_2, a_2), (b_2, a_3)\}.$$

In contrast to the axioms discussed in the previous section, which allowed us to build new sets from old, the next axiom simply asserts the existence of a set:

VII. AXIOM OF INFINITY $\exists S [\emptyset \in S \wedge \forall x \in S \ (x \cup \{x\} \in S)]$

In fact, this axiom guarantees the existence of a set which provides our precise definition of the natural numbers. Indeed, according to the Axiom of Infinity, the set S contains the following elements:

$$\begin{aligned}
&\emptyset,\\
&\{\emptyset\} = \emptyset \cup \{\emptyset\},\\
&\{\emptyset, \{\emptyset\}\} = \{\emptyset\} \cup \{\{\emptyset\}\},\\
&\{\emptyset, \{\emptyset\}, \{\emptyset, \{\emptyset\}\}\} = \{\emptyset, \{\emptyset\}\} \cup \{\{\emptyset, \{\emptyset\}\}\},\\
&\qquad\vdots
\end{aligned}$$

so we are constructing a collection of sets, each of which is an element of S, with zero, one, two, three, ... elements. This certainly should provide us with a reasonable model of the natural numbers, and indeed, it does as we shall see. Notice that the axiom only asserts that S contains these elements, but using the Schema of Separation together with the Axiom of Infinity, we can construct a set consisting of exactly these elements.

Our further discussion here depends upon the notion (which is made more precise in the next section) of a "choice function" for a set X of non-empty sets, namely, a *choice function* f for X is the specification of an element $f(S) \in S$ for each $S \in X$. For instance, on the set $X = \{\{0,1\}, \{1,2,3\}\}$ of sets, one possible choice function is

$$f(\{0,1\}) = 1 \text{ and } f(\{1,2,3\}) = 2,$$

and another possible choice function is

$$f(\{0,1\}) = 1 \text{ and } f(\{1,2,3\}) = 1.$$

The final axiom of set theory discussed in this section is

AXIOM OF CHOICE *If $X \neq \emptyset$ is a set whose elements are sets and $\emptyset \notin X$, then there is a choice function f for X.*

Here is the content of the Axiom of Choice. The set X itself has sets as elements, and we might imagine that an element of X is a "basket of fruit". We therefore think of X as consisting of a collection of such "baskets", where each basket is itself a set whose members are "pieces of fruit"; the set $\cup X$ is therefore the set whose elements are the pieces of fruit themselves. The Axiom of Choice asserts the existence of a choice function f for X which assigns to each basket (i.e., each element of X) some piece of fruit (i.e. an element of $\cup X$), where the piece of fruit associated to any given basket must itself lie in the given basket. Thus, in effect, the choice function f simultaneously "chooses a piece of fruit from each basket". It is worth pointing out that our "baskets of fruits" analogy only goes so far, since a given "piece of fruit" might occur in many "baskets", and the "same piece of fruit might be chosen from different baskets".

It is worth explicitly pointing out that there are obvious epistemological issues concerning the Axiom of Choice when the set X is infinite: How can I, a finite being, make infinitely many choices at once? It is also worth mentioning that the Axiom of Choice can be proved as a consequence of the Zermelo-Fraenkel axioms in case the set X is finite, but if X is infinite, then it cannot, even if each element of X contains only two elements. Further aspects of the Axiom of Choice and its colorful history will be discussed in the next section.

We shall actually require the Axiom of Choice in a few later proofs. Furthermore, we shall also rely on the Schema of Separation to employ the set-builder notation as well as rely on the various theorems and definitions (of union, intersection, Cartesion product, etc.) presented in this section. Other than these applications, the sequel will not explicitly depend upon the axioms. For completeness, the enumeration of the basic axioms of Zermelo-Fraenkel set theory is concluded in the next section following a discussion of their development.

EXERCISES

3.2 THE BASICS OF SET THEORY

1. Describe each of the following sets by listing its members.

(a) $\{\text{integers } n : 3 < n \leq 10\}$

(b) $\{\text{integers } n : n = 1 \vee n = 2\}$

(c) $\{\text{integers } n : n = m^2 \text{ for some natural number less than } 12\}$

(d) $\{\text{integers } n : n = m^3 - m \text{ for some even natural number less than } 12\}$

(e) $\{\text{integers } n : n^2 = 9\}$

(f) $\{\text{natural numbers } n : n^2 = 9\}$

2. For each of the following pairs of sets, determine whether the two sets are equal and justify your answers.

(a) $\{1, 2, 3, 4\}$ and $\{1, 4, 3, 3, 2\}$

(b) $\{\text{integers } n : n^2 - 3x = 18\}$ and $\{3, -6\}$

(c) $\{\text{natural numbers } n : 2 < x < 8 \text{ and } n \text{ is even}\}$ and $\{4, 6, 8\}$

(d) $\{\text{natural numbers } n : (n-2)^2 = -1\}$ and $\{-3\}$

(e) $\{\text{natural numbers } n : n^3 - n = n(n^2 - 1)\}$ and $\{0, 1, -1\}$

3. For each of the following sets, specify a predicate over a suitable universe of discourse and describe the set using set-builder notation.

(a) The set of all natural numbers strictly between 3 and 1000

(b) The set of all even integers

(c) The set of all integer multiples of 12

(d) The set of all cubes of integers which are negative

(e) The set of all prime natural numbers

4. Let $A = \{a, b, c, d\}$ and $B = \{d, e, f\}$. Find the following sets.

(a) $A \cup B$ (b) $A \cap B$

(c) $\cap\{A, B, \emptyset\}$ (d) $\cup\{A, B, \emptyset\}$

5. Let $A = \{a, b, c, d\}$, $B = \{c, d, e\}$, and $C = \{e, f\}$. Find the following sets.

(a) $\cup\{A, B, C\}$ (b) $\cap\{A, B, C\}$

(c) $\cup\{A \cup B, C\}$ (d) $\cup\{A, B \cap C\}$

(e) $\cap\{A \cup B, A \cup C, A\}$ (f) $\cup\{A \cap B, A \cap C, A\}$

6. Prove each of the following for sets A and B.

(a) $A \cap B \subseteq A$ (b) $A \subseteq A \cup B$

(c) $A \cap B \subseteq A \cup B$ (d) $(A \cup B) \cap A = A$

7. Prove or disprove each of the following for sets A, B, C.

(a) If $A \neq B$ and $B \neq C$, then $A \neq C$.

(b) At least one of the following holds: $A = B$, $A \subseteq B$, $B \subseteq A$.

(c) At most one of the following holds: $A = B$, A is a proper subset of B, B is a proper subset of A.

8. A set A is said to be *transitive* if each element of A is also a subset of A.

(a) Show that $\emptyset$ is transitive.

(b) Show that each of $\{\emptyset\}$ and $\{\emptyset, \{\emptyset\}\}$ is transitive.

(c) Find a transitive set with exactly three elements.

(d) Find a transitive set with infinitely many elements.

9. Let A, B, C be sets. Prove each of the following.

(a) If $C \subseteq A$ and $C \subseteq B$, then $C \subseteq A \cap B$; that is, the intersection of two sets is the largest set contained in both of them.

(b) If $A \subseteq C$ and $B \subseteq C$, then $A \cup B \subseteq C$; that is, the union of two sets is the smallest set containing both of them.

10. Suppose that $\mathcal{A}$ and $\mathcal{B}$ are each non-empty families of sets. Prove or disprove each of the following.

(a) $\cap\mathcal{A} \subseteq \cup\mathcal{A}$ (b) If $\cap\mathcal{A} = \cup\mathcal{A}$, then $\mathcal{A}$ is a singleton.

(c) If $\cup\mathcal{A} = \cup\mathcal{B}$, then $\mathcal{A} = \mathcal{B}$. (d) If $\cap\mathcal{A} = \cap\mathcal{B}$, then $\mathcal{A} = \mathcal{B}$.

(e) $\cup(\mathcal{A} \cup \mathcal{B}) = (\cup\mathcal{A}) \cup (\cup\mathcal{B})$ (f) $\cap(\mathcal{A} \cap \mathcal{B}) = (\cap\mathcal{A}) \cap (\cap\mathcal{B})$

11. Suppose that $\mathcal{A}$ and $\mathcal{B}$ are each sets of sets and suppose that $\mathcal{A} \subseteq \mathcal{B}$.

(a) Show that $\cup\mathcal{A} \subseteq \cup\mathcal{B}$. (b) Show that $\cap\mathcal{A} \supseteq \cap\mathcal{B}$.

12. Suppose that $\mathcal{A}$ is a set of sets. Prove each of the following.

(a) If B is a set so that $B \subseteq A$ for each $A \in \mathcal{A}$, then $B \subseteq \cap\mathcal{A}$.

(b) If B is a set so that $A \subseteq B$ for each $A \in \mathcal{A}$, then $\cup\mathcal{A} \subseteq B$.

13. Compute the following power sets.

(a) $\mathcal{P}(\emptyset)$ (b) $\mathcal{P}(\{a\})$

(c) $\mathcal{P}(\{a, b\})$ (d) $\mathcal{P}(\mathcal{P}(\{a\}))$

14. Let $A = \{a, b, c\}$, $B = \{c, d, e\}$, and $C = \{c, e\}$. List the elements of the following Cartesian products.

(a) $A \times B$ (b) $B \times A$

(c) $A \times (B \cap C)$ (d) $(A \times B) \cap (A \times C)$

(e) $A \times (B \cup C)$ (f) $A \times (B \times C)$

15. Prove or disprove each of the following for sets A and B.

(a) $\mathcal{P}(A \cap B) = \mathcal{P}(A) \cap \mathcal{P}(B)$ (b) $\mathcal{P}(A \cup B) = \mathcal{P}(A) \cup \mathcal{P}(B)$

(c) $\mathcal{P}(A \times B) = \mathcal{P}(A) \times \mathcal{P}(B)$ (d) If $A \subseteq B$, then $\mathcal{P}(A) \subseteq \mathcal{P}(B)$.

16. Define the set

$$\mathcal{A} = \{\{a, b, c\}, \{b, c, d, e\}, \{a, b, c, e\}, \{a, b, d\}\}.$$

(a) Describe the set $\cup\mathcal{A}$.

(b) Describe the set $\cap\mathcal{A}$.

(b) Exhibit two different choice functions for the family $\mathcal{A}$.

17. Define the sets $A_i = \{1, 2, \ldots, i\}$ for each natural number i, and let $\mathcal{A}$ be the set whose members are these sets A_i for $i \geq 1$.

(a) Describe the set $\cup\mathcal{A}$.

(b) Describe the set $\cap\mathcal{A}$.

(c) Exhibit two different choice functions for the family $\mathcal{A}$.

(d) Answer part (a)-(c) above for the sets $A_i = \{\text{natural numbers } n : n \geq i\}$.

(*) *3.3 ZFC SET THEORY*

This section is dedicated to a discussion of the various axioms, the "Zermelo-Fraenkel Axioms" which together with "The Axiom of Choice" form the foundations of modern set theory and define a "set", which is truly the fundamental object in all of mathematics. It is not a vast oversimplification, therefore, to assert that the axioms we present here are the basic assumptions of all mathematics, and it is remarkable that the totality of these assumptions may be stated quite concisely and explicitly as a collection of ten carefully chosen predicate forms. We have already seen the Axiom of Choice as well as the first seven Axioms I-VII in the previous section, and we discuss the remaining two axioms here after first discussing the history.

The development of these standard axioms of set theory was begun by Zermelo and Russell in 1908 and brought into its present form with further work by Fraenkel in 1922; this model of set theory is called "Zermelo-Fraenkel set theory" and is often abbreviated simply "ZF". The importance of the Axiom of Choice discussed in the previous section was recognized already at the time of Zermelo, but it became well accepted only more recently. The ZF axioms together with the Axiom of Choice is often abbreviated simply "ZFC", and we begin with some remarks on the Axiom of Choice.

In fact, the history of this axiom is most colorful, and the acceptance or rejection of the Axiom of Choice was fervently debated during the early part of the twentieth century. When formulated as in the previous section, the axiom and its epistemological ramifications seem pretty clear: Given any non-empty set of non-empty sets, there exists a choice function. Even with this formulation, though, various schools accepted this axiom, and various schools rejected it. The situation became much more confusing, however, when other equivalent formulations of the Axiom of Choice came to light. By "other equivalent formulations" here we mean other statements which were shown to be logically equivalent to the Axiom of Choice where one assumes only the axioms of ZF. We shall discuss these other formulations of the Axiom of Choice only later, but historically, the several equivalent formulations which arose served only to confuse matters. In part, this derives from the fact that one of these equivalent formulations (called the "Zermelo Well Ordering Principle" to be discussed in §5.10) was disturbing in that it posited (that is, asserted) the existence of a certain reasonably concrete mathematical structure which even today has not been explicitly constructed. The third main formulation of the axiom called "Zorn's Lemma" is discussed in §5.10 and is a standard tool in modern algebra for instance.

Despite the rocky history of the Axiom of Choice, it is fair to say that most working mathematicians today accept it in any of its equivalent formulations as a valid axiom. We also mention that at one time, mathematicians felt that it was important to examine each application of the Axiom of Choice to determine whether the application actually *required* the Axiom of Choice or whether it could be proved without this axiom. In fact, it turns out that many proofs in traditional mathematics do require the Axiom of Choice, and indeed, many results are actually equivalent to the Axiom of Choice in the sense discussed above.

Another axiom of ZFC with a fascinating history is the Schema of Separation. At one time, this axiom was replaced in ZF by the following:

> **AXIOM OF COMPREHENSION** *For any predicate $P(u)$ defined over any universe U, there is a set Y consisting of those elements u of U so that $P(u)$ is true.*

The reader should notice the subtle distinction between the Schema of Separation and the Axiom of Comprehension: Whereas in the former, we only claim that the collection of elements of a known set satisfying the predicate P is itself a set, in the latter, we claim that the elements of any universe (for instance, a universe not a priori known to be a set) satisfying P is actually a set. Thus, the Axiom of Comprehension is more general than the Schema of Separation in the sense that it allows the construction of many more sets.

There are, however, logical difficulties with the Axiom of Comprehension which were pointed out by Bertrand Russell and led to the replacement of this axiom with the Schema of Separation in ZF. Indeed, Russell considered the predicate $P(X) = [X \notin X]$ on the universe consisting of all sets. According to the Axiom of Comprehension, then, the collection of all sets so that $P(X)$ is true is itself a set, and we call this set S. The problem with the Axiom of Comprehension becomes apparent when one observes that if $S \in S$, then $P(S)$ must be true, so $S \notin S$, whereas if $S \notin S$, then $P(S)$ is true, so $S \in S$. Thus, neither $S \in S$ nor $S \notin S$ can be true, and this contradicts the law of the excluded middle. This contradiction among the earlier axioms is called "Russell's Paradox".

One concludes that there must be something fishy about the Axiom of Comprehension, and, over time, the replacement of the Axiom of Comprehension by the Schema of Separation was seen to resolve Russell's Paradox. Indeed, one cannot apply the Schema of Separation as in Russell's Paradox unless one knows in advance that the collection of all sets is itself a set. As a consequence of this discussion, we may conclude that the collection of all sets cannot be a set in ZF or ZFC, for otherwise, we could derive a paradox like Russell's using the Schema of Separation.

On the other hand, it is convenient to have a name for the collection of all sets, and the logicians Gödel and Bernays introduced the notion of a "class" for such large would-be sets. Thus, the collection of all sets is a class but is not a set, and the Schema of Separation is applied only to sets, not to classes. Indeed, if we allowed classes in the Schema of Separation to build sets, then we would run into the analogue of Russell's Paradox for classes. One can introduce the obvious analogue of the Schema of Separation for classes so as to "build" classes, but then the collection of all classes had better not be a class again so as to avoid the analogue of Russell's Paradox. Thus, one introduces the notion of a "super-class" of all classes, and so on. This paragraph describes what is called "Gödel-Bernays set theory" in distinction to Zermelo-Fraenkel set theory, and it is highly studied. We need not be so careful to distinguish sets from classes in the sequel and will sometimes refer to a class simply as a "collection".

Having made these remarks, we can finally be precise about what exactly is a "universe of discourse" for a predicate. The predicate calculus is regarded as completely syntactical, in the sense that it is regarded as simply formal manipulations of formal strings of symbols, and we use it to put forward the ZF or ZFC axioms and thereby axiomatically define sets. We can then go back and and reconsider the predicate calculus where the universe of discourse of a predicate is always just a set. This is called "first order predicate logic", and, in light of our remarks about the Gödel-Bernays theory, the reader can probably guess what is meant by "second order predicate logic" (where a universe of discourse is allowed to be a set or a class), "third order predicate logic", and so on.

The reader may have the feeling that mathematics has just squeaked by in avoiding Russell's Paradox with ZF by syntactical trickery, but this is the contemporary state of affairs.

More generally, the reader might wonder whether other paradoxes lurk hidden within ZF or ZFC waiting to be uncovered, and the unfortunate truth is that this is unknown at present: The consistency of ZF or ZFC is unknown, but other axioms can be introduced whose truth can be shown to imply the consistency of ZF or ZFC. Specifically, there is the notion of an "inaccessible cardinal", which, in a precise sense, is a really huge number, and the existence of an inaccessible cardinal has been shown to imply the consistency of ZFC. On the other hand, the consistency of ZF or ZFC together with an axiom positing the existence of an inaccessible cardinal is unknown. It is intriguing to imagine some physical experiment which "discovers" an inaccessible cardinal, thereby leading to a empirical "proof" of the consistency of ZFC.

Actually, it is worth mentioning here that there is another axiom, called the "(Generalized) Continuum Hypothesis", which is sometimes added to ZF or ZFC to give yet another model of set theory. We shall briefly discuss the Continuum Hypothesis in §7.5. Happily, the axioms of ZFC are known to be independent, and indeed, ZFC together with the Continuum Hypothesis are likewise known to be independent.

We close our historical and philosphical overivew of ZFC with a very brief discussion of the fundamental and shocking results of Gödel which are relevant to the current discussion. One such result, called Gödel's First Incompleteness Theorem (which has been touted in the popular press), roughly says that given any system of axioms, there is a true statement, that is, one that is not false, for which one cannot write down a finite proof. We mentioned before that this result dealt the death blow to the formalist Hilbert school and the program of Russell and Whitehead initiated with the *Principia Mathematica.* The interested reader can refer to the book *Gödel's Proof*, NYU Press, 1958, by Nagel and Newman.

Less well-known, Gödel's Second Incompleteness Theorem roughly states that it is impossible to give a finite proof of the consistency of any finite system of axioms. On the other hand, one can of course give a finite proof of inconsistency by exhibiting a contradiction such as Russell's Paradox. Thus, though the existence of an inaccessible cardinal implies the consistency of ZF or ZFC, one cannot prove this existence from ZF or ZFC alone. In the same way, one cannot then prove the consistency of ZF or ZFC together with the assumption that there is some inaccessible cardinal. Thus, in light of Gödel's Second Incompleteness Theorem, consistency of families of axioms in mathematics is most problematic. This is again, alas, a fact of the human condition.

Nevertheless, we continue now to finally complete the enumeration of the axioms of ZF. Most of the previous axioms allow us to "build new sets from old" and therefore, in effect, assert the existence of sets. In contrast, the next axiom prohibits certain behavior by sets and is never actually used in conventional mathematics except when studying the foundations of set theory:

VIII. AXIOM OF REGULARITY $\forall X[X \neq \emptyset \Rightarrow \exists x(x \in X \wedge x \cap X = \emptyset)]$

This axiom rules out the existence of a non-empty set X so that $X \in X$, for instance, since if there were such a set X, then $\{X\}$ would violate the Axiom of Regularity as the reader may check. Another somewhat deeper possibility which this axiom rules out is the existence of an "infinite chain"

$$\ldots, u_{n+1} \in u_n, \ldots, u_2 \in u_1, u_1 \in u_0.$$

In effect, this guarantees that "when building a set, one must start somewhere", and this is a reasonable rough interpretation of the Axiom of Regularity.

In order to present the final axiom, we must introduce the notion of a "function", which the reader has undoubtedly seen in elementary school. We shall systematically study functions only much later in Chapter 6 but give the precise definition here primarily in order to discuss this final axiom. Before giving this definition, we first complete the development of some background material.

Suppose that A, B are sets, $a \in A$, and $b \in B$. As in the previous section, we define the *ordered pair*

$$(a, b) = \{\{a\}, \{a, b\}\},$$

which is guaranteed to be a set by two applications of the Axiom of Pairs. The point is that this definition formalizes the notion of an "ordered pair" given in the previous section in the following sense.

3.3.1 THEOREM *Suppose that (a, b) and (a', b') are ordered pairs with $a, a' \in A$ and $b, b' \in B$. Then $(a, b) = (a', b')$ if and only if $a = a'$ and $b = b'$.*

Proof Notice that if $a = a'$ and $b = b'$, then of course

$$(a, b) = \{\{a\}, \{a, b\}\} = \{\{a'\}, \{a', b'\}\} = (a', b'),$$

so it remains only to prove the converse. To this end, first observe that if $a = b$, then

$$(a, b) = (a, a) = \{\{a\}, \{a, a\}\} = \{\{a\}, \{a\}\} = \{\{a\}\},$$

so the ordered pair (a, b) has only one element. Conversely, if (a, b) has only one element, then $\{a\} = \{a, b\}$, so $b \in \{a\}$, whence $a = b$.

Now suppose that indeed $(a, b) = (a', b')$. If $a = b$, then both (a, b) and (a', b') have only one element, so in particular $a' = b'$; since $\{a'\} \in (a, b)$ and $\{a\} \in (a', b')$, it follows that a, b, a' and b' must all be equal, and hence $(a, b) = (a', b')$. If $a \neq b$, then each of (a, b) and (a', b') contain exactly one set with a single element, namely $\{a\}$ and $\{a'\}$ respectively, so $a = a'$. Furthermore, each of (a, b) and (a', b') contain exactly one set with two elements, namely $\{a, b\}$ and $\{a', b'\}$ respectively, so we conclude that also $\{a, b\} = \{a', b'\}$. In particular, we have $b \in \{a', b'\}$, and since $b \neq a'$ (for otherwise, we find $b = a' = a$), we conclude that also $b = b'$. *q.e.d.*

Thus, we have paid some dues left over from the previous section. Proceeding to pay some further dues, we claim that if A and B are sets, then there is another set whose elements are exactly the collection of all ordered pairs (a, b), where $a \in A$ and $b \in B$. To see this, first observe that if $a \in A$ and $b \in B$, then we have $\{a\} \subseteq A$ and $\{b\} \subseteq B$, and so $\{a, b\} \subseteq A \cup B$. Since also $\{a\} \subseteq A \cup B$, we conclude that both $\{a\}$ and $\{a, b\}$ are elements of $\mathcal{P}(A \cup B)$. Thus, by definition $(a, b) = \{\{a\}, \{a, b\}\}$ is a subset of $\mathcal{P}(A \cup B)$, so in fact, (a, b) is an element of $\mathcal{P}(\mathcal{P}(A \cup B))$, whenever $a \in A$ and $b \in B$. It is then an easy exercise using the Schema of Separation and the Axiom of Extensionality to prove that there is a unique set consisting of all the ordered pairs (a, b) for $a \in A$ and $b \in B$. This set of all ordered pairs is called the *Cartesian product* of A and B or simply the *product* of A and B and is denoted

$$A \times B = \{z \in \mathcal{P}(\mathcal{P}(A \cup B)) : z = (a, b) \text{ for some } a \in A \text{ and } b \in B\}.$$

Armed with these remarks, we need never again resort to being so formal and simply regard $A \times B$ as the set of all such ordered pairs (a, b), where $a \in A$ and $b \in B$.

Now we turn to the precise definition of a "function". If A and B are fixed sets, then a *function from A to B* is a subset $f \subseteq A \times B$ so that

$$\forall a \in A \ \exists! b \in B \ [(a, b) \in f].$$

In a notation presumably more familiar to the reader, we shall write $f(a) = b$ if there is an ordered pair $(a, b) \in f$. According to the definition, there must be exactly one ordered pair $(a, b) \in f$ for each $a \in A$, and we write $f(a) = b$ in this case.

To relate this to the "functions" with which the reader is surely familiar, we consider a simple example: Define the sets $A = \{-2, -1, 0, 1, 2\}$, $B = \{0, 1, 2, 3, 4\}$, and let $f(n) = n^2$ be the function from A to B which returns the square n^2 of the integer $n \in A$; thus, we have $f(2) = 4 = f(-2)$, $f(1) = 1 = f(-1)$, and $f(0) = 0$. Using the definition of functions in terms of Cartesian products, we regard f as the collection of ordered pairs

$$f = \{(-2, 4), (-1, 1), (0, 0), (1, 1), (2, 4)\} \subseteq A \times B.$$

Thus, whereas one informally thinks of a function f from A to B as a "black box" which returns a value $f(a) \in B$ when given an element $a \in A$, the formal treatment requires regarding f as a collection of ordered pairs

$$f = \{(a, f(a)) \in A \times B : a \in A\}.$$

In order that a subset $f \subseteq A \times B$ be a function, our definition demands that the "black box returns exactly one value in B for each element of A". For another example, the "choice function" on X considered in the previous section is simply a function $f \subseteq X \times (\cup X)$ so that $f(S) \in S$ for each $S \in X$.

There is actually an enormous amount to say about the basic theory of functions, and we postpone this discussion until Chapter 6. We remind the reader that we have carefully defined functions here primarily in order to complete our discussion of the axioms of set theory, and we finally return to this task.

The final axiom states that given a function f from a set A to a set B, the collection $\{f(a) \in B : a \in A\}$ of values $f(a)$ taken by f as a varies over A is itself a set:

IX. SCHEMA OF REPLACEMENT

$$\forall A \forall f \exists C \forall u (u \in C \Leftrightarrow \exists a [a \in A \wedge (a, u) \in f]),$$

where f varies over the universe of all functions from A to another set.

For instance, in the example above where $A = \{-2, -1, 0, 1, 2\}$, $B = \{0, 1, 2, 3, 4\}$, and $f(n) = n^2$ returns the square of each element of A, the Schema of Replacement allows us to conclude directly that $\{0, 1, 4\}$ is itself a set.

This completes our discussion of the ZF Axioms I-IX and the Axiom of Choice, which together give the contemporary axiomatic definition ZFC of set theory

Just as one must exercise care in making definitions, so too care must be exercised in positing axioms. The axioms of ZFC have evolved over time as a suitable definition of sets, and great care by cadres of mathematicians and philosophers has been taken in their formulation. It is truly remarkable that essentially all the assumptions of modern mathematics can be so simply and concisely stated. Thus, the reader should have a healthy respect for the important foundational character of the axioms presented here.

At the risk of again waxing too philosophical, suppose that some extraterrestrial beings were to land on Earth; if we could agree with these beings on the ZFC axioms and on the Predicate Calculus, then we would presumably share the common body of all traditional mathematics. This argument has been made to include a rendering of ZFC in probes sent into outer space.

On the other hand, in light of our general discussion above about axiomatic systems and their incompletenesses, this remark perhaps overly aggrandizes the axioms of ZFC, which are in any case interesting historically and philosophically. The reader wishing to learn more about set theory might refer to Halmos' book *Naive Set Theory*, Springer-Verlag, 1974, to Kuratowski's book *Introduction to Set Theory and Topology*, Pergamon Press, 1961, or to the more advanced book *Introduction to Axiomatic Set Theory*, Springer-Verlag, 1971, by Takeuti and Zaring.

EXERCISES

3.3 ZFC SET THEORY

1. Define an *ordered triple* with entries a, b, c to be the set

$$(a, b, c) = \{\{a\}, \{a, b\}, \{a, b, c\}\}.$$

Prove that $(a, b, c) = (a', b', c')$ if and only if $a = a'$, $b = b'$, and $c = c'$.

2. Enumerate all of the functions from $\{a, b, c\}$ to $\{d, e, f\}$.

3. Which of the following are functions from $\{1, 2, 3\}$ to $\{1, 2, 3\}$? Prove your assertions.

(a) $\{(1, 1), (2, 1), (3, 3)\}$

(b) $\{(1, 1), (1, 2), (3, 3)\}$

(c) $\{(1, 1), (2, 3), (3, 3)\}$

(d) $\{(1, 1), (1, 2), (2, 2), (3, 2)\}$

4. Use the Axiom of Regularity to show that there is no infinite sequence $u_1, u_2, u_3, \ldots$ of sets so that $u_{n+1} \in u_n$.

5. Use the Axiom of Regularity to prove that there is no set of all sets.

6. Can there be sets A and B so that $A \in B$ and $B \in A$? Prove your assertion.

3.4 BINARY OPERATIONS ON SETS

We have already discussed the "binary operations" of union, intersection, and product of sets, namely, given sets A, B, we have constructed new sets $A \cup B$, $A \cap B$, and $A \times B$ respectively as described in the previous sections. There is actually another standard binary operation on sets A, B defined by

$$A - B = \{u \in A : u \notin B\}$$

called the *relative complement* of B in A. Thus, $A - B$ is the set of elements of A which are not elements of B, so for instance, if $A = \{a_1, a_2, a_3\}$ and $B = \{a_1, b_1, b_2\}$, then we have $A - B = \{a_2, a_3\}$ and $B - A = \{b_1, b_2\}$ provided a_1, a_2, a_3, b_1, b_2 are all distinct.

This section is dedicated to a discussion of various properties of $\cup$, $\cap$, and $-$ while the discussion of $\times$ is postponed to the next section. It is worthwhile to proceed somewhat formally, as follows: Given a collection $\mathcal{A}$ of objects, a *binary operation* $\triangle$ on $\mathcal{A}$ is a rule for producing a new object $X \triangle Y$ in $\mathcal{A}$ from the objects X, Y in $\mathcal{A}$. We have of course seen the binary operations $\wedge$ and $\vee$ on the collection of all predicate forms as well as the binary operations $\cup, \cap, \times$, and $-$ on the collection of sets, and we also mention the binary operations of addition, multiplication, and subtraction on the collection of all integers. We say that the binary operation $\triangle$ on $\mathcal{A}$ is *commutative* if

$$X \triangle Y = Y \triangle X \text{ for all } X, Y \text{ in } \mathcal{A},$$

and we say that $\triangle$ is *associative* if

$$(X \triangle Y) \triangle Z = X \triangle (Y \triangle Z) \text{ for all } X, Y, Z \text{ in } \mathcal{A}.$$

For instance, addition and multiplication of integers are both commutative and associative as the reader knows from elementary school, and we proved by truth table in Chapter 2 that $\wedge$ and $\vee$ are likewise both commutative and associative. On the other hand, subtraction of integers is neither commutative nor associative since

$$2 - 3 = -1 \neq 1 = 3 - 2 \quad \text{and} \quad (1 - 2) - 3 = -4 \neq 2 = 1 - (2 - 3)$$

for instance. For another negative example, we show that relative complement of sets is also neither commutative nor associative: Let $X = \{a_1, a_2, a_3\}$, $Y = \{a_2, a_3\}$ and $Z = \{a_2\}$ where $a_1 \neq a_2$, so we have

$$\begin{aligned} X - Y &= \{a_1\} \neq \emptyset = Y - X \\ (X - Y) - Z &= \{a_1\} \neq \{a_1, a_2\} = X - (Y - Z). \end{aligned}$$

We remark that the discussion of commutativity and associativity of $\wedge$ and $\vee$ in §2.8 applies more generally to any commutative and associative binary operation: If $\triangle$ is an associative binary operation on $\mathcal{A}$ and $X_1, X_2, \ldots, X_n$ are elements of $\mathcal{A}$ for some $n \geq 3$, then we may write $X_1 \triangle X_2 \triangle \cdots \triangle X_n$ with no parentheses to mean $(\cdots (X_1 \triangle X_2) \triangle \cdots \triangle X_n)$, and we may apply the operations $\triangle$ in the expression $X_1 \triangle X_2 \triangle \cdots \triangle X_n$ in any order so as to produce the same element $(\cdots (X_1 \triangle X_2) \triangle \cdots \triangle X_n)$ of $\mathcal{A}$. If $\triangle$ is both associative and commutative, then for any re-ordering $X_{i_1}, X_{i_2}, \ldots, X_{i_n}$ of $X_1, X_2, \ldots, X_n$ where $\{i_1, i_2, \ldots, i_n\} = \{1, 2, \ldots, n\}$, we have

$$X_{i_1} \triangle X_{i_2} \triangle \cdots \triangle X_{i_n} = X_1 \triangle X_2 \triangle \cdots \triangle X_n.$$

The proofs of these results follow verbatim from the arguments in §2.8.

3.4.1 THEOREM *Union and intersection of sets are each commutative and associative, that is, given any sets A, B, C, we have*

(a) $A \cup B = B \cup A$

(b) $A \cap B = B \cap A$

(c) $(A \cup B) \cup C = A \cup (B \cup C)$

(d) $(A \cap B) \cap C = A \cap (B \cap C)$

Proof For part (a), let u be arbitrary and observe that

$$\begin{aligned} u \in A \cup B &\Leftrightarrow (u \in A \vee u \in B) \text{ by definition of } \cup \\ &\Leftrightarrow (u \in B \vee u \in A) \text{ by commutativity of } \vee \\ &\Leftrightarrow u \in B \cup A \text{ by definition of } \cup. \end{aligned}$$

Since u was arbitrary, it follows from universal generalization that

$$\forall u[u \in A \cup B \Leftrightarrow u \in B \cup A],$$

as desired. The proof of part (b) follows similarly from commutativity of $\wedge$ and is left as an exercise.

For part (c), let u be arbitrary and observe that

$$\begin{aligned} u \in (A \cup B) \cup C &\Leftrightarrow (u \in A \cup B) \vee u \in C \quad \text{by definition of } \cup \\ &\Leftrightarrow (u \in A \vee u \in B) \vee u \in C \quad \text{by definition of } \cup \\ &\Leftrightarrow u \in A \vee (u \in B \vee u \in C) \quad \text{by associativity of } \vee \\ &\Leftrightarrow u \in A \vee (u \in B \cup C) \quad \text{by definition of } \cup \\ &\Leftrightarrow u \in A \cup (B \cup C) \quad \text{by definition of } \cup . \end{aligned}$$

Since u was arbitrary, we conclude by universal generalization that

$$\forall u[u \in (A \cup B) \cup C \Leftrightarrow u \in A \cup (B \cup C)],$$

so $(A \cup B) \cup C = A \cup (B \cup C)$, as desired. Part (d) follows similarly from associativity of $\wedge$ and is left as an exercise. *q.e.d.*

Regarding this proof, we remark that one usually just tacitly invokes universal generalization, one may perform several symbolic manipulations in each line of such a calculation (so there is a practical need to restrain oneself and perform just a few such manipulations in each line as with the predicate calculus), and one typically assumes a certain level of sophistication on the part of the reader and does not explicitly justify each line as we did above. We shall usually take this more informal approach in subsequent proofs.

Returning to the general setting of abstract binary operations $\triangle$ and $\triangledown$ defined on $\mathcal{A}$, we say that $\triangle$ *left-distributes* over $\triangledown$ if

$$X \triangle (Y \triangledown Z) = (X \triangle Y) \triangledown (X \triangle Z) \quad \text{for all } X, Y, Z \text{ in } \mathcal{A},$$

and we say that $\triangle$ *right-distributes* over $\triangledown$ if

$$(Y \triangledown Z) \triangle X = (Y \triangle X) \triangledown (Z \triangle X) \quad \text{for all } X, Y, Z \text{ in } \mathcal{A}.$$

In case $\triangle$ both left- and right-distributes over $\triangledown$, then we say simply that $\triangle$ *distributes* over $\triangledown$. In particular, if $\triangle$ is commutative, then left-distributivity, right-distributivity, and distributivity of $\triangle$ over $\triangledown$ are all logically equivalent as the reader should check. For instance, on the collection of integers, multiplication distributes over addition:

$$x \cdot (y + z) = x \cdot y + x \cdot z, \quad \text{for all integers } x, y, z,$$

but addition does not distribute over multiplication, e.g.,

$$1 + (2 \cdot 3) = 7 \neq 12 = (1 + 2) \cdot (1 + 3).$$

3.4.2 THEOREM *Union distributes over intersection, and intersection distributes over union, that is, for any sets A, B, C, we have*

(a) $A \cup (B \cap C) = (A \cup B) \cap (A \cup C)$

(b) $(B \cap C) \cup A = (B \cup A) \cap (C \cup A)$

(c) $A \cap (B \cup C) = (A \cap B) \cup (A \cap C)$

(d) $(B \cup C) \cap A = (B \cap A) \cup (C \cap A)$

Proof For part (a), let u be arbitrary and compute

$$\begin{aligned} u \in A \cup (B \cap C) &\Leftrightarrow u \in A \vee u \in B \cap C \\ &\Leftrightarrow u \in A \vee (u \in B \wedge u \in C) \\ &\Leftrightarrow (u \in A \vee u \in B) \wedge (u \in A \vee u \in C) \\ &\Leftrightarrow (u \in A \cup B) \wedge (u \in A \cup C) \\ &\Leftrightarrow u \in (A \cup B) \cap (A \cup C), \end{aligned}$$

where the third logical equivalence follows from distributivity of $\vee$ over $\wedge$, so indeed, $A \cup (B \cap C) = (A \cup B) \cap (A \cup C)$. As was mentioned above, part (b) then follows from commutativity of $\cup$. The proofs of parts (c) and (d) similarly rely on distributivity of $\wedge$ over $\vee$ and are left as exercises. *q.e.d.*

In fact, relative complement right-distributes but does not left-distribute over unions or intersections as we next prove.

3.4.3 THEOREM *For any sets A, B, C, we have*

(a) $(B \cup C) - A = (B - A) \cup (C - A)$

(b) $A - (B \cup C) = (A - B) \cap (A - C)$

(c) $(B \cap C) - A = (B - A) \cap (C - A)$

(d) $A - (B \cap C) = (A - B) \cup (A - C)$

Proof For part (a), we compute

$$\begin{aligned} u \in (B \cup C) - A &\Leftrightarrow u \in (B \cup C) \wedge u \notin A \\ &\Leftrightarrow (u \in B \vee u \in C) \wedge u \notin A \\ &\Leftrightarrow (u \in B \wedge u \notin A) \vee (u \in C \wedge u \notin A) \\ &\Leftrightarrow (u \in B - A) \vee (u \in C - A) \\ &\Leftrightarrow u \in (B - A) \cup (C - A), \end{aligned}$$

so indeed, $(B \cup C) - A = (B - A) \cup (C - A)$. Turning to part (b), we compute

$$\begin{aligned} u \in A - (B \cup C) &\Leftrightarrow u \in A \wedge u \notin (B \cup C) \\ &\Leftrightarrow u \in A \wedge \neg(u \in B \cup C) \\ &\Leftrightarrow u \in A \wedge \neg(u \in B \vee u \in C) \\ &\Leftrightarrow u \in A \wedge (u \notin B \wedge u \notin C) \\ &\Leftrightarrow (u \in A \wedge u \notin B) \wedge (u \in A \wedge u \notin C) \\ &\Leftrightarrow (u \in A - B) \wedge (u \in A - C) \\ &\Leftrightarrow u \in (A - B) \cap (A - C), \end{aligned}$$

where we have used DeMorgan's law in the fourth logical equivalence above.

The proofs of parts (c) and (d) are similar and left as exercises. *q.e.d.*

Our next result is an "omnibus theorem" which is intended to summarize various properties of union, intersection, and relative complement.

3.4.4 THEOREM *For any sets A, B, C, D, we have*

(a) $A \cup A = A$ *and* $A \cap A = A$

(b) $A \cup \emptyset = A$ *and* $A \cap \emptyset = \emptyset$

(c) $A \subseteq A \cup B$, $A \cap B \subseteq A$, *and* $A - B \subseteq A$

(d) *If* $A \subseteq B$ *and* $C \subseteq D$, *then* $(A \cup C) \subseteq (B \cup D)$ *and* $(A \cap C) \subseteq (B \cap D)$

(e) *If* $A \subseteq B$, *then* $A \cup B = B$ *and* $A \cap B = A$

(f) $A \cup (B - A) = A \cup B$ *and* $A \cap (B - A) = \emptyset$

Proof Beginning with part (a), for arbitrary u, we have

$$\begin{aligned} u \in A \cup A &\Leftrightarrow u \in A \vee u \in A \\ &\Leftrightarrow u \in A, \end{aligned}$$

while

$$\begin{aligned} u \in A \cap A &\Leftrightarrow u \in A \wedge u \in A \\ &\Leftrightarrow u \in A, \end{aligned}$$

completing the proof.

For part (b) with u arbitrary, we have

$$\begin{aligned} u \in A \cup \emptyset &\Leftrightarrow u \in A \vee u \in \emptyset \\ &\Leftrightarrow u \in A \vee 0 \\ &\Leftrightarrow u \in A, \end{aligned}$$

so indeed, $A \cup \emptyset = A$. Furthermore, we have

$$\begin{aligned} u \in A \cap \emptyset &\Leftrightarrow u \in A \wedge u \in \emptyset \\ &\Leftrightarrow u \in A \wedge 0 \\ &\Leftrightarrow 0, \end{aligned}$$

so $u \in A \cap \emptyset$ is false for each u, whence $A \cap \emptyset = \emptyset$.

Turning to part (c), the first inequality follows from the law of addition since for any u, we have $u \in A \Rightarrow (u \in A \wedge u \in B)$, and the second inequality follows from the law of simplification since for any u, we have $(u \in A \wedge u \in B) \Rightarrow u \in A$. Similarly, the third inequality again follows from the law of simplification since for any u, we have $(u \in A \wedge u \notin B) \Rightarrow u \in A$.

For part (d), suppose that $A \subseteq B$ and $C \subseteq D$, so for any u, we have $u \in A \Rightarrow u \in B$ and $u \in C \Rightarrow u \in D$. Now, given some arbitrary $x \in A \cup C$, we must have $x \in A$ or $x \in C$; in the former case, we conclude by Theorem 3.1.2 that $x \in B$, while in the latter case, we conclude again by Theorem 3.1.2 that $x \in D$. Furthermore by part (c), we have $B \subseteq B \cup D$ and $D \subseteq B \cup D$, so in either case, we conclude that $x \in B \cup D$. Thus, $x \in A \cup C \Rightarrow x \in B \cup D$, so $A \cup C \subseteq B \cup D$, as was claimed. For the second inclusion in part (d), given some arbitrary $x \in A \cap C$, we must have $x \in A$ and $x \in C$, so we conclude that $x \in B$ and $x \in D$, whence $x \in B \cap D$. Thus, $x \in A \cap C \Rightarrow x \in B \cap D$, so $A \cap C \subseteq B \cap D$, as desired.

In part (e), we assume that $A \subseteq B$ and begin with the first equality. Since $B \subseteq B$ as well, it follows from part (d) that $A \cup B \subseteq B \cup B$ and from part (a), we have $B \cup B = B$. Thus, we conclude that $A \cup B \subseteq B$, and by Theorem 3.1.1, it remains only to prove the reverse inclusion $B \subseteq A \cup B$, which follows from part (c). For the second equality, we have $A \subseteq B$ and $A \subseteq A$, so by part (d), we conclude that $A \cap A \subseteq A \cap B$. Since $A \cap A = A$ by part (a), we have proved the inclusion $A \subseteq A \cap B$, and again by Theorem 3.1.1, it remains to prove the reverse inclusion $A \cap B \subseteq A$, which was proved in part (c).

For the first equality in part (f), let u be arbitrary and compute

$$\begin{aligned} u \in A \cup (B - A) &\Leftrightarrow u \in A \vee u \in B - A \\ &\Leftrightarrow u \in A \vee (u \in B \wedge u \notin A) \\ &\Leftrightarrow (u \in A \vee u \in B) \wedge (u \in A \vee u \notin A) \\ &\Leftrightarrow (u \in A \vee u \in B) \wedge 1 \\ &\Leftrightarrow u \in A \vee u \in B \\ &\Leftrightarrow u \in A \cup B, \end{aligned}$$

as desired. For the final equality, again let u be arbitrary and compute

$$\begin{aligned} u \in A \cap (B - A) &\Leftrightarrow u \in A \wedge u \in B - A \\ &\Leftrightarrow u \in A \wedge (u \in B \wedge u \notin A) \\ &\Leftrightarrow (u \in A \wedge u \notin A) \wedge u \in B \\ &\Leftrightarrow 0 \wedge u \in B \\ &\Leftrightarrow 0, \end{aligned}$$

so $A \cap (B - A)$ has no elements, i.e., $A \cap (B - A) = \emptyset$. *q.e.d.*

Suppose that $X_1, X_2, \ldots X_n$ is a collection of sets for some $n \geq 1$. Since $\cap$ and $\cup$ are associative, we may write $X_1 \cap X_2 \cap \cdots \cap X_n$ or $X_1 \cup X_2 \cup \cdots \cup X_n$ in keeping with our conventions for parentheses. There is a particular situation which occurs in practice fairly frequently where the sets in question satisfy the inclusions

$$X_1 \subseteq X_2, X_2 \subseteq X_3, \ldots, X_{n-1} \subseteq X_n.$$

In this case, the family of sets $X_1, X_2, \ldots, X_n$ is said to be *nested*, and we have the simplified notation

$$X_1 \subseteq X_2 \subseteq \cdots \subseteq X_n$$

in this case. For example, we have the following nested family of sets

$$\{a_1\} \subseteq \{a_1, a_2\} \subseteq \{a_1, a_2, a_3\} \subseteq \{a_1, a_2, a_3, a_4, a_5\}.$$

3.4.5 THEOREM *Suppose that $X_1 \subseteq X_2 \subseteq \cdots \subseteq X_n$ is a nested family of sets for some $n \geq 1$. Then we have*

$$X_1 \cup X_2 \cup \cdots \cup X_n = X_n,$$
$$X_1 \cap X_2 \cap \cdots \cap X_n = X_1.$$

Proof For the first equality, the inclusion $X_n \subseteq X_1 \cup X_2 \cup \cdots \cup X_n$ follows immediately from Theorem 3.4.4c. An extension by induction of Theorem 3.1.1 shows that $X_i \subseteq X_n$ for each $i = 1, 2, \ldots, n$, and another extension by induction of Theorem 3.4.4d proves that $X_1 \cup X_2 \cup \cdots \cup X_n \subseteq X_n$. The first equality thus follows from Theorem 3.2.1.

For the second equality, the inclusion $X_1 \cap X_2 \cap \cdots \cap X_n \subseteq X_1$ again follows immediately from Theorem 3.4.4c. As before, it follows from Theorem 3.1.1 that $X_1 \subseteq X_i$ for each $i = 1, 2, \ldots, n$ and from Theorem 3.4.4d that $X_1 \subseteq X_1 \cap X_2 \cap \cdots \cap X_n$. The second equality finally follows from Theorem 3.2.1. *q.e.d.*

EXERCISES

3.4 BINARY OPERATIONS ON SETS

1. Define sets $A = \{a, b, c, d\}$, $B = \{a, c, e\}$, and $C = \{c, d, e, f\}$. Compute each of the following sets.

   (a) $A \cap B$ (b) $(A - B) \cap (A - C)$

   (c) $A \cup C$ (d) $(A \cup B) - (A \cap C)$

   (e) $A - B$ (f) $[A \cap (B \cup C)] - (B \cap C)$

2. Let A denote the set of all natural numbers, B denote the set of all even integers, and C denote the set of all odd natural numbers. For each part of Problem 1 above, describe the corresponding set in this case.

3. Prove the following results from the text.

(a) Theorem 3.4.1 parts (b) and (d)

(b) Theorem 3.4.2 parts (c) and (d)

(c) Theorem 3.4.3 parts (c) and (d)

4. Let A, B, C be sets, where $A \neq \emptyset$ and $A \cup B = A \cup C$. Show that it is not necessarily the case that $B = C$. If we assume in addition that $A \cap B = A \cap C$, then is it true that $B = C$? Prove your assertions.

5. Under what conditions on sets A and B is it true that $A - B = B - A$? Prove your assertions.

6. Let A, B, C be sets. Prove or disprove each of the following.

(a) $A \cap (A \cup B) = A$

(b) $A \cap (A - B) = B \cap (B - A)$

(c) $A - B = A \cap ((A \cup B) - B)$

(d) $A \cap (B \cup C) = (A \cap B) \cup C$

(e) $[A \cap (B \cup C)] - (B \cap C)$

(f) $A \cup ((A \cup B) - A) \cap B) = A \cup B$

(g) $\mathcal{P}(A - B) \subseteq \mathcal{P}(A) - \mathcal{P}(B)$

(h) $\mathcal{P}(A) - \mathcal{P}(B) \subseteq \mathcal{P}(A - B)$

7. Let A, B, C be sets. Prove or disprove each of the following.

(a) $A - (B - C) \subseteq (A - B) - C$

(b) $(A - B) - C \subseteq A - (B - C)$

(c) $A - B = A \Leftrightarrow B - A = B$

(d) $[(A - B) \cap C] \cap [(A - C) \cap B] = \emptyset$

(e) $A \cap B = A \cap C \Rightarrow B = C$

(f) $A \subseteq B \Rightarrow A \cup (B \cap C) = B \cap (A \cup C)$

(g) $A \cup (B \cap C) = (A \cup B) \cap C$

(h) $A \cap B = A \cup B \Rightarrow A = B$

8. Prove that for any sets A, B, we have $A \subseteq B$ if and only if $A \cap B = A$.

9. Let A, B, and C be arbitrary sets. Prove each of the following identities.

(a) $(A \cap B) \cup (A - B) = A$

(b) $(A \cap B) \cup (C - A) \cup (B \cap C) = (A \cap B) \cup (C - A)$

10. Let $A, B_1, B_2, \ldots, B_n$ be sets for some $n \geq 1$. Prove each of the following identities.

(a) $A \cap (B_1 \cup B_2 \cup \cdots \cup B_n) = (A \cap B_1) \cup (A \cap B_2) \cup \cdots \cup (A \cap B_n)$

(b) $A \cup (B_1 \cap B_2 \cap \cdots \cap B_n) = (A \cup B_1) \cap (A \cup B_2) \cap \cdots \cap (A \cup B_n)$

11. Let A, B, and C be sets. Prove or disprove each of the following.

(a) $A \cap B \subseteq C \Rightarrow A \subseteq C$

(b) $\mathcal{P}(A) \subseteq \mathcal{P}(B) \Rightarrow A \subseteq B$

(c) $C - A \subseteq C - B \Leftrightarrow B \subseteq A$ (d) $B - C = A - C \Rightarrow A \cap C = B \cap C$

(e) $A - (A - B) = A \cap B$ (f) $A \cap C = B \cap C \Rightarrow B - C = A - C$

12. Let A, B, and C be sets. Prove or disprove each of the following.

(a) $A \subseteq B \Rightarrow B - (B - A) = A$

(b) $B - (B - A) = A \Rightarrow A \subseteq B$

(c) If $A \cap C \subseteq C - B$ and $A \cap B \subseteq C$, then A and B are disjoint.

(d) $(A - B) \cap C$ and $(A - C) \cap B$ are disjoint.

(e) $C \cup (A \cap B) = C \cap (A \cup B)$ if and only if $C \subseteq A \cup B$ and $A \cap B \subseteq C$

13. Another standard binary operation operation on sets is the *symmetric difference* $\oplus$ defined by $A \oplus B = (A - B) \cup (B - A)$. Prove each of the following results for arbitrary sets A, B, C.

(a) $A \oplus B = (A \cup B) - (A \cap B)$ (b) $A \oplus B = B \oplus A$

(c) $(A \oplus B) \oplus C = A \oplus (B \oplus C)$ (d) $A \cap (B \oplus C) = (A \cap B) \oplus (A \cap C)$

(e) $A \oplus \emptyset = A$ and $A \oplus A = \emptyset$ (f) $A \oplus B = A \cup B \Leftrightarrow A \cap B = \emptyset$

14. Consider the binary operations $\cup$, $\cap$, $-$, and $\oplus$ on sets. For which pairs of operations among these can one express the other two operations in terms of the specified pair?

15. Prove that if a commutative binary operation $\triangle$ left-distributes over another binary operation ∇, then $\triangle$ also right-distributes over ∇.

16. Define the *disjoint union* $A \sqcup B$ of two sets A, B as follows. Set $A' = A \times \{1, 2\}$ and $B' = B \times \{1, 2\}$, and define $A \sqcup B = A \times \{1\} \cup B \times \{2\} \subseteq (A \cup B) \times \{1, 2\}$, so $A \sqcup B$ should be thought of as *two* disjoint copies of A and B, the former labeled "1" and the latter labeled "2". Is $\sqcup$ commutative or associative?

3.5 CARTESIAN PRODUCTS

In this section, we turn our attention to the Cartesian product $\times$. Let us first remark that there is a good intuitive way to think of the product $A \times B$ of sets A and B, where we imagine "listing" the elements of A horizontally and those of B vertically; for each specification of $a \in A$, $b \in B$, there is a corresponding point determined in the plane, and we imagine the ordered pair (a, b) as occurring at that point. For instance, taking the sets $A = \{a_1, a_2, a_3\}$ and $B = \{b_1, b_2\}$, we thus imagine $A \times B$ as the following collection of points in the plane:

$$
\begin{array}{cccc}
B\left\{\begin{array}{l} b_2 \\ \\ b_1 \end{array}\right. & \begin{array}{c}(a_1, b_2) \\ \\ (a_1, b_1)\end{array} & \begin{array}{c}(a_2, b_2) \\ \\ (a_2, b_1)\end{array} & \begin{array}{c}(a_3, b_2) \\ \\ (a_3, b_1)\end{array} \\
\\
 & \multicolumn{3}{c}{\underbrace{a_1 \qquad\qquad\qquad a_2 \qquad\qquad\qquad a_3}_{A}}
\end{array}
$$

whereas we imagine $B \times A$ as the following collection of points in the plane:

$$
\begin{array}{ccc}
A\left\{\begin{array}{l} a_3 \\ \\ a_2 \\ \\ a_1 \end{array}\right. & \begin{array}{c}(b_1, a_3) \\ \\ (b_1, a_2) \\ \\ (b_1, a_1)\end{array} & \begin{array}{c}(b_2, a_3) \\ \\ (b_2, a_2) \\ \\ (b_2, a_1)\end{array} \\
\\
 & \multicolumn{2}{c}{\underbrace{b_1 \qquad\qquad\qquad b_2}_{B}}
\end{array}
$$

Of course, this intuition is in keeping with the view of the plane as the Cartesian product of the set of real numbers with themselves, where we specify a point in this plane by giving the ordered pair (x, y) of real numbers:

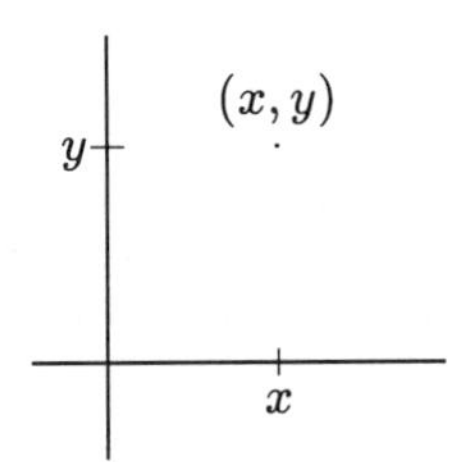

This interpretation of the plane as the set of all ordered pairs of real numbers is due to Descartes and explains the terminology "Cartesian product". (It seems remarkable that we can attribute what now seems so basic and almost innate to a single individual! Indeed, it is almost hard to accept that monkeys and porpoises, for instance, are not also aware of this unless Descartes informed them as well.)

This description of the Cartesian product shows that if A and B are finite sets with m and n elements respectively, then the product $A \times B$ is again finite with mn elements. In the extreme case that one of A or B is empty and has $m = 0$ or $n = 0$ elements, the same formula holds since $A \times B$ is itself empty (according to its definition) and has $0 = mn$ elements.

We first show that "Cartesian products respect inclusions" in the following sense.

3.5.1 THEOREM *Suppose that A, B, C, D are sets where $A \subseteq C$ and $B \subseteq D$. Then $A \times B \subseteq C \times D$.*

Proof Let (x, y) be an arbitrary element of $A \times B$, so $x \in A$ and $y \in B$. Insofar as $A \subseteq C$ and $B \subseteq D$, we conclude that in fact $x \in C$ and $y \in D$ by Theorem 3.1.2. Thus, we find $(x, y) \in C \times D$ as was claimed. *q.e.d.*

In fact, Cartesian product distributes across both union and intersection as we next show.

3.5.2 THEOREM *If A, B, and C are sets, then we have*

(a) $A \times (B \cup C) = (A \times B) \cup (A \times C)$

(b) $(B \cup C) \times A = (B \times A) \cup (C \times A)$

(c) $A \times (B \cap C) = (A \times B) \cap (A \times C)$

(d) $(B \cap C) \times A = (B \times A) \cap (C \times A)$

Proof For part (a), let (x, y) be an arbitrary ordered pair and compute

$$\begin{aligned}(x, y) \in A \times (B \cup C) &\Leftrightarrow x \in A \wedge y \in B \cup C \\ &\Leftrightarrow x \in A \wedge (y \in B \vee y \in C) \\ &\Leftrightarrow (x \in A \wedge y \in B) \vee (x \in A \wedge y \in C) \\ &\Leftrightarrow (x, y) \in A \times B \vee (x, y) \in A \times C \\ &\Leftrightarrow (x, y) \in (A \times B) \cup (A \times C)\end{aligned}$$

as desired. The proof of part (b) is similar and left as an exercise.

Turning to part (c), again let (x, y) be an arbitrary ordered pair and compute

$$\begin{aligned}(x, y) \in A \times (B \cap C) &\Leftrightarrow x \in A \wedge y \in B \cap C \\ &\Leftrightarrow (x \in A \wedge x \in A) \wedge (y \in B \wedge y \in C) \\ &\Leftrightarrow (x \in A \wedge y \in B) \wedge (x \in A \wedge y \in C) \\ &\Leftrightarrow (x, y) \in A \times B \wedge (x, y) \in A \times C \\ &\Leftrightarrow (x, y) \in (A \times B) \cap (A \times C)\end{aligned}$$

as desired. Again the proof of part (d) is similar and left as an exercise. *q.e.d.*

Our next result illustrates some further relationships between Cartesian products and unions or intersections.

3.5.3 THEOREM *If A, B, C, and D are sets, then we have*

(a) $(A \times B) \cap (C \times D) = (A \cap C) \times (B \cap D)$

(b) $(A \times B) \cup (C \times D) \subseteq (A \cup C) \times (B \cup D)$

Proof Beginning with part (a), let (x, y) be an arbitrary ordered pair and compute

$$\begin{aligned}(x,y) \in (A \times B) \cap (C \times D) &\Leftrightarrow (x,y) \in A \times B \wedge (x,y) \in C \times D \\ &\Leftrightarrow (x \in A \wedge y \in B) \wedge (x \in C \wedge y \in D) \\ &\Leftrightarrow (x \in A \wedge x \in C) \wedge (y \in B \wedge y \in D) \\ &\Leftrightarrow (x \in A \cap C) \wedge (y \in B \cap D) \\ &\Leftrightarrow (x,y) \in (A \cap C) \times (B \cap D)\end{aligned}$$

as desired.

For part (b), we compute using Theorem 3.5.2a that

$$\begin{aligned}(A \cup C) \times (B \cup D) &= A \times (B \cup D) \cup C \times (B \cup D) \\ &= (A \times B) \cup (A \times D) \cup (C \times B) \cup (C \times D) \\ &= [(A \times B) \cup (C \times D)] \cup [(A \times D) \cup (C \times B)],\end{aligned}$$

and the result then follows from the first inclusion in Theorem 3.4.4c. *q.e.d.*

■ **Example 3.5.1** Let us here illustrate some of the previous results and show that the reverse inclusion in part (b) of Theorem 3.5.3 does not hold. In general, given real numbers $a < b$, define the *interval*

$$[a, b] = \{\text{real numbers } t : a \leq t \leq b\}.$$

According to Theorem 3.5.2a, we have

$$\begin{aligned}([1,2] \cup [4,5]) \times ([1,2] \cup [4,5]) &= [1,2] \times ([1,2] \cup [4,5]) \cup [4,5] \times ([1,2] \cup [4,5]) \\ &= ([1,2] \times [1,2]) \cup ([1,2] \times [4,5]) \\ &\quad \cup ([4,5] \times [1,2]) \cup ([4,5] \times [4,5]),\end{aligned}$$

so the inclusion in Theorem 3.5.3b is a proper inclusion in this case.

We have seen in previous examples that Cartesian product is not commutative, and in fact, Cartesian product is also not associative: For instance if $a \in A$, $b \in B$, and $c \in C$, then we have $((a, b), c) \in (A \times B) \times C$ whereas we have $(a, (b, c)) \in A \times (B \times C)$. On the other hand, it is only for formal reasons (i.e., the distribution of parentheses) that we regard $((a, b), c)$ and $(a, (b, c))$ as different, and it is tempting to simply omit all but the first and last parentheses and write the corresponding element of $(A \times B) \times C$ or $A \times (B \times C)$ simply as (a, b, c), and indeed, this is often done in practice.

More precisely, given a collection of sets A_i for $i = 1, 2, \ldots, n$ where $n \geq 2$, we define an *ordered n-tuple* to be a collection of symbols of the form $(a_1, a_2, \ldots, a_n)$, where $a_i \in A_i$ for each $i = 1, 2, \ldots, n$. Two ordered n-tuples $(a_1, a_2, \ldots, a_n)$ and $(a'_1, a'_2, \ldots, a'_n)$ are regarded as identical if and only if $a_i = a'_i$ for each $i = 1, 2, \ldots, n$. We furthermore define the *(n-fold) Cartesian product* of the sets $A_1, A_2, \ldots, A_n$ to be

$$A_1 \times A_2 \times \cdots \times A_n = \{(a_1, a_2, \ldots, a_n) : a_i \in A_i \text{ for each } i = 1, 2, \ldots, n\}.$$

For an example of such a Cartesian product, if $A = \{a_1, a_2\}$, $B = \{b_1, b_2\}$ and $C = \{c_1, c_2\}$, then

$$\begin{aligned} A \times B \times C = \{&(a_1, b_1, c_1), (a_1, b_1, c_2), (a_1, b_2, c_1), (a_1, b_2, c_2), \\ &(a_2, b_1, c_1), (a_2, b_1, c_2), (a_2, b_2, c_1), (a_2, b_2, c_2)\}. \end{aligned}$$

A standard notation for the n-fold Cartesian product of sets $A_1, A_2, \ldots, A_n$ (in this order) is to write simply

$$\times_{i=1}^{n} A_i \quad = \quad A_1 \times A_2 \times \cdots \times A_n,$$

and we shall often rely on this notation below.

Our definition of ordered n-tuple naturally generalizes the notion of ordered pair and illustrates how one in practice avoids the debauch of parentheses associated with multiple Cartesian products.

As another notational point, it sometimes happens that we wish to take the n-fold Cartesian product of a set with itself, and in this case there is a special notation: If A is a set and $n \geq 2$, then we define

$$\begin{aligned} A^n &= \underbrace{A \times A \times \cdots \times A}_{n \text{ times}} \\ &= \{(a_1, a_2, \cdots, a_n) : a_i \in A \text{ for each } i = 1, 2, \ldots, n\}, \end{aligned}$$

and we extend this definition by setting $A^1 = A$.

Finally, we have observed above that if A and B are finite sets with m and n elements respectively, then $A \times B$ has mn elements. A standard extension by induction of this observation (which is left as an exercise) establishes

3.5.4 THEOREM *Suppose that $A_1, A_2, \ldots A_n$ is a collection of finite sets for $n \geq 1$, where A_i has $r_i \geq 0$ elements for each $i = 1, \ldots, n$. Then $\times_{i=1}^{n} A_i$ has the following number of elements:*

$$\prod_{i=1}^{n} r_i \quad = \quad r_1 \cdot r_2 \cdot \cdots \cdot r_n.$$

Just as the previous result enriches our understanding of Cartesian products, it is not uncommon in mathematics when confronted with a new definition to simply "enumerate" (that is, list) or even just "count" the new objects being defined, for one can often gain insight simply from counting!

EXERCISES

3.5 CARTESIAN PRODUCTS

1. Prove Theorem 3.5.2 parts (b) and (d).

2. Prove Theorem 3.5.4.

3. Let A, B, and C be sets. Prove or disprove each of the following

(a) $a \notin A \Rightarrow (a,b) \notin A \times B$

(b) $A = \emptyset \Rightarrow A \times B = \emptyset$

(c) If $A \times B = \emptyset$, then either $A = \emptyset$ or $B = \emptyset$.

(d) If $(a,b) \notin A \times B$, then $a \notin A$ and $b \notin B$.

(e) $A \cap B = \emptyset \Rightarrow (A \times B) \cap (B \times A) = \emptyset$

(f) $(A \times B) \cap (B \times A) = \emptyset \Rightarrow A \cap B = \emptyset$

(g) There are sets A and B so that

$$A \times B = \{(a,a),(a,b),(b,a),(c,a),(c,b),(b,b)\}.$$

(h) There are sets A and B so that

$$A \times B = \{(c,b),(c,g),(d,c),(c,d)\}.$$

(i) $A \times B = A \times C \Rightarrow B = C$

4. Suppose that A and B are sets. Write $A^2 - B^2$ as a union of two sets each of which is the Cartesian product of two sets.

5. Let $A, B_1, B_2, \ldots B_n$ be sets, where $n \geq 1$. Prove each of the following.

(a) $A \times (B_1 \cap B_2 \cap \cdots \cap B_n) = (A \times B_1) \cap (A \times B_2) \cap \cdots \cap (A \times B_n)$

(b) $A \times (B_1 \cup B_2 \cup \cdots \cup B_n) = (A \times B_1) \cup (A \times B_2) \cup \cdots \cup (A \times B_n)$

6. Give a precise definition of ordered n-tuple from the ZF Axioms and prove that two n-tuples $(a_1, a_2, \ldots, a_n)$ and $(a'_1, a'_2, \ldots, a'_n)$ are equal if and only if $a_i = a'_i$ for all $i = 1, \ldots, n$.

3.6 ABSOLUTE COMPLEMENTS AND DEMORGAN'S LAWS

It often happens that all of the sets under consideration in a given situation are themselves subsets of a fixed set $\mathcal{U}$, which we refer to as the "universe of discourse".

In fact, one can avoid essentially all foundational subtleties in set theory by making this assumption. As opposed to the "axiomatic set theory" discussed in §3.2 and §3.3, this version of set theory is called "naive set theory"; notice how such an assumption completely ducks the issue of carefully defining sets and instead manipulates just subsets of some given set whose existence is posited *a priori.*

In particular in this situation, we have $A \cup \mathcal{U} = \mathcal{U}$ and $A \cap \mathcal{U} = A$ for any set A (which is tacitly assumed to be a subset of $\mathcal{U}$) by Theorem 3.4.4e. When the universe of discourse is fixed, there is a "unary operation on sets" which produces from a set A another set

$$\overline{A} = \mathcal{U} - A = \{u \in \mathcal{U} : u \notin A\}$$

called the *absolute complement* of A. Thus, the absolute complement is just a special case of the relative complement of a set in its universe of discourse. For instance, if the universe is $\mathcal{U} = \{a_1, a_2, a_3, a_4, a_5\}$ and $A = \{a_1, a_2, a_4\}$, then we have $\overline{A} = \{a_3, a_5\}$, and if the universe $\mathcal{U}$ is the collection of all natural numbers and $A = \{n \in \mathcal{U} : n \geq 3\}$, then $\overline{A} = \{1, 2\}$. For another example, if

$$A = \{z \in \mathcal{U} : P(z)\}$$

for some predicate $P(z)$ defined on $\mathcal{U}$, then we have

$$\overline{A} = \{z \in \mathcal{U} : \neg P(z)\}.$$

It follows from the definitions of absolute and relative complement that for any two sets $A, B \subseteq \mathcal{U}$, we have

$$A - B = A \cap \overline{B},$$

and it is often useful to apply this basic formula to express relative complements in terms of absolute complements when appropriate.

3.6.1 THEOREM *Let A be an arbitrary subset of some universe $\mathcal{U}$. Then we have*

(a) $A \cup \overline{A} = \mathcal{U}$

(b) $A \cap \overline{A} = \emptyset$

Proof Both parts follow easily from Theorem 3.4.4f. *q.e.d.*

3.6.2 THEOREM *Suppose that A and B are subsets of a universe $\mathcal{U}$. Then $B = \overline{A}$ if and only if $A \cup B = \mathcal{U}$ and $A \cap B = \emptyset$.*

Proof That $B = \overline{A}$ satisfies the conditions is the content of the previous result. For the converse, suppose that $A \cup B = \mathcal{U}$, $A \cap B = \emptyset$ and compute

$$\begin{aligned} B &= \mathcal{U} \cap B \\ &= (A \cup \overline{A}) \cap B \\ &= (A \cap B) \cup (\overline{A} \cap B) \\ &= \emptyset \cup (\overline{A} \cap B) \\ &= (\overline{A} \cap A) \cup (\overline{A} \cap B) \\ &= \overline{A} \cap (A \cup B) \\ &= \overline{A} \cap \mathcal{U} \\ &= \overline{A} \end{aligned}$$

as desired. *q.e.d.*

3.6.3 THEOREM *Let A be an arbitrary set in the universe $\mathcal{U}$. Then we have $\overline{\overline{A}} = A$, that is, the complement of the complement of A is A.*

Proof By the previous result, $\overline{A} \cup A = \mathcal{U}$ and $\overline{A} \cap A = \emptyset$. It follows from Theorem 3.6.1 that A is the complement of $\overline{A}$, that is $\overline{\overline{A}} = A$. *q.e.d.*

A basic result called *DeMorgan's Laws* describe the absolute complement of a union or intersection as follows:

3.6.4 THEOREM *If A and B are sets in the universe $\mathcal{U}$, then we have*

(a) $\overline{A \cup B} = \overline{A} \cap \overline{B}$

(b) $\overline{A \cap B} = \overline{A} \cup \overline{B}$

Proof The proofs follow directly from the definition of absolute complement and the identities in Theorems 3.4.3b and 3.4.3d respectively. *q.e.d.*

A standard inductive argument which is left as an exercise extends DeMorgan's Laws to the setting of finite unions and intersections:

$$\overline{(A_1 \cup A_2 \cup \cdots \cup A_n)} = \overline{A}_1 \cap \overline{A}_2 \cap \cdots \cap \overline{A}_n$$

$$\overline{(A_1 \cap A_2 \cap \cdots \cap A_n)} = \overline{A}_1 \cup \overline{A}_2 \cup \cdots \cup \overline{A}_n$$

for any $n \geq 2$. On the other hand, DeMorgan's Laws hold in still greater generality as we next discuss after introducing some notation.

If $\mathcal{A}$ is a collection of subsets of the universe $\mathcal{U}$, then we have briefly discussed the unary union

$$\cup\mathcal{A} = \{z \in \mathcal{U} : \exists X(X \in \mathcal{A} \wedge z \in X)\}$$

and the unary intersection

$$\cap\mathcal{A} = \{z \in \mathcal{U} : \forall X(X \in \mathcal{A} \Rightarrow z \in X)\}$$

in §3.2. Let us observe the following peculiarities of these definitions in the extreme case that $\mathcal{A} = \emptyset$: The predicate $\exists X(X \in \mathcal{A} \wedge z \in X)$ is automatically false, so $\cup\mathcal{A} = \emptyset$ if $\mathcal{A} = \emptyset$ as one might expect; on the other hand, the predicate $\forall X(X \in \mathcal{A} \Rightarrow z \in X)$ is vacuously true for any $z \in \mathcal{U}$, so $\cap\mathcal{A} = \mathcal{U}$ if $\mathcal{A} = \emptyset$. It may seem perverse that an empty intersection is necessarily the entire universe, but this is the consistent definition.

We introduce here the alternate notation

$$\cup_{X \in \mathcal{A}} X = \cup\mathcal{A},$$
$$\cap_{X \in \mathcal{A}} X = \cap\mathcal{A},$$

for these unary operations. For a related notation, if Ω is a set, called the *index set*, and to each element $\alpha \in \Omega$ there corresponds a set X_α, then we refer to the set $\mathcal{A} = \{X_\alpha : \alpha \in \Omega\}$ of sets as *family of sets indexed by* Ω, and write

$$\cup_{\alpha \in \Omega} X_\alpha = \cup\mathcal{A},$$
$$\cap_{\alpha \in \Omega} X_\alpha = \cap\mathcal{A}.$$

Thus, by definition, we have the identities

$$\cup\mathcal{A} = \cup_{X_\alpha \in \mathcal{A}} X_\alpha = \cup_{\alpha \in \Omega} X_\alpha,$$
$$\cap\mathcal{A} = \cap_{X_\alpha \in \mathcal{A}} X_\alpha = \cap_{\alpha \in \Omega} X_\alpha.$$

Furthermore, the index set Ω is often a set of integers, and there is a special notation in this case, as follows: If $\Omega = \{\text{integers } n : a \leq n \leq b\}$ for instance, where $a \leq b$ are integers, then we may write

$$\cup_{n=a}^{b} X_n = X_a \cup X_{a+1} \cup \cdots \cup X_b = \cup_{n \in \Omega} X_n,$$
$$\cap_{n=a}^{b} X_n = X_a \cap X_{a+1} \cap \cdots \cap X_b = \cap_{n \in \Omega} X_n.$$

■ **Example 3.6.1** Let the universe $\mathcal{U}$ be the set $\{0, 1, 2, 3, 4\}$ and define $A_1 = \{0\}$, $A_2 = \{0, 1\}$, and $A_3 = \{0, 1, 2\}$. If $\mathcal{A} = \{\{0\}, \{0, 1\}, \{0, 1, 2\}\}$, then

$$\cup_{X \in \mathcal{A}} X = \cup_{i \in \{1,2,3\}} A_i = \cup_{i=1}^{3} A_i = \{0\} \cup \{0, 1\} \cup \{0, 1, 2\} = \{0, 1, 2\},$$
$$\cap_{X \in \mathcal{A}} X = \cap_{i \in \{1,2,3\}} A_i = \cap_{i=1}^{3} A_i = \{0\} \cap \{0, 1\} \cap \{0, 1, 2\} = \{0\}.$$

For further examples of the notation we have

$$\cup_{i \in \{1,2,3\}} \overline{A}_i = \cup_{i=1}^{3} \overline{A}_i = \{1, 2, 3, 4\} \cup \{2, 3, 4\} \cup \{3, 4\} = \{1, 2, 3, 4\},$$
$$\cap_{i \in \{1,2,3\}} \overline{A}_i = \cap_{i=1}^{3} \overline{A}_i = \{1, 2, 3, 4\} \cap \{2, 3, 4\} \cap \{3, 4\} = \{3, 4\}.$$

■ **Example 3.6.2** Suppose that the universe $\mathcal{U}$ is the set of all real numbers, let $[-n, n]$ denote the closed interval

$$[-n, n] = \{\text{real numbers } x : -n \leq x \leq n\}, \quad \text{for each } n \geq 1,$$

and set $\mathcal{I} = \{[-n, n] : n \geq 1\}$. We have

$$\cup\mathcal{I} = \cup_{X\in\mathcal{I}}X = \cup_{n\geq 1}[-n, n] = \{\text{real numbers}\},$$
$$\cap\mathcal{I} = \cap_{X\in\mathcal{I}}X = \cap_{n\geq 1}[-n, n] = [-1, 1],$$

where {real numbers} denotes the set of all real numbers.

Over the same universe, define the intervals

$$I_n = [-\frac{1}{n}, \frac{1}{n}] = \{\text{real numbers } x : -\frac{1}{n} \leq x \leq \frac{1}{n}\}, \quad \text{for each } n \geq 1.$$

We have

$$\cap_{n\geq 1}I_n = \{0\} \quad \text{and} \quad \cup_{n\geq 1} I_n = [-1, 1].$$

■ **Example 3.6.3** Suppose that the universe $\mathcal{U}$ is the set of all real numbers, and let the index set Ω also denote the set of all real numbers. For each $\alpha \in \Omega$, let $X_\alpha = \{\alpha\}$, so that

$$\cup_{\alpha\in\Omega}X_\alpha = \Omega,$$
$$\cap_{\alpha\in\Omega}X_\alpha = \emptyset,$$

This debauch of notation for unions and intersections may seem disconcerting, but we are obliged to mention all the possibilities here since they are all in common usage.

Returning finally to the promised extension of DeMorgan's Laws to arbitrary intersections and unions, we have

3.6.5 THEOREM *If $\mathcal{A}$ is a collection of sets, then we have*

(a) $\overline{(\cap_{X\in\mathcal{A}}X)} = \cup_{X\in\mathcal{A}}\overline{X}$

(b) $\overline{(\cup_{X\in\mathcal{A}}X)} = \cap_{X\in\mathcal{A}}\overline{X}$

Proof Beginning with part (a), first observe that in the extreme case that $\mathcal{A} = \emptyset$, we have

$$\overline{(\cap_{X\in\mathcal{A}}X)} = \overline{\mathcal{U}} = \emptyset = \cup_{X\in\mathcal{A}}\overline{X},$$

so DeMorgan's Law holds in this extreme case. More generally, we compute

$$\begin{aligned}
\overline{(\cap_{X\in\mathcal{A}}X)} &= \overline{\{z \in \mathcal{U} : \forall X(X \in \mathcal{A} \Rightarrow z \in X\}} \\
&= \{z \in \mathcal{U} : \neg\forall X(X \in \mathcal{A} \Rightarrow z \in X\} \\
&= \{z \in \mathcal{U} : \exists X\neg(X \notin \mathcal{A} \vee z \in X\} \\
&= \{z \in \mathcal{U} : \exists X(X \in \mathcal{A} \wedge z \notin X)\} \\
&= \{z \in \mathcal{U} : \exists X(X \in \mathcal{A} \wedge z \in \overline{X}\} \\
&= \cup_{X\in\mathcal{A}}\overline{X}
\end{aligned}$$

For part (b) in the extreme case that $\mathcal{A} = \emptyset$, we find

$$\overline{(\cup_{X\in\mathcal{A}} X)} = \overline{\emptyset} = \mathcal{U} = \cap_{X\in\mathcal{A}}\overline{X},$$

so DeMorgan's Law again holds in this extreme case. More generally, we again compute

$$\begin{aligned}
\overline{(\cup_{X\in\mathcal{A}} X)} &= \overline{\{z \in \mathcal{U} : \exists X(X \in \mathcal{A} \wedge z \in X)\}} \\
&= \{z \in \mathcal{U} : \neg\exists X(X \in \mathcal{A} \wedge z \in X)\} \\
&= \{z \in \mathcal{U} : \forall X \neg(X \in \mathcal{A} \wedge z \in X)\} \\
&= \{z \in \mathcal{U} : \forall X(X \notin \mathcal{A} \vee z \notin X)\} \\
&= \{z \in \mathcal{U} : \forall X(\neg(X \in \mathcal{A}) \vee z \in \overline{X}\} \\
&= \{z \in \mathcal{U} : \forall X(X \in \mathcal{A} \Rightarrow z \in \overline{X}\} \\
&= \cap_{X\in\mathcal{A}}\overline{X}
\end{aligned}$$

q.e.d.

Let us take a step back from the material being discussed here and make the following completely general comment: As was mentioned at the beginning of §2.7, one can often extend a result about a binary operation by induction to a result about arbitrary finite applications of the binary operation just as DeMorgan's Laws for finite unions and intersections follows by induction from Theorem 3.6.4. It is quite another matter to extend the result to include generalizations of the binary operation: Induction is typically of no avail in this endeavor, and other proof techniques must be found just as we gave an independent argument for the general case in Theorem 3.6.5. Indeed, the general result may not even be valid in certain situations.

To close this section, we comment briefly on an issue which has presumably occurred to the reader: There are obvious similarities between certain of our current manipulations with sets and the earlier manipulations with propositional forms. These similarities are due to the fact that there is a common underlying mathematical structure here called a "Boolean algebra", which we briefly discuss in passing. Precisely, let A be a set together with two binary operations $+$ and $\cdot$, one unary operation $^{-}$, and two distinguished elements $0, 1 \in A$; thus, given any elements $a, b \in A$, there are corresponding elements $a + b$, $a \cdot b$, $\bar{a} \in A$. We say that $(A, +, \cdot, ^{-}, 0, 1)$ is a *Boolean algebra* provided that the following ten identities hold for each $a, b, c \in A$:

- $a + b = b + a$
- $a \cdot b = b \cdot a$
- $(a + b) + c = a + (b + c)$
- $(a \cdot b) \cdot c = a \cdot (b \cdot c)$
- $a \cdot (b + c) = a \cdot b + a \cdot c$
- $a + (b \cdot c) = (a + b) \cdot (a + c)$
- $a + 0 = a$
- $a \cdot 1 = a$
- $a + \bar{a} = 1$
- $a \cdot \bar{a} = 0$

■ **Example 3.6.4** Suppose that $\mathcal{U}$ is some set which we take to be the universe of discourse. Then $(\mathcal{P}(\mathcal{U}), \cup, \cap, ^{-}, \emptyset, \mathcal{U})$ is a Boolean algebra (where $^{-}$ denotes the absolute complement), as we leave it for the reader to verify.

■ **Example 3.6.5** Fix a collection of propositional variables, and let Π denote the set of all propositional forms over these variables. Then $(\Pi, \vee, \wedge, \neg, 0, 1)$ is *not* a Boolean algebra; however, if we identify two elements of Π whenever they are logically equivalent to get a set Π' of logical equivalence classes with corresponding binary operations $\vee, \wedge, \neg$ and elements $0, 1$, then $(\Pi', \vee, \wedge, \neg, 0, 1)$ *is* indeed a Boolean algebra, as we again leave it for the reader to check.

The apparent similarities between certain results in set theory and in the propositional calculus are then explained by the fact that basic results and arguments about general Boolean algebras specialize in each of the previous examples to particular results which we have seen before; for instance, DeMorgan's Law is such an example. We shall not undertake a serious study of Boolean algebras here, but see §5.16 for a related discussion. In fact, Boolean algebras are an appropriate formalism for analyzing electrical "gate-networks" which underlie the hardware of modern computers, but we shall not take this up here.

EXERCISES

3.6 ABSOLUTE COMPLEMENTS AND DEMORGAN'S LAWS

1. Let the universe of discourse be the set $\mathcal{U}$ of all natural numbers, let A be the set of all natural numbers which are multiples of either 2 or 5, and let B be the set of all natural numbers which are multiples of either 2 or 3. Compute the following sets.

(a) $\overline{A} \cup \overline{B} = \overline{(A \cap B)}$ (b) $\overline{A} \cap \overline{B} = \overline{(A \cup B)}$

(c) $\overline{A - B}$ (d) $\overline{B - A}$

2. Prove Theorem 3.6.1.

3. Suppose that A and B are sets defined over a universe $\mathcal{U}$. Prove that $A - B = A \cap \overline{B} = \overline{B} - \overline{A}$.

4. Prove that the following three conditions on sets A and B defined in the universe $\mathcal{U}$ are logically equivalent to one another: $A \subseteq B$, $\overline{A} \cup B = \mathcal{U}$, and $A \cap \overline{B} = \emptyset$.

5. Let A and B be sets. Prove each of the following.

(a) $A \subseteq B \Leftrightarrow \overline{B} \subseteq \overline{A}$ (b) $(A \cup B) - B = A \cap \overline{B}$

(c) $(A \cup B) \cap (A \cup \overline{B}) = A$ (d) $\overline{(A - B)} = \overline{A} \cup B$

6. Let A, B, C, D be sets defined over a universe $\mathcal{U}$. Prove each of the following.

(a) If $A \cup C = \emptyset$ and $B \cup C = \mathcal{U}$, then $A \subseteq B$.

(b) If $A \cup B = \mathcal{U} = C \cup D$, then either $B \cap C \neq \emptyset$ or $A \cup D = \mathcal{U}$.

(c) If $(B \cup C) \cap (A \cup D) = \mathcal{U}$ and $(A \cap C) \cup (B \cup D) = \emptyset$, then we have the equalities $A = B$ and $C = D$.

(d) If $B \cap D = \emptyset$, $B \cup C = \mathcal{U}$, and $A \cap C = \emptyset$, then $A \cap D = \emptyset$.

(e) If $A \cup B = \mathcal{U} = C \cup D$ and $A \cap C = \emptyset$, then $B \cap D = \mathcal{U}$.

7. Let A_α for $\alpha \in \Omega$ be an indexed family of sets and suppose that A is another set. Prove the following identities.

(a) $A \cap (\cup_{\alpha\in\Omega} A_\alpha) = \cup_{\alpha\in\Omega}(A \cap A_\alpha)$ (b) $A \cup (\cap_{\alpha\in\Omega} A_\alpha) = \cap_{\alpha\in\Omega}(A \cup A_\alpha)$

(c) $A - (\cup_{\alpha\in\Omega} A_\alpha) = \cap_{\alpha\in\Omega}(A - A_\alpha)$ (d) $A - (\cap_{\alpha\in\Omega} A_\alpha) = \cup_{\alpha\in\Omega}(A - A_\alpha)$

8. Let B be a set and let $\mathcal{A}$ be a family of sets. Prove the following identities.

(a) $B \times (\cup\mathcal{A}) = \cup_{A\in\mathcal{A}}(B \times A)$ (b) $B \times (\cap\mathcal{A}) = \cap_{A\in\mathcal{A}}(B \times A)$

9. Let $A_1, A_2, A_3, \ldots$ be a family of sets indexed by the natural numbers. Prove each of the following inclusions.

(a) For each natural number m, we have the inclusion $A_m \subseteq \cup_{n\geq 1} A_n$.

(b) For each natural number m, we have the inclusion $\cap_{n\geq 1} A_n \subseteq A_m$.

(c) For each natural numbers ℓ and m where $\ell \leq m$, we have the inclusion $\cup_{n=1}^{\ell} A_n \subseteq \cup_{n=1}^{m} A_n$.

(d) For each natural numbers ℓ and m where $\ell \leq m$, we have the inclusion $\cap_{n=1}^{\ell} A_n \supseteq \cap_{n=1}^{m} A_n$.

10. Let $A_1, A_2, A_3, \ldots$ and $B_1, B_2, B_3, \ldots$ be families of sets indexed by the natural numbers and assume that $A_i \subseteq B_i$ for each $i \geq 1$. Prove each of the following inclusions.

(a) $\cup_{i\geq 1} A_i \subseteq \cup_{i\geq 1} B_i$ (b) $\cap_{i\geq 1} A_i \subseteq \cap_{i\geq 1} B_i$

11. (a) Verify that Example 3.6.4 is indeed an example of a Boolean algebra.

(b) Verify that Example 3.6.5 is indeed an example of a Boolean algebra.

(*) 3.7 *THE SET OF NON-NEGATIVE INTEGERS*

In this optional section, we carefully treat the set of non-negative integers and formally develop some of their basic properties. In particular, we shall prove the Basic Fact in §A.5 of Chapter 1 (that any non-empty set of natural numbers has a least element), upon which relies our previous argument regarding the validity of proof by induction. The starting point for this discussion is the Axiom of Infinity from Zermelo-Fraenkel set theory which asserts the existence of a set S containing $\emptyset$ and satisfying the predicate $\forall x[x \in S \Rightarrow (x \cup \{x\}) \in S]$. Thus, the set S contains the following elements

$$\begin{aligned}&\emptyset,\\&\{\emptyset\},\\&\{\emptyset,\{\emptyset\}\},\\&\{\emptyset,\{\emptyset\},\{\emptyset,\{\emptyset\}\}\},\\&\quad\vdots\end{aligned}$$

and we introduce the usual numerical symbols $0, 1, 2, 3, \ldots$ for these various sets. The symbol $n = 0, 1, 2, 3, \ldots$ therefore represents the set above with n elements. It follows from this definition that

$$\begin{aligned}0 &= \emptyset,\\1 &= \{0\},\\2 &= \{0,1\},\\3 &= \{0,1,2\},\\&\vdots\end{aligned}$$

which hopefully makes our definition seem more natural or palatable. It is worth commenting immediately that we shall "pay some dues" in this section to formally develop the basic theory from the axioms, and, having accomplished this, we shall then treat the non-negative integers and their arithmetic properties in the usual way. The situation is entirely analogous to our treatment of ordered pairs in §§3.2 and 3.3, where we carefully developed the basics from the axioms and could then essentially forget the underlying formalism. We also remark that it is more convenient here to describe the construction of the set of all non-negative integers rather than just the set of all natural numbers.

We refer to $x \cup \{x\}$ as the *successor* of x and introduce the notation x' for the successor of x. Thus, the successor of 0 is 1, the successor of 1 is 2, and so on. We shall call a set A a *successor set* if $0 \in A$ and $(x \cup \{x\}) \in A$ whenever $x \in A$, so the Axiom of Infinity simply guarantees the existence of a successor set S. Since the intersection of every non-empty family of successor sets contained in S is itself a successor set (as the reader should prove), the intersection of *all* the successor sets contained in S is again a successor set. We let ω denote this intersection, and refer to an element of ω as a *non-negative integer.*

We claim that ω is actually a subset of any successor set and refer to this property of ω as the *minimality property.* To see this, suppose that A is an arbitrary successor set, so $A \cap S$ is also a successor set. Since $A \cap S \subseteq S$, we find that $A \cap S$ is one of the sets in the intersection which defined ω, so $\omega \subseteq A \cap S$. In particular, we have $\omega \subseteq A$, which establishes the minimality property.

Historically, the development of the non-negative integers actually involved an axiomatic description of the set ω, and we next describe these axioms which are called the *Peano Postulates*:

PEANO POSTULATES

(a) $0 \in \omega$.

(b) *For any* $n \in \omega$, *we have also* $n' \in \omega$.

(c) *If* $X \subseteq \omega$, $0 \in X$, *and* $n' \in X$ *whenever* $n \in X$, *then we have* $X = \omega$.

(d) *For any* $n \in \omega$, *we have* $0 \neq n'$.

(e) *If* $m, n \in \omega$ *and* $m' = n'$, *then* $m = n$.

As with any axiomatic definition, the Peano Postulates were simply taken for granted historically and were the starting point for developing the theory of arithmetic. From our point of view of axiomatic set theory, however, each of the Peano Postulates is a theorem to be proved from our definition of ω above, and we turn next to proving these results. Postulates (a) and (b) follow from the fact that ω is a successor set, and (c) follows from the minimality property of ω discussed above. Postulate (c) itself is sometimes called the "Principle of Mathematical Induction", and we observe that the validity of the Principle of Mathematical Induction as discussed in §A.5 of Chapter 1 already follows from this result. Postulate (d) is trivial since $n \in n'$ for any $n \geq 1$ whereas $0 = \emptyset$ has no elements. The proof of the last Peano Postulate actually requires some work, and to this end, we must prove

3.7.1 LEMMA (i) *No non-negative integer is a subset of any of its elements, i.e., we have*

$$\forall n \in \omega \ \forall x (x \in n \Rightarrow n \not\subseteq x).$$

(ii) *If* n *is a non-negative integer and* $x \in n$, *then also* $x \subseteq n$, *i.e., we have*

$$\forall n \in \omega \ \forall x (x \in n \Rightarrow x \subseteq n).$$

As a point of terminology, a set X satisfying $\forall x (x \in X \Rightarrow x \subseteq X)$ is said to be *transitive*, so (ii) simply asserts that any non-negative integer is a transitive set.

Proof For part (i), define the set

$$X = \{n \in \omega : \forall x (x \in n \Rightarrow n \not\subseteq x)\}$$

so we must prove that $X = \omega$. We prove this equality by showing that X satisfies the hypotheses of Peano Postulate (c), and to this end first observe that 0 is indeed not a subset of any of its elements (since it has no elements), so $0 \in X$. It remains to prove that $n' \in X$ if $n \in X$, and we suppose that $n \in X$. Since $n \subseteq n$, we conclude that $n \notin n$ since $n \in X$, so $n' \not\subseteq n$ since $n \subseteq n'$. Turn attention now to n' and ask what might n' be a subset of: If $n' \subseteq x$, then $n \subseteq x$ since $n \subseteq n'$, so $x \notin n$ since $n \in X$. Thus, n' cannot be a subset of any element of n, and n' cannot be a subset of n itself as we saw above, so n' cannot be a subset of any element of n'. This proves that $n' \in X$, so Peano Postulate (c) implies that $X = \omega$ proving part (i).

Turning to part (ii), define the set

$$\begin{aligned} X &= \{n \in \omega : \forall x(x \in n \Rightarrow x \subseteq n)\} \\ &= \{n \in \omega : n \text{ is transitive}\}. \end{aligned}$$

We must again prove that $X = \omega$ and shall apply Peano Postulate (c) to this end. Of course, $0 \in X$ holds vacuously, and we suppose that $n \in X$ and must prove that also $n' \in X$. To see this, suppose that $x \in n'$ so that either $x \in n$ or $x = n$. In the former case, we have $x \subseteq n$ since $n \in X$, and since also $n \subseteq n'$, we conclude that $x \subseteq n'$. In the latter case, we find $x = n \subseteq n'$, so in either case, $n' \in X$, as desired. *q.e.d.*

Armed with the previous lemma, we turn to the proof of Peano Postulate (e) and suppose that $m, n \in \omega$ and $m' = n'$. Since $n \in n'$, we conclude that also $n \in m'$, so either $n \in m$ or $n = m$; likewise, we conclude that either $m \in n$ or $m = n$. Thus, if $m \neq n$, then we must conclude that $n \in m$ and $m \in n$. By Lemma 3.7.1ii, n is transitive, so $n \in n$. On the other hand, since $n \subseteq n$, this contradicts Lemma 3.7.1i and establishes Peano Postulate (e).

We say that two non-negative integers m, n are *comparable* provided that $m \in n$, $m = n$, or $n \in m$, and have

3.7.2 LEMMA *Any two non-negative integers are comparable, i.e., we have*

$$\forall m \in \omega \; \forall n \in \omega \; [m \in n \quad \vee \quad m = n \quad \vee \quad n \in m].$$

Proof For each $n \in \omega$, define the set

$$X(n) = \{m \in \omega : m \text{ and } n \text{ are comparable}\}$$

and let

$$X = \{n \in \omega : X(n) = \omega\},$$

so we must show that $X = \omega$. As usual, we proceed by showing that X satisfies the hypotheses of Peano Postulate (c), and to this end, begin by showing that $X(0) = \omega$. We prove this, in turn, by showing that $X(0)$ itself satisfies the hypotheses of Peano Postulate (c), and it is obvious that in fact $0 \in X(0)$. Suppose, then, that $m \in X(0)$. Since $m \in 0 = \emptyset$ is impossible, we conclude that either $m = 0$ or $0 \in m$, and in either case, we find $m' \in X(0)$. Applying Peano Postulate (c), we conclude that indeed $X(0) = \omega$, so $0 \in X$

To finish the proof, we must show that $n' \in X$ if $n \in X$, that is, we must show that $X(n') = \omega$ if $X(n) = \omega$. First of all, we have $0 \in X(n')$ since $n' \in X(0)$ by the previous paragraph. In order to apply Peano Postulate (c), it remains to show that $m' \in X(n')$ if $m \in X(n')$, and to this end, we suppose that indeed $m \in X(n')$. There are thus the three possibilities $n' \in m$, $n' = m$, or $m \in n'$, and in the first two cases, we conclude that $n' \in m'$ as well. In the third case $m \in n'$, we conclude that either $m = n$ (so $m' = n'$) or $m \in n$. On the other hand, since $m' \in X(n)$, we conclude that $n \in m'$, $n = m'$, or $m' \in n$. The first possibility is not compatible with the hypothesis that $m \in n$ (since $n \in m'$ implies that $n \in m$ or $n = m$, whence we have $n \subseteq m$, in contradiction to Lemma 3.7.1i); the second and third possibilities imply that $m' \in n'$. Thus, we find in all cases that m' and n' are comparable, i.e., we have $m' \in X(n')$, and application of Peano Postulate (c) finally shows that $X = \omega$. *q.e.d.*

Now, if $m, n \in \omega$ and $m \in n$, then m is a proper subset of n by Lemma 3.7.1, and we write $m < n$ and say that m is *less than* n in this case. Furthermore, if $m < n$ or $m = n$, then we write $m \leq n$ and say that m is *less than or equal to* n. Notice that if $m \leq n$ and $n \leq m$, then $m = n$ by Lemma 3.7.1; this property that

$$(m \leq n \wedge n \leq m) \Rightarrow m = n$$

is called *symmetry* of $\leq$. Furthermore, if $k \leq m$ and $m \leq n$, then $k \subseteq m \subseteq n$, so $k \subseteq n$ by Theorem 3.1.1, whence $k \leq n$ again by Lemma 3.7.1; this property that

$$(k \leq m \wedge m \leq n) \Rightarrow k \leq n$$

is called *transitivity* of $\leq$. The use of the term "transitivity" here is unrelated to the term "transitive" set discussed before; actually, the former term will be generalized later while the use of the latter term is isolated to this section.

Given a subset $X \subseteq \omega$ and $x \in X$, we say that x is a *least element* of X if $\forall y \in X\ (x \leq y)$.

3.7.3 THEOREM *If $X \subseteq \omega$ is non-empty, then X has a unique least element $x \in X$.*

Proof Since X is non-empty, we may choose some $m_0 \in X$ and consider the subset

$$X_0 = \{x \in X : x \leq m_0\} \subseteq X.$$

Thus, X_0 is a finite set, say with distinct elements $m_0, m_1, \ldots, m_n$. Since any two of these elements are comparable by Lemma 3.7.2, we may relabel these elements so that

$$m_n \leq m_{n-1} \leq \cdots \leq m_0.$$

We claim that $x = m_n \in X_0 \subseteq X$ is a least element of X. To prove this, first observe that $x \leq y$ for each $y \in X_0$ by construction and transitivity of $\leq$. On the other hand, if $y \in X - X_0$, then $m_0 \leq y$ by Lemma 3.7.2, so again $x \leq y$ by transitivity of $\leq$ since $x \leq m_0$ as proved above. It follows that x is indeed a least element of X completing the proof of existence.

To see that X has a unique least element, suppose that both x and y are least elements of X. Insofar as x is least and $y \in X$, we conclude that $x \leq y$, and by the same logic, we conclude that also $y \leq x$. Symmetry of $\leq$ finally implies that $x = y$ as desired. *q.e.d.*

Of course, the Basic Fact in §A.5 of Chapter 1 follows directly from the previous result completing this part of our discussion. To conclude this section, we briefly discuss the formal development of the usual arithmetic operations of addition and multiplication on ω and must first discuss some background material.

Recall that we have defined in §3.2 a function f from a set A to a set B to be a subset $f \subseteq A \times B$ so that $\forall a \in A\ \exists! b \in B\ [(a,b) \in f]$, and if $(a,b) \in f$, then we write $f(a) = b$. In particular, we consider a function f from a set X to itself and suppose that a is some specified element of X. It seems natural to attempt to define a function u from ω to X in such a way that $u(0) = a$, $u(1) = f(a)$, $u(2) = f(f(a))$, and so on, but to make this definition precise requires some further work. One might try to give an "inductive definition" as follows:

Basis: Define $u(0) = a$.

Induction: Define $u(n+1) = f(u(n))$, for each $n \geq 1$.

This would-be definition is imprecise in that we have not carefully proved that u is actually a function, and our next result, which is called the "Recursion Theorem", yields a careful proof of this fact.

3.7.4 THEOREM *Suppose that a is an element of a set X and f is a function from X to X. Then there exists a function u from ω to X so that $u(0) = a$ and for each $1 \leq n$, we have $u(n') = f(u(n))$.*

Proof We construct u explicitly as a collection of ordered pairs in $\omega \times X$. To this end, let $\mathcal{A}$ denote the collection of all those subsets A of $\omega \times X$ so that $(0,a) \in A$ and $(n', f(x)) \in A$ whenever $(n,x) \in A$. Since $\omega \times X$ itself has these properties, we have $\mathcal{A} \neq \emptyset$ and take the intersection $u = \cap\mathcal{A}$ of all these subsets. Since u evidently belongs to $\mathcal{A}$, it remains only to prove that u is actually a function from ω to X, and the proof is by induction. Let X denote the set of all $n \in \omega$ so that it is indeed the case that $(n,x) \in u$ for exactly one $x \in X$. We shall prove that $0 \in X$ and that if $n \in X$, then also $n' \in X$, and the result then follows from Peano Postulate (c).

For the basis step, we proceed by contradiction and suppose that $(0,b) \in u$ for some $b \in X$ distinct from a. Consider the set $u^* = u - \{(0,b)\}$ which must still contain $(0,a)$ since $a \neq b$. Furthermore, if $(n,x) \in u^*$, then also $(n', f(x)) \in u^*$ since $0 \neq n'$ for any $n \in \omega$ by Peano Postulate (d). Thus, $u^* \in \mathcal{A}$, contradicting that $u = \cap\mathcal{A}$, and we therefore conclude that $0 \in X$.

To verify the second hypothesis of Peano Postulate (c), suppose that $n \in X$, so there exists a unique element $x \in X$ so that $(n,x) \in u$. Since $(n,x) \in u$, it follows that $(n', f(x)) \in u$, and if $n' \notin X$, then $(n', y) \in u$ for some $y \in X$

with $y \neq f(x)$. Consider the set $u^* = u - \{(n', y)\}$, so $(0, a) \in u^*$ since $n' \neq 0$. Furthermore, if $(m, t) \in u^*$ for some $m \in \omega$ and $t \in X$, then also $(m', f(t)) \in u^*$. Indeed, if $m = n$, then we must have $t = x$ and u^* must contain $(n', f(x))$ since $f(x) \neq y$; on the other hand, if $m \neq n$, then $(m', f(t)) \in u^*$ since $m' \neq n'$ by Peano Postulate (e). We again conclude that $u^* \in \mathcal{A}$ contradicting the fact that $u = \cap\mathcal{A}$, and we conclude that $n' \in X$.

Thus, $X = \omega$, and it follows that indeed u is the desired function from ω to X. *q.e.d.*

Turning to the definition of addition and multiplication of natural numbers, it follows from the Recursion Theorem that for $m \in \omega$, there exists a function s_m from ω to ω so that $s_m(0) = m$ and $s_m(n') = s_m(n)'$, that is, $s_m(n)$ is the *sum*

$$m + n = s_m(n).$$

The usual arithmetic properties of addition are then proved by induction. For instance, to prove that addition is associative

$$(k + m) + n = k + (m + n) \text{ for any } k, m, n \in \omega,$$

we proceed by induction on n as follows. For the basis step, we have

$$(k + m) + 0 = k + m = k + (m + 0),$$

and for the induction, we compute

$$\begin{aligned}(k + m) + n' &= ((k + m) + n)' \\ &= (k + (m + n))' \\ &= k + (m + n)' \\ &= k + (m + n'),\end{aligned}$$

where the second equality follows from the inductive hypothesis and the others from the definition of addition.

To define multiplication, apply the Recursion Theorem to produce functions p_m from ω to ω for each $m \in \omega$ so that $p_m(0) = 0$ and $p_m(n') = p_m(n) + m$ for every $n \in \omega$. The value of $p_m(n)$ is by definition the *product*

$$m \cdot n = p_m(n).$$

All of the standard properties of arithmetic are then proved by induction starting from these definitions, but we shall not undertake this here and have included the proof of associativity of addition above just to give the flavor of such arguments. Owing to the familiarity of this material from previous math courses and everyday experience, this omission should cause the reader no discomfort. We systematically omit such a careful derivation of elementary arithmetic facts here simply because of constraints of time and energy and an eye towards other goals.

The interested reader could consult Halmos' book *Naive Set Theory*, Springer-Verlag, 1974.

We also remark that the careful application of the Recursion Theorem as in this section is often omitted, and an imprecise "definition" similar to the one given before Theorem 3.7.4 typically suffices in practice; see also the related discussion in §8.3. We have been quite formal here in order to carefully describe the foundations of arithmetic, but we shall typically take this more informal approach in the sequel.

EXERCISES

3.7 THE SET OF NON-NEGATIVE INTEGERS

1. Prove that if X is a non-empty subset of some non-negative integer, then there is an element $x \in X$ so that $x \in n$ for each element n of X different from x.

2. Prove that addition of non-negative integers is commutative. [HINT: First prove by induction on n that $0 + n = n$, then prove that $m' + n = (m + n)'$, and finally prove that $m + n = n + m$ by induction on m using the previous facts.]

3. (a) Prove that multiplication of non-negative integers is both commutative and associative.

(b) Prove that multiplication distributes over addition of non-negative integers.

4. Give a recursive definition of the nth power m^n of the non-negative integer m. [HINT: Use the recursion theorem to produce functions q_m, where $q_m(0) = 1$ and $q_m(n') = q_m(n) \cdot m$.]

5. Prove that if $m < n$ for non-negative integers m, n, then $m + k < n + k$ for each non-negative integer k.

6. Prove that if $m < n$ and $k \neq 0$, then $m \cdot k < n \cdot k$.

3.8 INDUCTIVE DEFINTIONS

This section discusses an "inductive" method of defining sets and other mathematical objects. We concentrate first on the inductive definition of a set S, and such a definition consists of the following three pieces of data:

- A *basis clause* which establishes that S contains certain specified elements.
- An *inductive clause* which describes ways in which given elements of S can be combined to produce other elements of S.
- An *extremal clause* which serves to limit the elements of S.

The basis clause thus determines a subset $S_0 \subseteq S$, which should be regarded as the "starting point" for the definition. Apply the inductive clause using S_0 to produce a collection S_1' of elements of S and define $S_1 = S_0 \cup S_1'$, then apply the inductive clause using S_1 to produce a collection S_2' of elements of S and define $S_2 = S_1 \cup S_2'$, and so on, in order to define a nested collection

$$S_0 \subseteq S_1 \subseteq S_2 \subseteq \cdots$$

of sets. The inductive clause should thus be regarded as a method of "building" the new elements S_{i+1}' of S from the old elements S_i, for each $i = 0, 1, 2, \ldots$. A typical extremal clause in this context might be: "The only elements of S are those that can be constructed from the basis clause by a finite number of applications of the inductive clause"; this serves to completely specify the set

$$S = \cup_{i \geq 0} S_i$$

determined by the inductive definition.

We mention that the term "recursive definition" is sometimes used in this context, but this term usually refers to such a definition of a function (as in the previous optional section); we shall discuss recursive definitions further in Chapter 8. The term "recursive routine" is also used in computer science to indicate that a sub-routine or function calls itself; though this is clearly related to our current use of the term, it really means something rather different. We also remark that the extremal clause is usually the "standard" one described above and is sometimes not stated explicitly; this rarely leads to confusion in practice.

Let us first give several examples of inductive definitions.

■ **Example 3.8.1** We give an inductive definition for the set $\mathcal{O}$ of all odd natural numbers as follows:

Basis: $1 \in \mathcal{O}$.

Induction: If $n \in \mathcal{O}$, then $n + 2 \in \mathcal{O}$.

Extremal: The only elements of $\mathcal{O}$ are those that arise from 1 by a finite number of applications of the inductive clause.

The basis clause determines the subset $\mathcal{O}_0 = \{1\} \subseteq \mathcal{O}$. The first application of the inductive clause to $\mathcal{O}_0$ produces the subsets $\mathcal{O}_1' = \{3\}$ and $\mathcal{O}_1 = \mathcal{O}_0 \cup \mathcal{O}_1' = \{1, 3\}$ of $\mathcal{O}$. The second application of the inductive clause to $\mathcal{O}_1$ produces the subsets $\mathcal{O}_2' = \{3, 5\}$ and $\mathcal{O}_2 = \mathcal{O}_1 \cup \mathcal{O}_2' = \{1, 3, 5\}$ of $\mathcal{O}$, and so on. The extremal clause stipulates that $\mathcal{O} = \cup_{i \geq 0} \mathcal{O}_i$ and therefore uniquely determines the set $\mathcal{O}$ being defined.

We hasten to add that the previous example presupposes that addition of natural numbers has been defined and is intended simply to illustrate the mechanics of an inductive definition. In the same way, the reader may be tempted to give the following "inductive definition" of the set of natural numbers:

Basis: 1 is a natural number.

Induction: If n is a natural number, then so too is $n + 1$.

Extremal: The only natural numbers are those that arise from 1 by a finite number of applications of the inductive clause.

This supposed definition is unsatisfactory because it implicitly assumes that we already know what is meant by "$n + 1$" given the natural number n. This is a perfectly legitimate *characterization* of the natural numbers but falls short of providing a suitable *definition*; one must resort to a more detailed discussion as in the previous optional section in order to carefully define the set of natural numbers together with the usual arithmetic operations on them.

We turn next to some more serious examples of inductive definitions and must begin by describing some background material.

Let A be a finite set which we refer to as an *alphabet* and whose elements are called *letters*. A finite ordered collection of elements of A is called a *string* over A. For instance if $A = \{0, 1\}$, then 00111, 0101, 000, and 1 are each examples of strings over A. The *length* of a string is the number of letters comprising it, so for instance, the previous examples of strings have respective lengths 5,4,3, and 1. Given two strings u and v over A, we define their *concatenation* to be the string uv over A gotten by simple juxtaposition. For instance, if $u = 01$ and $v = 10$ are strings over $\{0, 1\}$, then their concatenations are $uv = 0110$ and $vu = 1001$. We next give an inductive definition of the set A^+ of all nonempty strings over A.

■ **Example 3.8.2** Let A be an alphabet and inductively define the set A^+ as follows:

Basis: If $a \in A$, then also $a \in A^+$.

Induction: If $u \in A^+$ and $a \in A$, then also the concatenation $au \in A^+$.

Extremal: The only elements of A^+ are those that arise from A by a finite number of applications of the inductive clause.

The basis clause specifies that all strings of length 1 are elements of A^+, the first application of the inductive clause produces all the strings of length 2, the second application of the inductive clause produces all the strings of length 3, and so on. The extremal clause stipulates in particular that each element of A^+ has finite length.

For instance over the alphabet $A = \{0, 1\}$, our inductive definition produces the following elements of A^+:

$$0, 1, 00, 01, 11, 000, 001, 010, 011, 100, 101, 110, 111, 0000, \ldots.$$

In fact, the previous example is the starting point for the study to be undertaken in Chapter 9 of "languages" in the sense of computer science. Of course, the collection A^+ of *all* strings over an alphabet A is a rather uninteresting "language", and one is more typically interested in particular subsets of A^+. For instance, over the usual English alphabet, one might be interested in the set of English words, or over the usual ASCII character set in computer science, one might be interested in the set of all FORTRAN programs. A string over an alphabet which makes sense in some particular mathematical context is called a *well-formed formula*, which is often abbreviated simply *wff*. As was promised in §2.2, we next give an inductive definition of the wff's in the propositional calculus.

■ **Example 3.8.3** Fix some finite set $X = \{P, Q, R, \ldots\}$ whose elements will represent propositional variables, and suppose that X does not contain as elements any of the symbols $($,$)$,$\wedge$,$\vee$,$\Rightarrow$, $\Leftrightarrow$, $\neg$, 0, or 1. Inductively define the *propositional forms* over X as follows:

Basis: 0 is a propositional form.
1 is a propositional form.
If $x \in X$, then x is a propositional form.

Induction: If x and y are propositional forms, then so too are

$$(\neg x)$$
$$(x \vee y)$$
$$(x \wedge y)$$
$$(x \Rightarrow y)$$
$$(x \Leftrightarrow y)$$

Extremal: The only propositional forms are those that arise from $X \cup \{0, 1\}$ by a finite number of applications of the inductive clause.

Notice that this definition of propositional forms requires enclosing the result of each logical operation in a pair of parentheses, that is, this definition does not take account of the various conventions for omitting parentheses which were discussed in §2.3.

A standard problem in computer science called the "parsing problem" involves being given a string of letters over a given alphabet and determining whether that string is a wff relative to a given inductive definition. For instance, the string $((P \vee Q) \Leftrightarrow R)$ is indeed a propositional form since according to the basis clause above, each of P, Q and R are propositional forms, hence by a first application of the inductive clause so too is $(P \vee Q)$ and by a second application of the inductive clause so too is $((P \vee Q) \Leftrightarrow R)$. We shall postpone a careful treatment of the parsing problem to Chapter 9 and mention it here just in passing.

In fact, one may use inductive definitions to describe mathematical objects other than sets. A typical such situation is to rely on the inductive characterization above of the set of natural numbers in order to define new mathematical objects, and we next give two examples of such definitions. Actually, these examples fall a bit short in terms of rigor (cf. Theorem 3.7.4 and the subsequent discussion if this comment is unsettling) but are quite typical of such definitions in practice.

■ **Example 3.8.4** Suppose that a is a real number and let n be a non-negative integer. Letting $\cdot$ denote multiplication of real numbers, we give the following inductive definition of the *powers* a^n of a:

Basis: Define $a^0 = 1$.

Induction: Define $a^{n+1} = a^n \cdot a$ for $n \geq 0$.

Thus, we have

$$2^3 = 2^2 \cdot 2 = 2^1 \cdot 2 \cdot 2 = 2^0 \cdot 2 \cdot 2 \cdot 2 = 1 \cdot 2 \cdot 2 \cdot 2 = 8$$

for instance. Notice the absence of an extremal clause in this definition: The extremal clause is implicit in the fact that we are here defining a^n only for non-negative integers n.

■ **Example 3.8.5** Let n be a non-negative integer and give the following inductive definition of the natural number $n!$:

Basis: Define $0! = 1$.

Induction: Define $n! = n \cdot [(n-1)!]$ for $n \geq 1$.

Thus, we have

$$3! = 3 \cdot (2!) = 3 \cdot 2 \cdot (1!) = 3 \cdot 2 \cdot 1 \cdot (0!) = 3 \cdot 2 \cdot 1 \cdot 1 = 6$$

for instance.

We turn now from inductive definitions to discuss the utility of proofs by induction for mathematical objects which are defined inductively, and the basic idea is as follows. Suppose, for instance, that S is a set which is defined inductively and we wish to prove the quantified assertion $\forall s \in S\ P(s)$ for some particular predicate $P(s)$ defined on a universe containing S. We verify that $P(s)$ is valid for each element s of S which is specified in the basis clause of the definition of S. The inductive clause allows the construction of a "new" element s' of S from "old" ones, and we assume that $P(s)$ is valid for each old element and prove that $P(s')$ is also valid; that is, we assume $\forall s \in S_n\ P(s)$ and prove $\forall s \in S'_{n+1}\ P(s)$ for each $n \geq 0$ in the previous notation. Assuming the usual extremal clause in the inductive definition of S and arguing as for proofs by induction over the natural numbers, this provides a legitimate proof that $\forall s \in S\ P(s)$ is valid. Of course, our usual proofs by induction over the natural numbers are just a special case of this where we rely on the inductive characterization of the natural numbers discussed above, and as before, the proof breaks up into two steps called the "basis step" and the "induction step". Furthermore, the supposition in the induction step that the predicate is valid for each old element of S is again called the "inductive hypothesis".

The remainder of this section is dedicated to examples of such proofs by induction for inductively defined sets and other mathematical objects.

■ **Example 3.8.6** Suppose that x is a propositional form as defined inductively in Example 3.8.3 above and define $L[x]$ and $R[x]$ respectively to be the number of left and right parentheses which occur in the string x of letters. For instance, if x is the propositional form $((P \vee Q) \Leftrightarrow R)$, then $L[x] = 2 = R[x]$. We prove by induction that for any propositional form x, we similarly have $L[x] = R[x]$. For the basis step, the basis clause specifies that 0, 1 and any element of $X = \{P, Q, R, \ldots\}$ are propositional forms, and the assertion holds in this case since for any such propositional form x, we have $L[x] = 0 = R[x]$. For the induction step, let x and y be propositional forms as in the inductive clause and suppose by the induction hypothesis that $L[x] = R[x]$ and $L[y] = R[y]$. We then have

$$\begin{aligned}
L[(\neg x)] &= 1 + L[x] = 1 + R[x] = R[(\neg x)], \\
L[(x \vee y)] &= 1 + L[x] + L[y] = 1 + R[x] + R[y] = R[(x \vee y)], \\
L[(x \wedge y)] &= 1 + L[x] + L[y] = 1 + R[x] + R[y] = R[(x \wedge y)], \\
L[(x \Rightarrow y)] &= 1 + L[x] + L[y] = 1 + R[x] + R[y] = R[(x \Rightarrow y)], \\
L[(x \Leftrightarrow y)] &= 1 + L[x] + L[y] = 1 + R[x] + R[y] = R[(x \Leftrightarrow y)],
\end{aligned}$$

so in any case, the propositional form constructed from x and y again satisfies the required property. Since every propositional form is constructed in this way by a finite number of applications of the inductive clause to the elements specified in the basis clause, we conclude that indeed $L[x] = R[x]$ for any propositional form x.

■ **Example 3.8.7** We prove here from the inductive definition of factorials in Example 3.8.5 above that

$$n! = \prod_{i=1}^{n} i, \text{ for each } n \geq 0.$$

For the basis step, we have $0! = \prod_{i=1}^{0} i$, and it is a standard convention that the "empty product is unity", that is, a product with an empty index set such as $\prod_{i=1}^{0}$ has the value 1 by convention, so the basis step is valid. For the induction step with $n \geq 1$, we compute

$$n! = n \cdot [(n-1)!] \quad = \quad n \cdot \left(\prod_{i=1}^{n-1} i \right) \quad = \quad \prod_{i=1}^{n} i,$$

where the second equality holds by the induction hypothesis.

■ **Example 3.8.8** We prove here by induction two results about powers of real numbers, namely, for any real number a, we prove

$$(*) \qquad \forall m \forall n[a^m \cdot a^n = a^{m+n}]$$

and

$$(\dagger) \qquad \forall m \forall n[(a^m)^n = a^{mn}],$$

where the universe of discourse in each case is the set of all non-negative integers. For the equality $(*)$, fix an arbitrary non-negative integer m and proceed by induction on the non-negative integer n. For the basis step $n = 0$, we have

$$a^m \cdot a^0 = a^m \cdot 1 = a^m = a^{m+0},$$

and for the induction step, we compute

$$\begin{aligned} a^m \cdot a^{n+1} &= a^m \cdot (a^n \cdot a) \\ &= (a^m \cdot a^n) \cdot a \\ &= a^{m+n} \cdot a \\ &= a^{m+n+1}, \end{aligned}$$

where the first and last equalities follow from the definition of powers, the second from associativity of multiplication, and the third from the inductive hypothesis.

Turning to the latter equality $(\dagger)$, we again fix an arbitrary non-negative integer m and induct on the non-negative integer n. For the basis step $n = 0$, we have

$$(a^m)^0 = 1 = a^0 = a^{m \cdot 0},$$

where the first two equalities follow from the definition of powers. For the induction step, we compute

$$\begin{aligned} (a^m)^{n+1} &= (a^m)^n \cdot a^m \\ &= a^{mn} \cdot a^m \\ &= a^{mn+m} \\ &= a^{m(n+1)}, \end{aligned}$$

where the first equality follows from the definition of powers, the second from the inductive hypothesis, and the third from $(*)$.

We mention that the argument in the previous example is actually a paradigm for several later arguments where we give inductive definitions analogous to our definition here of powers of real numbers, and we shall have occasion later to recall the arguments in Example 3.8.8.

■ **Example 3.8.9** We prove by induction that if A is a finite set with n elements, then $\mathcal{P}(A)$ is finite with 2^n elements. For the basis step $n = 0$, suppose that A is a set with 0 elements, so $A = \emptyset$ and $\mathcal{P}(A) = \{\emptyset\}$. Since $2^0 = 1$, the basis step is established.

For the induction, suppose the assertion is true for sets with n elements and let $A = \{a_1, a_2, \ldots, a_n\}$ be such a set. Suppose that $a \notin A$, and define $A^* = A \cup \{a\}$. Given any subset $A' \subseteq A^*$, either $a \in A'$ and $A' - \{a\} \subseteq A$ or $a \notin A$ and $A' \subseteq A$. We conclude that $\mathcal{P}(A^*)$ can be written as the disjoint union

$$\mathcal{P}(A^*) = \mathcal{P}(A) \cup \{A' \cup \{a\} : A' \subseteq A\},$$

and by the inductive hypothesis $\mathcal{P}(A)$ has 2^n elements. Since there is exactly one corresponding element $A' \cup \{a\}$ for each $A' \in \mathcal{P}(A)$, the set $\{A' \cup \{a\} : A' \subseteq A\}$ also has 2^n elements, whence $\mathcal{P}(A^*)$ has $2^n + 2^n = 2 \cdot 2^n = 2^{n+1}$ elements.

EXERCISES

3.8 INDUCTIVE DEFINITIONS

1. Give inductive definitions of the following sets.

(a) The set of non-negative integers in their decimal representation, i.e., the set of wff's of the form $x_1 x_2 \cdots x_n$, where each x_i is a decimal digit $0, 1, \cdots, 9$ and n is some natural number.

(b) The set of real numbers in their decimal representation with terminating "fractional part", i.e., the set of wff's of the form

$$x_1 x_2 \cdots x_n . x_{n+1} x_{n_2} \cdots x_{n+m},$$

where $n \geq 0$ and $m \geq 0$.

(c) The set of even natural numbers in their binary representation, i.e., the set of wff's of the form $x_1 x_2 \cdots x_n$, where each $x_i \in \{0, 1\}$, representing an even natural number in base two.

2. Does there exist a set with exactly 13 distinct subsets? Does there exist a set with exactly 256 distinct subsets? Prove your assertions.

3. (a) Give an inductive definition of the product $\prod_{i=1}^{n} a_i$ of the real numbers $a_1, a_2, \ldots, a_n$.

(b) Prove that for all $0 \leq m \leq n$ we have the identity

$$\prod_{i=1}^{n} a_i = (\prod_{i=1}^{m} a_j)(\prod_{k=m+1}^{n} a_k).$$

4. (a) Give an inductive definition of the sum $\sum_{i=1}^{n} a_i$ of the real numbers $a_1, a_2, \ldots, a_n$.

(b) Prove that for all $n \geq 2$ we have the identity

$$\sum_{i=1}^{n} a_i = a_1 + a_n + \sum_{j=2}^{n-1} a_j.$$

5. (a) Give an inductive definition of $\cap_{i=1}^{n} A_i$ and $\cup_{i=1}^{n} A_i$, where $A_1, A_2, \ldots, A_n$ is a collection of sets and $n \geq 1$.

(b) Let $\mathcal{A} = \{A_1, A_2, \ldots, A_n\}$ and prove by induction that your inductive definitions of $\cup_{i=1}^{n} A_i$ and $\cap_{i=1}^{n} A_i$ agree with the usual definitions of $\cup\mathcal{A}$ and $\cap\mathcal{A}$ respectively.

6. (a) Give an inductive definition of the powers a^n of a real number a, where n is a non-positive integer.

(b) Prove that $a^m \cdot a^n = a^{m+n}$ for any real number a and any non-positive integers m, n.

(c) Prove that $(a^m)^n = a^{mn}$ for any real number a and any non-positive integers m, n.

(d) Prove the identities in parts (b) and (c) for any two integers m and n.

7. By a *number symbol*, we mean a string of letters built from natural numbers using the binary operations of addition $+$, multiplication $\cdot$, subtraction $-$ and the unary operation $-$ of taking the negative. We furthermore require that the result of each operation be enclosed in a pair of parentheses. For instance, $(-(((2+3)\cdot 5) - (4 \cdot 6)))$ is a well-formed number symbol. Give an inductive definition of a well-formed number symbol.

8. By a number in *scientific notation*, we mean an optional plus-or-minus sign, followed by a decimal digit, followed by an optional decimal point and a string of digits the last of which is non-zero, followed by the string *exp*, followed by a plus-or-minus sign, finally followed by a positive integer. For instance, $+0.031415exp + 2$ is a well-formed number in scientific notation. Give an inductive definition of a well-formed expression in scientific notation.

9. A polygon P in the plane is said to be *convex* if for any pair u, v of consecutive vertices, P lies entirely on one side of the line determined by u, v. For instance, any triangle is necessarily convex.

(a) Give an example of a polygon which is not convex.

(b) Give an inductive definition of a convex polygon. [HINT: Give an induction on the number $n \geq 3$ of sides of the polygon.]

(c) Assume that the sum of the interior angles of a triangle is π radians. Use this result to prove by induction that the sum of the interior angles of an n-sided convex polygon is $(n-2)\pi$ radians for each $n \geq 3$.

(d) Assume the usual *triangle inequality*: Given any triangle with edge lengths a, b, c, we have the inequalities $a < b+c$, $b < a+c$, and $c < a+b$. Use this result to prove by induction that for any n-sided convex polygon with edge lengths $a_1, a_2, \dots a_n$ and for any $i = 1, 2, \dots, n$, we have the *generalized triangle inequality* $a_i < a_1 + a_2 + \cdots a_{i-1} + a_{i+1} + \cdots + a_n$.

3.9 SETS OF NUMBERS

Our purpose here is simply to introduce the standard notation for the various sets of numbers, and the most basic such set is the collection $\mathbb{N}$ of all natural numbers, so we have $\mathbb{N} = \{1, 2, 3, \dots\}$; we have discussed the axiomatic development of $\mathbb{N}$ in §3.7 but shall never again need to be that formal. The set $\mathbb{N} \cup \{0\}$ is called the set of *whole numbers*, and there is no specialized standard notation for this set (which was denoted ω in §3.7). $\mathbb{Z}$ denotes the set $\{0, \pm 1, \pm 2, \dots\}$ of all integers, whose axiomatic development is easily pursued starting from $\mathbb{N}$ as the reader can imagine. $\mathbb{Q}$ denotes the set of all rational numbers, that is, the set of all expressions of the form p/q, with $p \in \mathbb{Z}$, $q \in \mathbb{Z} - \{0\}$, where we regard two such expressions p/q and r/s as identical if $ps = rq$; we shall comment in §5.12 about a more formal treatment of the definition of $\mathbb{Q}$. $\mathbb{R}$ denotes the set of all real numbers, whose careful definition is rather involved and will not be given here. (The rough idea, though, is that a real number is defined to be a collection of suitable sequences of rational numbers; see [W. Rudin, *Principles of Mathematical Analysis*, Mc-Graw Hill (1964)] for a treatment which is accessible to the reader at this point.) A real number is said to be irrational if it is not rational; there is no special standard notation for the set $\mathbb{R} - \mathbb{Q}$ of all irrational numbers. Finally, $\mathbb{C}$ denotes the set of all complex numbers, that is, the set of all expressions of the form $x + y\sqrt{-1}$, where $x, y \in \mathbb{R}$, and where two such expressions $x + y\sqrt{-1}$ and $u + v\sqrt{-1}$ are regarded as identical if $x = u$ and $y = v$. Thus, we have

$$\mathbb{N} \subseteq \mathbb{N} \cup \{0\} \subseteq \mathbb{Z} \subseteq \mathbb{Q} \subseteq \mathbb{R} \subseteq \mathbb{C},$$

where we regard $\mathbb{Z} \subseteq \mathbb{Q}$ by setting $n = n/1$ for each $n \in \mathbb{Z}$, and we regard $\mathbb{R} \subseteq \mathbb{C}$ by setting $x = x + 0\sqrt{-1}$ for each $x \in \mathbb{R}$.

We shall not actually require complex numbers in the sequel but include the discussion here just for notational completeness. We shall, however, require the sets $\mathbb{N}$, $\mathbb{N} \cup \{0\}$, $\mathbb{Q}$, and $\mathbb{R}$ together with their usual arithmetic properties in the sequel, and we take the formal development of this material for granted from now on. Owing to the familiarity of these properties from elementary school and from everyday experience, this omission should cause the reader no discomfort.

To close, we repeat the following famous remark of the mathematician Felix Klein: *Plato said "God is a geometer". Jacobi changed this to "God is an arithmetician". Then came Kronecker and fashioned the memorable expression "God created the natural numbers, and all the rest is the work of man".*

Chapter 4

Elementary Number Theory

This chapter is dedicated to a quick discussion of certain algebraic properties of the natural numbers, and in effect, this represents an application of the material of the preceding chapters. Indeed, we have spent substantial energy developing the predicate calculus and set theory in the previous chapters, and here we take a step back from such foundational material in order to apply it to the study of arithmetic. The first part of our discussion is surely familiar to the reader from previous algebra courses as well as from everyday experience with the natural numbers. In contrast, our final topics are reasonably deep and lead directly to research-level mathematical questions.

The primary intent in presenting this material here is to convince the reader of the real power and efficacy of what has been learned before. Indeed, we apply the methods and constructions of the previous chapters here to a topic, namely arithmetic, which was presumably "mastered" by the reader long ago, and in these few pages, we uncover facts and insights which are surely new. We would thus hope that the reader comes away from this chapter with the conclusion that the methods already covered in this book are of great power and utility. At the same time, the reader might rightly conclude that the passage from basic to research-level mathematics can be quite direct: Beginning from the foundations of set theory and the predicate calculus, this chapter concludes with interesting and unsolved research questions!

Our methods here rely on the Basic Fact considered in Chapter 1 and proved as Theorem 3.7.3 that any non-empty subset of $\mathbb{N}$ has a unique least element.

The reader willing to assume this fact as well as the basic definitions and properties of addition and multiplication should be able to tackle this chapter with no discomfort. A reader demanding more rigor should first study §3.7.

4.1 COMMON MULTIPLES

If $a, b \in \mathbb{Z}$, then we say that b is *divisible* by a and write $a|b$, which is read in English as "a divides b", if there is some $c \in \mathbb{Z}$ so that $b = ac$. In this case, we say that a is a *divisor* of b and that b is a *multiple* of a. In particular, $0 \in \mathbb{Z}$ is a multiple of any $a \in \mathbb{Z}$ since $0 = 0 \cdot a$, and $1 \in \mathbb{N}$ is a divisor of any $a \in \mathbb{Z}$ since $a = a \cdot 1$. As before, we shall also write $a \nmid b$ if a does not divide b.

We shall usually restrict our attention to the case that $a, b \in \mathbb{N}$ in this chapter, but essentially all of our results here apply (in slightly modified form) for pairs of integers. Notice that if $a, b \in \mathbb{N}$ and $a|b$, then $b = ac$, where c is necessarily a natural number rather than just an integer. In fact, we have

4.1.1 LEMMA *If $a, b \in \mathbb{N} \cup \{0\}$ and $a|b$, then either $b = 0$ or $a \leq b$. In particular, if $a, b \in \mathbb{N}$ and $a|b$, then $a \leq b$.*

Proof There is an integer $c \in \mathbb{Z}$ so that $b = ac$ since $a|b$, and if $b \neq 0$, then also $a \neq 0$. Thus, either $b = 0$ or $a, b \in \mathbb{N}$. In the latter case, $c \in \mathbb{Z}$ is actually a natural number as was mentioned above, so $c \geq 1$, and hence $b = ac \geq a$. *q.e.d.*

It follows that if $a \in \mathbb{N}$ and $a|1$, then $a \leq 1$, so in fact $a = 1$, that is, the only natural divisor of 1 is 1 itself.

4.1.2 PROPOSITION: *Suppose that $a, b, c \in \mathbb{N}$.*

(a) *If $a|b$ and $b|c$, then $a|c$.*

(b) *If $a|b$, then $a|(kb)$ for any $k \in \mathbb{N}$.*

(c) *If $c|a$ and $c|b$, then $c|(a + b)$ and $c|(a - b)$.*

Proof For part (a), since $a|b$ and $b|c$ by hypothesis, there are $m, n \in \mathbb{N}$ so that $b = ma$ and $c = nb$. Thus, $c = (mn)a$, so $a|c$. Part (b) follows from part (a) since $a|b$ and $b|(kb)$ for any $k \in \mathbb{N}$. As to part (c), since $c|a$ and $c|b$, there are $m, n \in \mathbb{N}$ so that $a = mc$ and $b = nc$. Thus, $a \pm b = (m \pm n)c$, so $c|(a \pm b)$. *q.e.d.*

Given $a \in \mathbb{N}$, it is easy to find the natural multiples of a, for these multiples are simply $\{a, 2a, 3a, \ldots\}$. In contrast, it is difficult to find the divisors of natural numbers. As we have observed, if $d|a$ where $a, d \in \mathbb{N}$, then $d \leq a$, so to find the divisors of a, we might simply enumerate the numbers $b \leq a$ and check whether $b|a$ or $b \nmid a$. There is thus an obvious finite algorithm for finding the set of divisors of a natural number. On the other hand, this algorithm is inefficient and entirely impractical (even on a big and fast computer) when the number a is very large. In fact, finding divisors of large natural numbers is sufficiently difficult in practice that modern cryptography schemes are sometimes based on divisibility.

Now, suppose that $a_1, \ldots, a_n \in \mathbb{N}$ for some $n \geq 1$ and set $A = \{a_1, \ldots, a_n\}$. A natural number $x \in \mathbb{N}$ so that x is a multiple of a_i, for each $i = 1, 2, \cdots, n$, is called a *common multiple* of the set A. For instance, the product $a_1 \cdots a_n$ is a common multiple of A. Of course, if x is a common multiple of A, then so too is any multiple kx of x where $k \in \mathbb{N}$ by Proposition 4.1.2a.

Thus, the set of all common multiples of $A = \{a_1, \ldots, a_n\}$ is a non-empty subset of $\mathbb{N}$, and this set therefore has a least element by Theorem 3.7.3. This least element is called the *least common multiple* (to be abbreviated simply "*lcm*") of A and is denoted $[a_1, \ldots, a_n]$. For an example, we conclude that $[4, 3] = 12$ by simply observing that no natural number less than 12 is a common multiple of 4 and 3. For another example, consider $d = [3, 4, 18]$, and observe that $[3, 4, 18] = [3, 2^2, 2 \cdot 3^2]$, so some power of 2 divides d and some power of 3 divides d. To find out which powers, notice that 2^2 and 3^2 must divide d, and in fact, $d = 2^2 \cdot 3^2 = 36$ is a common multiple of $\{3, 4, 18\}$. Enumerating the natural numbers less than 36, we conclude that 36 is actually the *lcm*, so $36 = [3, 4, 18]$.

EXERCISES

4.1 COMMON MULTIPLES

1. Compute the least common multiple of each of the following sets.

(a) $\{2, 4, 8\}$ (b) $\{2, 6, 8\}$

(c) $\{5, 7, 12\}$ (d) $\{2, 3, 4, 5, 12\}$

2. Prove that if $c|a$ and $c|b$, then $c|(ma + nb)$ for any integers a, b, c, m, n.

3. Prove that the least common multiple of a non-empty set $A \subseteq \mathbb{N}$ is uniquely determined.

4. Show that if $a, b \in \mathbb{N}$ satsify $a|b$ and $b|a$, then $a = b$.

5. Prove or disprove each of the following for $a, b, c, d \in \mathbb{N}$.

(a) If $b|ac$, then $b|c$. (b) If $a|b$ and $c|d$, then $ac|bd$.

(c) If $a|b$, then $a|bc$. (d) If $a|(b + c)$, then $a|b$ or $a|c$.

6. Prove or disprove each of the following for $a, b, c, d \in \mathbb{N}$.

(a) If $a|c$ and $b|c$, then $ab|c$. (b) If $[a, b] = d = [b, c]$, then $[a, c] = d$.

(c) $a|b$ if and only if $ac|bc$. (d) $2|[a(a + 1)]$ and $3|[a(a + 1)(a + 2)]$

7. Prove each of the following by induction on $n \in \mathbb{N}$.

(a) $7|(2^{3n} - 1)$ (b) $8|(3^{2n} + 7)$ (c) $3|[2^n + (-1)^{n+1}]$

8. Suppose that $a, b \in \mathbb{N}$ and prove each of the following.

(a) $a!b!|(a+b)!$ [HINT: Argue by induction on $a+b$.]

(b) Use part (a) to show that $(a+1)(a+2)\cdots(a+k)$ is divisible by $k!$ for each $a, k \in \mathbb{N}$.

4.2 THE DIVISION ALGORITHM

The main result of this section is of fundamental importance both in our current discussion of divisors and multiples and for later applications. It is also presumably intuitively evident to the reader from everyday experience with natural numbers.

4.2.1 DIVISION ALGORITHM *If $a, b \in \mathbb{N}$, then there are unique $q, r \in \mathbb{N} \cup \{0\}$ so that $b = aq + r$ where $0 \leq r < a$.*

Proof We must prove existence and uniqueness of q, r, and we begin with existence. To this end, define the set

$$S = \{b - ax : x \in \mathbb{N} \cup \{0\} \text{ and } b - ax \geq 0\},$$

and observe that $S \neq \emptyset$ since, for instance, if $x = 0$, then $b - ax = b \geq 0$, so $b \in S$. Applying Theorem 3.7.3 again, we conclude that S has a least element, and we let $r \in S$ denote this least element. By construction, there is some $q \geq 0$ so that $b - aq = r$, where $r \geq 0$.

To finish the proof of existence, we must show that $r < a$. To derive a contradiction, we suppose that $r \geq a$, so $r = a + c$ for some $0 \leq c = r - a < r$. Thus, $b - aq = a + c$, so $c = b - a(q-1) \in S$. Finally, since $c < r$, this contradicts that r is the least element of S.

Turning to the uniqueness of q, r, suppose that $b = aq + r = aq' + r'$, where $0 \leq r, r' < a$. From these equations, $a(q' - q) = (r - r')$, and taking absolute values (letting $|x| \in \mathbb{N} \cup \{0\}$ denote the absolute value of $x \in \mathbb{Z}$), we find $a|(|r - r'|)$, whence either $a \leq |r - r'|$ or $|r - r'| = 0$ by Lemma 4.1.1. On the other hand, $0 \leq r, r' < a$ implies that $|r - r'| < a$, so we have reached a contradiction unless $|r - r'| = 0$. Thus, $r = r'$ and $aq = b - r = b - r' = aq'$, whence we have $q = q'$ as well. *q.e.d.*

The integer $r \geq 0$ in the expression $b = aq + r$, where $0 \leq r < a$ is called the *remainder* on division of b by a, and the letter "q" in this expression stands for "quotient".

As a first application of the Division Algorithm, we have the following fact about common multiples.

4.2.2 THEOREM *Each common multiple of the set $\{a_1, \ldots, a_n\}$ of natural numbers is divisible by the lcm $[a_1, \ldots, a_n]$.*

Proof Let M be a common multiple of $\{a_1, \ldots, a_n\}$ and set $N = [a_1, \ldots, a_n]$. Applying the Division Algorithm, we find $M = qN + r$, where $0 \leq r < N$, so $r = M - qN$. Furthermore, since M and N are common multiples, for each $i = 1, \ldots, n$, there are natural numbers $x_i, y_i \in \mathbb{N}$ so that $M = x_i a_i$ and $N = y_i a_i$. Combining these last equations, we find

$$r = M - qN = (x_i - qy_i)a_i, \text{ for } i = 1, \ldots, n.$$

From these equations, we conclude that $a_i | r$, for $i = 1, \ldots, n$, so r is actually a common multiple of $\{a_1, \ldots, a_n\}$. Finally, $r < N$ by construction, so we have contradicted that N is the *least* common multiple unless $r = 0$ Finally, if $r = 0$, then $M = qN$, so $N|M$ as desired. *q.e.d.*

EXERCISES

4.2 THE DIVISION ALGORITHM

1. Use the Division Algorithm to write $b = aq + r$ where $a, b \in \mathbb{N}$ for each of the following pairs a, b.

(a) $a = 56$ and $b = 11$ (b) $a = 963$ and $b = 657$

(c) $a = 11$ and $b = 56$ (d) $a = 1109$ and $b = 4999$

2. Prove that if $a, b \in \mathbb{N}$ and $a \geq 2b$, then there are unique $q, r \in \mathbb{N}$ satisfying $a = qb + r$ where $b \leq r < 2b$.

3. (a) Show that any odd natural number n may be written either in the form $n = 4k + 1$ or in the form $n = 4k + 3$ for some $k \in \mathbb{N} \cup \{0\}$.

(b) Show that the square of any odd natural number n may be written in the form $n^2 = 8k + 1$ for some $k \in \mathbb{N} \cup \{0\}$.

(c) Show that if $a \in \mathbb{N}$ is odd, then $24|a(a^2 - 1)$.

(d) Show that if $a, b \in \mathbb{N}$ are odd, then $8|(a^2 - b^2)$

4. (a) Show that any natural number n which may be written in the form $n = 6k + 5$ for some $k \in \mathbb{N} \cup \{0\}$ may also be written in the form $n = 3\ell + 2$ for some $\ell \in \mathbb{N} \cup \{0\}$.

(b) Does the converse of part (a) hold? Prove your assertion.

5. Show that the square of any natural number n may be written either in the form $n^2 = 3k$ or in the form $n^2 = 3k + 1$ for some $k \in \mathbb{N} \cup \{0\}$.

6. Show that the cube of any natural number n may be written in one of the forms $n^3 = 9k$, $n^3 = 9k + 1$, or $n^3 = 9k + 8$ for some $k \in \mathbb{N} \cup \{0\}$.

7. Prove that for any $a \in \mathbb{N}$, one of a, $a + 2$, or $a + 4$ is divisible by 3.

8. Prove that for each $n \in \mathbb{N}$, the quantity $n(n+1)(n+2)/6$ is a natural number. [HINT: Any natural n admits one of the expressions $6k, 6k+1, \ldots, 6k+5$.]

4.3 COMMON DIVISORS

Let $S \subseteq \mathbb{N}$ be a given non-empty subset. Every $d \in \mathbb{N}$ so that $d|s$ for each $s \in S$ is called a *common divisor* of S. In particular, 1 is a common divisor of any subset $S \subseteq \mathbb{N}$ since each natural number is divisible by 1. Of course, if d is a common divisor of S and $s \in S$, then $d \leq s$ by Lemma 4.1.1, so the set of all common divisors of S is finite and hence has a greatest element. This greatest element is called the *greatest common divisor* (to be abbreviated simply "*gcd*") of S.

Now, suppose that d is a common divisor of $S \subseteq \mathbb{N}$, and let $d(S)$ denote the *gcd* of S. If $a \in S$, then $d|a$ as well as $d(S)|a$, and therefore a is itself a common multiple of $\{d, d(S)\}$. By Theorem 4.2.2, $[d, d(S)]|a$, so $N = [d, d(S)]$ is a common divisor of S. Thus, $d(S) \geq N$ since $d(S)$ is the *greatest* common divisor.

On the other hand, we have $d(S)|N$ (since $N = [d, d(S)]$ by definition), so $N \geq d(S)$ again by Lemma 4.1.1. We conclude that $d(S) = N = [d, d(S)]$, so $d|d(S)$, and we summarize this discussion with

4.3.1 THEOREM *Suppose $S \subseteq \mathbb{N}$ where $S \neq \emptyset$. There is then a gcd $d(S)$ of S, and if d is a common divisor of S, then $d|d(S)$.*

In particular, if S is a finite set $S = \{a_1 \ldots, a_n\} \subseteq \mathbb{N}$, then we let $(a_1, \ldots, a_n) = d(S)$ denote the *gcd* of S. For some examples (which are again pursued by enumerating the various divisors), we have $(3, 4, 18) = (3, 2^2, 2 \cdot 3^2) = 1$, $(4, 18) = (2^2, 2 \cdot 3^2) = 2$, and $(3, 18) = (3, 2 \cdot 3^2) = 3$.

Our next result relates the *gcd* and *lcm* of a pair $\{a, b\}$ of natural numbers.

4.3.2 THEOREM *For any natural numbers $a, b \in \mathbb{N}$, we have the equality $ab = (a, b)[a, b]$.*

Proof According to Theorem 4.2.2, $[a, b]|ab$, and we set

$$ab = d[a, b], \quad [a, b] = ka, \quad \text{and} \quad [a, b] = \ell b.$$

Thus, $ab = d[a, b] = dka = d\ell b$, so $a = d\ell$ and $b = dk$, and we conclude that d is a common divisor of $\{a, b\}$.

We must show that d is the *greatest* common divisor of a and b, and to this end, suppose that t is another common divisor of a and b. We may write $a = ta_1$, $b = tb_1$, so ta_1b_1 is a common multiple of a and b. By Theorem 4.2.2 again, $[a, b]|ta_1b_1$, so $ta_1b_1 = u[a, b]$ for some natural number u.

Combining the previous equations, we find

$$d[a, b] = ab = t^2a_1b_1 = tu[a, b],$$

so, in fact, $d = tu$. Thus, $t|d$, so $t \leq d$ by Lemma 4.1.1, and d is indeed greatest, as desired. *q.e.d.*

EXERCISES

4.3 COMMON DIVISORS

1. Compute the least common multiple and greatest common divisor of each of the following sets.

(a) $\{16, 36\}$ (b) $\{2, 3, 4, 9\}$

(c) $\{18, 56\}$ (d) $\{4, 6, 12\}$

2. Show that if $a, b \in \mathbb{N}$, then $(a, b) = [a, b]$ if and only if $a = b$.

3. Prove that for all $a, n \in \mathbb{N}$, we have $(a, a + n)|n$.

4. For each natural number n, evaluate $(n, n + 1)$ and $[n, n + 1]$.

5. Suppose that $a, b, c \in \mathbb{N}$ and prove that if $(a, b) = 1$ and $c|(a + b)$, then $(a, c) = (b, c) = 1$.

6. Suppose that $a, b \in \mathbb{N}$ and prove each of the following implications.

(a) If $(a, b) = b$, then $b|a$. (b) If $[a, b] = b$, then $a|b$.

7. Given $a, b, c \in \mathbb{N}$, define $d = (a, b, c)$, and prove the following identities

$$d = ((a, b), c) = (a, (b, c)) = ((a, c), b).$$

8. Suppose that $a, b, c \in \mathbb{N}$ satisfy $a = bx + cy$ for some $x, y \in \mathbb{N}$ and prove that $(b, c)|(a, b)$.

4.4 RELATIVELY PRIME PAIRS

Suppose that a, b are natural numbers. If $(a, b) = 1$, then we say that a and b are *relatively prime*, so for instance, 1 is relatively prime to any natural number. This notion is of fundamental importance in our ensuing discussions, and as an immediate consequence of Theorem 4.3.2, we have

4.4.1 COROLLARY *Given natural numbers a and b, if $(a, b) = 1$, then $ab = [a, b]$.*

The next result is of basic importance and is originally due to Euclid. It is called the "fundamental theorem of arithmetic" by some authors, but we reserve this term for a later theorem, whose proof is based on it.

4.4.2 LEMMA *Given natural numbers $a, b, c \in \mathbb{N}$, if $b|ac$ and $(a, b) = 1$, then $b|c$.*

Proof Since $b|ac$ by hypothesis and obviously $a|ac$, we conclude that $[a, b]|ac$ by Theorem 4.2.2. Corollary 4.4.1 then gives that $ab = [a, b]$, so $ab|ac$, or in other words, $ac = tab$ for some natural number t. Dividing by a, we find $c = tb$, so indeed $b|c$. *q.e.d.*

Prefatory to our final result about relatively prime pairs, observe that given natural numbers a and b, we may compute their *gcd* $d = (a, b)$; there are thus natural numbers $a_1, b_1 \in \mathbb{N}$ satisfying $a = da_1, b = db_1$, and we can write these natural numbers a_1 and b_1 as the respective quotients $a_1 = a/d$ and $b_1 = b/d$.

4.4.3 THEOREM *Given natural numbers a and b, the natural numbers $a/(a, b)$ and $b/(a, b)$ are relatively prime.*

Proof Let $d = (a, b)$ and define $a_1 = a/d, b_1 = b/d$, and let $d_1 = (a_1, b_1)$ and define $a_2 = a_1/d_1, b_2 = b_1/d_1$. Thus, we must show that $d_1 = 1$, and to this end, we compute

$$a = dd_1a_2 \text{ and } b = dd_1b_2,$$

so dd_1 is a common divisor of a and b. Thus, $dd_1|d$, whence $dd_1 \leq d$ by Lemma 4.1.1, so finally $d_1 = 1$. *q.e.d.*

We mention parenthetically that the analogous proof shows that given natural numbers $a_1, a_2, \ldots, a_n$, where $n \geq 2$, we have

$$\left(\frac{a_1}{(a_1, a_2, \ldots, a_n)}, \frac{a_2}{(a_1, a_2, \ldots, a_n)}, \ldots, \frac{a_n}{(a_1, a_2, \ldots, a_n)}\right) = 1.$$

To close this section, we take a break from the main developments and pursue some sample problems.

■ **Example 4.4.1** We prove that if $(a, b) = 1$ and $c|a$, then $(b, c) = 1$ for any natural numbers a, b, c. To see this, set $d = (b, c)$ so $d|b$ and $d|c$. We assume that $c|a$, and it therefore follows that $d|a$. Thus, d is a common divisor of $\{a, b\}$, so $d|(a, b)$ by Theorem 4.3.1. Finally, we also assume that $(a, b) = 1$, so $d|1$, whence $d = 1$, as was claimed.

■ **Example 4.4.2** Given natural numbers a, b, c, we show that if $a|c$, $b|c$, and $(a, b) = 1$, then $ab|c$. Indeed, according to the first two hypotheses, c is a common multiple of a and b, so $[a, b]|c$ by Theorem 4.2.2. On the other hand using the third hypothesis, $[a, b] = ab$ by Corollary 4.4.1, so indeed $ab|c$.

■ **Example 4.4.3** We prove here that for any natural number n, $n!+1$ is relatively prime to $(n+1)!+1$. To prove this, let d be some common divisor of $n!+1$ and $(n+1)!+1$. Thus, d divides any multiple of $n!+1$, so in particular, d divides $(n!+1)(n+1) = ((n+1)!+1)+n$. Since d also divides $(n+1)!+1$ by hypothesis, d divides $n = (n!+1)(n+1) - ((n+1)!+1)$, that is, $d|n$. Since $n|n!$, we conclude that $d|n!$ as well.

Finally, since $d|n!$ and $d|(n!+1)$, d divides the difference $1 = (n!+1) - n!$, that is, $d|1$ so $d = 1$, as was to be shown.

EXERCISES

4.4 RELATIVELY PRIME PAIRS

1. Prove each of the following assertions about products of consecutive natural numbers.
   (a) The product of three consecutive natural numbers is divisible by 6.
   (b) The product of four consecutive natural numbers is divisible by 24.
   (c) The product of five consecutive natural numbers is divisible by 120.
2. Suppose that $a, b, m \in \mathbb{N}$ and prove that if $(a, b) = 1$, then $(a + mb, b) = 1$.
3. Exhibit three natural numbers $a, b, c \in \mathbb{N}$ so that any two are relatively prime, yet $\{a, b, c\}$ has no common divisors other than 1.
4. Suppose that $a, b, c \in \mathbb{N}$ and prove that if $(a, b) = 1$ and $c|(a + b)$, then $(a, c) = 1 = (b, c)$.
5. Suppose that $a, b, c \in \mathbb{N}$ and prove that if $(a, b) = 1$ and $c|a$, then $(b, c) = 1$.

4.5 LINEAR DIOPHANTINE EQUATIONS

Given natural numbers $a, b, c \in \mathbb{N}$, we consider here the equation $ax + by = c$ in the variables x, y. More explicitly, given $a, b, c \in \mathbb{N}$, we wish to find *integers* (not necessarily natural numbers) $x, y \in \mathbb{Z}$ satisfying the equation $ax + by = c$. Of course, we might solve for $x = (c - by)/a$ in terms of y, but there is no guarantee of an integral choice of y so that x is integral as well. In fact, the requirement that x, y be integral is a restrictive one. In general, equations whose solution are required to be integral are called "Diophantine equations" after the Greek (actually, probably Helenized Babylonian) mathematician Diophantus who first studied them; Diophantus is sometimes called the "father of notation" and dates from about 250 AD. The term "linear" refers to the particular form of the equation $ax + by = c$.

4.5.1 THEOREM *If $a, b \in \mathbb{N}$, then there are integers $m_0, n_0 \in \mathbb{Z}$ so that*

$(a,b) = m_0a + n_0b$. Furthermore, $c = (a,b)$ is the least natural number which can be expressed in the form $c = ma + nb$ for integral m, n.

Proof To begin, define the set

$$\mathcal{M} = \{ma + nb : m, n \in \mathbb{Z}\}.$$

Notice that $x = ma + nb \in \mathcal{M}$ if and only if $-x = (-m)a + (-n)b \in \mathcal{M}$, and we therefore consider instead the subset $\mathcal{M}_+ = \mathcal{M} \cap \mathbb{N}$. $\mathcal{M}_+$ is evidently non-empty, so by Theorem 3.7.3, it has a unique least element $c = m_0a + n_0b \in \mathcal{M}_+$.

Now, suppose that $d|a$ and $d|b$. Thus, d divides $c = m_0a + n_0b$, so we have the inclusion

$$\{d \in \mathbb{N} : d|a \text{ and } d|b\} \subseteq \{d \in \mathbb{N} : d|c\},$$

or in other words, any common divisor of a and b is a divisor of c. We shall prove that $c|a$ and $c|b$, so c is itself a common divisor of a and b. Thus,

$$\{d \in \mathbb{N} : d|c\} \subseteq \{d \in \mathbb{N} : d|a \text{ and } d|b\},$$

and we conclude from the previous pair of inset equations that the set of common divisors of a and b is exactly the set of divisors of c. Since these sets are equal, they must have the same greatest element, yet (a,b) is the greatest of the common divisors of a and b by definition, while c is obviously the greatest divisor of c. Thus, $(a,b) = c = m_0a + n_0b$, which would complete the proof.

For the first part of the theorem, it therefore remains only to prove that $c|a$ and $c|b$. To this end, observe that $a = 1 \cdot a + 0 \cdot b \in \mathcal{M}_+$ and $b = 0 \cdot a + 1 \cdot b \in \mathcal{M}_+$, so to complete the proof that $c|a$ and $c|b$, it suffices to prove that $c|x$ for each $x \in \mathcal{M}_+$.

To see this, choose some $x = ma + nb \in \mathcal{M}_+$ and use the Division Algorithm to write $x = tc + r$, where $0 \leq r < c$ and $t \geq 0$. Thus,

$$ma + nb = t(m_0a + n_0b) + r, \quad \text{so} \quad r = (m - tm_0)a + (n - tn_0)b,$$

and we conclude that $r \in \mathcal{M}_+ \cup \{0\}$. On the other hand $c > r > 0$ violates that c is least, so $r = 0$, and it follows that $x = tc$, i.e., $c|x$. Thus, c does indeed divide x for each $x \in \mathcal{M}_+$, so in particular $c|a$ and $c|b$.

This completes the proof that there are integers m_0, n_0 satisfying $(a,b) = m_0a + n_0b$. For the final assertion in the theorem, just observe that $(a,b) = c$ is by definition the least element of $\mathcal{M}_+$. *q.e.d.*

In the special case that a and b are relatively prime, we have

4.5.2 COROLLARY *If $a, b \in \mathbb{N}$, then a and b are relatively prime if and only if $\exists m, n \in \mathbb{Z}\,(ma + nb = 1)$.*

Proof We must prove that if $ma + nb = 1$ for some integers m, n, then a and b are relatively prime, for the converse follows immediately from the previous theorem. In the notation of the proof of the previous theorem, we have $1 = ma + nb \in \mathcal{M}_+$, so 1 must be the least element of $\mathcal{M}_+$. On the other hand, we found above that (a, b) is the least element of $\mathcal{M}_+$, so indeed $(a, b) = 1$, as was asserted. *q.e.d.*

The previous result can be most useful in applications, since the hypothesis that two natural numbers a, b are relatively prime can be replaced by the equation $ma+nb = 1$, where $m, n \in \mathbb{Z}$, and such an equation can be used in computational manipulations. As a simple example of this technique, we prove that if a and b are relatively prime natural numbers, then for any $c \in \mathbb{Z}$, there are $m, n \in \mathbb{Z}$ so that $c = ma + nb$. Indeed, since $(a, b) = 1$, there are $x, y \in \mathbb{Z}$ so that $1 = xa + yb$ by Corollary 4.5.2. Multiplying this equation by c, we find $c = (cx)a + (cy)b$, as required.

We have the following refinement of the previous results.

4.5.3 LEMMA *If $a, b \in \mathbb{N}$ and $(a, b) = 1$, then there are $u, v \in \mathbb{N}$ so that $au - bv = 1$.*

The distinction between Corollary 4.5.2 and Lemma 4.5.3 is that $m, n \in \mathbb{Z}$ in the former while $m, n \in \mathbb{N}$ in the latter.

Proof According to Corollary 4.5.2, there are integers $x_0, y_0 \in \mathbb{Z}$ so that $ax_0 + by_0 = 1$. Choose any $t_0 \in \mathbb{Z}$ satisfying $t_0 > -x_0/b$ and $t_0 > y_0/a$, and define

$$u = x_0 + bt_0 > 0 \quad \text{and} \quad v = -(y_0 - at_0) > 0.$$

Thus, $u, v \in \mathbb{N}$ and

$$au - bv = ax_0 + abt_0 + by_0 = abt_0 = ax_0 + by_0 = 1,$$

as desired. *q.e.d.*

Given $a, b \in \mathbb{N}$, we now consider the various c for which there is a solution $c = ax + by$ for $x, y \in \mathbb{N}$ and have

4.5.4 THEOREM *Suppose $a, b \in \mathbb{N}$ and $(a, b) = 1$. If $c > ab$, then $c = ax + by$ for some $x, y \in \mathbb{N}$. Furthermore, there are no $x, y \in \mathbb{N}$ satisfying $ab = ax + by$.*

Proof We may choose $u, v \in \mathbb{N}$ so that $au - bv = 1$ by the previous lemma. Thus, for $n > ab$, we have $anu - bnv = n > ab$, and so

$$\frac{nu}{b} - \frac{nv}{a} > 1.$$

We may therefore choose an integer $t \in \mathbb{Z}$ so that $nv/a < t < nu/b$, namely, we may take t to be the greatest integer less that nu/b. Define $x - nu - bt > 0$ and $y = at - nv > 0$, so

$$ax + by = a(nu - bt) + b(at - nv) = n,$$

as required.

For the final assertion, if $ab = ax + by$, then $ax = (a - y)b$. Since $(a, b) = 1$, we have $b|x$ by Lemma 4.4.2, so $x \geq b$. Thus, $ab = ax + by \geq ab + by > ab$, which is absurd. *q.e.d.*

We close this section with a reasonably sophisticated result, which is included to communicate the fact that our discussion here just scratches the surface of extremely deep and complicated phenomena.

4.5.6 THEOREM *If $a > 1$ is a natural number and m, n are natural numbers, then we have $(a^m - 1, a^n - 1) = a^{(m,n)} - 1$.*

Proof Define the quantities

$$\delta = (m, n) \quad \text{and} \quad d = (a^m - 1, a^n - 1),$$

so the assertion $d = a^\delta - 1$ follows if we can show that $(a^\delta - 1)|d$ (so $a^\delta - 1 \leq d$) and $d|(a^\delta - 1)$ (so $d \leq a^\delta - 1$). To prove each of these divisibility results, we first require the

4.5.7 CLAIM *For any real number $r \neq 1$ and any natural numbers p and k, we have*

$$(r^p - 1)|(r^{pk} - 1).$$

To prove the claim, simply notice that

$$\begin{aligned}
&(r^p - 1)(r^{p(k-1)} + r^{p(k-2)} + \cdots + r^p + 1) \\
&= r(r^{p(k-1)} + r^{p(k-2)} + \cdots + r^p + 1) \\
&\quad - (r^{p(k-1)} + r^{p(k-2)} + \cdots + r^p + 1) \\
&= r^{pk} + r^{p(k-1)} + \cdots \ + r^p \\
&\qquad\quad - r^{p(k-1)} - \cdots - r^p - 1 \\
&= r^{pk} - 1.
\end{aligned}$$

A sum such as this one, where all but the first and last terms cancel in pairs, is called a "telescoping sum" (suggesting an obvious visual imagery). Of course, one can give a more formal inductive proof of the previous claim, but we prefer the proof above (in keeping with the imagery).

Armed with the claim, we return to the divisibility proofs promised before. Since $\delta | m$ and $\delta | n$ by definition of $\delta = (m, n)$, it follows from the claim that

$$(a^\delta - 1)|(a^m - 1) \text{ and } (a^\delta - 1)|(a^n - 1),$$

so $(a^\delta - 1)|d$ by Theorem 4.3.1 (and the definition of $d = (a^m - 1, a^n - 1)$), whence $a^\delta - 1 \le d$.

For the other inequality and again by definition of δ, there are $m_1, n_1 \in \mathbb{N}$ so that $m = \delta m_1$ and $n = \delta n_1$. Furthermore, $(m_1, n_1) = 1$ by Theorem 4.4.3. By Lemma 4.5.3, there are natural numbers $u, v \in \mathbb{N}$ so that $m_1 u - n_1 v = 1$, and multiplying this equation by δ, we find that $\delta = mu - nv$. Since $d|(a^m - 1)$ and $d|(a^n - 1)$ by definition of $d = (a^m - 1, a^n - 1)$, we find that

$$d|(a^{mu} - 1) \text{ and } d|(a^{nv} - 1)$$

again applying the claim. Thus, d must divide the difference

$$(a^{mu} - 1) - (a^{nv} - 1) = a^{mu} - a^{nv} = a^{nv}(a^{mu-nv} - 1) = a^{nv}(a^\delta - 1).$$

Finally, as we noted above, $d|(a^{nv} - 1)$, so there is some $k \in \mathbb{N}$ so that $kd = a^{nv} - 1$. Thus, $1 = a^{nv} - kd$, so $(d, a^{nv}) = 1$ by Corollary 4.5.2. Since $d|a^{nv}(a^\delta - 1)$ and $(d, a^{nv}) = 1$, it follows from Lemma 4.4.2 that $d|(a^\delta - 1)$, whence $d \le a^\delta - 1$. *q.e.d.*

EXERCISES

4.5 LINEAR DIOPHANTINE EQUATIONS

1. Determine whether each of the following Diophantine equations has an integral solution $x, y \in \mathbb{Z}$, and if so, give a solution.

(a) $243x + 198y = 9$ (b) $56x + 138y = 18$

(c) $71x - 50y = 12$ (d) $84x - 438y = 156$

2. Determine whether each of the following Diophantine equations has a natural solution $x, y \in \mathbb{N}$, and if so, give a solution.

(a) $123x + 360y = 99$ (b) $93x - 81y = 3$

(c) $158 - 57y = 7$ (d) $84x - 438y = 156$

3. Suppose that $a, b, c \in \mathbb{N}$ and prove that if $(a, b) = 1 = (a, c)$, then $(a, bc) = 1$.

4. Suppose that $a, b, c \in \mathbb{N}$ and prove that if $(a, b) = 1$, then $(ac, b) = (b, c)$.

5. Suppose that $a, b, c \in \mathbb{N}$ and prove that if $a|bc$, then $a|(a, b)c$.

6. Prove that $(ma, mb) = m(a, b)$ for each $a, b, m \in \mathbb{N}$; conclude that furthermore $[ma, mb] = m[a, b]$.

7. Prove that $(n-1)|(n^k-1)$ for each $k, n \in \mathbb{N}$.

8. Suppose that $k, n \in \mathbb{N}$. Prove that $(n-1)^2|(n^k-1)$ if and only if $(n-1)|k$. [HINT: Write $n^k = [(n-1)+1]^k$.]

9. Let $a, m, n \in \mathbb{N}$ and prove each of the following.

(a) If $m > n$, then $(a^{2n}+1)|(a^{2m}-1)$.

(b) If $m \neq n$, then $(a^{2m}+1, a^{2n}+1)$ has value 1 if a is even and has value 2 if a is odd.

4.6 THE FUNDAMENTAL THEOREM OF ARITHMETIC

We say that a natural number $p \in \mathbb{N}$ is *prime* if it has exactly two natural divisors, namely 1 and p. In contrast, a natural number $n \in \mathbb{N}$ is said to be *composite* if it may be written $n = ab$ for some $a, b \in \mathbb{N}$, where $\{a, b\} \neq \{1, p\}$. In particular, 1 is neither prime (since it has only one divisor), nor composite. On the other hand, by definition every natural number other than 1 is either prime or composite, but not both.

4.6.1 LEMMA *If n is a natural number and p a prime, then either $p|n$ or $(p, n) = 1$.*

Proof Let $d = (p, n)$, so, in particular, $d|p$ and $d|n$. Since p is prime and $d|p$, either $d = p$ or $d = 1$. If $d = p$, then we have $p|n$, whereas if $d = 1$, then we have $(p, n) = 1$. *q.e.d.*

4.6.2 COROLLARY *Suppose that p is prime, $a_1, \ldots a_n \in \mathbb{N}$ for $n \geq 2$, and $p|(a_1 a_2 \cdots a_n)$. Then $p|a_i$ for some $i = 1, \ldots, n$.*

Proof The proof is by induction on n. For the basis step $n = 2$, we have $p|a_1a_2$, so by the previous lemma, either $p|a_1$ or $(p, a_1) = 1$. In the former case, the corollary holds trivially (i.e., the implication is trivial), and in the latter case, we apply Lemma 4.4.2 to conclude that $p|a_2$. This completes the basis step.

For the induction step, suppose that $p|a_1 \cdots a_n a_{n+1}$, so $p|(a_1 \cdots a_n)a_{n+1}$. Applying the case $n = 2$, we conclude that either $p|a_{n+1}$ or $p|a_1 \cdots a_n$, and in the latter case, the induction hypothesis implies that $p|a_i$ for some $i = 1, \ldots, n$, completing the proof. *q.e.d.*

We have seen in Example A.5.8 of Chapter 1 the proof (using strong induction) that any natural number $a > 1$ can be written as a product $a = p_1 p_2 \cdots p_m$ of primes $p_1, p_2, \ldots, p_m$. Our main goal for this section is to show that this expression is "essentially" unique in a sense which we next make precise. Since multiplication of natural numbers is commutative, this expression is certainly not unique in general, for we may simply re-order the primes (writing for instance $12 = 2 \cdot 3 \cdot 2$ or $12 = 3 \cdot 2 \cdot 2$). On the other hand, we may re-order the primes p_i once and for all in such a way that $p_1 \leq p_2 \leq \cdots \leq p_m$ (writing for instance $12 = 2 \cdot 2 \cdot 3$). Requiring this ordering on the primes evidently uniquely determines the ordered set of primes $p_1 \leq \cdots \leq p_m$.

Our main result in this section is the following theorem, which is often called the "Fundamental Theorem of Arithmetic".

4.6.3 THEOREM *Every natural number $a > 1$ can be expressed uniquely as a product $a = p_1 p_2 \cdots p_m$ of primes $p_1 \leq \cdots \leq p_m$. That is, if $a = q_1 q_2 \cdots q_n$ is another expression of a as a product of primes, where $q_1 \leq \cdots \leq q_n$, then $m = n$ and $p_1 = q_1, \ldots, p_m = q_m$.*

Proof As remarked above, we have proved existence in Example A.5.8 in Chapter 1, and we prove uniqueness here by induction on m. For the basis step $m = 1$, we find $p_1 = q_1 q_2 \cdots q_n$. By the previous corollary, $p_1 = q_j$ for some $j = 1, \ldots, n$, whence $p_1 \leq q_n$. On the other hand, $q_n \leq p_1$ (since $q_n | p_1$), so we conclude that $p_1 = q_n$. Therefore, we have $1 = q_1 q_2 \cdots q_{n-1}$, so each $q_j | 1$, whence $q_j = 1$, for each $j = 1, \ldots, n-1$. This contradicts that each q_j is prime unless $n = 1$ and $p_1 = q_1$.

For the induction step, we assume that if $p_1 p_2 \cdots p_m = q_1 q_1 \cdots q_n$, then $m = n$ and $p_1 = q_1, \ldots, p_m = q_n$. Letting $p_1 p_2 \ldots p_{m+1} = q_1 q_2 \cdots q_n$, we conclude as before that $p_{m+1} = q_j$ for some j. Since $q_j \leq q_n$, we find that $p_{m+1} \leq q_n$. In the same way, we conclude that $q_n \leq p_{m+1}$, so $p_{m+1} = q_n$. It follows, therefore, that $p_1 p_2 \cdots p_m = q_1 q_2 \cdots q_{n-1}$, so the induction hypothesis gives that $m = n - 1$ and $p_1 = q_1, \ldots p_m = q_{n-1}$. Thus, $m+1 = n$ and $p_1 = q_1, \ldots, p_{m+1} = q_n$. *q.e.d.*

The unique expression $a = p_1 p_2 \cdots p_m$, where $p_1 \leq \cdots \leq p_m$, is called the *prime factorization* of $a \geq 1$. An equivalent but often more convenient notation for the prime factorization is to write $a = q_1^{\alpha_1} q_2^{\alpha_2} \cdots q_r^{\alpha_r}$, where the primes are distinct $q_1 < \cdots < q_r$, each exponent $\alpha_i \geq 1$, and, of course, $\{p_1, \ldots, p_m\} = \{q_1, \ldots, q_r\}$. As before, the expression $a = q_1^{\alpha_1} q_2^{\alpha_2} \cdots q_r^{\alpha_r}$ is unique, where $q_1 < \cdots < q_r$ and $\alpha_i \geq 1$ for each $i = 1, \ldots, r$.

We turn finally to the problem of computing the *gcd* and *lcm* of natural numbers a and b given their prime factorizations. Combining the primes occurring as factors of a with those occurring for b, we may write

$$a = p_1^{\alpha_1} p_2^{\alpha_2} \cdots p_n^{\alpha_n},$$
$$b = p_1^{\beta_1} p_2^{\beta_2} \cdots p_n^{\beta_n},$$

where we take $\alpha_i = 0$ (so $p_i^{\alpha_i} = 1$) if $p_i \nmid a$ and $\beta_i = 0$ (so $p_i^{\beta_i} = 1$) if $p_i \nmid b$, for $i = 1, \ldots, n$.

If p is a prime dividing (a, b), then (since $(a, b)|a$ and $(a, b)|b$) we have $p = p_i$ for some i with $\alpha_i, \beta_i > 0$. Furthermore, for each prime p_i, we evidently have

$p_i^{\gamma_i}|d$, where γ_i is the minimum of α_i and β_i, and, in fact, γ_i is the largest power of p_i which divides both a and b, for $i = 1, \ldots, n$. We conclude that

$$(a, b) = p_1^{\gamma_1} p_2^{\gamma_2} \cdots p_n^{\gamma_n}.$$

It follows in a similar way that

$$[a, b] = p_1^{\delta_1} p_2^{\delta_2} \cdots p_n^{\delta_n},$$

where δ_i is the maximum of α_i and β_i. Notice, in particular, that these formulas for (a, b) and $[a, b]$ give another proof that $ab = (a, b)[a, b]$.

Thus, (a, b) and $[a, b]$ are easily computed from the prime factorizations of a and b. For instance, given

$$a = 73500 = 2^2 \cdot 3 \cdot 5^3 \cdot 7^2 \text{ and } b = 6174 = 2 \cdot 3^2 \cdot 5^0 \cdot 7^3,$$

we find

$$(a, b) = 2 \cdot 3 \cdot 5^0 \cdot 7^2 = 294 \text{ and } [a, b] = 2^2 \cdot 3^2 \cdot 5^3 \cdot 7^3 = 1543500.$$

At the risk of dispelling possible enthusiasm for this method of computing (a, b) and $[a, b]$, we remind the reader that computing the prime factorizations of a and b is a formidable problem (even on the computer), yet this data is the starting point for our computation of (a, b) and $[a, b]$ above. The next section is dedicated to a method of computing (a, b) and $[a, b]$ when the prime factorizations of a and b are not known.

EXERCISES

4.6 THE FUNDAMENTAL THEOREM OF ARITHMETIC

1. Compute the greatest common divisor (a, b) and least common multiple $[a, b]$ for each of the following pairs a, b.

(a) $a = 56$ and $b = 72$ (b) $a = 138$ and $b = 24$

(c) $a = 243$ and $b = 198$ (d) $a = 93$ and $b = 81$

2. Show that for each $n \in \mathbb{N}$, the sequence

$$(n+1)! + 2, (n+1)! + 3, \ldots, (n+1)! + (n+1)$$

of consecutive natural numbers are all composite. (Thus, there are necessarily larger and larger "gaps" among the natural numbers between primes.)

3. Suppose that $n \in \mathbb{N} - \{1\}$ has the prime decomposition $n = p_1^{\alpha_1} p_2^{\alpha_2} \cdots p_k^{\alpha_k}$. Prove that $n = m^2$ is the square of a natural number $m \in \mathbb{N}$ if and only if each α_i is even.

4. Let $a, b \in \mathbb{N}$ be natural numbers and prove each of the following.

(a) If $(a, b) = 1$, then $(a^n, b^n) = 1$.

(b) If $a^n | b^n$, then $a | b$.

5. Suppose that $p \in \mathbb{N}$ is prime and let $a \in \mathbb{N}$. Prove that $(a, p) = 1$ if and only if p does not divide a.

6. Suppose that $(a, b) = p$, where $a, b, p \in \mathbb{N}$ and p is prime. What are the possible values of (a^2, b), (a^3, b), and (a^2, b^3)? Prove your assertions.

7. Suppose that $a \in \mathbb{N}$ is a composite number, so it may be written in the form $a = bc$ for $b, c \in \mathbb{N}$, where $1 < b < a$ and $1 < c < a$. We may re-label b and c, if necessary, in order to insure that $b \leq c$, and therefore $b^2 \leq bc = a$, so $b \leq \sqrt{a}$. According to the Fundamental Theorem of Arithmetic, b must have at least one prime factor p, and $p \leq b \leq \sqrt{a}$. We conclude that a composite number a always has a prime divisor satisfying $p \leq \sqrt{a}$. These remarks lead to the following method of enumerating primes, called the *Sieve of Eratosthenes* after Eratosthenes of Cyrene (276-194 B.C.). To enumerate all of the prime numbers less than or equal to a given $n \in \mathbb{N}$, one writes down all of the numbers $2, 3, \ldots, n$ in order, and serially crosses out all the multiples $2p$, $3p$, $4p, \ldots$ of the primes $p \leq \sqrt{n}$. Those numbers which do not get crossed out (i.e., do not "fall through the sieve") are precisely the primes less than n. For instance, to enumerate the primes less than $n = 30$, we serially cross out multiples of the primes $2, 3, 5 \leq \sqrt{30} \approx 5.4$ as follows

	2	3	~~4~~	5	~~6~~
7	~~8~~	~~9~~	~~10~~	11	~~12~~
13	~~14~~	~~15~~	~~16~~	17	~~18~~
19	~~20~~	~~21~~	~~22~~	23	~~24~~
~~25~~	~~26~~	~~27~~	~~28~~	29	~~30~~

to find the primes $2, 3, 5, 7, 11, 13, 17, 19, 23, 29 < 30$.

(a) Enumerate the prime numbers less than 100.

(b) Enumerate the prime numbers less than 400.

8. Suppose that $a, b, c \in \mathbb{N}$ and $p \in \mathbb{N}$ is prime. Prove or disprove each of the following.

(a) If $(a, b) = (a, c)$, then also $[a, b] = [a, c]$.

(b) If $a^3 | b^3$, then $a | b$.

(c) If $(a, b) = (a, c)$, then $(a^2, b^2) = (a^2, c^2)$.

(d) If $a^3 | b^2$, then $a | b$.

(e) If $p | a$ and $p | (a^2 + b^2)$, then $p | b$.

(f) If $a^2 | b^3$, then $a | b$.

(g) If $a | (b^2 \pm 1)$, then $a | (b^4 \pm 1)$.

(h) If $p | a^5$, then $p | a$.

9. Suppose that $a, a_i \in \mathbb{N}$ satisfy $(a, a_i) = 1$ for each $i = 1, 2, \ldots, n$. Prove that $(a_1, a_2, \ldots, a_n, a) = 1$.

10. Give complete proofs of the formulas in the text for (a, b) and $[a, b]$ in terms of the prime factorizations of natural numbers a and b.

11. Prove that for any natural numbers a, b, c, we have the identities

$$[[a, b], c] = [[b, c], a] = [[c, a], b] = [a, b, c].$$

4.7 THE EUCLIDEAN ALGORITHM

We describe here a useful algorithm for computing the *gcd* (a, b) given the natural numbers $a, b \in \mathbb{N}$, and the theory underlying the algorithm is as follows. According to the Division Algorithm 4.2.1, we may uniquely write $a = qb + r$ for some $0 \leq r < b$. It follows that every common divisor of a and b is also a divisor of $r = a - qb$, and moreover, every common divisor of r and b is a divisor of $a = qb + r$. Thus, the set of common divisors of $\{a, b\}$ is identical with the set of common divisors of $\{b, r\}$, so these sets have the same greatest elements, or, in other words, we have proved that $(a, b) = (b, r)$.

Now, set $n_0 = a, n_1 = b, n_2 = r$, so we seek (n_0, n_1) and have proved that $(n_0, n_1) = (n_1, n_2)$. If $n_2 = 0$, then $n_1 | n_0$, so $(n_0, n_1) = n_1$, and we have computed the desired *gcd*. On the other hand, if $n_2 \neq 0$, we can again apply the Division Algorithm, this time to n_1, n_2. Continuing in this way as long as the remainder in the Division Algorithm is non-zero, we find the sequence

$$\begin{aligned} n_0 &= q_1 n_1 + n_2 \\ n_1 &= q_2 n_2 + n_3 \\ &\vdots \\ n_{k-1} &= q_k n_k + n_{k+1} \\ &\vdots \end{aligned}$$

of equations, where $(n_0, n_1) = (n_1, n_2) = (n_2, n_3) = \cdots = (n_{k-1}, n_k)$ as was proved above. Furthermore, $n_{i+1} < n_i$ (since n_{i+1} is a remainder on division by n_i), so $n_1 > n_2 > \cdots > n_k > \cdots$. It follows that the sequence above must terminate, so eventually some remainder must vanish. Let us assume that the sequence above terminates with the equation $n_{k-1} = q_k n_k + 0$, so $n_{k+1} = 0$. Thus, $(n_{k-1}, n_k) = n_k$, and so

$$(a, b) = (n_0, n_1) = \cdots = (n_{k-1}, n_k) = n_k.$$

This discussion gives an effective algorithm, called the *Euclidean Algorithm*, for computing (a, b) and hence $[a, b] = ab/(a, b)$ as well from a and b, and we next give some examples of its application. Notice that the Euclidean Algorithm computation of (a, b) is especially amenable to implementation on the computer.

■ **Example 4.7.1** We compute $(115, 25) = (5 \cdot 23, 5^2) = 5$ using the Euclidean Algorithm:

$$\begin{aligned} 115 &= 4 \cdot 25 + 15 & n_0 &= q_1 n_1 + n_2 \\ 25 &= 1 \cdot 15 + 10 & n_1 &= q_2 n_2 + n_3 \\ 15 &= 1 \cdot 10 + 5 & n_2 &= q_3 n_3 + n_4 \\ 10 &= 2 \cdot 5 \ \ + 0 & n_3 &= q_4 n_4 + \ 0, \end{aligned}$$

so $(115, 25) = n_4 = 5$, as expected.

Notice that in general, if $a < b$, then the first application of the Division Algorithm to the computation of (a, b) simply interchanges a and b. For instance, the first two steps of the Euclidean Algorithm applied to $(25, 115)$ give:

$$\begin{aligned} 25 &= 0 \cdot 115 + 25 & n_0 &= q_1 n_1 + n_2 \\ 115 &= 4 \cdot 25 + 15 & n_1 &= q_2 n_2 + n_3 \end{aligned}$$

■ **Example 4.7.2** We compute here $(31415, 10000)$ by the Euclidean Algorithm:

$$\begin{aligned} 31415 &= 3 \cdot 10000 + 1415 \\ 10000 &= 7 \cdot 1415 + 95 \\ 1415 &= 14 \cdot 95 + 85 \\ 95 &= 1 \cdot 85 + 10 \\ 85 &= 8 \cdot 10 + 5 \\ 10 &= 2 \cdot 5 + 0, \end{aligned}$$

so $(31415, 10000) = 5$.

In fact, the Euclidean Algorithm is quite an efficient method of computing (a, b). Indeed, according to a theorem of Lamé (whose proof is beyond the scope of our discussion here), the number of divisions in the Euclidean Algorithm applied to (a, b) is at most six times the number of digits in the decimal expansion of a or b, whichever is larger. In contrast to this efficiency of the Euclidean Algorithm, our next example gives an infinite sequence $f_1, f_2, \cdots$ of natural numbers with the property that the Euclidean Algorithm applied to (f_{n+2}, f_{n+1}) requires n divisions. Furthermore, $(f_{n+2}, f_{n+1}) = 1$, for each $n \geq 1$. These numbers f_n for $n \geq 1$ are called the *Fibonacci numbers* and are defined in the next example.

■ **Example 4.7.3** Begin by setting $f_1 = f_2 = 1$, and inductively define

$$f_n = f_{n-1} + f_{n-2}, \text{ for } n > 2,$$

thus, $f_1, f_2, f_3, \ldots = 1, 1, 2, 3, 5, 8, 13, 21 \ldots$. According to the very definition of these numbers, we have

$$
\begin{aligned}
f_{n+2} &= 1 \cdot f_{n+1} + f_n \\
f_{n+1} &= 1 \cdot f_n + f_{n-1} \\
&\vdots \\
f_4 &= 1 \cdot f_3 + f_2 \\
f_3 &= 2 \cdot f_2 + 0
\end{aligned}
$$

Thus, $(f_{n+2}, f_{n+1}) = f_2 = 1$ for each n, and the Euclidean Algorithm requires n divisions as was claimed.

The Fibonacci numbers are a remarkable sequence of numbers: Sunflower florets, pine cones, pineapples, and the mating dances of bees all exhibit properties described by the Fibonacci numbers. For instance, the seeds of sunflowers occur in patterns consisting of two sets of spirals, where one set of spirals turns clockwise and the other turns counter-clockwise; the number of clockwise and counter-clockwise spirals are usually consecutive Fibonacci numbers! In fact, the consecutive Fibonacci numbers occurring for sunflower seeds are usually 34 and 55.

These numbers were first studied in 1202 by Leonardo of Pisa, who is better known as Fibonacci, and the problem which Fibonacci posed that led him to these numbers is as follows: suppose that a pair of adult rabbits produce a pair of baby rabbits each month and after two months, the baby rabbits are themselves adults capable of producing another pair of babies. The reader may check that after n months, there are f_n adult pairs and f_{n+1} baby pairs for each $n \geq 1$ starting with one pair of adult rabbits, where f_n denotes the nth Fibonacci number. The sequence of Fibonacci numbers will be discussed further in Chapter 8.

In fact, the Euclidean Algorithm is also useful for solving the linear Diophantine equation $(a, b) = xa + yb$, that is, given $a, b \in \mathbb{N}$, one can effectively compute $x, y \in \mathbb{Z}$ satisfying the equation as we next indicate. Set $n_0 = a, n_1 = b$, and apply the Euclidean Algorithm to the computation of (a, b):

$$
(*) \qquad \begin{cases}
n_0 = q_1 n_1 + n_2 \\
n_1 = q_2 n_2 + n_3 \\
\qquad \vdots \\
n_{k-3} = q_{k-2} n_{k-2} + n_{k-1} \\
n_{k-2} = q_{k-1} n_{k-1} + n_k \\
n_{k-1} = q_k n_k + 0,
\end{cases}
$$

where $(n_0, n_1) = n_k$ and $n_1 > n_2 > \cdots > n_k > 0$.

The next-to-last equation in $(*)$ gives

$$(a, b) = n_k = n_{k-2} - q_{k-1} n_{k-1},$$

the previous equation gives

$$n_{k-1} = n_{k-3} - q_{k-2}n_{k-2},$$

and substituting this expression for n_{k-1}, we find

$$\begin{aligned} n_k &= n_{k_2} - q_{k-1}(n_{k-3} - q_{k-2}n_{k-2}) \\ &= -q_{k-1}n_{k-3} + (1 + q_{k-1}q_{k-2})n_{k-2}. \end{aligned}$$

We might then substitute n_{k-2} from the previous equation, and so on. After $k-2$ such substitutions, we find $(a, b) = n_k = xn_0 + yn_1 = xa + yb$, and this describes the required algorithm for solving the Diophantine equation.

Unlike the beautiful simplicity of computing (a, b) using the Euclidean Algorithm, this algorithm is rather awkward to use (though it is again easy to implement on the computer), and we next give an example.

■ **Example 4.7.4** To begin, we compute $(5, 12)$:

$$\begin{aligned} 5 &= 0 \cdot 12 + 5 \qquad & n_0 &= q_1n_1 + n_2 \\ 12 &= 2 \cdot 5 + 2 & n_1 &= q_2n_2 + n_3 \\ 5 &= 2 \cdot 2 + 1 & n_2 &= q_3n_3 + n_4 \\ 2 &= 2 \cdot 1 + 0 & n_3 &= q_4n_4 + 0 \end{aligned}$$

so $(5, 12) = 1$, and we wish to find $x, y \in \mathbb{Z}$ so that $xa + yb = 1$. To this end, we compute

$$\begin{aligned} 1 = n_4 &= n_2 - q_3n_3 \\ &= n_2 - q_3(n_1 - q_2n_2) \\ &= (1 + q_2q_3)n_2 - q_3n_1 \\ &= -q_3n_1 + (1 + q_2q_3)(n_0 - q_1n_1) \\ &= (-q_3 - q_1 - q_1q_2q_3)n_1 + (1 + q_2q_3)n_0. \end{aligned}$$

Substituting in the values computed earlier, we find

$$1 = (-2 - 0 - 0)n_1 + (1 + 4)n_0 = (-2)12 + (5)5,$$

as desired.

Yuck! A more elegant and simpler algorithm for the same purpose is as follows. Adopt the notation of (*) above, set $s_0 = 1, t_0 = 0, s_1 = 0, t_1 = 1$ and recursively define

$$s_n = s_{n-2} - q_{n-1}s_{n-1} \text{ and } t_n = t_{n-2} - q_{n-1}t_{n-1}, \quad \text{for } n \geq 2.$$

A solution to the linear Diophantine equation is then given by $(a, b) = s_ka + t_kb$. We have presented this second procedure in order to emphasize to the reader that for a given purpose there are typically many different algorithms, some of which are more or less awkward, economical, or elegant than others.

EXERCISES

4.7 THE EUCLIDEAN ALGORITHM

1. Use the Euclidean Algorithm to compute the greatest common divisor (a, b) of the following pairs a, b.

(a) $a = 172$ and $b = 20$ (b) $a = 2947$ and $b = 2464$

(c) $a = 1109$ and $b = 4999$ (d) $a = 2689$ and $b = 4001$

2. Find the following least common multiples using the Euclidean Algorithm.

(a) $[50, 52]$ (b) $[284, 1248]$

(c) $[184, 26]$ (d) $[94, 52]$

3. Determine an integral solution $x, y \in \mathbb{Z}$ to each of the following Diophantine equations

(a) $243x + 198y = 9$ (b) $71x - 50y = 1$

(c) $81x - 93y = 12$ (d) $46x + 64y = 8$

4. Compute $d = (1819, 3587)$ using the Euclidean Algorithm and find an integral solution $x, y \in \mathbb{Z}$ to the equation $1819x + 3587y = d$.

5. Compute $d = (1918, 3581)$ using the Euclidean Algorithm and find an integral solution $x, y \in \mathbb{Z}$ to the equation $1918x + 3581y = d$.

6. Let $a, b, k \in \mathbb{N}$ and apply the Euclidean Algorithm to the computation of (ka, kb) in order to prove that $(ka, kb) = k(a, b)$.

7. Prove that the Fibonacci numbers indeed solve the problem about rabbit populations described in the text.

8. Given natural numbers a, b, c, how might one use the Euclidean algorithm to compute (a, b, c)? Prove your assertions.

9. Prove that the algorithm presented after Example 4.7.4 does indeed give a solution to the required Diophantine equation.

(*) *4.8 CONTINUED FRACTIONS*

Here we briefly discuss another aspect of the Euclidean Algorithm which gives an interesting representation of numbers. In order to conform to standard notation in this setting, we modify our notation for the system (*) of equations in the previous section by setting $m = k - 1$ and $a_i = q_{i+1}$ for each $i = 0, 1, \ldots, m$, so the system $(*)$ becomes

$$(\dagger) \quad \begin{cases} n_0 = a_0 n_1 + n_2 \\ n_1 = a_1 n_2 + n_3 \\ \vdots \\ n_{m-2} = a_{m-2} n_{m-1} + n_m \\ n_{m-1} = a_{m-1} n_m + n_{m+1} \\ n_m = a_m n_{m+1} + 0, \end{cases}$$

Manipulating these equations, we find

$$\frac{n_m}{n_{m+1}} = a_m$$

$$\begin{aligned} \frac{n_{i-1}}{n_i} &= a_{i-1} + \frac{n_{i+1}}{n_i} \\ &= a_{i-1} + \frac{1}{n_i/n_{i+1}}, \text{ for } i = 1, \ldots, m, \end{aligned}$$

and substituting we find

$$(\ddagger) \quad \frac{n_0}{n_1} = a_0 + \cfrac{1}{a_1 + \cfrac{1}{\ddots + \cfrac{1}{a_{m-1} + \cfrac{1}{a_m}}}}.$$

This algebraic expression for the rational number n_0/n_1 is called the *continued fraction expansion* of $n_0/n_1 \in \mathbb{Q}$, but we shall informally refer to the sequence $a_0, a_1, \ldots, a_m$ itself simply as the "continued fraction expansion" of n_0/n_1. To simplify the notation, we shall write $\nu(a_0, a_1, \ldots, a_m) \in \mathbb{Q}$ to denote the rational number n_0/n_1 determined by the sequence $a_0, a_1, \ldots, a_m$ in accordance with equation $(\ddagger)$. Notice that whereas a_0 may vanish, each a_i is actually a natural number for each $i = 1, 2, \ldots, m$.

■ **Example 4.8.1** We compute the continued fraction expansion of the rational number 22/7, and the Euclidean Algorithm gives

$$\begin{aligned} 22 &= 3 \cdot 7 + 1 \\ 7 &= 7 \cdot 1 + 0, \end{aligned}$$

whence $22/7 = 3 + 1/7 = \nu(3, 7)$ has continued fraction expansion 3,7.

■ **Example 4.8.2** The rational number 31415/10000 has continued fraction expansion 3,7,14,1,8,2 by Example 4.7.2, so we have

$$\begin{aligned}\frac{31415}{10000} &= \nu(3,7,14,1,8,2) \\ &= 3 + \cfrac{1}{7 + \cfrac{1}{14 + \cfrac{1}{1 + \cfrac{1}{8 + \cfrac{1}{2}}}}}.\end{aligned}$$

■ **Example 4.8.3** We leave it for the reader to check that the quotient f_{n+1}/f_n of consecutive Fibonacci numbers discussed in Example 4.7.3 has continued fraction expansion $1, 1, \ldots, 1$ with a grand total of n 1's, that is, we have

$$f_{n+1}/f_n = \nu(\underbrace{1, 1, \ldots, 1}_{n \text{ times}}).$$

Turning things around, suppose that $a_0 \in \mathbb{Z}$ and $a_1, a_2, \ldots, a_m \in \mathbb{N}$ is an arbitrary sequence of numbers, and use equation (‡) to produce a corresponding rational number $\nu(a_0, a_1, \ldots, a_m) \in \mathbb{Q}$. Thus, any such sequence $a_0, a_1, \ldots, a_m$ itself determines a rational number, and we refer to such a sequence itself as a "continued fraction". The number a_i for $i = 0, 1, \ldots, m$ is called the ith *convergent* of the continued fraction $a_0, a_1, \ldots, a_m$. There are always two ways to represent a given rational number as a continued fraction, where one such representation has an even number of convergents, and the other has an odd number of convergents. Indeed, it follows directly from the definition (‡) that

$$\text{if } a_m \geq 2, \text{ then } \nu(a_0, a_1, \ldots, a_m) = \nu(a_0, a_1, \ldots, a_m - 1, 1),$$

while

$$\text{if } a_m = 1, \text{ then } \nu(a_0, a_1, \ldots, a_{m-1}, 1) = \nu(a_0, a_1, \ldots, a_{m-2}, a_{m-1} + 1).$$

In fact, there are always exactly two continued fractions representing a given rational number, but we shall not prove this here.

The situation becomes much more interesting and complicated if we consider an arbitrary *real* number and attempt to construct a sequence analogous to the continued fraction expansion of a rational number. To accomplish this, given $x \in \mathbb{R}$, we let $[x]$ denote the largest integer not exceeding x, so for instance, we have

$$[\frac{5}{2}] = 2, \quad [\frac{1}{2}] = 0, \quad \text{and} \quad [-\frac{3}{2}] = -2.$$

The notation $[x]$ is completely standard in mathematics, and $[x]$ is called simply the *integral part* of x. According to the definition, we have

$$x = [x] + f \quad \text{where} \quad 0 \leq f < 1$$

for any $x \in \mathbb{R}$.

To define the sequence for $x \in \mathbb{R}$ which is analogous to the continued fraction expansion of a rational number, we begin by setting $a_0 = [x] \in \mathbb{Z}$ and therefore have

$$x = a_0 + \xi_0, \quad 0 \leq \xi_0 < 1.$$

If $\xi_0 \neq 0$, then we may write

$$a_1' = \frac{1}{\xi_0}, \quad a_1 = [a_1'], \quad a_1' = a_1 + \xi_1, \quad 0 \leq \xi_1 < 1.$$

If $\xi_1 \neq 0$, then we may similarly write

$$\frac{1}{\xi_1} = a_2' = a_2 + \xi_2, \quad 0 \leq \xi_2 < 1$$

and so on to define the sequence $a_0, a_1, a_2, \ldots$. Notice that $a_n' = 1/\xi_{n-1} > 1$, so $a_n \geq 1$ for each $n \geq 1$. Of course, it may happen that this sequence terminates if ξ_N vanishes for some N, and in this case, the real number x is necessarily rational

$$x = \nu(a_0, a_1, \ldots, a_N).$$

The sequence $a_0, a_1, a_2, \ldots$ associated in this way with the real number x is called the *continued fraction expansion* of $x \in \mathbb{R}$, and it terminates, i.e., it is a finite sequence, if and only if x is actually a rational number.

Turning this around as before, suppose that $a_0 \in \mathbb{Z}$ and $a_1, a_2, \ldots \in \mathbb{N}$ is an arbitrary such sequence of numbers. We might ask whether there is some $x \in \mathbb{R}$ so that the given sequence is actually the continued fraction expansion of x. In fact, any sequence $a_0, a_1, a_2, \ldots$ is actually the continued fraction expansion of a real number x; furthermore, if the sequence does not terminate, i.e., if the corresponding $x \in \mathbb{R}$ is not rational, then it is the unique such sequence representing x, but we shall not prove these results here.

These continued fraction expansions are of basic utility when studying algebraic properties of real numbers. In particular, given the real number $x \in \mathbb{R}$ with continued fraction expansion $a_0, a_1, a_2, \ldots$, the rate at which the convergents a_i grow as i tends to infinity is a rough measure of how "complicated" the number x is. In fact, one can make precise the notion of "how well an irrational number is approximated by rational numbers" on the basis of how fast these convergents grow, and, at the risk of appearing irrational, we remark that "certain irrational numbers are more rational than others". We shall not take up these subtleties here, but next give some examples of continued fraction expansions.

■ **Example 4.8.4** Consider the infinite sequence $1, 2, 1, 2, \ldots$ of natural numbers and the corresponding continued fraction

$$x = 1 + \cfrac{1}{2 + \cfrac{1}{1 + \cfrac{1}{2 + \cfrac{1}{\ddots}}}}.$$

It follows from the definition that

$$\frac{1}{\frac{1}{x-1}-2} = 1 + \cfrac{1}{2+\cfrac{1}{1+\cfrac{1}{\ddots}}} = x,$$

or, in other words, $x = (x-1)/(3-2x)$. Clearing denominators, we find the quadratic $2x^2 - 2x - 1 = 0$, which has solutions $\frac{1}{2}(1 \pm \sqrt{3})$, and we evidently take the positive sign, since x is positive by definition. Thus, we have found that

$$x = \frac{1}{2}(1+\sqrt{3})$$

in this example.

■ **Example 4.8.5** Consider the infinite sequence $1, 1, 1, \ldots$ and its corresponding continued fraction

$$x = 1 + \cfrac{1}{1+\cfrac{1}{1+\cfrac{1}{1+\cfrac{1}{\ddots}}}}.$$

Computing as in the previous example, we find

$$x = \frac{1}{x-1},$$

so $x^2 - x - 1 = 0$, which has solutions $\frac{1}{2}(1 \pm \sqrt{5})$, and we again evidently take the positive sign. Thus, $x = \frac{1}{2}(1+\sqrt{5})$, which is called the "golden mean". The Greeks claimed that creations of art and nature owed their beauty to various underlying mathematical patterns and ascribed special such significance to this golden mean. For instance, most people supposedly regard a rectangle as most aesthetically pleasing when its sides are in the approximate ratio of the golden mean, which is roughly 5-by-3. It appears naturally in the symmetries of certain marine mammals, in the proportions of the human body, and so on.

■ **Example 4.8.6** Consider the infinite sequence $2, 1, 1, 1, \ldots$ and its corresponding continued fraction

$$z = 2 + \cfrac{1}{1+\cfrac{1}{1+\cfrac{1}{1+\cfrac{1}{\ddots}}}}.$$

Computing as in the previous examples, we find

$$\frac{1}{z-2} = 1 + \cfrac{1}{1 + \cfrac{1}{1 + \cfrac{1}{\ddots}}} = z - 1,$$

so $(z-1)(z-2) = 1$. By the quadratic formula, the solutions are $z = \frac{1}{2}(3 \pm \sqrt{5})$, and we again take the positive sign $z = \frac{1}{2}(3 + \sqrt{5})$ since $z \geq 2$ by definition.

In fact, the previous examples illustrate typical phenomena of continued fractions, namely, $x \in \mathbb{R}$ solves a quadratic equation (and such a real number is called a "surd" in the classical literature) if and only if its continued fraction expansion $a_0, a_1, a_2, \ldots$ is "eventually periodic", by which we mean that there are natural numbers N and M so that whenever $k \geq N$, we have $a_{k+M} = a_k$. We shall not prove this result here, but it should at least be plausible in light of the previous examples.

In the introduction to this chapter, we promised the reader that we would state an open research-level problem, and here it is: characterize the real solutions to cubic equations in terms of their continued fraction expansions. That is, we ask for the analogue of the result mentioned in the previous paragraph for solutions $x \in \mathbb{R}$ to equations of the form $ax^3 + bx^2 + cx + d = 0$, where $a, b, c, d \in \mathbb{R}$. We do not mean to suggest that this is a central problem in contemporary mathematics, for it is not (though there is a modest literature on it). Rather, it is characteristic of a large class of open and entirely natural problems seemingly just beyond the limits of existing mathematical theory about which absolutely nothing is known!

We have included this brief discussion of continued fraction expansions of real numbers in order to illustrate the more general context in which an analogue of the Euclidean Algorithm can be applied. Part of our intent is to convince the reader that research-level questions in mathematics can be easily appreciated (and even sometimes solved!) by beginners for they are just below the surface. Finally, we feel it is important to encounter continued fractions here since they appear in many guises in mathematics in branches as diverse as geometry, number theory, mathematical physics, dynamical systems, analysis, as well as others. The interested reader might profitably consult *An Introduction to the Theory of Numbers*, Oxford University Press, 1979, by Hardy and Wright, or *An Introduction to the Theory of Numbers*, John Wiley and Sons, 1991, by Niven, Zuckerman, and Montgomery.

EXERCISES

4.8 CONTINUED FRACTIONS

1. Compute the rational number $\nu \in \mathbb{Q}$ corresponding to each of the following finite continued fraction expansions.

(a) 2,1,4 (b) 0,2,1,4

(c) -3,2,12,3 (d) 0,2,2,100

2. Compute the continued fraction expansion of each of the following rational numbers.

(a) 17/3 (b) 3/17

(c) 7/56 (d) 9/56

3. Compute the real number $x \in \mathbb{R}$ corresponding to each of the following continued fraction expansions.

(a) $2, 2, 2, \ldots$ (b) $1, 2, 1, 2, 1, 2, \ldots$

(c) $2, 1, 2, 1, 2, 1, \ldots$ (d) $1, 2, 3, 1, 1, 1, \ldots$

4. Recall that $\nu(a_1, a_2, \ldots, a_m) \in \mathbb{Q}$ denotes the rational number with finite continued fraction expansion $a_1, a_2, \ldots, a_m$. Prove each of the following.

(a) If $a, b, c \in \mathbb{N}$, where $b > c$, then $\nu(a, b) < \nu(a, c)$.

(b) If $a, b, c, d \in \mathbb{N}$, where $c > d$, then $\nu(a, b, c) > \nu(a, b, d)$.

(c) If $c, a_1, a_2, \ldots, a_n \in \mathbb{N}$, then $\nu(a_1, a_2, \ldots, a_n) > \nu(a_1, a_2, \ldots, a_n + c)$ if n is even. On the other hand, this inequality does not hold for n odd.

5. Prove that a real number is a surd if and only if its continued fraction expansion is eventually periodic.

Chapter 5

Relations

One often in mathematics wishes to compare two elements of a set. For instance, given two natural numbers $n, m \in \mathbb{N}$, we might consider which of them is the smaller number, and if n is not larger than m, then we write $n \leq m$. This is a simple example of a "relation", the relation of being less than or equal to, on the set $\mathbb{N}$. More generally, this chapter is dedicated to the study of general relations on arbitrary sets, and we shall find that this material provides a convenient and useful formalism. Under suitable conditions, one can "draw a picture of a relation as a directed graph", and this "picture" is a useful way to intuit about the relation itself; thus, the abstract mathematical structure of a relation admits an essentially geometric realization.

In fact, the collection of all relations on a given set is quite rich: The usual set-theoretic operations of intersection, union, and so on, induce corresponding operations on relations, as we shall see. Furthermore, the collection of all relations on a given set also supports another binary operation called "composition" as well as another unary operation called "inversion", and we shall investigate how these new operations interact with the usual set-theoretic ones.

On the other hand, there is not an interesting theory of general relations, and we introduce various basic properties which a relation might satisfy. Certain combinations of these properties are used to define special classes of relations which do have substantial theories and which are of fundamental importance in mathematics, and we shall separately study two such classes called "partial orderings" and "equivalence relations"; these are truly among the most basic of mathematical structures, and there is much to say about the general theories of each of these classes. Another such basic subclass of partially ordered sets, called "lattices", are discussed in the final optional sections.

5.1 N-ARY RELATIONS

Recall from §3.5 the definition of the Cartesian product

$$\times_{i=1}^{n} A_i = A_1 \times \cdots \times A_n$$

as the set of all n-tuples $(a_1, \ldots, a_n)$, where $a_i \in A_i$ for each $i = 1, 2, \ldots, n$. By definition, an *(n-ary) relation* R on $A_1, \ldots, A_n$ is simply a subset $R \subseteq \times_{i=1}^{n} A_i$. In particular, recall the n-fold Cartesian product

$$A^n = \underbrace{A \times \cdots \times A}_{n \text{ times}}$$

of a set A with itself, and define an *(n-ary) relation* on A itself to be simply a subset $R \subseteq A^n$. In the special cases $n = 1, 2$, and 3 respectively, we refer to R as a *unary*, *binary*, and *ternary* relation.

For instance, for any collection $A_1, \ldots, A_n$ of sets where $n \geq 1$, we may take $R = \emptyset \subseteq \times_{i=1}^{n} A_i$, which is called the *empty relation* on $A_1, \ldots, A_n$. At the other extreme, we might take $R = \times_{i=1}^{n} A_i$, which is called the *universal relation* on $A_1, \ldots, A_n$.

Relations are regarded as being identical only under the most restrictive circumstances: Specifically, two relations $R \subseteq \times_{i=1}^{m} A_i$ and $S \subseteq \times_{j=1}^{n} B_j$ are regarded as the same if and only if $m = n$, $A_i = B_i$ for each $i = 1, \ldots, m$, and $R = S$ as subsets of $\times_{i=1}^{n} A_i$.

To emphasize the stringency of the notion of equality among relations, observe that the empty relation $\emptyset \subseteq \mathbb{N}^2$ is by definition different from the empty relation $\emptyset \subseteq \mathbb{Z}^2$ for instance. In the same way, a unary relation on $\times_{i=1}^{1} A_i = A_1$ is not simply a subset of A_1, rather, it is the specification of A_1 together with a subset of A_1.

To get a feeling for this very general definition, let us immediately give some further examples.

■ **Example 5.1.1** On the set $A = \{a, b, c, d, e\}$, take the ternary relation

$$R = \{(a, b, c), (b, c, d), (c, d, e), (e, a, b), (a, a, a), (a, c, e)\}.$$

■ **Example 5.1.2** Suppose that $P(a_1, \ldots, a_n)$ is a predicate defined on a universe $\times_{i=1}^{n} A_i$, and define a relation

$$R = \{(a_1, \ldots, a_n) \in \times_{i=1}^{n} A_i : P(a_1, \ldots, a_n)\}.$$

On the other hand, if R is any n-ary relation on $A_1, \ldots, A_n$, then we may define a predicate $P(a_1, \ldots, a_n)$ with universe $\times_{i=1}^n A_i$ by setting

$$P(a_1, \ldots, a_n) = 1 \Leftrightarrow (a_1, \ldots, a_n) \in R.$$

The reader might rightly infer from this example that the mathematical structure which relations formalize is as rich as the structure of the predicate calculus.

Consider the case now that each set A_i is finite and suppose that the set A_i has r_i elements for each $i = 1, \ldots, n$. In this case, the Cartesian product $\times_{i=1}^n A_i$ has $r = \prod_{i=1}^n r_i$ elements as we have seen in Theorem 3.5.4. The collection of all possible relations on $A_1, \ldots, A_n$ is by definition simply the power set of this Cartesian product, and we have seen in Example 3.8.9 that this power set has precisely 2^r elements. This proves the following

5.1.1 THEOREM *Suppose that A_i is finite with r_i elements for each $i = 1, 2, \ldots, n$ and let $r = \prod_{i=1}^n r_i$. Then there are exactly 2^r n-ary relations on $A_1, \ldots, A_n$.*

Again, we remark on the utility of simply counting a newly defined object in order to get a feeling for it.

Actually, we shall be primarily interested in binary relations in the sequel, and we shall typically adopt the notation that $R \subseteq A \times B$ is a binary relation. We say that R is a relation *from A to B* under these circumstances, and we call A the *domain* and B the *codomain* of the relation R. By definition of equality among relations, the binary relation R uniquely determines its domain A and codomain B. Introducing some notation, which is called "infix" notation in computer science, particular to this situation, we write aRb if $(a, b) \in R$ and $a\not{R}b$ if $(a, b) \notin R$, where $(a, b) \in A \times B$.

■ **Example 5.1.3** Define the binary relation

$$R = \{(m, n) \in \mathbb{N}^2 : m \leq n\} \subseteq \mathbb{N}^2$$

on $\mathbb{N}$. Thus, $(1, 2) \in R$ (since $1 \leq 2$) so $1R2$, while $(2, 1) \notin R$ (since $2 \not\leq 1$) so $2\not{R}1$ for instance. Of course, we might simply introduce the symbol "$\leq$" for the subset R itself so as to obtain a consistent notation. In this same way, we might define the binary relation

$$< \; = \; \{(m, n) \in \mathbb{N}^2 : m < n\} \subseteq \mathbb{N}^2$$

on $\mathbb{N}$, so $1 < 2$, while $2 \not< 1$ for instance. It is common to denote the relation R itself in this way by its corresponding symbol $<$ in the infix notation.

■ **Example 5.1.4** On any set A, we define the binary relation

$$E_A = \{(a, a) \in A^2 : a \in A\} \subseteq A^2,$$

which is called the *relation of equality* on A.

■ **Example 5.1.5** It is common to let the symbol "=" denote the relation of equality on a given set A provided that no confusion can result. To illustrate this, let $=$ denote the relation of equality on $\mathbb{N}$, let $\leq$ denote the usual relation of weak inequality on $\mathbb{N}$ discussed in Example 5.1.3, so that $\leq$ is a binary relation on $\mathbb{N}$ and $(m,n) \in \leq$ if and only if $m \leq n$, and let $<$ denote the usual relation of strict inequality on $\mathbb{N}$. According to the definitions, we have

$$\leq \; = \; < \; \cup \; =$$

While this equation is slightly abusive (since the same symbol "=" occurs with two different meanings), such equations are fairly common in practice.

Let $R \subseteq \times_{i=1}^{n} A_i$ be a relation on the sets $A_1, A_2, \ldots, A_n$, where $n \geq 1$. For any collection

$$A_i' \subseteq A_i \quad \text{for } i = 1, \ldots, n$$

of subsets, there is a corresponding subset

$$\begin{aligned} R' &= \{(a_1, a_2, \ldots, a_n) \in R : a_i \in A_i' \text{ for all } i = 1, \ldots, n\} \\ &= R \cap (\times_{i=1}^{n} A_i') \\ &\subseteq R. \end{aligned}$$

Regarding $R' \subseteq \times_{i=1}^{n} A_i'$ as an n-ary relation on $\times_{i=1}^{n} A_i' \subseteq \times_{i=1}^{n} A_i$, we refer to R' as the *restriction* of the relation R to $\times_{i=1}^{n} A_i'$.

■ **Example 5.1.6** The restriction of the usual relation $\leq$ on $\mathbb{R}^2$ to the subset $\mathbb{N}^2 \subseteq \mathbb{R}^2$ is the usual relation $\leq$ on $\mathbb{N}^2$.

■ **Example 5.1.7** Given any subset $A \subseteq B$, the restriction of the relation of equality on B to A is again the relation of equality on A.

■ **Example 5.1.8** Define the binary relation

$$D = \{(a,b) \in \mathbb{N} \times \mathbb{Z} : b \text{ is an integral multiple of } a\} \subseteq \mathbb{N} \times \mathbb{Z},$$

so $2D(-6)$ for instance. Of course, we have considered at length the related relation

$$| = \{(a,b) \in \mathbb{N}^2 : a \text{ is a divisor of } b\} \subseteq \mathbb{N}^2$$

in the previous chapter, which is simply the restriction of D to $\mathbb{N} \times \mathbb{N} \subseteq \mathbb{N} \times \mathbb{Z}$.

EXERCISES

5.1 N-ARY RELATIONS

1. Let A denote the set $\{1,2,3,4,5,6\}$. Specify the following relations as explicit subsets of A^2.

(a) $\{x \in A : x \leq 2\}$

(b) $\{(x,y,z) \in A^3 : x < yz\}$

(c) $\{(x,y) \in A^2 : x < y\}$

(d) $\{(x,y,u,v) \in A^4 : x+y = u+v\}$

2. Give inductive definitions of each of the following relations.

(a) $R = \{(a,b) \in \mathbb{N}^2 : a \leq b\} \subseteq \mathbb{N}^2$

(b) $R = \{(a,b) \in \mathbb{N}^2 : a = 3b\} \subseteq \mathbb{N}^2$

(c) $R = \{(a,b,c) \in \mathbb{N}^3 : a+b = c\} \subseteq \mathbb{N}^3$

(d) $R = \{(a,b,c,d) \in \mathbb{N}^4 : a+b = c+d\} \subseteq \mathbb{N}^4$

3. Suppose that $R \subseteq A \times B$ is a relation from A to B. We may draw "pictures" of R as a subset of $A \times B$ for suitable sets A and B, where we imagine A along the horizontal, B along the vertical, and explicitly draw R as a subset of the Cartesian product $A \times B$. This is called the "graph" of the relation.

(a) Let $A = \{1,2,3\}$ and $B = \{2,4,6\}$, and draw the graph of the relation $R = \{(1,2),(2,4),(3,6),(1,1),(2,6)\}$.

(b) Let $A = B = \mathbb{R}$, and draw the graph of the relation $\{(x,y) \in \mathbb{R}^2 : x = (y-1)^2\}$.

(c) Let $A = \mathbb{R}^2$ and $B = \mathbb{R}$, and draw the graph of the relation $\{(x,y,z) \in \mathbb{R}^3 : x^2+y^2+z^2 = 1\}$

(d) Let $A = \mathbb{Z}$ and $B = \mathbb{R}$, and draw the graph of the relation $\{(n,x) \in \mathbb{N} \times \mathbb{R} : x = n \text{ or } x = n-1\}$

4. Let A denote the set $\{1,2,3\}$.

(a) List the various unary relations $R \subseteq A$.

(b) How many different binary relations $R \subseteq A^2$ are there on A?

(c) How many different n-ary relations $R \subseteq A^n$ are there on A?

5. Let A denote the set $\{1,2,\ldots,n\}$ for some $n \geq 1$.

(a) Prove that there are 2^n different unary relations on A.

(b) Prove that there are 2^{n^2} different binary relations on A.

(c) How many different n-ary relations $R \subseteq A^n$ are there on A?

6. Suppose that $A' \subseteq A$, $B' \subseteq B$, and let $R \subseteq A \times B$ be a relation. In each of the following examples, describe the restriction of R to $A' \times B'$.

(a) Set $A = B = \mathbb{R}$ and consider the relation $R = \{(x,y) \in \mathbb{R}^2 : y = x^2\}$; take subsets $A' = \mathbb{N}$ and $B' = \mathbb{R}$.

(b) Set $A = B = \mathbb{N}$ and consider the relation $|$, where $m|n$ if and only if m divides n; take subsets $A' = B' = \{2, 3, 6, 9, 30\}$.

(c) Set $A = B = \mathbb{R}$ and consider the relation $R = \{(x, y) \in \mathbb{R}^2 : x = y^2\}$; take subsets $A' = \{x \in \mathbb{R} : -3 \leq x \leq -1\}$ and $B' = \mathbb{R}$.

(d) Set $A = \mathbb{R}^2$, $B = \mathbb{R}$, and consider the relation $R = \{(x, y, z) \in \mathbb{R}^3 : x^2 + y^2 + z^2 = 1\}$; take subsets $A' = \{(x, y) \in \mathbb{R}^2 : x^2 + y^2 \geq 1/2\}$ and $B' = \mathbb{R}$.

5.2 BINARY RELATIONS AND DIGRAPHS

In the special case of a binary relation $R \subseteq A^2$, there is a pictorial description of R which is most useful as we next explain. For simplicity, we assume for now that the set A is finite, and hence so too is the set $R \subseteq A^2$ finite. One associates a figure G in the plane $\mathbb{R}^2 \supseteq G$ to R by first drawing a point in the plane for each element a of the set A and labeling the point associated to $a \in A$ with the symbol "a". Next, for each element (a_1, a_2) in the finite set R, we draw an oriented segment in the plane beginning at a_1 and ending at a_2, and we indicate the orientation on this segment by putting an arrow on it pointing from a_1 to a_2.

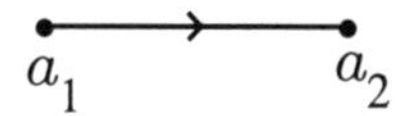

A figure $G \subseteq \mathbb{R}^2$ which we draw in this way is called a *(labeled) digraph* (which is short for "directed graph") associated to R. The points of G corresponding to the elements of A are called the *vertices* and the (oriented) segments corresponding to elements of R are called the *(oriented) edges* of G. By definition, if $(a, a) \in R$ for some $a \in A$, then there is an edge of G beginning and ending at the vertex of G labeled a. An edge of this type (whose endpoints correspond to the same vertex) is called a *loop* of G, and we say that the loop is *based* at the vertex a in this case. Notice that the orientation on the loop corresponding to $(a, a) \in R$ is actually of no consequence (since if we reverse the entries in the ordered pair (a, a) we obtain the same ordered pair), and, in fact, arrows on the loops in a digraph are typically omitted.

We next give examples of digraphs associated to various binary relations.

■ **Example 5.2.1** On the set $A = \{a, b, c, d, e\}$, consider the binary relation

$$R = \{(a, b), (d, c), (e, a), (a, a), (b, a)\}.$$

In this case, a corresponding digraph is given by

so $G(R)$ has exactly one loop in this example.

■ **Example 5.2.2** On the set $A = \{a, b, c, d\}$, consider the binary relation

$$R = \{(a, a), (a, b), (b, a), (b, c), (b, d), (a, c), (c, d), (d, a)\}$$

with corresponding digraph

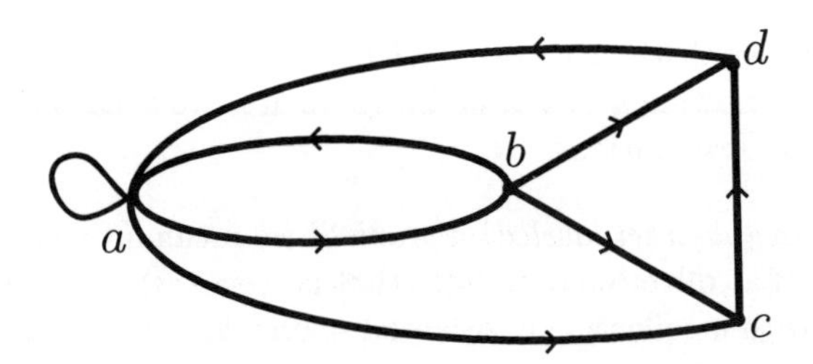

■ **Example 5.2.3** On the set $A = \{1, \ldots, 8\} \subseteq \mathbb{N}$, define the relation $R = \{(m, n) \in A^2 : (m|n)\}$ with corresponding digraph

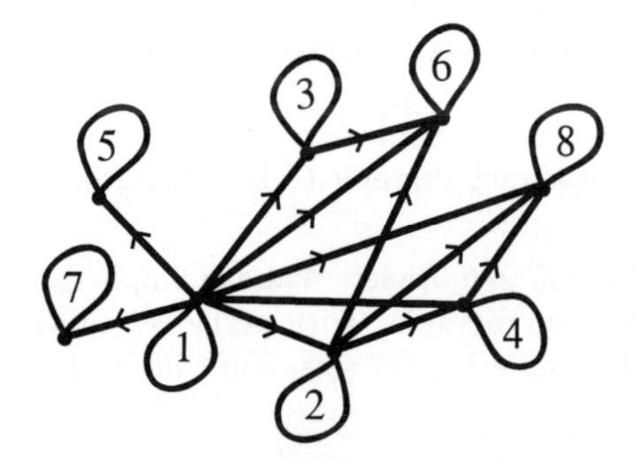

■ **Example 5.2.4** Consider the binary relation $R = \{(x_i, x_j) \in A^2 : i < j\}$ on the set $A = \{x_1, \ldots, x_5\}$ with corresponding digraph

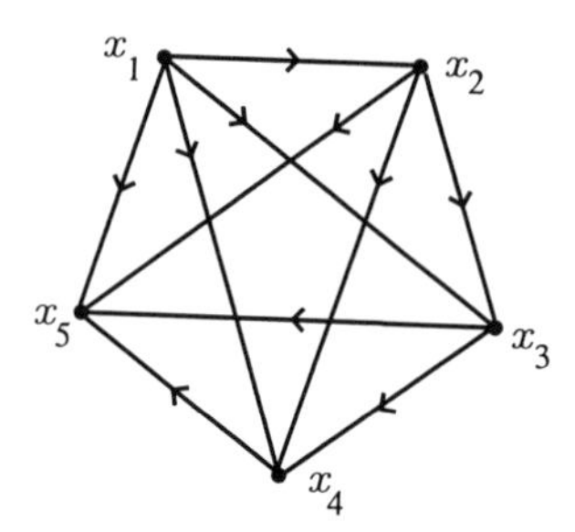

Notice that we draw the digraph in this example in such a way that various pairs of distinct edges actually cross. (Actually, there is no way to draw the digraph in this example in the plane without such "crossings"; we shall discuss this fact from "graph theory" in §10.4.)

Thus, given the binary relation $R \subseteq A^2$ on the finite set A, we can associate a digraph G whose vertices are labeled by the elements of A. By construction, the digraph G possesses the following properties.

(i) There is at most one loop based at each vertex of G.

(ii) For any two distinct vertices a, b of G, there are either zero, one, or two edges connecting a and b, and if there are two such edges, then they have opposite orientations.

Conversely, by an *(abstract labeled) digraph G* we mean an abstract figure in the plane consisting of a collection of points (that is, *vertices*) in the plane labeled by a set A together with a collection of oriented segments (that is, oriented *edges*) connecting these points, where we assume that conditions (i) and (ii) above hold. In this case, we may uniquely construct a binary relation $R(G) \subseteq A^2$ in the natural way from G, where

$$R(G) = \{(a, b) \in A^2 : \text{there is an oriented edge in } G \text{ from } a \text{ to } b\},$$

and of course the original abstract digraph G is a digraph associated to this relation $R(G)$.

We have restricted attention to *finite* sets A above simply for convenience, and one can clearly associate labeled digraphs (with perhaps infinitely many vertices and edges) to binary relations on arbitrary sets, and conversely. For instance, we have

■ **Example 5.2.5** A digraph corresponding to the relation $\leq$ on $\mathbb{N}$ is as follows

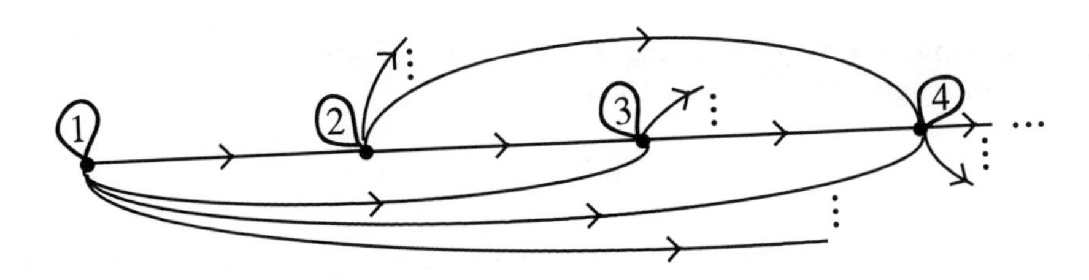

We conclude that any binary relation is represented by some abstract labeled digraph, and this is really the main thrust of our discussion here. Indeed, we have introduced this material on digraphs now primarily in order to develop this intuitively appealing geometric representation of a binary relation. This representation as a digraph is a useful way to imagine a binary relation in the mind's eye, for one can often "visualize" a proof about binary relations on the level of digraphs, and digraphs can therefore be useful in the discovery of proofs about binary relations. Such an operative "visual" picture of an abstract structure can often be a valuable tool in mathematics.

Part of the subsequent utility of this pictorial representation of a binary relation depends on the following concept. Suppose that G is an abstract digraph whose vertices are labeled by a set A and let $R = R(G)$ denote the corresponding binary relation on A. If $a, b \in A$, then we define a *(directed) edge-path P in G from a to b* to be a finite sequence $P = (v_0, v_1, \ldots, v_n)$, where $n \geq 0$, of vertices of G so that $v_0 = a$, $v_n = b$, and $v_i R v_{i+1}$ for each $i = 0, \ldots, n-1$. Thus, one may imagine a directed edge-path in a digraph G as a possible "trajectory" which a particle might take in traversing the edges of G, where the particle is allowed to travel along an edge of G only in a direction determined by a specified orientation on the edge. The vertex a is called the *initial point*, the vertex b is called the *terminal point* of P, and we say simply that "P is a path from a to b" in this case. The *length* of the edge-path P is defined to be n. If all the vertices $v_0, \ldots, v_n$ are distinct *except* perhaps that $v_0 = v_n$ (that is, if $v_0, \ldots, v_{n-1}$ are all distinct and $v_1, \ldots, v_n$ are all distinct), then we say that P is *simple*. Finally, if the initial and terminal points of an edge-path coincide, then we call the edge-path a *cycle*. In particular, a loop $(a, a) \in R$ gives rise to a simple cycle of length 1.

■ **Example 5.2.6** There are no cycles whatsoever on the digraph in Example 5.2.4, there is a unique longest path, and it is $(x_1, x_2, x_3, x_4, x_5)$.

■ **Example 5.2.7** On the digraph in Example 5.2.3, there are the paths $(1, 8)$, $(1, 4, 8)$, $(1, 2, 4, 8)$ from 1 to 8. The only simple cycles in this example are the loops of length 1, and there is one such loop based at each vertex $i = 1, 2, \cdots 8$. The only cycles in the example are of the form

$$\underbrace{(i, i, \ldots, i)}_{n \text{ times}}$$

for any $i = 1, 2, \ldots, 8$, and $n \geq 1$.

■ **Example 5.2.8** On the digraph in Example 5.2.2, there are paths from a to b of any length: For instance, the path

$$(\underbrace{a, a, \ldots, a}_{n-1 \text{ times}}, b)$$

for $n \geq 1$ has initial point a, terminal point b, and length n; among these, only the path (a, b) is simple. If $n = 2k - 1 \geq 1$ is an odd number, then another path from a to b of length n is

$$(\underbrace{a, b, a, b, \ldots, a, b}_{k \text{ copies of } a,b});$$

among these, again only (a, b) is simple. We furthermore have the simple cycles (a, a), (a, b, a), (b, d, a, b), for instance, while

$$(b, d, a, a, b, c, d, a, b) \quad \text{and} \quad (a, b, a, b, a, b, a, a, a, c, d, a)$$

furnish examples of non-simple cycles.

It is evident that different abstract labeled digraphs can give rise to the same binary relation. For instance, one might simply move the digraph in the plane to obtain what is properly speaking a "different" digraph, and the reader may easily imagine still more drastic examples. In fact, one simply regards as identical two abstract labeled digraphs which give rise to the same binary relation. In this way, one *defines* a labeled digraph to be simply an ordered pair (A, R), where A is a set and $R \subseteq A^2$ is a binary relation on A; the elements of A are themselves called vertices and the elements of R are themselves called edges. The figures in the plane which we have discussed above are simply a convenient representation for the data type of labeled digraphs, and this data type is entirely equivalent to the data type of binary relations by definition.

EXERCISES

5.2 BINARY RELATIONS AND DIGRAPHS

1. Let A denote the set $\{1, 2, \ldots, 6\}$, and draw the digraph corresponding to each of the following binary relations R on A.

(a) aRb if and only if $a \geq b$

(b) aRb if and only if a divides b

(c) aRb if and only if $a \neq b$

(d) aRb if and only if $a + b$ is even

(e) aRb if and only if $a + b$ is odd

(f) aRb if and only if 3 divides $a + b$

2. Let A denote the set $\{1, 2, \ldots, 6\}$, and describe the binary relation R on A corresponding to each of the following digraphs.

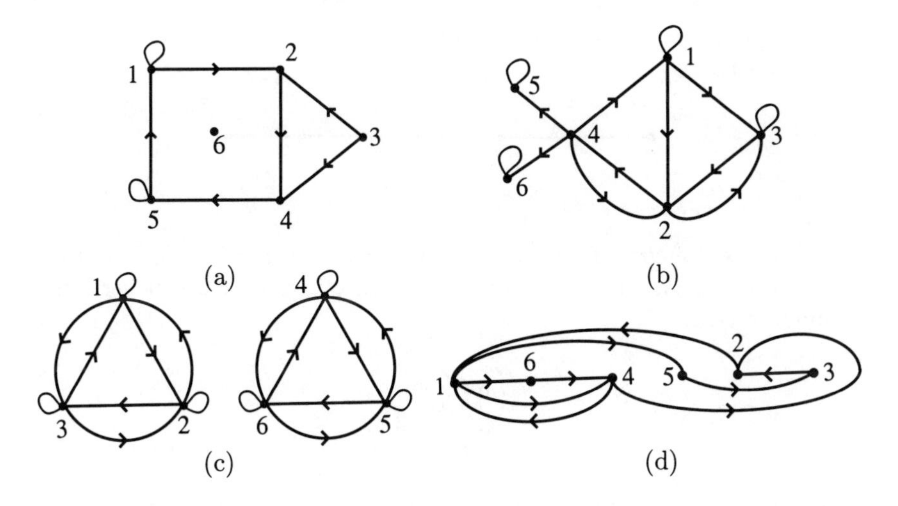

3. For each of the six digraphs in Problem 1

(a) List all the simple edge-paths from 1 to 2. (b) from 1 to 3.

4. For each of the six digraphs in Problem 2

(a) List all the simple edge-paths from 1 to 2. (b) from 1 to 3.

5.3 PROPERTIES OF RELATIONS

Here we define and discuss several properties which a binary relation might satisfy. Various combinations of these properties are used to define certain classes of relations which are among the most basic of mathematical structures.

Suppose that $R \subseteq A^2$ is a binary relation on a set A, and let G be a corresponding digraph. We say that

> R is *reflexive* if aRa for each $a \in A$. That is, there is a loop based at each vertex a of G:

> R is *irreflexive* if $a \not R a$ for each $a \in A$. That is, there is no loop based at any vertex a of G:

> R is *symmetric* if $\forall a, b \in A(aRb \Rightarrow bRa)$. That is, if there is an edge of

G running from a to b, then there must also be an edge in G running from b to a:

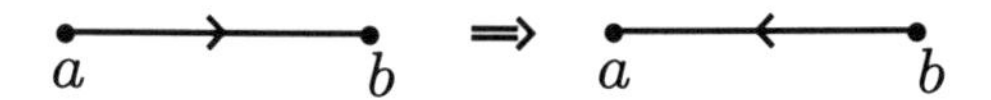

R is *antisymmetric* if $\forall a, b \in A([(aRb \wedge bRa) \Rightarrow a = b]$. That is, there can never be two distinct edges in G connecting two distinct vertices a and b of G:

R is *asymmetric* if $\forall a, b \in A(aRb \Rightarrow b \not R a)$. That is, there are no loops and there can be no two distinct edges connecting distinct vertices of G:

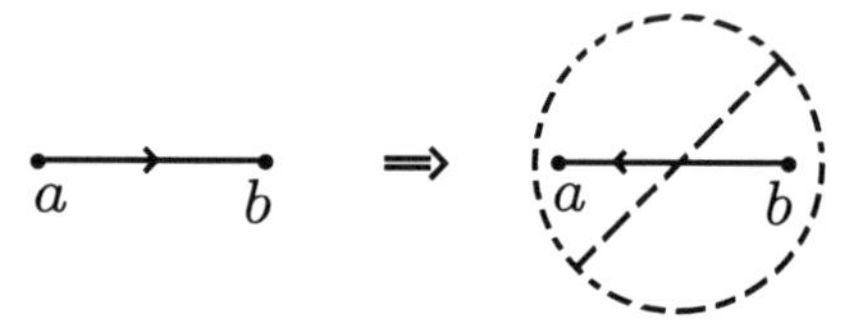

R is *transitive* if $\forall a, b, c \in A[(aRb \wedge bRc) \Rightarrow aRc]$. That is, if there are edges in G from a to b and from b to c (so there is a path of length two in G from a to c), then there is an edge in G from a to c:

We next make several elementary observations about the logical dependence among these definitions. Notice first that if a relation is not reflexive, then it is not necessarily irreflexive since there may be loops based at some vertices and not at others. Also, observe that an asymmetric relation is automatically antisymmetric. Furthermore, a relation might be neither symmetric nor antisymmetric and hence not asymmetric either. Finally, suppose that R is a transitive binary relation on A. If there is a directed edge-path in a corresponding digraph G with initial point a and terminal point b, then there is an edge in G from a to b, as the reader should prove (by induction on the length of a path from a to b).

Let us give some examples of relations satisfying various combinations of these properties.

■ **Example 5.3.1** The relation $<$ of strict inequality on $\mathbb{N}$ is irreflexive, transitive, and (vacuously) antisymmetric and asymmetric.

■ **Example 5.3.2** The relation $\leq$ of weak inequality on $\mathbb{N}$ is reflexive, antisymmetric, and transitive, but it is not asymmetric.

■ **Example 5.3.3** The relation of equality on any non-empty set is reflexive, symmetric, antisymmetric, and transitive, but it is not asymmetric.

■ **Example 5.3.4** The universal relation on any non-empty set is reflexive, symmetric, and transitive, but it is not antisymmetric if the set has more than one element.

■ **Example 5.3.5** The relation corresponding to the digraph

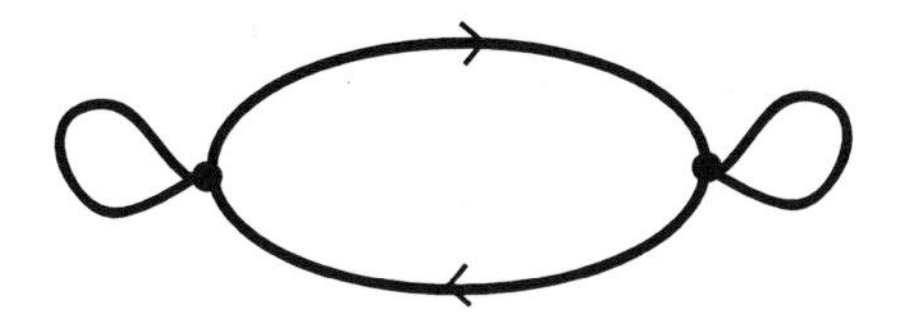

is reflexive, symmetric and transitive.

■ **Example 5.3.6** The relation corresponding to the digraph

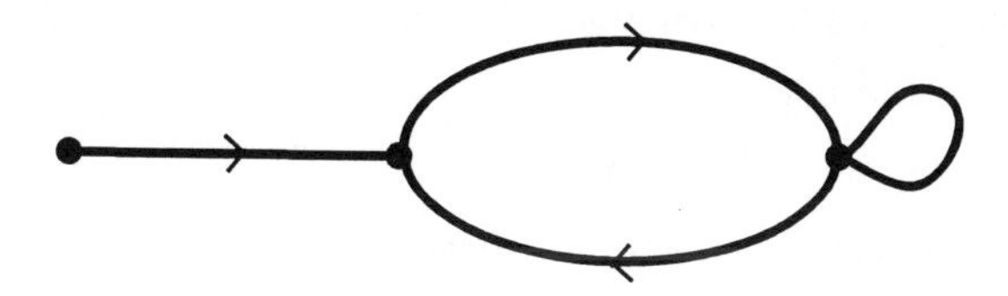

satisfies none of the conditions.

As a set-theoretic characterization of reflexivity, we have

5.3.1 THEOREM *A binary relation $R \subseteq A^2$ on a set A is reflexive if and only if $E_A \subseteq R$, where E_A is the relation of equality on A.*

Proof Suppose first that R is reflexive, so $(a, a) \in R$ for all $a \in A$, i.e.,

$E_A \subseteq R$, proving the first implication. Conversely, if $E_A \subseteq R$ and $a \in A$, then $(a,a) \in E_A \subseteq R$, so $(a,a) \in R$ and R is indeed reflexive. *q.e.d.*

We mentioned at the beginning of this section that two particular combinations of these properties lead to fundamental mathematical objects. Specifically, a *partial ordering* is a reflexive, antisymmetric and transitive relation, and an *equivalence relation* is a reflexive, symmetric and transitive relation. We shall study both partial orderings and equivalence relations in detail later. For now, let us just observe that the relation $\leq$ on $\mathbb{N}$ is a partial ordering and equality on any set is an equivalence relation by Examples 5.3.2 and 5.3.3 above.

EXERCISES

5.3 PROPERTIES OF RELATIONS

1. Let A denote the set $\{1,2,3,4\}$. Find a binary relation on A satisfying each of the following combinations of properties.

(a) reflexive, symmetric, but not transitive

(b) reflexive, transitive, but not symmetric

(c) symmetric, transitive, but not reflexive

(d) reflexive, symmetric, and transitive

(e) reflexive, antisymmetric, and transitive

2. Let A denote the set $\{1,2,3,4\}$. Find a binary relation on A satisfying each of the following combinations of properties.

(a) transitive but neither reflexive nor irreflexive

(b) reflexive but neither asymmetric nor antisymmetric

(c) antisymmetric but not asymmetric

(d) reflexive, symmetric, but not transitive

(e) irreflexive, antisymmetric, but not transitive

3. Sketch the graphs of the following relations $R \subseteq \mathbb{R}^2$ and determine which properties hold in each case.

(a) $R = \{(x,y) \in \mathbb{R}^2 : y^2 = x^2\}$

(b) $R = \{(x,y) \in \mathbb{R}^2 : x \leq 1 \text{ or } x \leq y\}$

(c) $R = \{(x,y) \in \mathbb{R}^2 : x^2 \leq 1 \text{ and } y \geq 0\}$

(d) $R = \{(x,y) \in \mathbb{R}^2 : x^2 + y^2 = 1\}$

4. For each of the following digraphs, determine which properties hold for the corresponding binary relation.

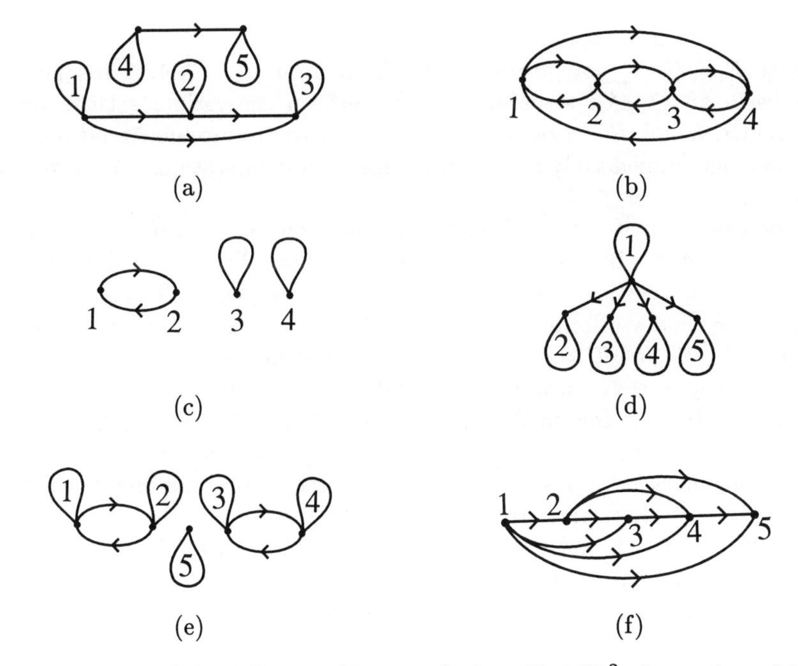

5. For each of the following binary relations $R \subseteq \mathbb{Z}^2$, determine which properties hold.

   (a) mRn if and only if $mn > 0$.

   (b) mRn if and only if $|m - n|$ equals 0,3, or 6.

   (c) mRn if and only if $|m + n|$ equals 0,3, or 6.

   (d) mRn if and only if $yx^2 = xy^2$.

6. Prove that if $R \subseteq A^2$ is an equivalence relation as well as a partial ordering on $A \neq \emptyset$, then R is the relation of equality on A.

7. Suppose that A is a finite set with $n \geq 0$ elements.

   (a) Show that among the 2^{n^2} distinct binary relations on A, exactly $(2^n - 1)^n$ relations R satisfy the condition that for each $a \in A$, there is some ordered pair $(a, a') \in R$.

   (b) Show that among the $(2^n - 1)^n$ relations on A satisfying the condition in part (a), exactly 2^{n^2-n} are reflexive.

8. Suppose that $R \subseteq A^2$ is a transitive relation with corresponding digraph G, and let P be an oriented edge-path on G with initial point u and terminal point v. Prove that uRv.

9. Can a non-empty relation be both symmetric and anti-symmetric? Prove your assertions.

5.4 SET-THEORETIC OPERATIONS

Suppose that $R, S \subseteq \times_{i=1}^{n} A_i$ are general relations on $A_1, \ldots, A_n$. Insofar as R, S are subsets of $\times_{i=1}^{n} A_i$, we might simply perform the various set-theoretic operations $R \cup S$, $R \cap S$, $R - S$ or $\overline{R} = (\times_{i=1}^{n} A_i) - R$ to produce new relations on $\times_{i=1}^{n} A_i$. Thus, one immediately finds a rich collection of operations on the set of all relations on $A_1, \ldots, A_n$.

For some examples, if R is the binary relation $\leq$ on $\mathbb{N}$, then $\overline{R}$ is simply the binary relation $>$ on $\mathbb{N}$. If S is the binary relation of equality on $\mathbb{N}$, then $R - S$ is simply the binary relation $<$ on $\mathbb{N}$. For a final example, let $T = \{(m, n) \in \mathbb{N}^2 : m|n\}$ be the binary relation of divides on $\mathbb{N}$. Since $mTn \Rightarrow mRn$, we have $T \subseteq R$, so $R \cap T = T$, $R \cup T = R$, and $T - R = \emptyset$. Furthermore, $R - T$ is the set of ordered pairs $(m, n) \in \mathbb{N}^2$ where $m \leq n$ and $m \nmid n$.

We restrict attention now to the case that $R, S \subseteq A^2$ are binary relations on a set A. In this case, we have considered six special properties (reflexivity, transitivity, etc.) of binary relations in the previous section, and one might wonder whether these special properties are invariant under the four set-theoretic operations $R \cup S$, $R \cap S$, $R - S$ and $\overline{R}$; for instance, if R and S are reflexive, then is $R \cup S$ reflexive?. There are thus 24 sensible questions here we might ask, and we summarize the corresponding 24 answers in the following table.

	$R \cup S$	$R \cap S$	$R - S$	$\overline{R}$
reflexive	Y	Y	N	N
irreflexive	Y	Y	Y	N
symmetric	Y	Y	Y	Y
antisymmetric	N	Y	Y	N
asymmetric	N	Y	Y	N
transitive	N	Y	N	N

For instance, the entry "Y" in the entry of the table corresponding to "reflexive" and "$R \cup S$" indicates that if R and S are reflexive, then so too is $R \cup S$. To prove this, we observe that if R is reflexive, then there is a loop based at each vertex of digraph corresponding to R, so the same is true for the digraph of $R \cup S$. More formally (that is, as one should write down in a formal exposition of the proof), suppose that R is reflexive, so for each $a \in A$, we have aRa, i.e., $(a, a) \in R$. Since $R \subseteq R \cup S$, we have $(a, a) \in R \cup S$, so $R \cup S$ is indeed reflexive. Notice that we did not even need to assume here that S is reflexive.

We next prove that if R and S are symmetric, then so too is $R \cap S$. Suppose that $(a, b) \in R \cap S$, so

$$\begin{aligned} (a, b) \in R \cap S \ &\Rightarrow [(a, b) \in R \ \wedge \ (a, b) \in S] \\ &\Rightarrow [(b, a) \in R \ \wedge \ (b, a) \in S] \\ &\Rightarrow (b, a) \in R \cap S, \end{aligned}$$

where we have used the hypothesis that R and S are symmetric in the second implication. We leave it to the reader to visualize this proof on the level of digraphs.

Next, we prove that if R and S are transitive, then so too is $R \cap S$. To see this, suppose that $(a,b),(b,c) \in R \cap S$, so that

$$\begin{aligned}(a,b),(b,c) \in R \cap S &\Rightarrow [(a,b),(b,c) \in R] \wedge [(a,b),(b,c) \in S] \\ &\Rightarrow [(a,c) \in R] \wedge [(a,c) \in S] \\ &\Rightarrow (a,c) \in R \cap S,\end{aligned}$$

where we have used transitivity of R and S in the second implication.

Finally, we give a counter-example to the claim that if R and S are asymmetric, then so too is $R \cup S$, that is, we justify the corresponding entry "N" in the table above. In our counter-example, $R \cup S$ must not be asymmetric so we should be able to find vertices, say the vertices a, b, of the corresponding digraph with $(a,b),(b,a) \in R \cup S$. Of course, one such edge, say (a,b) must lie in R and the other (b,a) in S, for otherwise R or S would fail to be asymmetric (and we would not produce the desired counter-example). We are led to define the simple counter-example on the set $A = \{a,b\}$, where we take $R = \{(a,b)\}$ and $S = \{(b,a)\}$.

We leave the rest of the proofs (or at least their visualizations in terms of digraphs as above) of the entries in the table above to the untiring reader.

EXERCISES

5.4 SET-THEORETIC OPERATIONS

1. We have proved 4 of the 24 entries in the table discussed in this section.

(a) Prove 5 further entries in the table.

(b) Prove 10 further entries in the table.

(c) Prove all 20 remaining entries in the table.

2. Recall that an equivalence relation is a reflexive, symmetric and transitive binary relation, whereas a partial ordering is a reflexive, antisymmetric, and transitive binary relation.

(a) Construct a table of Y's and N's analogous to the table in this section which describes whether unions, intersections, and relative complements of equivalence relations and partial orderings are again equivalence relations and partial orderings.

(b) Give proofs of 3 of the entries in each table constructed in part (a).

(c) Give proofs of all 6 entries in each table constructed in part (a).

5.5 INVERSION

We have considered the set-theoretic operations on relations in the previous section. Here, we are given a binary relation R from a set A to a set B and construct a new binary relation from B to A, which represents the "inverse" of R.

Precisely, the unary operation of *inversion* is defined as follows. If $R \subseteq A \times B$, then we define the *inverse* or *converse* of R by

$$R^{-1} = \{(b,a) \in B \times A : (a,b) \in R\} \subseteq B \times A.$$

Thus, the binary relation R^{-1} from B to A is derived from R by simply changing the order of the ordered pairs from $(a,b) \in R$ to $(b,a) \in R^{-1}$. For instance, if $R = \{(1,a),(2,a)\} \subseteq \{1,2\} \times \{a\}$, then $R^{-1} = \{(a,1),(a,2)\} \subseteq \{a\} \times \{1,2\}$. For another example, if R is the relation $\leq$ on $\mathbb{N}$, then R^{-1} is simply the relation $\geq$.

We summarize the effects of inversion on the set-theoretic operations in the following

5.5.1 THEOREM *Suppose that $R, S \subseteq A \times B$ are binary relations from A to B. Then we have*

(a) $(R^{-1})^{-1} = R$

(b) $(R \cup S)^{-1} = R^{-1} \cup S^{-1}$

(c) $(R \cap S)^{-1} = R^{-1} \cap S^{-1}$

(d) $(A \times B)^{-1} = B \times A$ and $\emptyset^{-1} = \emptyset$

(e) $(\overline{R})^{-1} = \overline{(R^{-1})}$

(f) $(R - S)^{-1} = R^{-1} - S^{-1}$

(g) $R \subseteq S \Rightarrow R^{-1} \subseteq S^{-1}$

Proof The proof of each part follows from standard manipulations in the predicate calculus and/or set theory. For instance, for part (a), we simply observe

$$(a,b) \in R \Leftrightarrow (b,a) \in R^{-1} \Leftrightarrow (a,b) \in (R^{-1})^{-1}.$$

For part (c), we calculate

$$\begin{aligned}(a,b) \in (R \cap S)^{-1} &\Leftrightarrow (b,a) \in R \cap S \\ &\Leftrightarrow [(b,a) \in R] \wedge [(b,a) \in S] \\ &\Leftrightarrow [(a,b) \in R^{-1}] \wedge [(a,b) \in S^{-1}] \\ &\Leftrightarrow (a,b) \in R^{-1} \cap S^{-1},\end{aligned}$$

as desired.

For part (e), we calculate

$$\begin{aligned}(a,b) \in (\overline{R})^{-1} &\Leftrightarrow (b,a) \in \overline{R} \\ &\Leftrightarrow (b,a) \notin R \\ &\Leftrightarrow (a,b) \notin R^{-1} \\ &\Leftrightarrow (a,b) \in \overline{R^{-1}},\end{aligned}$$

as was claimed.

For part (f), recall that $R - S = R \cap \overline{S}$, so we have

$$\begin{aligned}(R-S)^{-1} = (R \cap S)^{-1} &= R^{-1} \cap (\overline{S})^{-1} \\ &= R^{-1} \cap \overline{S^{-1}} \\ &= R^{-1} - S^{-1},\end{aligned}$$

where we have used part (c) in the second equality and part (e) in the third equality.

We leave the proofs of the remaining parts as exercises (which are similar to the arguments above). *q.e.d.*

In the special case of a binary relation $R \subseteq A^2$ from A to itself, the set-theoretic interpretation of symmetry is described by

5.5.2 THEOREM *If $R \subseteq A^2$ is a binary relation on A, then R is symmetric if and only if $R = R^{-1}$.*

Proof We suppose first that R is symmetric and prove separately that $R \subseteq R^{-1}$ and $R^{-1} \subseteq R$ in order to prove one implication of the theorem. Since R is symmetric, $(a,b) \in R \Rightarrow (b,a) \in R$. Thus, we have $(a,b) \in R^{-1}$ by definition of inversion, so $R \subseteq R^{-1}$. For the reverse inclusion, suppose that $(a,b) \in R^{-1}$, so by definition of inversion, we have $(b,a) \in R$. Since R is symmetric, we conclude that $(a,b) \in R$ as well, and this proves that $R^{-1} \subseteq R$.

For the reverse implication, we suppose that $R = R^{-1}$. Thus, $(a,b) \in R \Rightarrow (b,a) \in R^{-1} = R$, and we conclude that $(a,b) \in R \Rightarrow (b,a) \in R$, or in other words, R is symmetric. *q.e.d.*

EXERCISES

5.5 INVERSION

1. Prove parts (b), (d), and (g) of Theorem 5.5.1.

2. Suppose that $R \subseteq A^2$ is a binary relation on a set A.

(a) Prove that $R \cup R^{-1}$ is symmetric.

(b) Prove that $R \cap R^{-1}$ is symmetric.

3. Suppose that $R \subseteq A^2$ is a binary relation on a set A. Prove or disprove each of the following.

(a) If R is reflexive, then so is R^{-1}.
(b) If R is symmetric, then so is R^{-1}.
(c) If R is irreflexive, then so is R^{-1}.
(d) If R is antisymmetric, then so is R^{-}
(e) If R is transitive, then so is R^{-1}.
(f) If R is asymmetric, then so is R^{-1}.

5.6 COMPOSITION

We turn now to the definition of a certain "multiplication" of relations, and the setting in which this operation is applied is as follows. Suppose that $R \subseteq A \times B$ is a binary relation from A to B and $S \subseteq B \times C$ is a binary relation from B to C, or in other words, the codomain of R is the same set B as the domain of S. In this case, we define a new binary relation $R \circ S \subseteq A \times C$, which is called the *composition* of R and S, from A to C by

$$R \circ S = \{(a, c) \in A \times C : \exists b \in B \text{ so that } aRb \wedge bSc\}.$$

■ **Example 5.6.1** If $A = \{1, 2\}$, $B = \{3, 4, 5\}$, $C = \{6, 7\}$, and we define

$$R = \{(1, 3), (2, 4), (2, 5)\} \text{ and } S = \{(3, 6), (3, 7), (4, 6)\},$$

then

$$R \circ S = \{(1, 6), (1, 7), (2, 6)\}.$$

Notice in particular that though $R \circ S$ is defined since the codomain of R is the same as the domain of S, the composition $S \circ R$ is *not* defined since the codomain of S is C and the domain of R is $A \neq C$.

We make the useful convention that given binary relations R and S, we shall only write the collection $R \circ S$ of symbols if the codomain of R agrees with the domain of S. Thus, by simply writing $R \circ S$, we make this tacit assumption. In the same way, given binary operations R, S, we shall tacitly agree to write a binary set-theoretic operation $R \cup S$, $R \cap S$, or $R - S$ only when R and S have the same domain and the same codomain (for we have defined the binary operations $\cap$, $\cup$, and $-$ only under these conditions).

For an extended and indeed very illuminating example of composition of relations we consider the following analysis of "relations amongst relations".

■ **Example 5.6.2** Let A denote the set of all people who have ever lived, and introduce the binary relations M, F, B on A, where

$$aMb \Leftrightarrow a \text{ is the mother of } b$$
$$aFb \Leftrightarrow a \text{ is the father of } b$$
$$aBb \Leftrightarrow a \text{ is the brother of } b$$

for any $a, b \in A$. Since each of M, F, B are binary relations from A to A, every possible composition of these relations is actually defined. For instance, $a(F \circ M)b$ if and only if a is the maternal grandfather of b, since $a(F \circ M)b \Leftrightarrow \exists c(aFc \wedge cMb)$, so a is the father of c and c is the mother of b, whence a is indeed the maternal grandfather of b. On the other hand, $a(M \circ F)b$ if and only if a is the paternal grandmother of b, so even though both $M \circ F$ and $F \circ M$ are defined, they are not equal. Thus, composition of relations is definitely not commutative even if both orders of composition are defined. For another example, $a(B \circ M)b$ if and only if a is the maternal uncle of b. For a more complicated example, notice that $a[B \circ (M \circ F)]b$ if and only if a is the paternal great uncle of b on the grandmother's side, and moreover $a[B \circ (M \circ F)]b \Leftrightarrow a[(B \circ M) \circ F]b$.

The previous example faithfully illustrates this binary operation of composition of relations in familiar familial terms and again illustrates that composition is not commutative in general. For another example, we have

■ **Example 5.6.3** Let $R \subseteq \mathbb{N}^2$ denote the binary relation $|$ of divides on $\mathbb{N}$, and let $S \subseteq \mathbb{N}^2$ be defined by $(a, b) \in S$ if and only if $a = b^2$. Then $a(R \circ S)b$ if and only if there is some $c \in \mathbb{N}$ so that $a|c$ and $c = b^2$, i.e., if $a|(b^2)$. On the other hand, $a(S \circ R)b$ if and only if there is some $c \in \mathbb{N}$ so that $a = c^2$ and $c|b$, i.e., if $\sqrt{a} \in \mathbb{N}$ and $(\sqrt{a})|b$.

We turn next to various basic results on composition of relations.

5.6.1 THEOREM *Composition of relations is associative. That is, given relations R, S, T, we have $R \circ (S \circ T) = (R \circ S) \circ T$.*

Proof By our convention, the codomain of R agrees with the domain of S, and the codomain of S agrees with the domain of T, and we suppose that $R \subseteq A \times B$, $S \subseteq B \times C$, and $T \subseteq C \times D$. By definition,

$$\begin{aligned} (a, d) \in R \circ (S \circ T) &\Leftrightarrow \exists b \in B[aRb \wedge b(S \circ T)d] \\ &\Leftrightarrow \exists b \in B[aRb \wedge \exists c \in C(bSc \wedge cTd)] \\ &\Leftrightarrow \exists b \in B \exists c \in C[aRb \wedge bSc \wedge cTd] \qquad (*). \end{aligned}$$

An entirely analogous manipulation shows that $(a, d) \in (R \circ S) \circ T$ is also logically equivalent to the predicate $(*)$ above, which proves the theorem. *q.e.d.*

In light of the associativity of composition, we shall as usual often omit parentheses in an iterated composition, writing simply $R \circ S \circ T$ rather than $R \circ (S \circ T)$ for instance.

5.6.2 THEOREM *Suppose that $R \subseteq A \times B$ is a binary relation from A to B, and let $E_A \subseteq A^2$, $E_B \subseteq B^2$ denote the relations of equality on A, B.*

Then

$$E_A \circ R = R = R \circ E_B.$$

Proof We concentrate first on proving the identity $E_A \circ R = R$, and to this end, compute

$$\begin{aligned}(a,b) \in R &\Rightarrow [(a,a) \in E_A] \wedge [(a,b) \in R] \\ &\Rightarrow \exists c\big[[(a,c) \in E_A] \wedge [(c,b) \in R]\big],\end{aligned}$$

so indeed $R \subseteq E_A \circ R$.

For the reverse inclusion, we find

$$(a,b) \in E_A \times R \Rightarrow \exists c\big[[(a,c) \in E_A] \wedge [(c,b) \in R]\big],$$

but $(a,c) \in E_A \Rightarrow a = c$ by definition of E_A, so in fact $(a,b) \in R$, as desired.

Thus, we have proved that $E_A \circ R = R$, and the proof that $R = R \circ E_B$ is substantially similar and left as an exercise. *q.e.d.*

Our next result is a basic one relating composition with the set-theoretic operations.

5.6.3 THEOREM *Given binary relations R, S, T, we have*

(a) $R \circ (S \cup T) = (R \circ S) \cup (R \circ T)$

(b) $(S \cup T) \circ R = (S \circ R) \cup (T \circ R)$

(c) $R \circ (S \cap T) \subseteq (R \circ S) \cap (R \circ T)$

(d) $(S \cap T) \circ R \subseteq (S \circ R) \cap (T \circ R)$

Proof We prove part (a) as follows

$$\begin{aligned}(a,c) \in R \circ (S \cup T) &\Leftrightarrow \exists b[(a,b) \in R \wedge (b,c) \in S \cup T)] \\ &\Leftrightarrow \exists b[(a,b) \in R \wedge [(b,c) \in S \vee (b,c) \in T]] \\ &\Leftrightarrow \exists b\big[[(a,b) \in R \wedge (b,c) \in S] \vee [(a,b) \in R \wedge (b,c) \in T]\big] \\ &\Leftrightarrow \exists b[(a,b) \in R \wedge (b,c) \in S] \vee \exists b[(a,b) \in R \wedge (b,c) \in T] \\ &\Leftrightarrow [(a,c) \in R \circ S] \vee [(a,c) \in R \circ T] \\ &\Leftrightarrow (a,c) \in [(R \circ S) \cup (R \circ T)].\end{aligned}$$

The proof of part (b) is entirely analogous and is left as an exercise.

To prove part (c), we compute

$$\begin{aligned}(a,c) \in R \circ (S \cap T) &\Leftrightarrow \exists b[(a,b) \in R \wedge (b,c) \in S \cap T] \\ &\Leftrightarrow \exists b[(a,b) \in R \wedge (b,c) \in S \wedge (b,c) \in T] \\ &\Leftrightarrow \exists b\big[[(a,b) \in R \wedge (b,c) \in S] \wedge [(a,b) \in R \wedge (b,c) \in T]\big] \\ &\Rightarrow \exists b[(a,b) \in R \wedge (b,c) \in S] \wedge \exists b[(a,b) \in R \wedge (b,c) \in T] \\ &\Leftrightarrow [(a,c) \in R \circ S] \wedge [(a,c) \in R \circ T] \\ &\Leftrightarrow (a,c) \in [(R \circ S) \cap (R \circ T)].\end{aligned}$$

For a simple counter-example to the reverse inclusion in part (c), just take $R = \{(a,b_1),(a,b_2)\} \subseteq A \times B$, $S = \{(b_1,c)\}, T = \{(b_2,c)\} \subseteq B \times C$, where $A = \{a\}$, $B = \{b_1, b_2\}$, and $C = \{c\}$. In this example, $S \cap T = \emptyset$, so

$$R \circ (S \cap T) = R \circ \emptyset = \emptyset.$$

On the other hand,

$$R \circ S = R \circ T = (R \circ S) \cap (R \circ T) = \{(a,c)\} \neq \emptyset,$$

so this is indeed a counter-example to the reverse inclusion.

As before, the proof of part (d) follows closely that of part (c) and is left as an exercise. *q.e.d.*

We close this section with a basic and useful result relating inversion and composition.

5.6.4 THEOREM *Suppose that $R \subseteq A \times B$ and $S \subseteq B \times C$ are binary relations, so the compositions $R \circ S$ and $S^{-1} \circ R^{-1}$ are defined. Then we have*

$$(R \circ S)^{-1} = S^{-1} \circ R^{-1}.$$

Proof We have

$$\begin{aligned}(c,a) \in (R \circ S)^{-1} &\Leftrightarrow (a,c) \in R \circ S \\ &\Leftrightarrow \exists b \in B\ (aRb \wedge bSc) \\ &\Leftrightarrow \exists b \in B\ [b(R^{-1})a \wedge c(S^{-1})b] \\ &\Leftrightarrow \exists b \in B\ [c(S^{-1})b \wedge b(R^{-1})a] \\ &\Leftrightarrow (c,a) \in S^{-1} \circ R^{-1},\end{aligned}$$

as desired. *q.e.d.*

Suppose that we have a binary relation $R \subseteq A^2$. In this case, we may evidently compose R with itself to get the binary relation $R \circ R$ on A, and this fact leads to a rich mathematical structure as we finally and briefly discuss. Make the following recursive definition of R^n for each natural number n:

$$\begin{aligned}R^0 &= E_A, \text{ the relation of equality on } A, \\ R^{n+1} &= (R^n) \circ R.\end{aligned}$$

Thus, $R^2 = R \circ R$ and $R^3 = R \circ R \circ R$ for instance. The relation $R^n \subseteq A^2$ is called the nth *iterate* of R (and the reader should prove by induction on n that each iterate R^n is actually a binary relation on A).

Essentially repeating the argument of Example 3.8.8, we conclude $R^m \circ R^n = R^{m+n}$ and $(R^m)^n = R^{mn}$ for any nonnegative integers m, n. We shall study iterates carefully in the next optional section.

EXERCISES

5.6 COMPOSITION

1. Prove parts (b) and (d) of Theorem 5.6.3.

2. Prove that if $R \subseteq A \times B$ and $E_B \subseteq B \times B$ is the relation of equality, then $R = R \circ E_B$.

3. Prove that for $n \geq 0$ and any $R \subseteq A^2$, we have $(R^n)^{-1} = (R^{-1})^n$.

4. Let A denote the set $A = \{1, 2, 3, 4\}$ and define the binary relations

$$R = \{(1,1), (1,2), (2,3), (3,4), (4,4)\}$$
$$S = \{(2,2), (2,1), (3,2), (3,3), (4,1), (4,2)\}$$

on A. Compute each of the following relations.

(a) $R \circ S$ and $S \circ R$ (b) R^2 and S^2

(c) $R \circ S^{-1}$ and $S \circ R^{-1}$ (d) $S \circ R \circ S^{-1}$ and $R \circ S \circ R^{-1}$

5. Let R denote the usual binary relation $\leq$ on $\mathbb{Z}$ and let S denote the binary relation on $\mathbb{Z}$ where mSn if and only if 3 divides $m - n$. Compute each of the following relations.

(a) $R \circ S$ and $S \circ R$ (b) R^2 and S^2

(c) $R \circ S^{-1}$ and $S \circ R^{-1}$ (d) $S \circ R \circ S^{-1}$ and $R \circ S \circ R^{-1}$

6. For each of the following digraphs representing a binary relation R, find the least $m, n \in \mathbb{N}$ so that $R^m = R^n$ where $m \neq n$.

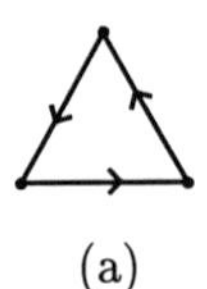

(a)

(b)

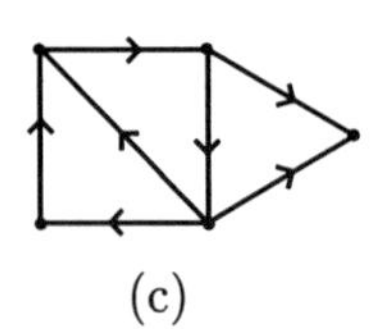

(c)

7. Prove that a binary relation R on a set A is transitive if and only if $R \circ R \subseteq R$.

8. (a) Prove that if R is either the empty relation or the universal relation on a set A, then $R^2 = R$.

(b) Give an example of a relation R on a non-empty set A where $R^2 = R$ and R is neither the empty nor the universal relation on A.

9. Suppose that R, S, T are binary relations on a set A. Prove or disprove each of the following.

(a) If $R \subseteq S$, then $R \circ T \subseteq S \circ T$.

(b) If $R \subseteq S$, then $T \circ R \subseteq T \circ S$.

(c) If $R, S \subseteq T$, then $R \circ S \subseteq T^2$.

(d) $E_B \subseteq R \circ R^{-1}$ and $E_A \subseteq R^{-1} \circ R$

10. Suppose that R, S are binary relations on a set A. Prove or disprove each of the following.

(a) If R and S are both reflexive, then so too is $R \circ S$.

(b) If R and S are both irreflexive, then so too is $R \circ S$.

(c) If R and S are both symmetric, then so too is $R \circ S$.

(d) If R and S are both antisymmetric, then so too is $R \circ S$.

(e) If R and S are both asymmetric, then so too is $R \circ S$.

(f) If R and S are both transitive, then so too is $R \circ S$.

(g) If $R \circ S = S \circ R$, then $R = S$.

(h) If R and S are both symmetric, then $(R \circ S)^{-1} = R \circ S$.

11. Let R be a binary relation on a set A and let E denote the relation of equality on A

(a) For each integer $n \geq 0$, prove that $(E \cup R)^n = \cup_{i=0}^n R^i$.

(b) Prove that $\cup_{i \geq 1} R^i$ is transitive.

(c) Prove that $\cup_{i \geq 0} R^i$ is both reflexive and transitive.

(*) 5.7 *ITERATIONS*

We restrict attention in this optional section to the setting of a binary relation $R \subseteq A^2$ on a set A. As observed in the last section, we may evidently compose R with itself to get the binary relation $R \circ R$ on A and define the *iterates* of R recursively by

$$R^0 = E_A, \text{ the relation of equality on } A,$$
$$R^{n+1} = (R^n) \circ R.$$

Essentially repeating a previous argument in the setting of natural powers of real numbers in Example 3.8.8, we have

5.7.1 THEOREM *Suppose that $R \subseteq A^2$. Then we have*

(a) $\forall m \geq 0 \ \ \forall n \geq 0 \ \ [R^m \circ R^n = R^{m+n}]$

(b) $\forall m \geq 0 \ \ \forall n \geq 0 \ \ [(R^m)^n = R^{mn}]$

Proof For part (a), let $m \geq 0$ be arbitrary and induct on n, so for the basis step, we must show $R^m \circ R^0 = R^m \circ E_A = R^m$, which follows from Theorem 5.6.2. For the induction step, we have

$$\begin{aligned} R^m \circ R^{n+1} &= R^m \circ (R^n \circ R) \\ &= (R^m \circ R^n) \circ R \\ &= R^{m+n} \circ R \\ &= R^{m+n+1} \end{aligned}$$

using associativity of composition.

For part (b), we again let $m \geq 0$ be arbitrary and induct on n, so the basis step is $(R^m)^0 = E_A = R^0 = R^{m \cdot 0}$. For the induction step, we have

$$\begin{aligned} (R^m)^{n+1} &= (R^m)^n \circ R^m \\ &= R^{mn} \circ R^m \\ &= R^{mn+m} \\ &= R^{m(n+1)}, \end{aligned}$$

where the second equality follows from the induction hypothesis and the third from part (a). *q.e.d.*

There is a pleasant geometric interpretation of iterates R^n in terms of the digraph of R as follows:

5.7.2 THEOREM *Suppose that $R \subseteq A^2$, let G be a corresponding digraph, and let $n \geq 0$. Then $(a, b) \in R^n$ if and only if there is an oriented edge-path of length n in G from a to b.*

Proof We argue by induction, and the basis step $n = 0$ asserts that $(a, b) \in R^0 = E_A$ (where E_A is the relation of equality on A) if and only if there is an oriented edge-path in D of length zero from a to b. To see this, observe that an edge-path of length zero is of the form (a), where a is a vertex of G, so we must have $a = b$, whence $(a, b) = (a, a) \in E_A = R^0$.

For the induction step, suppose that $(a, b) \in R^{n+1} = R^n \circ R$, so there is some $c \in A$ so that $aR^nc \ \wedge \ cRb$. By the inductive hypothesis, there is an oriented edge-path $u_0, \ldots, u_n$ of length n in G from $a = u_0$ to $c = u_n$, and of course cRb gives rise to an edge of G, which we think of as an oriented edge-path (c, b) of length one. Thus, $u_0, \ldots, u_n, b$ is an oriented edge-path in G from a to b, as was claimed.

Conversely (and still in the induction step), suppose that $u_0, \ldots, u_{n+1}$ is an oriented edge-path in G from a to b, so $u_0, \ldots, u_n$ is an oriented edge-path from

a to $c = u_n$ of length n. By the inductive hypothesis, $(a, c) \in R^n$ and $(c, b) \in R$ (since $u_0, \ldots, u_{n+1}$ is an oriented edge-path), so in fact, we have $(a, b) \in R^n \circ R = R^{n+1}$. *q.e.d.*

Thus, one may visualize the iterates of a relation directly on a corresponding digraph in terms of the edge-paths on the digraph. Armed with these insights, we turn to a study of the various iterates R^n for $n \geq 0$ of a binary relation $R \subseteq A^2$ and begin with

■ **Example 5.7.1** Let $E = R^0$ be the relation

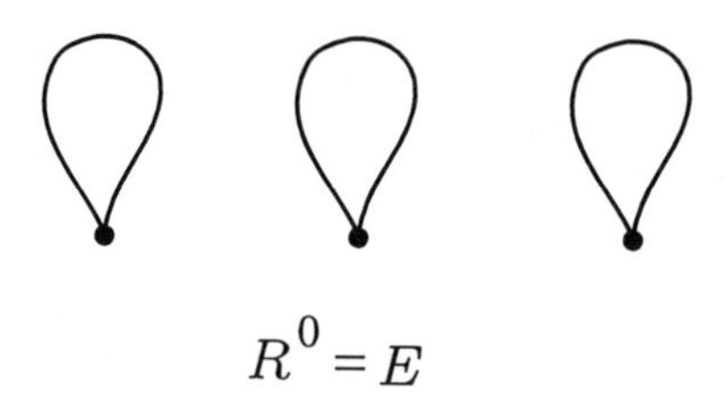

$R^0 = E$

of equality on a set A with three elements, and define R to be the binary relation

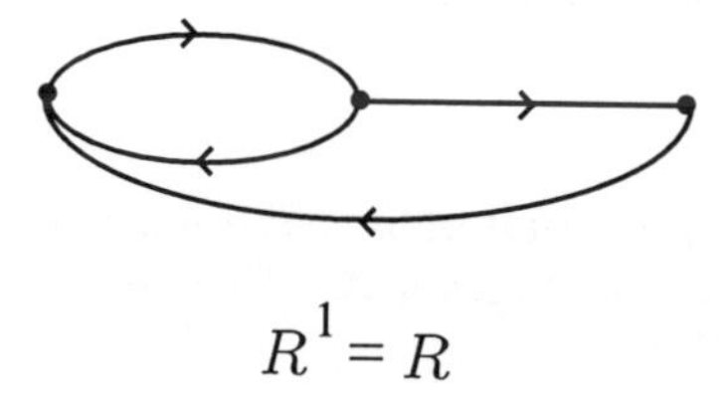

$R^1 = R$

Using the previous theorem and exhaustively considering the collection of oriented edge-paths of various lengths on the previous digraph, we can enumerate digraphs corresponding to the positive iterates of R:

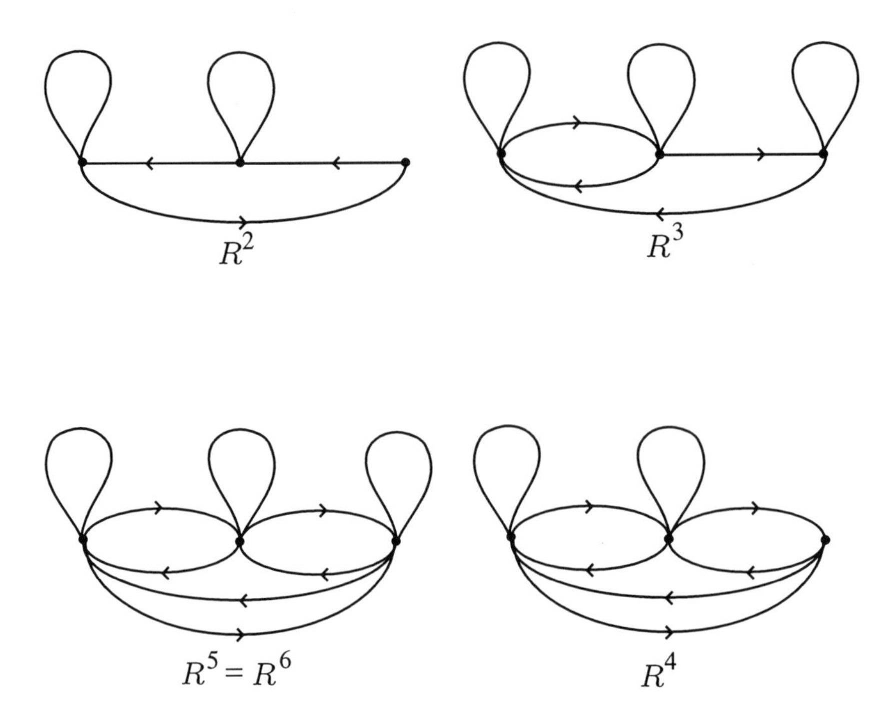

Thus, the relation R^5 is actually the universal relation, and in fact, R^6 is the universal relation as well. It follows that $R^7 = R^6 \circ R^1 = R^5 \circ R^1 = R^6$ for instance, and in fact, $R^k = R^5$, i.e., R^k is the universal relation, for each $k \geq 5$.

This example illustrates certain typical phenomena of iterates which we next make precise.

5.7.3 THEOREM *Suppose that A is a finite set with $n \geq 1$ elements, and let $R \subseteq A^2$ be a binary relation on A. Then there are $s, t \geq 0$ so that $R^s = R^t$ where $0 \leq s < t \leq 2^{n^2}$.*

Proof Since A has n elements, A^2 has n^2 elements by Theorem 3.5.4, so the power set $\mathcal{P}(A)$ has 2^{n^2} elements by Example 3.8.9. Thus, there are exactly 2^{n^2} distinct binary relations on A, and it follows that among the $2^{n^2}+1$ relations $R^0, R^1, \ldots R^{2^{n^2}}$, two must be identical. *q.e.d.*

In contrast to the previous result, if A is an infinite set, then no two iterates of R need be identical. For instance, let R denote the binary relation on $\mathbb{N}$ where mRn if and only if $n = m + 1$. In this case, $mR^q n$ if and only if $q = n - m$, so the relations $R^0, R^1, \ldots$ are all distinct.

Our final result here gives a more precise description of the set of iterates of a given relation which applies in particular to any binary relation on a finite set. Specifically, we show that if two distinct iterates of a relation R agree, then there are actually only finitely many distinct iterates of R.

5.7.4 THEOREM *Suppose that $R \subseteq A^2$ is a binary relation on a set A where $R^s = R^t$ for some $s < t$, and define $p = t - s$. Then we have*

(a) $\forall n \geq 0 \; (R^{s+n} = R^{t+n})$

(b) $\forall n, i \geq 0 \; (R^{s+np+i} = R^{s+i})$

(c) $\forall q \geq 0 \; (R^q \in \{R^0, \ldots, R^{t-1}\})$

Proof Part (a) is proved by induction, and the basis step $n = 0$ is the hypothesis that $R^s = R^t$. For the induction step,

$$\begin{aligned} R^{s+(n+1)} &= R^{s+n} \circ R^1 \\ &= R^{t+n} \circ R^1 \quad \text{by the induction hypothesis} \\ &= R^{t+n+1}. \end{aligned}$$

Part (b) is proved by induction on n for an arbitrary fixed $i \geq 0$, and the basis step $n = 0$ requires observing that $R^{s+0\cdot p+i} = R^{s+i}$. For the induction, we compute

$$\begin{aligned} R^{s+(n+1)p+i} = R^{s+np+p+i} &= R^{s+np+i} \circ R^p \quad \text{by Theorem 5.7.1} \\ &= R^{s+i} \circ R^p \quad \text{by the induction hypothesis} \\ &= R^{s+i} \circ R^{t-s} \quad \text{by definition of } p \\ &= R^{t+i} \quad \text{by Theorem 5.7.1 again} \\ &= R^{s+i} \quad \text{by part (a).} \end{aligned}$$

For part (c), if $q < t$, then clearly $R^q \in \{R^0, \ldots, R^{t-1}\}$. On the other hand, if $q \geq t$, then we may write $q = s + \ell$ for some $\ell > 0$ (since $s < t$ by hypothesis). Using the Division Algorithm 4.2.1, we then uniquely write $\ell = np + i$, where $i < p$. Thus,

$$R^q = R^{s+np+i} = R^{s+i},$$

where the last equality follows from part (b). Finally, we have $s + i < t$ since $i < p = t - s$, so indeed $R^q \in \{R^0, \cdots, R^{t-1}\}$. *q.e.d.*

EXERCISES

5.7 ITERATIONS

1. Let A denote the set $\{1, 2, \ldots, 6\}$. For each of the following digraphs, draw the digraphs corresponding to the first five iterates of the underlying binary relation on A.

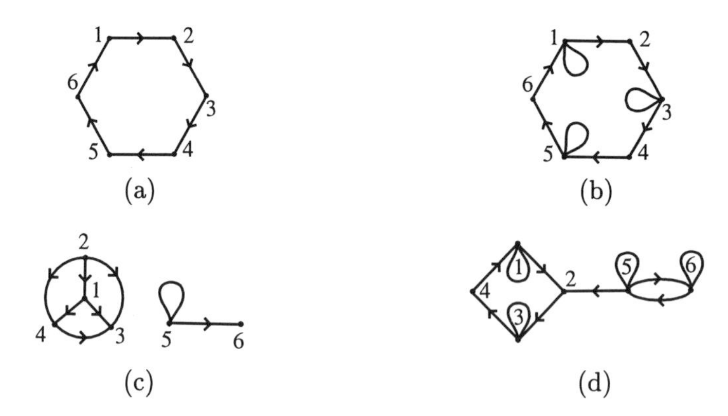

2. Verify Theorem 5.7.2 by listing all of the edge-paths of length at most three for each digraph in Problem 1.

3. Verify Theorem 5.7.3 by finding r and s for each digraph in Problem 1, and prove your assertions by induction.

4. Verify Theorem 5.7.4 for each digraph in Problem 1, and prove your assertions by induction.

5.8 POSETS

A relation $R \subseteq A^2$ is called a *partial ordering* if R is reflexive, antisymmetric and transitive, and in this case, we refer to the pair (A, R) as a *partially ordered set* or simply as a *poset*. As a point of notation, one might also refer to A itself simply as a poset when the relation R is understood. Posets are one of the fundamental mathematical structures, and this section is dedicated to their study. For some simple examples, the binary relation $\leq$ on $\mathbb{R}$ is evidently reflexive, antisymmetric and transitive, so $(\mathbb{R}, \leq)$ is actually a poset. In the same way, the binary relation $\leq$ is also a partial ordering on $\mathbb{N}$.

For a more interesting class of examples, let us fix some natural number $r \in \mathbb{N}$ and consider the finite set $A_r = \{1, \ldots, r\} \subseteq \mathbb{N}$ together with the binary relation $m|n$ (i.e., m divides n) on the set A_r. Using familiar properties of divisibility, the reader should check that each A_r is actually a poset with this partial ordering.

As a point of notation, we mention that one sometimes uses the symbol "$\leq$" or a variant such as "$\preceq$" to denote an abstract partial ordering on an arbitrary set A. For instance, let A denote the set $A = \mathcal{P}(\{a, b\})$, and define $\leq$ to be the relation $\{(x, y) \in A^2 | x \subseteq y\}$. Thus, $(A, \leq)$ is a poset (as the reader can check), where we use the symbol $\leq$ for the partial ordering on A even though the relation in this example has nothing to do with the usual relation of being "less than or equal to". Students sometimes find this slight abuse of notation confusing, but it is rather common in practice.

Our next result furnishes us with the prototypical examples of posets.

5.8.1 THEOREM *Suppose that A is a set of sets and let $\subseteq$ denote inclusion of sets. Then $(\mathcal{P}(A), \subseteq)$ is a poset.*

Proof We must show that $\subseteq$ is reflexive, antisymmetric and transitive. Reflexivity is obvious, antisymmetry follows immediately from Theorem 3.2.1, and transitivity was proved in Theorem 3.1.1. *q.e.d.*

Before giving some explicit examples of the previous theorem, we introduce a standard figure for a poset (A, R) which is akin to the digraph G of the underlying relation R. The basic point is that if we assume *a priori* that R is a partial ordering, then many of the edges of G are superfluous. For instance, since the relation R is reflexive by hypothesis, we know in advance that there is a loop based at each vertex of G. We might therefore simply omit these loops in a corresponding digraph $G_1 = G - \{\text{loops of } G\}$, or more precisely, we might draw the digraph of the relation $R_1 = R - E_A$ on A, where E_A is the relation of equality on A.

The other simplification of digraph we shall consider derives from the *a priori* fact that R is also transitive, i.e., $(aRb \wedge bRc) \Rightarrow aRc$ for any vertices $a, b, c \in A$. Thus, the existence of edges $(a, b), (b, c)$ of G implies the existence of the edge (a, c), and the latter edge of G is therefore superfluous. Supposing that A is finite, we define

$$G_2 = G_1 - \{\text{edges } (a, c) \in R : \exists b \in A[(a, b) \in R \ \wedge \ (b, c) \in R]\},$$

so the relation R is uniquely determined by the relation

$$\begin{aligned} R_2 &= R_1 - \{(a, c) \in R : \exists b \in A[(a, b) \in R \ \wedge \ (b, c) \in R]\} \\ &= R_1 - R_1 \circ R_1 \end{aligned}$$

on A, and the digraph of the relation R_2 is G_2.

The digraph G_2 associated in this way to G or R is called the *Hasse diagram* of the poset and is the standard diagrammatic representation of a poset. In fact, one also omits the orientations of edges of G_2 in the Hasse diagram employing the convention that all edges are drawn in such a way that all orientations point upwards in the diagram.

■ **Example 5.8.1** Let R denote the partial ordering $\subseteq$ of inclusion of subsets of $A = \{a, b, c\}$, where a, b, c are all distinct. The Hasse diagram of the poset (A, R) is

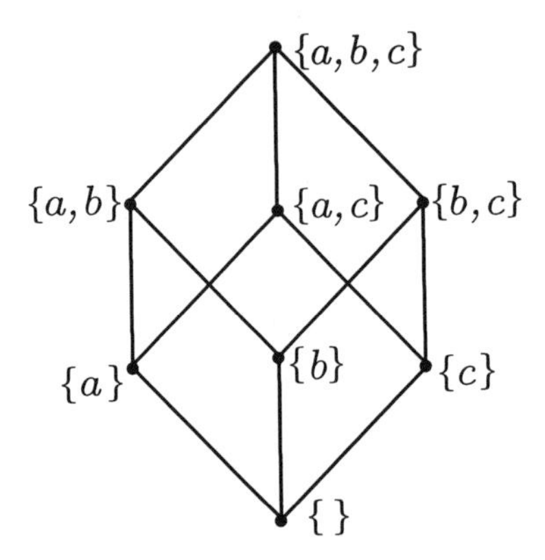

■ **Example 5.8.2** Consider the restriction R of the binary relation $|$ of "divides" to the set $A = \{1, 2, \ldots, 12\}$. The Hasse diagram of the poset (A, R) is

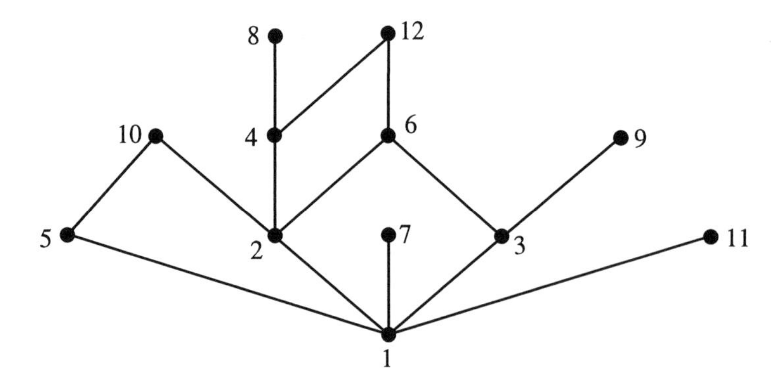

Actually, we shall never use Hasse diagrams in the sequel and include their definition here just for completeness. On the other hand, the Hasse diagram of a finite poset is a useful and common formalism.

Implicit in our condition on orientations in Hasse diagrams is the following result about the usual digraph of a partial ordering.

5.8.2 THEOREM *Suppose that G is a digraph associated to the binary relation $\leq$ on a set A where $\leq$ is a partial ordering. Then there are no simple cycles on G other than loops.*

Proof Suppose that $(a_0, a_1, \ldots, a_n)$ is a simple cycle which is not a loop so in particular $a_0 = a_n$. According to the definitions, we must have

$$a_0 \leq a_1 \leq \cdots \leq a_n \quad \text{and} \quad a_n = a_0,$$

so for each $j = 1, 2, \ldots, n$, we have

$$[(a_0 \leq a_j) \wedge (a_j \leq a_n = a_0)] \Rightarrow (a_0 = a_j)$$

by antisymmetry of $\leq$. This contradicts that the cycle is simple unless it is a loop. *q.e.d.*

A binary relation R on a set A is called a *strict partial ordering* if R is irreflexive, asymmetric and transitive. For instance, the binary relation $<$ of strict inequality on $\mathbb{R}$ or on $\mathbb{N}$ is a strict partial ordering. Just as for partial orderings, one often uses the symbol $<$ or a variant such as $\prec$ to denote an abstract strict partial ordering on an arbitrary set even though the particular relation has nothing to do with the usual relation of being "strictly less than". For an example of this, define the relation

$$< \;=\; \{(x, y) \in A^2 : x \subseteq y \wedge x \neq y\}$$

on the set $A = \mathcal{P}(\{a, b\})$, and notice that $<$ is indeed a strict partial ordering on A.

Observe that if a binary relation R on a set A is irreflexive and transitive, then R is automatically also asymmetric. To see this, suppose that aRb and bRa, so we have aRa by transitivity. On the other hand, aRa is impossible by irreflexivity, so $aRb \wedge bRa$ must be false, or in other words, we have $aRb \Rightarrow b\not{R}a$, so R is indeed asymmetric (and hence antisymmetric). Thus, we might simply define a strict partial ordering to be an irreflexive and transitive relation, and asymmetry of the relation follows automatically.

Let $\leq$, $<$, and $=$ denote the usual relations of being "less than or equal to", "strictly less than" and "equal to", respectively, on the set $\mathbb{R}$. Thus, the relation $<$ is none other than the relation $\leq - =$, and $\leq$ is none other than $< \cup =$. These are examples of general phenomena relating partial orderings and strict partial orderings which we next describe.

5.8.3 THEOREM *Suppose that A is a set and let E denote the relation of equality on A. If P is a partial ordering on A, then $S = P - E$ is a strict partial ordering on A. Conversely, if S is a strict partial ordering on A, then $P = S \cup E$ is a partial ordering on A.*

Proof To begin, suppose that P is a partial ordering on A and define the relation $S = P - E$ on A as in the theorem. We must show that S is irreflexive and transitive. According to the definition

$$(a, a) \in S \Leftrightarrow [(a, a) \in P] \wedge [(a, a) \notin E],$$

yet $(a, a) \in E$ by definition of equality. Thus, $(a, a) \in S$ is absurd, so S is indeed irreflexive.

To prove transitivity of S, suppose that aSb and bSc for $a, b, c \in A$. By definition of $S = P - E$, we conclude that $(aPb) \wedge (a \neq b)$ and $(bPc) \wedge (b \neq c)$. By transitivity of the partial ordering P, we conclude that aPc, and it remains only to show that $a \neq c$ (for then aSc by definition). In the contrary case $a = c$, we find $(aPb) \wedge (bPc) \wedge (a = c)$, so $a = b$ by antisymmetry of P. Finally, this contradicts our supposition that $a \neq b$.

For the second part, we suppose that S is a strict partial ordering on A and must prove that $P = S \cup E$ is a partial ordering, that is, P is reflexive, antisymmetric and transitive. Reflexivity is immediate since $[(a,a) \in E] \Rightarrow [(a,a) \in S \cup E] \Rightarrow [(a,a) \in P]$. As to antisymmetry of P, suppose that $aPb \wedge bPa$, so

$$[(aSb) \vee (a = b)] \wedge [(bSa) \vee (b = a)].$$

On the other hand, asymmetry of S shows that $a \not S b$ or $b \not S a$, so the previous inset equation can be valid only if $a = b$, and we conclude that P is indeed antisymmetric.

We turn finally to transitivity of P, and suppose that $aPb \wedge bPc$ for some $a, b, c \in A$. According to the definition of $P = S \cup E$, we conclude that

$$[(aSb) \vee (a = b)] \wedge [(bSc) \vee (b = c)],$$

so there four cases, as follows

$$\begin{aligned}
[(aSb) \wedge (bSc)] &\Rightarrow (aSc) \\
[(aSb) \wedge (b = c)] &\Rightarrow (aSc) \\
[(a = b) \wedge (bSc)] &\Rightarrow (aSc) \\
[(a = b) \wedge (b = c)] &\Rightarrow (a = c).
\end{aligned}$$

Thus, in any case $(aSc) \vee (a = c)$ so indeed aPc, as desired. *q.e.d.*

The previous result shows that partial orderings and strict partial orderings are essentially the same mathematical structures. In fact, one usually studies posets directly and regards strict partial orderings as a subsidiary concept.

To close, we recall from the table in §5.4 that the intersection of any two partial orderings on a common set A is again a partial ordering on A. On the other hand, the union of two partial orderings on A may fail to be a partial ordering A, as we leave it for the reader to verify (in fact, both antisymmetry and transitivity can fail).

EXERCISES

5.8 POSETS

1. Determine which of the following relations represented by digraphs are actually posets. Prove your assertions.

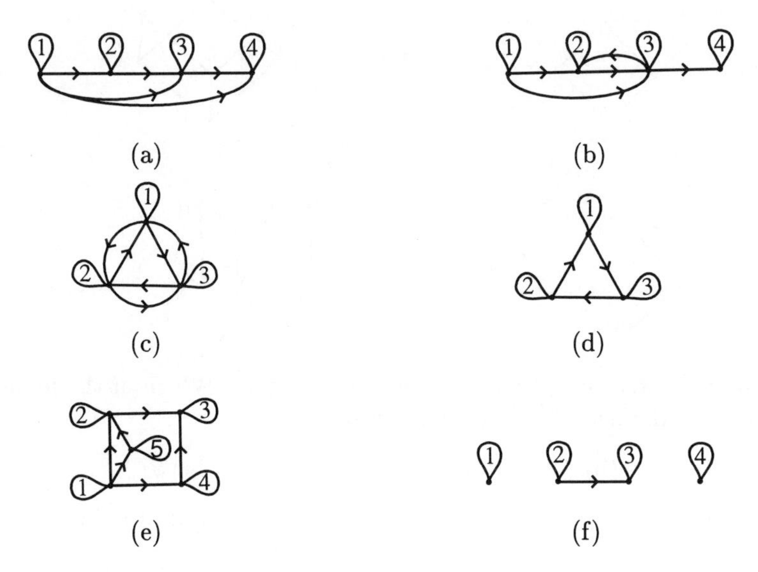

2. Determine whether each of the following relations is actually a partial ordering. Prove your assertions.

(a) The relation $\subseteq$ of inclusion on $\mathcal{P}(\{a,b,c\})$

(b) The relation of proper inclusion on $\mathcal{P}(\{a,b,c\})$

(c) The binary relation $|$ of "divides" on $\mathbb{N}$

(d) The binary relation $|$ of "divides" on the set $\{1,2,3,4,7,8,9\}$

(e) The binary relation

$$\{(a,a),(b,b),(c,c),(d,d),(a,b),(b,c),(c,d),(d,a)\}$$

on the set $\{a,b,c,d\}$

3. Suppose that A is a non-empty set and R is a binary relation on A. Prove that R is a partial ordering if and only if R^{-1} is a partial ordering.

4. Draw the Hasse diagram for each of the following posets.

(a) The relation $|$ of "divides" on $\{2,4,6,8,10,12\}$

(b) The relation $|$ of "divides" on the set $\{1,2,\ldots,15\}$

(c) The relation $\subseteq$ of inclusion on the power set $\mathcal{P}(\{1,2,\ldots,5\})$

5. For each of the following Hasse diagrams, explicitly describe the corresponding partial ordering and draw the usual digraph of this relation.

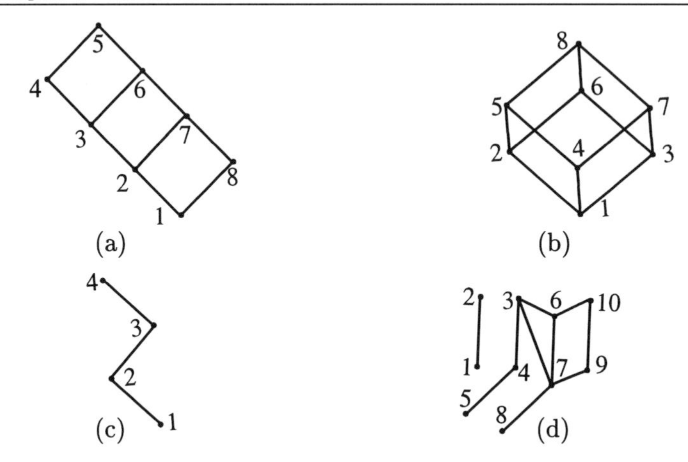

6. Suppose that R, S are partial orderings on a set A. Which of the following are partial orderings? Prove your assertions.

(a) $R \circ S$ (b) $(R - S) \cup (S - R)$ (c) $R^{-1} \cap S^{-1}$

7. Prove that the relation $\mid$ of divides restricts to a partial ordering on $A_r = \{1, 2, \ldots, r\}$ for any $r \geq 1$.

5.9 LINEAR AND WELL ORDERS

There is a marked difference between the posets $(\mathbb{R}, \leq)$ and $(\mathcal{P}(X), \subseteq)$ for a general set X: In the former poset, given any $x, y \in \mathbb{R}$ either $x \leq y$ or $y \leq x$, while in the latter poset (provided X has at least two elements), there are elements $Y, Z \in \mathcal{P}(X)$ so that $Y \not\subseteq Z$ and $Z \not\subseteq Y$.

To formalize these observations, suppose that $\leq$ is a partial ordering on a set A. We say that two elements $a, b \in A$ are *comparable* if either $a \leq b$ or $b \leq a$. Furthermore, we say that $\leq$ is a *linear ordering* (also called a *simple* or *total* ordering) if $\leq$ is a partial ordering on A so that if $a, b \in A$, then either $a \leq b$ or $b \leq a$, and we call $(A, \leq)$ a *linearly ordered set* or a *chain* in this case. In other words, a linear ordering is a partial ordering in which any two elements are comparable. Given elements $a, b \in A$, it may happen that $a \leq b$ and $b \leq a$, but in this case $a = b$ by antisymmetry of the linear ordering $\leq$.

Thus, $\leq$ is a linear ordering on $\mathbb{R}$, while $\subseteq$ is not a linear ordering on $\mathcal{P}(X)$ if X has at least two elements. Of course, $\leq$ is also a linear ordering on $\mathbb{N}$.

■ **Example 5.9.1** Consider a linear ordering $\leq$ on a finite set $A = \{a_1, \ldots, a_n\}$ with $n \geq 1$ elements. In fact, the elements of A can be re-ordered, say in the order $a_{i_1}, \ldots, a_{i_n}$ where $\{1, \ldots, n\} = \{i_1, \ldots, i_n\}$ so that

$$a_{i_1} \leq a_{i_2} \leq \cdots \leq a_{i_n},$$

and this "chain" of inequalities uniquely determines the linear ordering, which must be transitive and reflexive of course. The proof of this claim is left as an exercise for the reader (by induction on n). Thus, a linear ordering $\leq$ on a finite set simply corresponds to a chain $a_{i_1} \leq a_{i_2} \leq \cdots \leq a_{i_n}$ of inequalities.

Suppose that $(A, \leq)$ is a poset and $B \subseteq A$. An element $b \in B$ is called a *greatest* element of B if $\forall b' \in B(b' \leq b)$. Dually, an element $b \in B$ is called a *least* element of B, if $\forall b' \in B(b \leq b')$. For instance, in the finite linearly ordered set in Example 5.9.1, the least element of A itself is a_{i_1} and the greatest is a_{i_n}. As a point of terminology if X is a partially ordered set, then its least element, if it exists (and its greatest element, respectively), is written $\min X$ (and $\max X$) and is called the *minimum* or *infimum* (and the *maximum* or *supremum*) of X.

■ **Example 5.9.2** Let $A = \mathcal{P}(\{a, b\})$ denote the power set of $\{a, b\}$ with the usual partial ordering $\subseteq$ of inclusion of sets. If $B = \{\{b\}\}$, then b is both the greatest and least element of B. If $B = \{\{a\}, \{b\}\}$, then there is neither a greatest nor least element of B, since $\{a\}$ and $\{b\}$ are not comparable under $\subseteq$. Finally, if $B = \{\{b\}, \emptyset\}$, then $\emptyset$ is a least and $\{b\}$ a greatest element of B. Of course, $\{a, b\}$ is a greatest element of $B = A$ itself.

■ **Example 5.9.3** We have proved in Theorem 3.7.3 that any non-empty subset of $\mathbb{N}$ actually has a least element. On the other hand, the subset $\mathbb{N} \subseteq \mathbb{N}$ evidently has no greatest element. In contrast, any *finite* subset of $\mathbb{N}$ has a greatest element (a fact we used in defining the greatest common divisor). Of course, any finite linearly ordered set has both a least and greatest element.

■ **Example 5.9.4** Consider the usual linear ordering $\leq$ on $\mathbb{R}$. The subset $(0, 1) = \{x \in \mathbb{R} : 0 < x < 1\}$ has neither least nor greatest element, whereas the subset $[0, 1) = \{x \in \mathbb{R} : 0 \leq x < 1\}$ has a least but no greatest element.

The reader should infer from these examples that greatest and least elements of $B \subseteq A$, where A is partially ordered, may or may not exist. On the other hand, if a greatest or least element of B exists, then it is unique, as we next prove.

5.9.1 THEOREM *Let $(A, \leq)$ be a poset and suppose that $B \subseteq A$. If a and b are both greatest elements of B (or if a and b are both least elements of B), then $a = b$.*

Proof Suppose that a and b are both greatest elements of B, so in particular, $a, b \in B$. Since a is greatest and $b \in B$, we have $b \leq a$, and since b is greatest and $a \in B$, we have $a \leq b$. By antisymmetry of the partial ordering $\leq$, we conclude that therefore $a = b$. The proof that two least elements are identical is analogous and left as an exercise. *q.e.d.*

Suppose that $(A, \leq)$ is a linearly ordered set with corresponding strict ordering $<$. Generalizing the usual notation for intervals in the set $\mathbb{R}$ with its standard linear ordering, given $b, c \in A$, we define the *intervals* in A by

$$\begin{aligned}(b,c) &= \{a \in A : b < a < c\},\\ [b,c] &= \{a \in A : b \leq a \leq c\},\\ (b,c] &= \{a \in A : b < a \leq c\},\\ [b,c) &= \{a \in A : b \leq a < c\}.\end{aligned}$$

The first interval is said to be *open*, the second is said to be *closed*, and the last two are said to be *half-open* or *half-closed*. In particular, an interval $[b, c] = \emptyset$ unless $b \leq c$, and if $b = c$, then $[b, c] = \{b\}$ while $(b, c) = (b, c] = [b, c) = \emptyset$. We also mention that some authors use the notation $]b, c[$ for the open interval (b, c) and write $]b, c]$, $[b, c[$ for the half-open intervals, but we shall always use the notation above.

■ **Example 5.9.5** Consider the finite linearly ordered set $(A, \leq)$ determined by the chain

$$a_1 \leq a_2 \leq \cdots \leq a_{10}$$

of inequalities. We have the intervals

$$\begin{aligned}[a_1, a_1) &= \emptyset,\\ [a_1, a_1] &= \{a_1\},\\ [a_1, a_4] &= \{a_1, a_2, a_3, a_4\},\\ (a_2, a_5] &= \{a_3, a_4, a_5\},\\ (a_4, a_{10}) &= \{a_5, a_6, a_7, a_8, a_9\},\\ [a_2, a_1] &= \emptyset,\end{aligned}$$

for instance.

Suppose that $(A, \leq)$ is a linearly ordered set. We say that $\leq$ is a *well order* on A (and we call $(A, \leq)$ a *well ordered set*), if every non-empty subset $B \subseteq A$ of A has a least element.

■ **Example 5.9.6** Any finite linearly ordered set is actually well ordered.

■ **Example 5.9.7** We have seen above that $\leq$ is a linear ordering on $\mathbb{N}$, and in fact from Theorem 3.7.3, $(\mathbb{N}, \leq)$ is also a well ordered set.

■ **Example 5.9.8** For a negative example, $(\mathbb{Z}, \leq)$ is a linearly ordered set, but it is not well ordered: For instance, the collection of all negative integers has no least element.

■ **Example 5.9.9** For another negative example, the usual linear ordering $\leq$ on the set $\mathbb{R}$ is not a well order; for instance, any open interval $(a, b) \subseteq \mathbb{R}$ has no least element.

Recapitulating parts of the discussion above, we have defined the notion of a partial ordering $\leq$ on a set A, namely $\leq$ is a reflexive, antisymmetric, and transitive binary relation on A. More specially, $\leq$ is a linear ordering if any two elements of the poset A are comparable under $\leq$. More specially still, a linear ordering $\leq$ is a well ordering if any non-empty subset of A has a a least element for $\leq$. All three of these concepts are of basic importance in mathematics.

As a parenthetical note, we mention that the natural province of proofs by induction is actually for an assertion of the form $\forall x \in XP(x)$ over a universe X which is itself a well ordered set $(X, \leq)$. This general version of induction is called *transfinite induction* and is developed in the exercises.

Recall from §5.1 that if R is a binary relation on A and $A' \subseteq A$, then we may restrict R to $A' \times A' \subseteq A \times A$ to get a binary relation R' on the subset A'. Our next basic result states that partial, linear, and well orderings always restrict, respectively, to partial, linear and well orderings.

5.9.2 THEOREM *If $(A, \leq)$ is a poset and $A' \subseteq A$, then the restriction of $\leq$ to A' is again a partial ordering; if $\leq$ is a linear ordering, then its restriction to A' is again a linear ordering; if $\leq$ is a well ordering, then its restriction to A' is again a well ordering.*

Proof Each property required of the restriction of $\leq$ follows directly from the corresponding property of $\leq$ itself. A more complete proof is a good exercise in manipulating the definitions, which we therefore leave to the reader. *q.e.d*

We turn finally to an extended discussion of a useful method of producing a new binary relation $L \subseteq (A \times B)^2$ given binary relations R on A and S on B. Indeed, we define the relation $L \subseteq (A \times B)^2$ by setting

$$(a, b)L(c, d) \text{ if and only if } [aRc] \wedge [(a = c) \Rightarrow (bSd)].$$

The relation L is called the *lexicographic ordering* (or *dictionary ordering*) on $A \times B$ derived from $R \subseteq A^2$ and $S \subseteq B^2$. For instance, if $A = \{a, b, c\}$ with linear ordering given by $a \leq b \leq c$ and $B = \{c, d, e\}$ with linear ordering given by $c \leq d \leq e$, then in the derived lexicographic ordering L on $\{a, b, c\} \times \{c, d, e\}$, we have

$$(a, d)T(a, e), (a, e)T(b, c), (b, c)T(b, d), (b, d)T(c, d), \text{ and } (c, c)T(c, e).$$

Of course, this example already illustrates the rationale for the terminology "lexicographic" ordering.

■ **Example 5.9.10** Consider the lexicographic ordering on $\mathbb{R} \times \mathbb{R}$ induced by the usual linear ordering $\leq$ on each factor $\mathbb{R}$ in the Cartesian product. To analyze intervals in $\mathbb{R}^2$ with lexicographic ordering, suppose that $a, b, c, d \in \mathbb{R}$ with $(a, b)L(c, d)$; if $a < c$, then the interval $((a, b), (c, d))$ is

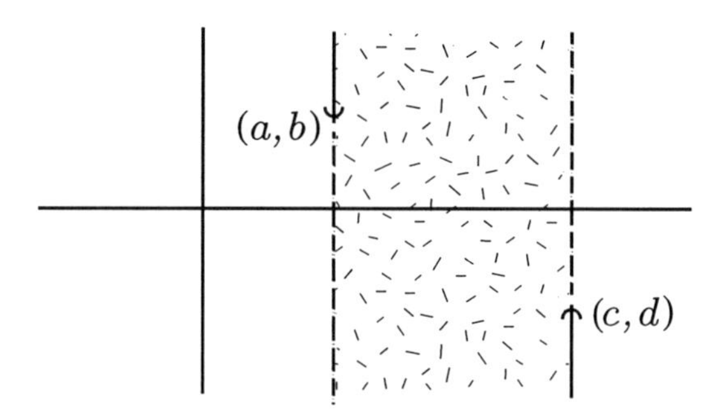

while if $a = c$ and $b < d$, then the interval $((a,b),(c,d))$ is

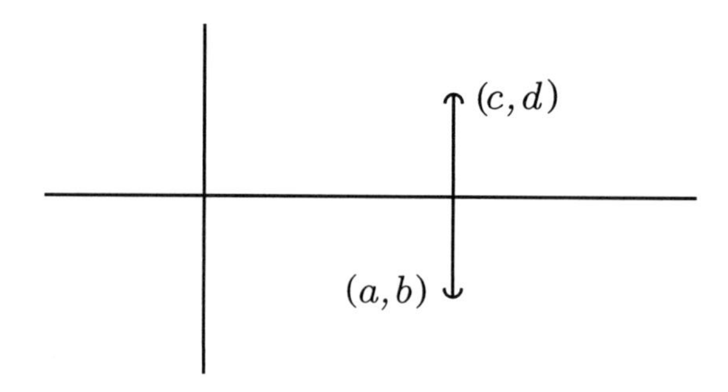

We have the following theorem which roughly states that the lexicographic order derived from partial, linear and well orderings is itself a partial, linear and well ordering, respectively

5.9.3 THEOREM *Suppose that $R \subseteq A^2$ and $S \subseteq B^2$ are partial orderings. If L is the derived lexicographic ordering on $A \times B$, then L is also a partial ordering. Furthermore, if R and S are each linear orderings, then so too is L. Finally, if R and S are each well orderings, then so too is L.*

Proof We first prove that L is reflexive, antisymmetric and transitive. For reflexivity, notice that if $(a,b) \in A \times B$, then aRa and bSb by reflexivity of R and S, and it follows from the definition of L that $(a,b)L(a,b)$, so L is indeed reflexive. Turning to antisymmetry, suppose that $(a,b)L(c,d)$ and $(c,d)L(a,b)$, so

$$[aRc] \wedge [(a = c) \Rightarrow (bSd)] \wedge [cRa] \wedge [(c = a) \Rightarrow (dSb)].$$

In particular, we have $aRc \wedge cRa$, so $c = a$ by antisymmetry of R. We conclude that $bSd \wedge dSb$, so $b = d$ by antisymmetry of S. Thus, we find that $(a,b) =$

(c,d), so L is indeed antisymmetric. Turning finally to transitivity, suppose that $(a,b)L(c,d)$ and $(c,d)L(e,f)$, so

$$[aRc] \wedge [(a=c) \Rightarrow (bSd)] \wedge [cRe] \wedge [(c=e) \Rightarrow (dSf)].$$

In particular, we have $aRc \wedge cRe$, so aRe by transitivity of R. If $a \neq e$, then indeed $(a,b)L(e,f)$, as desired. On the other hand, if $a = e$, then we conclude that $aRc \wedge cRa$, so $a = c$ by antisymmetry of R. Thus, we must have $bSd \wedge dSf$, so bSf by transitivity of S, and finally, then $(a,b)L(e,f)$. This completes the proof that L is transitive, and hence L is a partial ordering.

For the second assertion of the theorem, suppose now that R and S are actually linear orderings. Given $(a,b),(c,d) \in A \times B$, we must have either aRc or cRa (since R is a linear ordering), so if $a \neq c$, then either $(a,b)L(c,d)$ or $(c,d)L(a,b)$, respectively. On the other hand, we have either bSd or dSb (since S is a linear ordering), so if $a = c$, then either $(a,b)L(c,d)$ or $(c,d)L(a,b)$, respectively. Thus, $(a,b),(c,d) \in A \times B$ are indeed comparable, so L is a linear ordering.

Finally, suppose that R and S are each well orderings, and let $\emptyset \neq X \subseteq A \times B$. Define the "projection" $X_A = \{a \in A : (a,b) \in X\}$, so $\emptyset \neq X_A \subseteq A$ has a least element $a_0 \in A$ since A is well ordered. Now, consider the subset $X' = \{b \in B : (a_0, b) \in X\}$, so $\emptyset \neq X' \subseteq B$ has a least element $b_0 \in B$ since B is well ordered. By construction, $(a_0, b_0) \in X$ is the least element of X (as the reader should check) for the lexicographic ordering L, so L is indeed a well order.

q.e.d.

■ **Example 5.9.11** We conclude from the previous theorem that the lexicographic ordering L on $\mathbb{N} \times \mathbb{N}$ induced by the usual well ordering on each factor $\mathbb{N}$ in the Cartesian product is itself a well ordering. Let us briefly analyze this example. By definition, $(1,n)L(m,n)$ for any $m,n \in \mathbb{N}$, and furthermore, $(m,n)L(p,q)$ whenever $m < p$. Moreover, each (m,n) with $n \neq 1$ has an "immediate predecessor" (p,q), by which we mean

$$(p,q)L(m,n) \text{ and } \forall (x,y) \in \mathbb{N}^2[(x,y)L(m,n) \Rightarrow (x,y)L(p,q)],$$

or more informally, $(p,q)L(m,n)$ and there is no other element of $\mathbb{N}^2$ "between" (p,q) and (m,n). Indeed, the immediate predecessor of (m,n) is $(m,n-1)$ provided $n \neq 1$. In contrast, the elements $(m,1) \in \mathbb{N}^2$ do not have an immediate predecessor. To prove this, simply notice that $(p,q)L(m,1) \Rightarrow (p,q+1)L(m,1)$, so $(p,q+1)$ lies between (p,q) and $(m,1)$ for any $(p,q) \in \mathbb{N}^2$.

We have discussed the previous two examples in order to persuade the reader that the intuition that a well or linearly ordered set must "look like" $\mathbb{N}$ or $\mathbb{R}$ can be misleading.

EXERCISES

5.9 LINEAR AND WELL ORDERS

1. Give a complete proof of Theorem 5.9.2.

2. Determine whether each of the following binary relations is a strict partial ordering, partial ordering, linear ordering, or well ordering. Prove your assertions.

   (a) The relation $<$ of strict inequality on the set $\mathbb{N}$

   (b) The relation $\leq$ of weak inequality on the set $\mathbb{N}$

   (c) The relation $\leq$ of weak inequality on the set $\mathbb{Z}$

   (d) The relation $\leq$ of weak inequality on the set $\mathbb{R}$

   (e) The relation $\subseteq$ of containment on the set $\mathcal{P}(\mathbb{N})$

   (f) The relation of proper containment on the set $\mathcal{P}(\mathbb{N})$

3. Determine whether the each of the following digraphs represents a strict partial ordering, partial ordering, linear ordering, or well ordering.

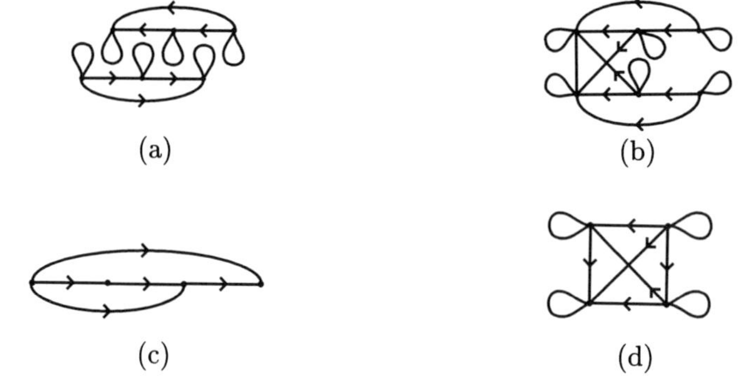

4. Consider the relation $R \subseteq \mathbb{R}^2$, where $(a, b)R(c, d)$ if and only if $(a \leq c) \wedge (b \leq d)$. Prove or disprove each of the following.

   (a) R is a partial ordering.

   (b) R is a linear ordering.

   (c) R is a well ordering.

5. Consider the relation $R \subseteq \mathbb{R}^2$, where $(a, b)R(c, d)$ if and only if $(a < c) \vee (a = c \wedge b \geq d)$. Prove or disprove each of the following.

   (a) R is a partial ordering

   (b) R is a linear ordering

   (c) R is a well ordering

6. Give an example of a well ordering on $\mathbb{Z} \times \mathbb{Z}$ other than the usual lexicographic ordering.

7. Let R be a binary relation on a set A. Prove or disprove each of the following.

(a) If R is a partial ordering, then so too is R^{-1}.

(b) If R is a linear ordering, then so too is R^{-1}.

(c) If R is a well ordering, then so too is R^{-1}.

8. Suppose that R and S are binary relations on a set A. Prove or disprove each of the following.

(a) If R and S are linear orderings, then so too is $R \cap S$.

(b) If R and S are linear orderings, then so too is $R \circ S$.

9. Suppose that R is a partial ordering on a finite set A. Show that there is some linear ordering S on A so that $R \subseteq S$.

10. Suppose that R is a linear ordering on a finite set A and prove that the intersection of two intervals in A is either empty or another interval in A.

11. The *length* of a chain (i.e., a linearly ordered subset) in a finite poset is simply the number of elements comprising it.

(a) Enumerate the chains of length two in the poset $(\mathcal{P}(\{1,2,3\}), \subseteq)$.

(b) How many chains of length 2 are there in the poset $(\mathcal{P}(\{1,2,\ldots,n\}), \subseteq)$?

(c) How many chains of length 3 are there in the poset $(\mathcal{P}(\{1,2,\ldots,n\}), \subseteq)$?

12. Suppose that $(X, \leq)$ is a well ordered set. An element $x \in X$ determines its *initial segment* $s(x) = \{y \in X : y \leq x \text{ and } y \neq x\}$; in other words, the set of elements of X strictly preceding x is the initial segment $s(x)$. The *principle of transfinite induction* asserts that if $S \subseteq X$ so that for each $x \in X$ we have the implication $s(x) \subseteq S \Rightarrow x \in S$, then we must have $S = X$. (See also Zermelo's Well Ordering Principle discussed in §5.11.)

(a) Consider the set $\mathbb{N}^+ = \mathbb{N} \cup \{\infty\}$ together with an order $\preceq$, where $m \preceq n$ if $m, n \in \mathbb{N}$ and $m \leq n$ as usual and where $n \preceq \infty$ for each $n \in \mathbb{N}$. Prove that $\preceq$ is a well ordering on $\mathbb{N}^+$.

(b) For each element $x \in \mathbb{N}^+$, describe the initial segment $s(x) \subseteq \mathbb{N}^+$.

(c) Prove the principle of transfinite induction.

13. Prove that if $\leq$ is a linear ordering on the finite set $A = \{a_1, a_2, \ldots, a_n\}$, then we may re-order the elements of A as $a_{i_1}, a_{i_2}, \ldots, a_{i_n}$, where $a_{i_1} \leq a_{i_2} \leq \cdots \leq a_{i_n}$.

14. Consider the line $\mathcal{L} = \{(x,y) \in \mathbb{R}^2 : x = y\}$. Describe the restriction to $\mathcal{L}$ of the lexicographic ordering on $\mathbb{R}^2$ induced by the usual linear ordering on each factor $\mathbb{R}$.

(*) 5.10 BOUNDS IN POSETS

Suppose that $B \subseteq A$, where $(A, \leq)$ is a poset. We have considered the notion of greatest and least elements of B in the previous section, and we here elaborate in this optional section on these ideas and discuss other related distinguished elements which are determined by B. We say that an element $b \in B$ is a *maximal* element of B (and a *minimal* element of B, respectively) if $b \in B$ and there is no element $b' \in B$ so that $b \neq b'$ and $b \leq b'$ (and $b' \leq b$, respectively). Thus, just as for greatest and least elements of B, a maximal or minimal element of B is necessarily also an element of B by definition.

In the next definitions, we describe elements of A (as opposed to elements of B as in the definitions above) which are determined by the subset $B \subseteq A$. We say $a \in A$ is an *upper bound* for B (and a *lower bound*, respectively) if for each $b \in B$, we have $b \leq a$ (and $a \leq b$, respectively). Furthermore, we say that an upper bound $a \in A$ for B is a *least upper bound* (often abbreviated simply "*lub*") for B if for every upper bound b of B, we have $a \leq b$. In the same way, $a \in A$ is a *greatest lower bound* (abbreviated simply "*glb*") if for every lower bound b of B, we have $b \leq a$.

Notice that by definition a greatest element of B is necessarily maximal and a least element of B is necessarily minimal.

We have no guarantee that any of these distinguished elements of A or B actually exist, and often they do not. Let us give some examples of both existence and non-existence of these distinguished elements.

■ **Example 5.10.1** Let A denote the poset $\mathcal{P}(\{a, b\})$ with the partial ordering given by inclusion $\subseteq$ of sets, which has been considered before. Taking $B = \{\{b\}\}$, we find that $\{b\}$ and $\{a, b\}$ are upper bounds, and $\{b\}$ is the least upper bound. Dually, the lower bounds of B are $\{b\}$ and $\emptyset$, and $\{b\}$ is also the greatest lower bound.

■ **Example 5.10.2** Consider the poset $\mathbb{R}$ with the usual linear ordering $\leq$, and take $B = \{t \in \mathbb{R} : 0 < t \leq 1\}$. In this case, B has no minimal or least element, but it does have common greatest and maximal element, namely $1 \in B$. The collection of upper bounds of B is the set $\{t \in \mathbb{R} : t \geq 1\}$, and the collection of lower bounds of B is the set $\{t \in \mathbb{R} : 0 \leq t\}$. Thus, 1 is the *lub* and 0 is the *glb* of B, and so B contains its *lub* but does not contain its *glb* in this case.

■ **Example 5.10.3** Consider the collection $A = \{1, \ldots, 6\}$ together with the binary relation $a|b$, i.e., a divides b, on A. We saw in the previous section that $A = A_6$ is indeed a poset, and we first let $B = A$, so $4, 5, 6$ are all maximal elements, B has no greatest element, and B has no upper bounds (hence no *lub*). In contrast, 1 is the least element, the minimal element, the unique lower bound, and hence the *glb* of B.

The subset $B = \{1, 2, 3\} \subseteq A_6$ has maximal elements 2 and 3, has no greatest element, and 1 is its least and minimal element. The unique upper bound of B in

A is 6 (so 6 is also the *lub* of B), and 1 is the unique lower bound and *glb* of B.

The Hasse diagram of the poset $(A, \mid)$ nicely illustrates these remarks

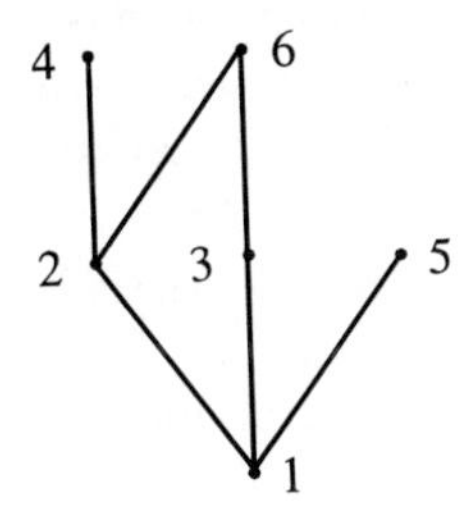

■ **Example 5.10.4** For any finite linearly ordered set A and any subset $B \subseteq A$, the *lub* and *glb* of B in A clearly exists and is an element of B.

Observe that we have given examples where many different upper bounds may exist. In contrast, we next prove that if *lub* or *glb* exists, then it must be unique.

5.10.1 THEOREM *Let $B \subseteq A$, where $(A, \leq)$ is a poset. If $a, b \in A$ are each least upper bounds for B (or if $a, b \in A$ are each greatest lower bounds for B), then $a = b$.*

Proof If a, b are both *lub*'s of B, then in particular a, b are both upper bounds of B. Since a is least, we have $a \leq b$, and since b is least, we have $b \leq a$, so $b = a$ by antisymmetry of $\leq$. The corresponding assertion for *glb*'s is proved similarly. *q.e.d.*

We next explicate some simple relationships among the concepts introduced above.

5.10.2 THEOREM *Let $B \subseteq A$, where $(A, \leq)$ is a poset. Then we have*

(a) *If b is a greatest element of B, then b is a maximal element of B; furthermore, b is also the lub of B. Likewise, for least and minimal elements.*

(b) *If $a \in B$ is an upper bound (i.e., if the set B contains an upper bound a), then a is the greatest element of B. Likewise, for lower bounds and least elements.*

Proof For the first assertion in part (a), we prove instead the contrapositive: if b is not maximal, then b is not greatest. To prove this, if b is not maximal, then there must exist $b' \in B$ so that $b' \neq b$ and $b \leq b'$. Using antisymmetry of $\leq$, we conclude that $b' \leq b$ must be false, so indeed, b is not greatest.

For the second assertion in part (a), suppose that b is greatest. By definition, b is automatically an upper bound of B, and we must prove that b is the least upper bound. To this end, if a is another upper bound of B, then $b \leq a$ (since the greatest element b lies in B by definition), so b is the least upper bound of B.

Finally for part (b), if $a \in B$ is an upper bound of B, then $b \leq a$ for all $b \in B$, and since $a \in B$, we conclude that a is the greatest element of B. *q.e.d.*

To close, we mention that a *lattice* is a partially ordered set A so that for any doubleton $\{a, b\} \subseteq A$, the *glb* and *lub* of $\{a, b\}$ each exist. For instance, a finite linearly ordered set is a lattice by Example 5.10.4. Lattices are among the most basic of mathematical objects, but we postpose presenting the prototypical examples and a further discussion to the optional §5.16 below. We just remark here that one regards each of *glb* and *lub* of doubletons as a binary operation on A in the natural way, and under suitable circumstances (for a so-called "Boolean lattice"), these operations are used to define a Boolean algebra (which was discussed before Examples 3.6.4 and 3.6.5). Thus, a lattice is a mathematical object more general than a Boolean algebra and, of course, more special than a poset.

EXERCISES

5.10 BOUNDS IN POSETS

1. Consider the relation $|$ of "divides" on $\mathbb{N}$, and let A denote the subset $A = \{28, 42, 70, 154, 2002\} \subseteq \mathbb{N}$.

(a) Find a lower bound for A in $\mathbb{N}$.

(b) Find an upper bound for A in $\mathbb{N}$.

(c) Does A have a greatest element or a least element? Prove your assertions.

(d) Does A have a *glb* or *lub* in $\mathbb{N}$? Prove your assertions.

2. Construct examples of the following.

(a) A non-empty linearly ordered set in which some subsets do not have a least element

(b) A non-empty partially ordered set which is not linearly ordered and in which some subsets do not have a greatest element

(c) A poset together with a subset which has a *lub* but no greatest element

(d) A poset together with a subset which has a lower bound but no *glb*

3. Prove each of the following assertions.

(a) A finite non-empty poset has at least one minimal and maximal element.

(b) In a linearly ordered set, a minimal element of a subset is least and a maximal element is greatest.

(c) A finite non-empty subset of a linearly ordered set has both a greatest and least element.

4. Take the partial ordering $\subseteq$ of inclusion on $A = \mathcal{P}(Z)$, and suppose that $X, Y \in A$. Find the *glb* and *lub* of $\{X, Y\}$.

5. On the set $A = \mathbb{N}$ take the usual binary relation $|$ of "divides". For each $m, n \in \mathbb{N}$, find the *glb* and *lub* of $\{m, n\}$.

(*) 5.11 AXIOM OF CHOICE REVISITED

We have discussed in Chapter 3 the following

AXIOM OF CHOICE (AC) *Every set of non-empty sets has a choice function. That is, if S is a set of sets and $\emptyset \notin S$, then there is a function $f : S \to \cup S$ so that $f(X) \in X$ for each $X \in S$.*

and we remarked that there were other equivalent formulations of this axiom. This optional section is dedicated to a brief discussion of these other formulations, and we shall abbreviate the Axiom of Choice here by writing simply "AC", as indicated above.

The simplest reformulation of AC is the following statement.

ZERMELO'S WELL ORDERING PRINCIPLE (ZWOP) *Given any set X, there exists some well order on X.*

A rather more complicated but extremely useful formulation of AC will be given below. In preparation for this result, suppose that A is a poset and $B \subseteq A$. In this case, the partial ordering $\leq$ on A restricts to a binary relation

$$\{(b_1, b_2) \in B^2 : (b_1, b_2) \in \leq\}$$

on B as in §5.1, and this binary relation is actually a partial ordering on B by Theorem 5.9.2. Thus, given the poset $(A, \leq)$ and the subset $B \subseteq A$, there is a corresponding partial ordering on B, and we abuse notation (slightly) and use the same symbol $\leq$ for the partial ordering induced on B.

KURATOWSKI-ZORN LEMMA (KZL) *Suppose that $(P, \leq)$ is a non-empty poset, and assume that the following condition holds*

If $L \subseteq P$ so that the partial ordering $\leq$ induced by restriction to L is actually a linear ordering, then L has an upper bound for $\leq$ in P.

Then P has a maximal element for $\leq$.

These various incarnations of AC are equivalent in the following precise sense, whose proof is beyond the scope of our discussion here (but see [P. R. Halmos, *Naive Set Theory*, Springer-Verlag (1974)] for a proof which should accessible to the interested reader).

5.11.1 THEOREM *Assuming the axioms of ZF set theory, we have the logical equivalences* $AC \Leftrightarrow ZWOP \Leftrightarrow KZL$.

We mentioned in Chapter 3 that a certain reformulation of AC seemed sufficiently bizarre to mathematicians early in the twentieth century that it led to the questioning of AC as a reasonable axiom, and the formulation we alluded to there is ZWOP. To get a sense of why ZWOP indeed seems bizarre, according to this axiom, there must be a well order on $\mathbb{R}$, yet nobody has successfully written down such a well order, and it is not for lack of trying! Thus, even in this very special case of ZWOP when $X = \mathbb{R}$, nobody has been able to exhibit the putative (that is, the supposed) well order, and this perhaps seems suspicious. At any rate, it seemed suspicious early in the twentieth century.

Let us emphasize that AC in any of its equivalent formulations is essentially universally accepted in mathematics today. In fact, the KZL formulation, which is usually referred to simply as "Zorn's Lemma", is perhaps the most useful formulation of AC. For instance, KZL is a standard tool in the field of algebra, and we shall see examples of its application in Chapter 7 below.

EXERCISES

5.12 AXIOM OF CHOICE REVISITED

1. Prove several of the implications in Theorem 5.11.1

2. Assuming Theorem 5.11.1, prove the logical equivalence of AC and each of the following conditions:

   (a) Every partial order R on A admits a chain $K \subseteq A$ which contains all its upper bounds.

   (b) [*Hausdorff maximum principle*] Every partial order R on a set A admits a maximal chain.

3. We say that $K \subseteq \mathcal{P}(A)$ has *finite character* over A if a subset B of A is a member of K if and only if every finite subset of B is a member of K. Assume Theorem 5.11.1 and prove the equivalence of AC and the following *Teichmüller-Tukey Lemma:* A non-empty set K with finite character over A has a maximal element.

4. Assume AC and prove that if A is a set of non-empty pairwise disjoint sets, then there exists a set containing one element from each member of A.

5.12 EQUIVALENCE RELATIONS

Just as our study of posets in the previous sections was motivated by the fact that they are one of the basic types of relations which arise most commonly in practice, we undertake here the study of another such basic mathematical structure. Precisely, we say that a binary relation $R \subseteq A^2$ on a set A is an *equivalence relation* if R is reflexive, symmetric, and transitive. In other words, a corresponding digraph G satisfies the following properties: there is a loop based at each vertex of G; for each edge e of G other than a loop, both orientations on e occur as oriented edges of G; for any oriented edge-path of length two with initial point a and terminal point b, there is an edge in G from a to b.

To get a feeling for this definition, we begin with some examples.

■ **Example 5.12.0** Let $A_n = \{1, \ldots, n\}$ and consider the universal relation $A_n^2 \subseteq A_n^2$. The corresponding digraph G_n is called the *complete digraph on n vertices.* For instance, the complete digraph on $n = 3$ vertices is

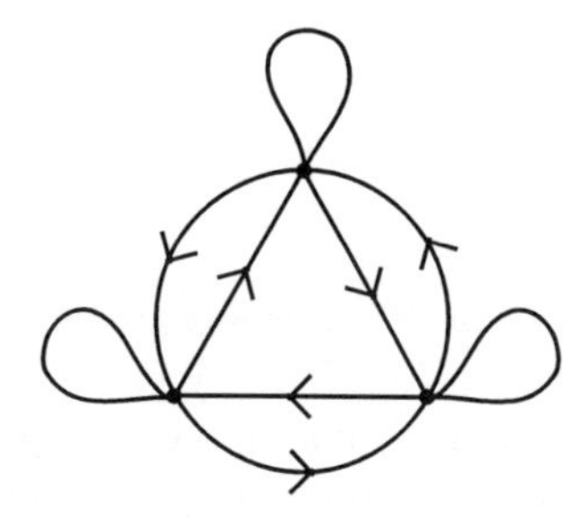

Reflexivity in this case is manifest (since each vertex of G_3 has a loop) as is symmetry (since for each edge which is not a loop, both orientations occur in G_3). Moreover, transitivity holds trivially since for any two vertices v_1, v_2, there is an edge of G_3 with initial point v_1 and terminal point v_2. Thus, the universal relation on A_3 is an equivalence relation.

More formally and in general, since $(a,b),(b,a) \in A_n^2$ for each $a, b \in A_n$, we have that A_n^2 is symmetric and (taking $a = b$) reflexive as well. As to transitivity, if $(a,b),(b,c) \in A_n^2$, then $a, b, c \in A_n$, so in particular $(a,c) \in A_n^2$. In effect, since each of the properties reflexive, symmetric, and transitive requires the existence of certain edges and the universal relation includes all possible edges, the universal relation is an equivalence relation.

■ **Example 5.12.1** On the set $A = \{a, b, \ldots, e\}$, take the relation $R = E_A \cup \{(a,b),(b,a),(d,e),(e,d)\}$, where E_A is as usual the relation of equality on A, so the digraph G is given by

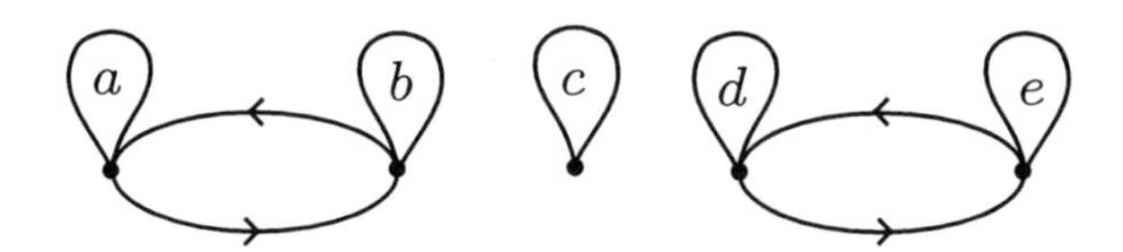

Notice how the vertices of G decompose into "blocks" $\{a,b\}$, $\{c\}$, $\{d,e\} \subseteq A$, where any two vertices in such a block can be joined by an edge-path in G; indeed, the restriction to each such block of the relation R on A is the corresponding universal relation, and the associated complete digraph which is a sub-digraph of G is called a "component" of G. There are thus three components in this example, namely, the digraphs $\{(a,a),(b,b),(a,b),(b,a)\}$, $\{(c,c)\}$, and $\{(d,d),(e,e),(d,e),(e,d)\}$. This is the typical behavior for an equivalence relation on a finite set: The digraph is comprised of a finite collection of disjoint complete sub-digraphs, one such sub-digraph for each component.

■ **Example 5.12.2** For any set A, the relation E_A of equality on A is reflexive by definition, symmetric (since $(a,b) \in E_A \Rightarrow (a=b)$ so $(b,a)=(a,a) \in E_A$), and transitive (since $[(a=b) \wedge (b=c)] \Rightarrow [a=c]$). Thus, the relation of equality on any set is an equivalence relation.

■ **Example 5.12.3** Suppose that P is a predicate defined over some universe $\mathcal{U}$, and define a binary relation $\sim$ on $\mathcal{U}$, where by definition $x \sim y$ if and only if $P(a) \Leftrightarrow P(b)$. Reflexivity of $\sim$ is trivial, symmetry follows from the definition of bi-implication (and commutativity of disjunction), and transitivity was proved in 2.5.11.

It is worth mentioning explicitly that just as one often uses the symbol $\leq$ to denote an arbitrary partial ordering, one often uses the symbol $\sim$ as in the example above to denote an arbitrary equivalence relation. Of course, these are simply notational conventions.

■ **Example 5.12.4** Suppose that $p \in \mathbb{N}$, and define a binary relation R_p on $\mathbb{Z}$, where $(m,n) \in R_p$ if and only if $p|(n-m)$. If $(m,n) \in R_p$, then we say that m and n are *equivalent modulo* p, and we write

$$m \equiv n(p) \Leftrightarrow p|(m-n)$$

using a specialized but completely standard notation.

In particular, equivalence modulo 1 is the universal relation on $\mathbb{Z}$. Turning to the consideration of equivalence modulo 2, observe that an integer n is even if and only if $n \equiv 0(2)$, while n is odd if and only if $n \equiv 1(2)$. Thus, each integer n satisfies either $n \equiv 0(2)$ or $n \equiv 1(2)$, and no integer satisfies both of these conditions as we proved in Example B.2.2 of Chapter 1. In fact, the notion of equivalence modulo $p \in \mathbb{N}$ is a natural generalization of the usual dichotomy of even/odd integers when $p = 2$.

5.12.1 THEOREM *For each $p \in \mathbb{N}$, equivalence modulo p is an equivalence relation. That is, R_p is an equivalence relation on $\mathbb{Z}$ for each $p \in \mathbb{N}$.*

Proof Let us fix $p \in \mathbb{N}$. R_p is reflexive since $p|0$, and R_p is symmetric since obviously $p|(m-n) \Rightarrow p|(n-m)$ for each $m, n \in \mathbb{N}$. As to transitivity, suppose that $m \equiv n(p)$ and $n \equiv r(p)$, so $p|(m-n)$ and $p|(n-r)$. Thus, there are $k, \ell \in \mathbb{N}$ so that $m - n = kp$ and $n - r = \ell p$. Adding these two equations, we find $m - r = (k + \ell)p$, so $m \equiv r(p)$. *q.e.d.*

■ **Example 5.12.5** Define the binary relation $\sim$ on $\mathbb{R}$, where

$$(x \sim y) \Leftrightarrow (x - y \in \mathbb{Z}),$$

which is obviously reflexive since $0 \in \mathbb{Z}$ and symmetric since $y - x = -(x - y)$. For transitivity, if $x \sim y$ and $y \sim z$, where $x, y, z \in \mathbb{R}$, then there are $k, \ell \in \mathbb{Z}$ so that $x - y = k$ and $y - z = \ell$. Adding these two equations, we find $x - z = k + \ell \in \mathbb{Z}$, so indeed $x \sim z$.

■ **Example 5.12.6** Define the binary relation $\sim$ on $\mathbb{Z} \times (\mathbb{Z} - \{0\})$, where

$$[(m, n) \sim (p, q)] \Leftrightarrow (mq = np).$$

Both reflexivity and symmetry of $\sim$ are again obvious. As to transitivity, if $(m, n) \sim (p, q)$ and $(p, q) \sim (r, s)$, then we have $mq = pn$ and $ps = rq$, or in other words, $m/n = p/q$ and $p/q = r/s$. Thus, $m/n = p/q = r/s$, so $ms = nr$, i.e., we have $(m, n) \sim (r, s)$.

Suppose that R is an equivalence relation on a set A. For each element $a \in A$, we define a corresponding subset

$$[a]_R = \{x \in A : xRa\} \subseteq A$$

which is called the *R-equivalence class of a*. When the relation is fixed or understood, then we write $[a] = [a]_R$ and refer to this set simply as the equivalence class of a. In other words, the equivalence class of an element $a \in A$ is the collection of elements of A which are related to a by R.

■ **Example 5.12.7** For each element $m \in A_n$ in Example 5.12.0, we have $[m] = A_n$ since any element of A_n is related to m by the universal relation. There is thus only one equivalence class in this example.

■ **Example 5.12.8** There are the three equivalence classes $[a] = [b] = \{a, b\}$, $[c] = \{c\}$, and $[d] = [e] = \{d, e\}$ for the relation in Example 5.12.1

It is important to emphasize that it often happens that $[a] = [b]$ even though $a \neq b$ as in the previous two examples. In contrast to this, however, we have

■ **Example 5.12.9** For each $a \in A$ in Example 5.12.2, we have $[a] = \{a\}$ since only a is related to a by equality. Thus, each element $a \in A$ gives rise to a distinct equivalence class $\{a\}$ in this case.

■ **Example 5.12.10** Suppose that a lies in the universe $\mathcal{U}$ of the predicate P in Example 5.12.3. If the specialization $P(a)$ is true, then the equivalence class is

$$[a] = \{b \in \mathcal{U} : P(b)\},$$

whereas if the specialization $P(a)$ is false, then the class is

$$[a] = \{b \in \mathcal{U} : \neg P(b)\},$$

and there are thus at most two equivalence classes.

■ **Example 5.12.11** In Example 5.12.4 we have considered equivalence modulo p on $\mathbb{Z}$, and there are exactly p equivalence classes in this case. Specifically, for $r = 0, \ldots, p-1$, we have

$$[r] = \{r + kp : k \in \mathbb{Z}\} \subseteq \mathbb{Z}.$$

Furthermore, given any $n \in \mathbb{N}$, we may apply the Division Algorithm 4.2.1 to write $n = qp + r$, where $0 \leq r < p$, so $[n] = [r]$ for some $r \in \{0, \ldots, p-1\}$. On the other hand, if $n \in \mathbb{Z} - (\mathbb{N} \cup \{0\})$, then we may use the Division Algorithm to write $-n = qp + r$ for some integers $q \geq 0$ and r, where $0 \leq r < p$. Thus, $n = -qp - r$, whence $[n] = [-r]$. Finally, notice that $[-r] = [p - r]$ for each $r \in \mathbb{Z}$, so there are indeed exactly the equivalence classes $[0], [1], \ldots, [p-1]$ for equivalence modulo p.

■ **Example 5.12.12** In Example 5.12.5, there are the distinct equivalence classes

$$[t] = \{t + n : n \in \mathbb{Z}\} \subseteq \mathbb{R},$$

where t satisfies $0 \leq t < 1$, and these are the only sets which arise as equivalence classes in this case.

■ **Example 5.12.13** In Example 5.12.6, there are the distinct equivalence classes $[(p, q)]$, one such equivalence class for each rational number p/q.

5.12.2 THEOREM *Suppose that R is an equivalence relation on a set $A \neq \emptyset$. Then*

(a) *For any* $a, b \in A$, *either* $[a] = [b]$ *or* $[a] \cap [b] = \emptyset$. *That is, two equivalence classes are either disjoint or identical.*

(b) *We have* $\cup_{a \in A}[a] = A$.

Proof For part (a), suppose that $[a] \cap [b] \neq \emptyset$, and choose some $c \in [a] \cap [b]$. Thus,

$$\begin{aligned}[(cRa) \wedge (cRb)] &\Rightarrow [(aRc) \wedge (cRb)] \\ &\Rightarrow aRb \\ &\Rightarrow [(aRb) \wedge (bRa)],\end{aligned}$$

where the first and third implications depend on symmetry and the second on transitivity of R. Thus, we have proved that $([a] \cap [b] \neq \emptyset) \Rightarrow [(aRb) \wedge (bRa)]$.

Now, suppose that $x \in [a]$ so xRa. In light of the previous remarks, we have also aRb, so it follows from transitivity of R that xRb, i.e., $x \in [b]$. We have therefore proved that $[a] \subseteq [b]$, and the analogous argument (interchanging the roles of a and b) proves the reverse inclusion $[b] \subseteq [a]$. We conclude that if $[a] \cap [b] \neq \emptyset$, then $[a] = [b]$, as required.

For part (b), simply observe that $a \in [a]$ for each $a \in A$ by reflexivity, so evidently

$$a \in [a] \subseteq \cup_{a \in A}[a].$$

Thus, we have shown that $A \subseteq \cup_{a \in A}[a]$. The reverse inclusion follows as well since by definition $[a] \subseteq A$ for each $a \in A$, so in fact, $\cup_{a \in A}[a] \subseteq A$. *q.e.d.*

Suppose that R is an equivalence relation on a set A, and define the set

$$A/R = \{[a] \in \mathcal{P}(A) : a \in A\},$$

so A/R is itself a set of sets. The set A/R is called the *quotient* of A by R (and is sometimes referred to simply as "A modulo R"). One thinks of producing A/R from A by "collapsing each equivalence class to a point"; that is, the elements of A/R are exactly the equivalence classes of R on A.

■ **Example 5.12.14** In Example 5.12.0, the quotient A_n/A_n^2 of A_n by the universal relation is simply the singleton $\{A_n\}$.

■ **Example 5.12.15** The quotient is $A/R = \{[a], [c], [e]\}$ in Example 5.12.1.

■ **Example 5.12.16** In Example 5.12.2, we have seen that $[a] = \{a\}$ for each $a \in A$, so

$$A/E_A = \{\{a\} \in \mathcal{P}(A) : a \in A\}.$$

■ **Example 5.12.17** In Example 5.12.3, there are the two possible equivalence classes

$$\{u \in \mathcal{U} : P(u)\} \text{ and } \{u \in \mathcal{U} : \neg P(u)\}$$

■ **Example 5.12.18** In Example 5.12.4, the various equivalence classes of equivalence modulo $p \in \mathbb{N}$ are simply $[r] = \{r + kp : k \in \mathbb{Z}\}$ where $0 \le r < p$, and the quotient

$$\mathbb{Z}/R_p = \{[r] \in \mathcal{P}(\mathbb{Z}) : 0 \le r < p\}$$

has exactly p elements.

There is a standard notation for the quotient in the previous example, where one writes $\mathbb{Z}/p = \mathbb{Z}/R_p$. The notation $\mathbb{Z}/p$ for the quotient of $\mathbb{Z}$ by equivalence modulo p is often read in English as "zee modulo pee" or "zee mod pee" and is completely standard.

■ **Example 5.12.19** In Example 5.12.5, the various equivalence classes were found to be $[t] = \{t + n : n \in \mathbb{Z}\}$, where $0 \le t < 1$, so in this case the quotient

$$\mathbb{R}/\sim \; = \; \{[t] \in \mathcal{P}(\mathbb{R}) : 0 \le t < 1\}$$

is not a finite set.

■ **Example 5.12.20** In Example 5.12.6, we have found that there is a distinct equivalence class $[(p,q)]$ for each rational number $p/q \in \mathbb{Q}$, so the quotient $(\mathbb{Z} \times (\mathbb{Z} - \{0\}))/\sim$ can be identified with the set $\mathbb{Q}$. In fact, one actually *defines* the set $\mathbb{Q}$ to be the quotient $(\mathbb{Z} \times (\mathbb{Z} - \{0\})/\sim$, which fulfills our promise in §3.9 to give a precise definition of $\mathbb{Q}$.

■ **Example 5.12.21** We cannot resist pointing out that the previous Examples 5.12.0-20 should be "taken modulo seven" in the sense that two examples numbered 5.12.i and 5.12.j above for $0 \le i, j \le 20$ refer to the same underlying example if and only if $i \equiv j(7)$.

We have the following immediate consequence of the previous theorem.

5.12.3 THEOREM *Suppose R is an equivalence relation on a set $A \neq \emptyset$, and consider the quotient A/R. Then*

(a) *If $X, Y \in A/R$, then either $X = Y$ or $X \cap Y = \emptyset$.*

(b) *We have $\cup(A/R) = A$.*

In order to become comfortable with the important notion of the quotient of a set by an equivalence relation, the reader should carefully verify that this result is simply a reformulation of the previous one.

To complete our discussion of quotients, we have

5.12.4 THEOREM *Suppose that R_i is an equivalence relation on the set A_i for $i = 1, 2$. Then $R_1 = R_2$ as relations if and only if $A_1/R_1 = A_2/R_2$ as sets.*

Proof Suppose first that $R_1 = R_2$ as relations, so necessarily $A_1 = A_2$. If $X \in A_1/R_1$, then $X = [a]_{R_1}$ for some $a \in A_1$ by definition, so in fact $X = [a]_{R_2} \in A_2/R_2$. Thus, we have shown that $A_1/R_1 \subseteq A_2/R_2$, and an analogous argument proves the reverse inclusion $A_2/R_2 \subseteq A_1/R_1$, as desired.

Conversely, suppose that $A_1/R_1 = A_2/R_2$, so $A_1 = \cup(A_1/R_1) = \cup(A_2/R_2) = A_2$ by Theorem 5.12.3b. Thus, R_1 and R_2 are actually equivalence relations on the common set $A = A_1 = A_2$. Suppose that $(a, b) \in R_1$ for some $(a, b) \in A^2$, so $b \in [a]_{R_1} \in A/R_1$. Now, $[a]_{R_1} = [c]_{R_2}$ for some $c \in A$ since $A/R_1 = A/R_2$, and since $a \in [a]_{R_1} = [c]_{R_2}$ and $a \in [a]_{R_2}$, we conclude that $[a]_{R_2} = [c]_{R_2}$ by Theorem 5.12.3a, so $[a]_{R_1} = [a]_{R_2}$.

Finally, $b \in [a]_{R_1} = [a]_{R_2}$ implies that $(a, b) \in R_2$ as desired. *q.e.d.*

As a final note, we remark that the the intersection $R_1 \cap R_2$ of two equivalence relations is itself an equivalence relation, as the reader may easily check directly or by using the table in §5.4. In contrast, it is *not* true that the union of two equivalence relations is actually an equivalence relation (in fact, the union of two transitive relations is not necessarily transitive), and we leave it to the reader to construct an appropriate counter-example.

EXERCISES

5.12 EQUIVALENCE RELATIONS

1. Prove Theorem 5.12.3.

2. Prove that the behavior described in Example 5.12.1 is indeed typical for equivalence relations on finite sets.

3. Prove that the intersection of two equivalence relations on a common set is again an equivalence relation, and prove that the union of two equivalence relations on a common set is not necessarily an equivalence relation.

4. Determine whether each of the following binary relations on $\mathbb{Z}$ is an equivalence relation, and prove your assertions.

(a) The relation $<$ of weak inequality

(b) $\{(m, n) \in \mathbb{Z}^2 : (m < 0 \wedge n < 0) \vee (m > 0 \wedge n > 0)\}$

(c) $\{(m, n) \in \mathbb{Z}^2 : (m \geq 0 \wedge n \leq 0) \vee (m \geq 0 \wedge n \geq 0)\}$

(d) $\{(m, n) \in \mathbb{Z}^2 : (m \leq 0 \wedge n < 0) \vee (m > 0 \wedge n \geq 0)\}$

(e) $\{(m, n) \in \mathbb{Z}^2 : (m \geq 0 \wedge n \geq 0) \vee (m \leq 0 \wedge n \leq 0)\}$

(f) $\{(m, n) \in \mathbb{Z}^2 : |m - n| \leq 6\}$, $\{(m, n) \in \mathbb{Z}^2 : |m - n| = 6\}$, and $\{(m, n) \in \mathbb{Z}^2 : |m - n| \geq 6\}$

5. Prove that if R is a partial ordering, then $R \cap R^{-1}$ is an equivalence relation.

6. Prove that if R is both a partial ordering and an equivalence relation, then R is the relation of equality.

7. Prove that the empty relation is an equivalence relation on the empty set.

8. Define a binary relation $\sim$ on $\mathbb{R}^2$, where $(a, b) \sim (c, d)$ if and only if $a^2 + b^2 = c^2 + d^2$.

(a) Prove that $\sim$ is an equivalence relation.

(b) What is the equivalence class $[(0,0)]$ of $(0,0) \in \mathbb{R}^2$?

(c) What is the equivalence class $[(0,1)]$ of $(0,1) \in \mathbb{R}^2$?

9. Define a binary relation R on $\mathbb{Q}$, where xRy if and only if when expressed as fractions $x = p/q$, $y = r/s$ in lowest terms (i.e., 1 is the only common divisor of p and q and similarly for r and s), we have $q = s$.

(a) Prove that R is an equivalence relation.

(b) Prove the equality $[1/6] = [5/6]$ of equivalence classes.

(c) Are the equivalence classes $[4/6]$ and $[5/6]$ disjoint? Prove your assertion.

10. Consider the binary relation $R = \{(a, b) \in \mathbb{R}^2 : a - b = k\sqrt{2} \text{ for some } k \in \mathbb{Z}\}$ on $\mathbb{R}$.

(a) Prove that R is an equivalence relation.

(b) Enumerate the elements of the equivalence class $[\sqrt{2}/4]$.

(c) Enumerate the elements of the equivalence class $[1]$.

11. Prove that each equivalence class of an equivalence relation is necessarily non-empty.

12. Suppose that R, S are equivalence relations on a set A. Which of the following are equivalence relations? Prove your assertions.

(a) $R \circ S$ $\qquad$ $(R - S) \cup (S - R)$ $\qquad$ (c) $R^{-1} \cap S^{-1}$

5.13 PARTITIONS

We introduce here a mathematical object, called a "partition", which we shall find is simply another manifestation of the mathematical structure underlying equivalence relations. Just as with digraphs and binary relations, it is useful here to have different but equivalent manifestations of an abstract mathematical structure.

Roughly, a partition is a decomposition of a set $A \neq \emptyset$ into disjoint non-empty subsets. Imagining that the set A is a pie (the edible kind, not the trigonometric kind!), a partition of the set A is simply a way of "cutting the pie into pieces".

More precisely, a *partition* of a set $A \neq \emptyset$ is a collection $\pi \subseteq \mathcal{P}(A)$ of non-empty subsets $S \subseteq A$ so that

(i) We have $A = \cup\pi = \cup_{S\in\pi} S$.

(ii) For each $S, T \in \pi$, either $S = T$ or $S \cap T = \emptyset$.

A collection $\pi \subseteq \mathcal{P}(A)$ is said to *exhaust* (or *cover*) A if condition (i) holds, so a partition is simply a cover of A by disjoint non-empty sets. In case π is a partition, the elements of π (which are subsets of A) are called the *blocks* of π. The number of blocks in a partition π is called the *rank* of π.

In terms of our image of A as pie, one should not imagine actually cutting the pie into pieces, but rather think of the whole pie itself decomposed into what would become pieces after cutting. These subsets of the pie A are the blocks of the corresponding partition. We must be careful to include each point of the pie in some block (so there are no crumbs!) in order that these subsets exhaust A, and, of course, these subsets must all be pairwise disjoint by condition (ii).

■ **Example 5.13.1** On the set $A = \{a, b, \ldots, e\}$, we have the partition $\pi = \{\{a,b\},\{c\},\{d,e\}\}$ of rank 3.

■ **Example 5.13.2** Consider the set of all points A inside and on an equilateral triangle in the plane, let a_1, a_2, a_3 be the line segments (together with their endpoints) connecting the midpoints of the edges of the equilateral triangle, and let A_0, A_1, A_2, A_3 denote the regions of $A - (a_1 \cup a_2 \cup a_3)$ as indicated in the following figure.

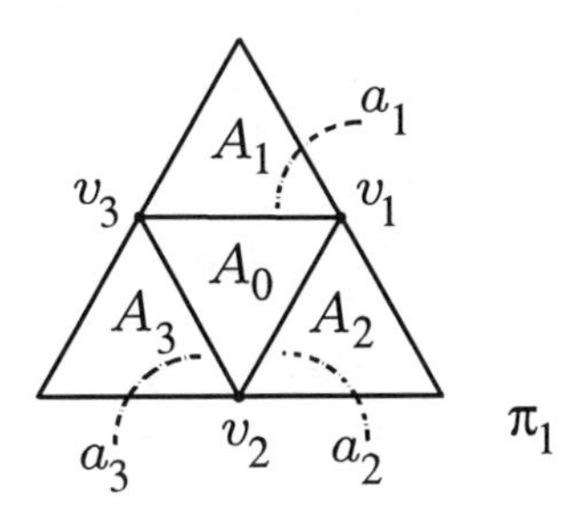

where we have also set $v_i = a_i \cap a_{i+1}$ for $i = 1, 2, 3$ (defining $a_4 = a_1$ for convenience here). We define also $\hat{a}_i = a_i - \{v_1, v_2, v_3\}$ for $i = 1, 2, 3$, so the v_i are the vertices of the inscribed equilateral triangle, and the $\hat{a}_i$ are the line segments (minus their endpoints) of the sides of this inscribed triangle. The collection

$$\pi_1 = \{A_0, A_1, A_2, A_3\} \cup \{\hat{a}_1, \hat{a}_2, \hat{a}_3\} \cup \{v_1, v_2, v_3\}$$

is a partition of A of rank 10, as the reader should check.

This gives a precise example of our image of a partition of a set as a method of cutting it into pieces. Notice that if we had taken instead

$$\{A_0, A_1, A_2, A_3, a_1, a_2, a_3\},$$

then we would have a cover of A but not a partition of A since the putative blocks would not be pairwise disjoint.

■ **Example 5.13.3** Let $p \in \mathbb{N}$, and consider equivalence modulo p on $\mathbb{Z}$. According to Theorem 5.12.1, the collection of equivalence classes $\{[0], \ldots, [p-1]\}$ is itself a partition of $\mathbb{Z}$, and we imagine decomposing $\mathbb{Z}$ into the disjoint subsets $[0], \ldots, [p-1]$ which cover $\mathbb{Z}$.

■ **Example 5.13.4** Given a non-empty set A, the collection $\mathcal{P}(A) - \{\emptyset\}$ is evidently a collection of non-empty subsets of A which covers A. On the other hand, $\mathcal{P}(A) - \{\emptyset\}$ is not a partition of A unless A is a singleton.

The basic result relating partitions and equivalence relations is as follows.

5.13.1 THEOREM *Let A be a non-empty set. If R is an equivalence relation on A, then*

$$A/R = \{[a]_R : a \in A\}$$

is a partition of A. Conversely, if π is a partition of A, then the binary relation R_π on A defined by

$$aR_\pi b \Leftrightarrow \exists S \in \pi[(a \in S) \wedge (b \in S)]$$

is an equivalence relation.

Proof As we observed in Example 5.13.3 above, Theorem 5.12.1 guarantees that if R is an equivalence relation, then A/R is indeed a partition. For the second part, given the partition π, we must show that R_π is an equivalence relation, and symmetry follows essentially from commutativity of conjunction:

$$[aR_\pi b] \Leftrightarrow [\exists S \in \pi[(a \in S) \wedge (b \in S)]] \Leftrightarrow [\exists S \in \pi[(b \in S) \wedge (a \in S)]] \Leftrightarrow [bR_\pi a].$$

To see that R_π is reflexive, suppose that $a \in A$. Insofar as π exhausts A, there must be some $S \in \pi$ so that $a \in S$, and it follows that $(a \in S) \wedge (a \in S)$, so indeed $aR_\pi a$. Turning finally to transitivity, suppose $aR_\pi b$ and $bR_\pi c$. Thus, there are $S, T \in \pi$ so that $a, b \in S$ and $b, c \in T$, so $b \in S \cap T$. We conclude that $S \cap T$ is non-empty, so $S = T$ since π is a partition. It follows that $a, b, c \in S$, so in particular $a, c \in S$, whence $aR_\pi c$. *q.e.d.*

The next result explains the sense in which we may regard partitions and equivalence relations as different manifestations of the same mathematical structure.

5.13.2 THEOREM *Suppose that A is a non-empty set. If π is a partition of A, then*

$$A/R_\pi = \pi.$$

On the other hand, if R is an equivalence relation on A with corresponding partition $\pi = A/R$, then

$$R = R_\pi = R_{A/R}$$

Proof For the first part, given the partition π of A, we have

$$R_\pi = \{(a, b) \in A^2 : \exists S \in \pi(a, b \in S)\}$$

and a block of A/R_π is simply an element S of π. Thus, the partitions π and A/R_π have the same blocks and are therefore identical.

For the second part, observe that

$$\begin{aligned} S \in A/R_\pi &\Leftrightarrow S \text{ is an } R_\pi - \text{equivalence class} \\ &\Leftrightarrow S \text{ is a block of } \pi = A/R \\ &\Leftrightarrow S \text{ is an } R - \text{equivalence class} \\ &\Leftrightarrow S \in A/R \end{aligned}$$

The result then follows from Theorem 5.12.4. *q.e.d.*

The previous two results show that an equivalence relation on a set A uniquely determines a corresponding partition on A and conversely. In this sense, partitions are entirely equivalent to equivalence relations, and results in one setting translate immediately to results in the other. In certain contexts, one formalism may be more convenient than the other, so it is valuable to have both formulations. The remainder of this section is dedicated to a further study of partitions.

Suppose that π_1, π_2 are partitions of a set $A \neq \emptyset$. We say that π_1 *refines* π_2 if every block of π_1 is contained in some block of π_2, and in this case, we write $\pi_1 \leq \pi_2$. Of course, a block of π_1 can be contained in at most one block of π_2 by condition (ii) above on the partition π_2. In keeping with our image of the partition π_2 of A as a method of cutting A into pieces, a refinement π_1 of π_2 is simply a method of cutting A into still smaller pieces.

■ **Example 5.13.5** Recall Example 5.13.2 above, where we gave a partition π_1 of an equilateral triangle of unit edge-length into four triangles (with certain combinations of their edges included), three line segments (without their endpoints), and three points. There is another partition π_2 into four triangles (with none of their edges included), nine line segments (without their endpoints), and six points as indicated

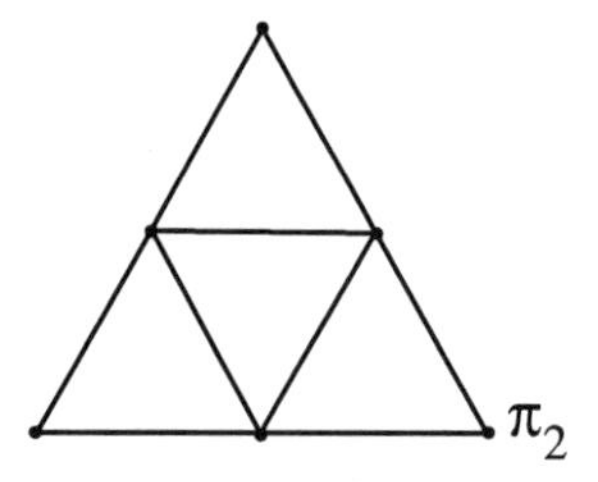

and the reader may check that $\pi_2 \leq \pi_1$.

For another example, consider the partition π_3 of the equilateral triangle into sixteen triangles (with none of their edges included), thirty line segments (without their endpoints), and 15 points as indicated

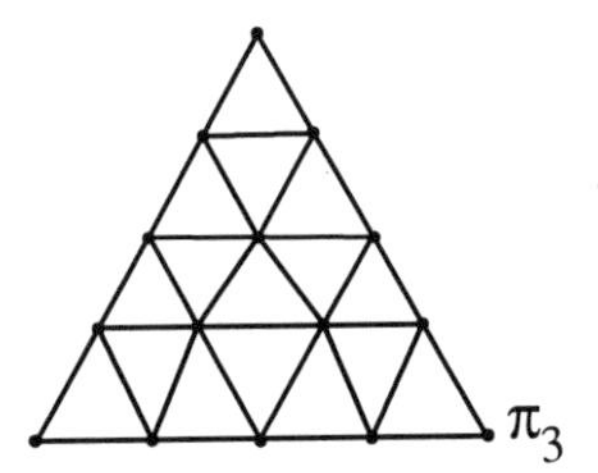

and the reader may check that $\pi_3 \leq \pi_2$, $\pi_2 \leq \pi_1$, and $\pi_3 \leq \pi_1$.

■ **Example 5.13.6** Consider equivalence R_p and R_q modulo p and q, respectively, on $\mathbb{Z}$, where $p, q \in \mathbb{N}$. Let $\mathbb{Z}/p = \mathbb{Z}/R_p$ and $\mathbb{Z}/q = \mathbb{Z}/R_q$ denote the corresponding partitions of $\mathbb{Z}$ as before, and let $[x]_p$ (and $[x]_q$, respectively) denote the block of $\mathbb{Z}/p$ (and $\mathbb{Z}/q$) containing $x \in \mathbb{Z}$. We claim that $\mathbb{Z}/q$ refines $\mathbb{Z}/p$ if and only if p divides q. To see this, suppose first that $p|q$ and choose some $x \in \mathbb{Z}$. We have

$$y \in [x]_q \Leftrightarrow q|(x-y) \Rightarrow p|(x-y) \Rightarrow y \in [x]_p,$$

so for each $x \in \mathbb{Z}$, we find $[x]_q \subseteq [x]_p$. Thus, if $p|q$, then $\mathbb{Z}/q$ does refine $\mathbb{Z}/p$, as was asserted. Indeed, writing $q = kp$ for some $k \in \mathbb{Z}$, we have the following containments among the blocks of $\mathbb{Z}/q$ and $\mathbb{Z}/p$

$$\begin{aligned}
[0]_q, [p]_q, \ldots, [(k-1)p]_q &\subseteq [0]_p, \\
[1]_q, [p+1]_q, \ldots, [(k-1)p+1]_q &\subseteq [1]_p, \\
\vdots \qquad\qquad \vdots \qquad & \quad \vdots \\
[p-1]_q, [2p-1]_q, \ldots, [(k-1)p+p-1]_q &\subseteq [p-1]_p.
\end{aligned}$$

Conversely, suppose that $\mathbb{Z}/q$ refines $\mathbb{Z}/p$, so in particular the block $[0]_q$ of $\mathbb{Z}/q$ is contained in some block $[x]_p$ of $\mathbb{Z}/p$. Since $0 \in [0]_q \subseteq [x]_p$, we must have $0 \in [x]_p$, whence $p|x$. On the other hand, $q \in [0]_q \subseteq [x]_p$, so $q \in [x]_p$, whence $p|(x-q)$. Thus, we find $p|x$ and $p|(x-q)$, so $p|[x-(x-q)]$, i.e., $p|q$, as desired.

The notion of refinement of partitions has a simple translation into the setting of equivalence relations, as follows.

5.13.3 THEOREM *Suppose that π_1 and π_2 are partitions of a set $A \neq \emptyset$, and let $R_1 = R_{\pi_1}$ and $R_2 = R_{\pi_2}$ denote the corresponding equivalence relations on A. Then π_1 refines π_2 if and only if $R_1 \subseteq R_2$.*

Proof Suppose that π_1 refines π_2. If aR_1b, then there is some $S_1 \in \pi_1$ so that $a, b \in S_1$. Since π_1 refines π_2, there is some $S_2 \in \pi_2$ so that $a, b \in S_2$, and we conclude that aR_2b. Thus, $[(a, b) \in R_1] \Rightarrow [(a, b) \in R_2]$, so $R_1 \subseteq R_2$ as desired.

Conversely, suppose that $R_1 \subseteq R_2$. If S_1 is a block of π_1, and $a \in S_1$, then

$$S_1 = [a]_{R_1} = \{x \in A : xR_1a\} \subseteq \{x \in A : xR_2a\} = [a]_{R_2},$$

and of course $[a]_{R_2}$ is a block $S_2 \in \pi_2$ by definition. Thus, the block S_1 is contained in the block S_2, so π_1 indeed refines π_2. *q.e.d.*

Suppose that A is some non-empty set, and let $\Pi(A)$ denote the collection of all partitions of A. Recall that we have written $\pi_1 \leq \pi_2$ if $\pi_1 \in \Pi(A)$ refines $\pi_2 \in \Pi(A)$, and we may thus regard $\leq$ as a binary relation on $\Pi(A)$ itself. In fact, the notation $\leq$ was chosen because of the following result.

5.13.4 THEOREM *For any set A, refinement $\leq$ is a partial ordering on the set $\Pi(A)$ of all partitions of A.*

Proof Suppose that $\pi_i \in \Pi(A)$ and let $R_i = R_{\pi_i}$ denote the corresponding equivalence relation on A for $i = 1, 2, 3$. By the previous theorem, one partition π_1 refines another π_2 if and only if $R_1 \subseteq R_2$. It follows immediately that $\pi_1 \leq \pi_1$ since $R_1 \subseteq R_1$, so $\leq$ is indeed reflexive. As to antisymmetry, suppose that $\pi_1 \leq \pi_2$ and $\pi_2 \leq \pi_1$, so that $R_1 \subseteq R_2$ and $R_2 \subseteq R_1$ again using the previous theorem. Thus, $R_1 = R_2$, so $\pi_1 = \pi_2$ by Theorem 5.12.4, and $\leq$ is seen to be antisymmetric. Turning finally to transitivity, suppose that $\pi_1 \leq \pi_2$ and $\pi_2 \leq \pi_3$, so $R_1 \subseteq R_2$ and $R_2 \subseteq R_3$. It follows from transitivity of inclusion Theorem 3.1.1 that $R_1 \subseteq R_3$, so $\pi_1 \leq \pi_3$ by the previous theorem, whence $\leq$ is indeed transitive. *q.e.d.*

The set $\Pi(A)$ of partitions of a set $A \neq \emptyset$ actually has a still richer structure, that of a "lattice", as we shall discuss in the optional section §5.16 below.

EXERCISES

5.13 PARTITIONS

1. Find the equivalence relation corresponding to each of the following partitions of $A = \{0, 1, 2, 3, 4\}$.

   (a) $\pi_a = \{\{0, 1, 2, 3, 4\}\}$
   (b) $\pi_b = \{\{0\}, \{1\}, \{2\}, \{3\}, \{4\}\}$
   (c) $\pi_c = \{\{0\}, \{1\}, \{2, 3, 4\}\}$
   (d) $\pi_d = \{\{0, 1\}, \{2, 3\}, \{4\}\}$

2. For each $c \in \mathbb{R}$, there is a corresponding subset

$$A_c = \{(x, y) \in \mathbb{R}^2 : y - x = c\} \subseteq \mathbb{R}^2,$$

   and we define the set $\mathcal{A} = \{A_c \in \mathcal{P}(\mathbb{R}^2) : c \in \mathbb{R}\}$ of sets.

   (a) Prove that $\mathcal{A}$ is a partition of $\mathbb{R}^2$.

   (b) Describe the blocks of $\mathcal{A}$.

   (c) Describe the equivalence relation on $\mathbb{R}^2$ corresponding to $\mathcal{A}$.

3. Define the binary relation R on $\mathbb{N}^2$, where we have $(a, b)R(c, d)$ if and only if $a + d = b + c$.

   (a) Prove that R is an equivalence relation on $\mathbb{N}$.

   (b) Describe the equivalence classes of R.

   (c) Describe the partition of $\mathbb{N}$ corresponding to the equivalence relation R.

4. For each of the following binary relations R on a set A, check that R is an equivalence relation and describe the corresponding partition of A.

   (a) On the set $A = \mathbb{R}$, define xRy if and only if $x^2 = y^2$.

   (b) On the set $A = \mathbb{R}^2$, define $(a, b)R(c, d)$ if and only if $a^2 + b^2 = c^2 + d^2$.

   (c) Let A be the set of all propositional forms over a fixed set of propositional variables, and take the usual relation $\Leftrightarrow$ of bi-implication on propositional forms; see Example 3.6.5.

   (d) On the set $A = \{1, 2, \ldots, n\}$, define aRb if and only if a and b end in the same decimal digit.

   (e) On the set $A = \{1, 2, \ldots, n\}$, define aRb if and only if a and b have the same number of decimal digits.

5. Consider the set P consisting of the partitions $\pi_a, \pi_b, \pi_c, \pi_d$ in Problem 1 together with the partial ordering $\leq$ of refinement of partitions.

   (a) Draw the digraph of the poset $(P, \leq)$.

   (b) Draw the Hasse diagram of the poset $(P, \leq)$.

6. Let A denote the set $A = \{1, 2, \ldots, n\}$, for some $n \geq 1$, and let $\Pi(A)$ denote the set of all partitions of A together with its partial ordering $\leq$ of refinement.

(a) Does $(\Pi(A), \leq)$ have a least element? Prove your assertions.

(b) Does $(\Pi(A), \leq)$ have a greatest element? Prove your assertions.

7. Suppose that π_1 and π_2 are partitions of a set A. Which of the following are themselves partitions of A? Prove your assertions.

(a) $\pi_1 \cup \pi_2$ (b) $\pi_1 - \pi_2$

(c) $\pi_1 \cap \pi_2$ (d) $(\pi_1 - \pi_2) \cup (\pi_2 - \pi_1)$

8. Recall that we denote by $\mathbb{Z}/p$ the partition of $\mathbb{Z}$ corresponding to the relation of equivalence modulo $p \in \mathbb{N}$, i.e., the relation $\{(a, b) \in \mathbb{Z}^2 : m \equiv n(p)\}$. Let $\leq$ denote the relation of refinement on partitions of $\mathbb{Z}$.

(a) Draw the digraph for $\leq$ on the set $\{\mathbb{Z}/2, \mathbb{Z}/3, \mathbb{Z}/6\}$.

(b) Draw the digraph for $\leq$ on the set $\{\mathbb{Z}/1, \mathbb{Z}/2, \ldots, \mathbb{Z}/10\}$.

(c) Draw the Hasse diagram for $\leq$ on the set $\{\mathbb{Z}/1, \mathbb{Z}/2, \ldots, \mathbb{Z}/14\}$

9. Let A denote the set $\{1, 2, \ldots, n\}$ and suppose that $\pi_1, \pi_2, \ldots, \pi_{N+1}$ is a collection of distinct partitions of A so that π_{i+1} refines π_i for each $i = 1, \ldots, N$.

(a) Show that restriction of $\leq$ to $\{\pi_1, \pi_2, \ldots, \pi_{N+1}\}$ is a linear ordering.

(b) What is the maximum possible value of N in terms of n, i.e., what is the longest possible length of a chain of partitions of A?

(*) 5.14 CLOSURE OPERATIONS

Suppose that $R \subseteq A^2$ is a binary relation on A. We discuss in this optional section three "closure operations", each of which produces from R another binary relation on A. The three operations are denoted $r(R)$, $s(R)$, $t(R)$ and are called the *reflexive, symmetric, transitive closure* of R, respectively. Each relation $r(R), s(R), t(R)$ contains R as a subset, has the corresponding attribute reflexive, symmetric, transitive, and is "as small as possible". Thus, even if R does not have these various attributes, we can construct from it a relation that does.

Actually, it is easy to define these closures directly:

$$r(R) = \cap\{R' \subseteq A^2 : (R \subseteq R') \text{ and } R' \text{ is reflexive}\},$$
$$s(R) = \cap\{R' \subseteq A^2 : (R \subseteq R') \text{ and } R' \text{ is symmetric}\},$$
$$t(R) = \cap\{R' \subseteq A^2 : (R \subseteq R') \text{ and } R' \text{ is transitive}\}.$$

We must check, for instance, that $t(R)$ is indeed transitive, and to this end, suppose that $a, b, c \in A$, and $a[t(R)]b \wedge b[t(R)]c$, so (by definition of $t(R)$) for each transitive relation $R' \supseteq R$, we have $aR'b \wedge bR'c$. Since R' is transitive by assumption, we find $aR'c$, and since R' was arbitrary, we conclude that $a[t(R)]c$, as required. The proofs that $r(R)$ is reflexive and $s(R)$ is symmetric are similar (and a tad simpler) than the argument here that $t(R)$ is transitive and are left as exercises.

Thus, as defined above each closure $r(R), s(R), t(R)$ indeed has the corresponding attribute reflexive, symmetric, and transitive, and we now explain the sense in which these closures are "as small as possible". Suppose, for instance, that R' is a transitive relation containing R; according to the definition of $t(R)$ (since $t(R)$ is an intersection of sets, one of which is R'), we must have $t(R) \subseteq R'$, and it is in this sense that $t(R)$ is as small as possible.

Summarizing our discussion here, we have proved

5.14.1 THEOREM *Suppose that R is a binary relation on a set A, and let $r(R), s(R), t(R)$ be defined as above. Then*

(1) *We have $R \subseteq r(R), R \subseteq s(R), R \subseteq t(R)$.*

(2) *The relations $r(R), s(R), t(R)$ are reflexive, symmetric, and transitive, respectively.*

(3) *Suppose that R' is a binary relation on A so that $R \supseteq R'$. If R' is reflexive, then $R' \supseteq r(R)$, if R' is symmetric, then $R' \supseteq s(R)$, and if R' is transitive, then $R' \supseteq t(R)$.*

In fact, properties (1)-(3) in the previous theorem actually uniquely determine the closure operations. For instance, if $t(R)$ and $t'(R)$ satisfy these properties, then in particular $t(R)$ satisfies properties (1) and (2) so by property (3) for $t'(R)$, we conclude that $t'(R) \subseteq t(R)$. Of course, an analogous argument shows that $t(R) \subseteq t'(R)$, so $t(R) = t'(R)$ and the properties (1)-(3) indeed uniquely determine $t(R)$. Similar arguments apply to show that these properties uniquely determine $r(R)$ and $s(R)$.

Some comments are in order here regarding mathematical exposition under the current circumstances where a collection of properties is found to uniquely characterize an object being defined. It is sometimes regarded as an "elegant" mathematical treatment to use the properties themselves to define the object. For instance, we might have *defined* $t(R)$ to be the relation satisfying properties (1)-(3). One can often prove various facts directly from this "elegant" definition, and we shall do so presently. Such arguments are perfectly rigorous, but one must take care to prove that there is actually an object satisfying the characterizing properties, for otherwise, one is proving valid theorems about objects that do not exist! Such a definition may seem like so much hocus-pocus to the beginner (and indeed it is hocus-pocus unless an object satisfying the characterizing properties does exist), but such "elegant" expositions under these circumstances are rather common in practice.

Of course, in our current situation, we have proved the existence of the closure relations $r(R), s(R), t(R)$, and we next give several results which follow more or less directly from properties (1)-(3) above.

5.14.2 THEOREM *Suppose that R is a binary relation on a set A. R is reflexive if and only if $r(R) = R$, R is symmetric if and only if $s(R) = R$, and R is transitive if and only if $t(R) = R$.*

Proof We concentrate on transitivity, leaving reflexivity and symmetry as similar exercises. Suppose that R is transitive, so R satisfies properties (1)-(2), i.e., $R \subseteq R$ and R is transitive. By property (3) for $t(R)$, we have $t(R) \subseteq R$, while the reverse inclusion $R \subseteq t(R)$ follows from property (1) of $t(R)$. Thus, if R is transitive, then $t(R) = R$. Conversely, if $t(R) = R$, then R is transitive since $t(R)$ is transitive. *q.e.d.*

Actually, there are much more down-to-earth descriptions of the closures, as we next describe.

5.14.3 THEOREM *Suppose that R is a binary relation on a set A, and let E denote the relation of equality on A. Then we have the identities*

$$\begin{aligned} r(R) &= R \cup E, \\ s(R) &= R \cup R^{-1}, \\ t(R) &= \cup_{i \geq 1} R^i, \end{aligned}$$

where R^{-1} is the inverse relation of R, and R^i for $i \geq 1$ are the iterates (i.e., the iterated compositions) of R.

Proof Beginning with reflexivity, we show that $R \cup E$ satisfies the characterizing properties (1)-(3). Of course, $R \subseteq R \cup E$ and $R \cup E$ is reflexive, so properties (1)-(2) are obvious. If R' is a reflexive relation containing R, then $R \subseteq R'$ and $E \subseteq R'$, so $R \cup E \subseteq R'$. Thus, $R \cup E$ satisfies the characterizing properties (1)-(3), so $r(R) = R \cup E$.

Turning to symmetry, we show that $R \cup R^{-1}$ satisfies the characterizing properties. Again, property (1) is obvious. Using Theorem 5.5.2, we prove that $R \cup R^{-1}$ is symmetric by verifying that $R \cup R^{-1} = (R \cup R^{-1})^{-1}$, and to this end, we compute

$$(R \cup R^{-1})^{-1} = R^{-1} \cup (R^{-1})^{-1} = R^{-1} \cup R = R \cup R^{-1},$$

so $R \cup R^{-1}$ satisfies properties (1)-(2). As to property (3), suppose that R' is symmetric and $R \subseteq R'$. Of course, $(a, b) \in R \cup R^{-1}$ if and only if $(a, b) \in R \subseteq R'$ or $(a, b) \in R^{-1}$. In the latter case, we have $(b, a) \in R \subseteq R'$, so $(a, b) \in (R')^{-1}$. On the other hand, $R' = (R')^{-1}$ since R' is symmetric again using Theorem 5.5.2, so in either case we find $(a, b) \in R'$. Thus, $R \cup R^{-1} \subseteq R'$, so property (3) holds as well, whence $s(R) = R \cup R^{-1}$.

It remains to prove that $t(R) = \cup_{i \geq 1} R^i$, and we begin by proving that $\cup_{i \geq 1} R^i \subseteq t(R)$. To this end, we prove that $R^i \subseteq t(R)$ by induction on $i \geq 1$, and the basis step $i = 1$ follows from property (1) for $t(R)$. For the induction step, suppose that $R^n \subseteq t(R)$ and let $(a,b) \in R^{n+1} = R^n \circ R$. By definition of composition, there is some $c \in A$, so that aR^nc and cRb, and since $R^n \subseteq t(R)$ by the induction hypothesis, we find $a[t(R)]c$. Since $R \subseteq t(R)$, we also find $c[t(R)]b$, so by transitivity of $t(R)$, we conclude $a[t(R)]b$. We have shown that $[(a,b) \in R^{n+1}] \Rightarrow [(a,b) \in t(R)]$ completing the induction step and hence the proof that $\cup_{i\geq 1} R^i \subseteq t(R)$.

For the reverse inclusion $t(R) \subseteq \cup_{i\geq 1} R^i$, we use the characterizing property (3) of $t(R)$ and must therefore show that $\cup_{i\geq 1} R^i$ satisfies properties (1)-(2), and property (1) follows since $R = R^1$ as before. As to property (2) (that is, transitivity of $\cup_{i\geq 1} R^i$), suppose that $(a,b),(b,c) \in \cup_{i\geq 1} R^i$, so $(a,b) \in R^p$ and $(b,c) \in R^q$, for some $p,q \in \mathbb{N}$. By definition of composition, we have $(a,c) \in R^p \circ R^q = R^{p+q}$, where we have used Theorem 5.7.1 in the last equality. Thus, $(a,c) \in R^{p+q} \subseteq \cup_{i\geq 1} R^i$, so $\cup_{i\geq 1} R^i$ is indeed transitive. By property (3) of $t(R)$, we conclude that $t(R) \subseteq \cup_{i\geq 1} R^i$, as required. *q.e.d.*

Let us here briefly reflect on these closure operations in terms of corresponding digraphs. Suppose that R is a binary relation on a set A, and let G be a corresponding digraph. Of course, there is a digraph corresponding to $r(R) = R \cup E$ which arises from G by simply adding a loop based at each vertex (which does not already have such a loop):

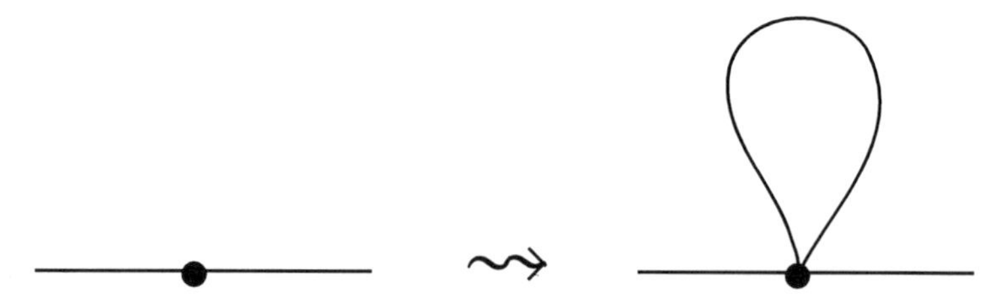

Similarly, we may construct a digraph for R^{-1} from G by simply reversing the arrow on each edge of G. Furthermore, we produce a digraph for $s(R) = R \cup R^{-1}$ from G by adding an edge (b,a) for each edge (a,b) of G (actually only for each edge (a,b) so that (b,a) is not already an edge of G):

Finally, consider $t(R) = \cup_{i\geq 1} R^i$. The edges of a digraph for $t(R)$ thus correspond to the oriented edge-paths of various lengths on G, so one builds a digraph for $t(R)$ from G by adding an oriented edge from vertex a to vertex b if and only if there is an oriented edge-path in G with initial point a and terminal point b.

One thus starts with G, then adds edges corresponding to oriented edge-paths of length two, then adds edges corresponding to oriented edge-paths of length three, and so on:

These remarks hopefully give the reader a reasonably vivid mental image of these closure operations.

■ **Example 5.14.1** The reflexive closure of the usual relation $<$ of strict inequality on the integers is the usual relation $\leq$ of weak inequality. The symmetric closure of the usual relation of strict inequality $<$ on the integers is the complement $\mathbb{Z}^2 - E$ of the relation E of equality on the integers, that is, it is the relation $\neq$ on $\mathbb{Z}$.

■ **Example 5.14.2** The symmetric closure of the usual relation of weak inequality $\leq$ on the integers is the universal relation on $\mathbb{Z}$.

■ **Example 5.14.3** If the relation $R \subseteq \mathbb{Z}^2$ is defined by $mRn \Leftrightarrow (n = m+1)$, then the transitive closure of R is the usual relation $<$ of strict inequality on $\mathbb{Z}$.

■ **Example 5.14.4** If R is the relation "is the parent of" defined on the set of all people who have ever lived, then the transitive closure of R is the relation "is a forbear of".

We have thus seen three perfectly suitable possible definitions of the closure relations: as an abstract intersection at the beginning of this section, in terms of characterizing properties in Theorem 5.14.1, and in explicit set-theoretic terms in Theorem 5.14.3. Each of these formulations serves a purpose in our exposition here: the first definition to guarantee existence, the characterizing properties as a tool for proving theorems, and the third explicit description we next use to analyze the interaction of the various closure operations. This state of affairs (several equivalent formulations each with its own utility) occurs reasonably often as a paradigm for various mathematical structures.

5.14.4 THEOREM *Suppose that R is a binary relation on a set A.*

(a) *If R is reflexive, then $s(R)$ and $t(R)$ are reflexive.*

(b) *If R is symmetric, then $r(R)$ and $t(R)$ are symmetric.*

(c) *If R is transitive, then $r(R)$ is transitive.*

Proof , Let E denote the relation of equality on A. For part (a), suppose that R is reflexive, so $R = R \cup E$ by Theorem 5.3.1, whence $E \subseteq R$. Since $R \subseteq s(R)$, we conclude that $E \subseteq s(R)$, so $s(R) = s(R) \cup E$ is also reflexive again by Theorem 5.3.1. Similarly, $E \subseteq R \subseteq t(R)$, so $t(R) = t(R) \cup E$ is also reflexive.

For part (b), we begin by showing that $r(R)$ is symmetric if R is symmetric. By Theorem 5.5.2, symmetry of R shows that $R = R^{-1}$, so

$$r(R) = R \cup E = R^{-1} \cup E = R^{-1} \cup E^{-1} = (R \cup E)^{-1} = [r(R)]^{-1},$$

so $r(R)$ is indeed symmetric. Turning now to $t(R)$, we compute

$$[t(R)]^{-1} = \left(\cup_{i\geq 1} R^i\right)^{-1} = \cup_{i\geq 1}(R^i)^{-1},$$

where the second equality follows directly from the definitions (and is left as an exercise). Another elementary argument by induction (which is also left as an exercise) shows that $(R^i)^{-1} = (R^{-1})^i$ for each $i \geq 1$. Furthermore, since R is symmetric by hypothesis, we have $R = R^{-1}$, so in fact, we find $(R^i)^{-1} = R^i$. Plugging this back into the previous inset equation, we find

$$[t(R)]^{-1} = \cup_{i\geq 1}(R^i)^{-1} = \cup_{i\geq 1} R^i = t(R),$$

so $t(R)$ is indeed symmetric by Theorem 5.5.2.

For part (c), we suppose that R is transitive, so that $R = t(R)$ by Theorem 5.14.2. To prove that $r(R)$ is transitive, we must similarly show that $tr(R) = r(R)$. To this end, we first observe

$$tr(R) = t(R \cup E) = \cup_{i\geq 1}(R \cup E)^i.$$

An elementary induction (left as an exercise) shows that $(R \cup E)^i = \cup_{j=0}^{i} R^j$ for each $i \geq 1$, so we find

$$\begin{aligned} tr(R) &= \cup_{i\geq 1} \cup_{j=0}^{i} R^j \\ &= \cup_{i\geq 0} R^i \\ &= E \cup \left(\cup_{i\geq 1} R^i\right) \\ &= E \cup t(R) \\ &= E \cup R \\ &= r(R), \end{aligned}$$

whence $r(R)$ is indeed transitive. *q.e.d.*

As an application of the previous material, we consider the reflexive symmetric transitive closure $tsr(R)$ of a binary relation $R \subseteq A^2$. Of course, $r(R)$ is reflexive, so $sr(R)$ is symmetric and reflexive by part (a) of the previous theorem. Finally, $tsr(R)$ is transitive, symmetric, and reflexive by parts (a) and (b) of the previous theorem. Thus, $tsr(R)$ is actually an equivalence relation, and it is called the *equivalence relation induced by* R. Thus, an arbitrary binary relation R gives rise to a well-defined equivalence relation $tsr(R)$.

In fact, the equivalence relation $tsr(R)$ is uniquely determined by characterizing properties analogous to properties (1)-(3) above, as we next show.

5.14.5 THEOREM *Let R be a binary relation on a set A, and suppose that R' is a binary relation on A satisfying the properties*

(1) *We have $R \subseteq R'$.*

(2) *R' is an equivalence relation.*

(3) *If R'' is an equivalence relation containing R, then $R' \subseteq R''$.*

Then $R' = tsr(R)$.

Proof As before, we must show that $tsr(R)$ satisfies properties (1)-(3) in the theorem, and properties (1)-(2) are obvious. For property (3), if R'' is an equivalence relation containing R, then R'' is reflexive and symmetric, hence $R'' \supseteq R \cup R^{-1} \cup E = sr(R)$. Finally, since R'' is transitive as well and contains $sr(R)$, we conclude that $R'' \supseteq tsr(R)$. *q.e.d.*

EXERCISES

5.14 CLOSURE OPERATIONS

1. Draw the digraphs of $r(R), s(R)$ and $t(R)$ for each of the following binary relations.

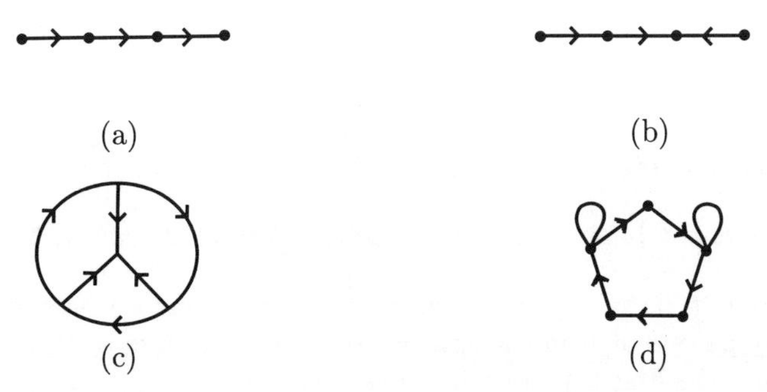

2. Let R be a binary relation and define $S = tsr(R)$. Prove that $tsr(S) = S$.

3. Prove Theorem 5.14.1 for $r(R)$ and $s(R)$.

4. Prove Theorem 5.14.2 for $r(R)$ and $s(R)$.

5. Let R and S be binary relations on a set A and suppose that $R \subseteq S$. Prove each of the following inclusions.

(a) $r(R) \subseteq r(S)$ (b) $s(R) \subseteq s(S)$

(c) $t(R) \subseteq t(S)$ (d) $st(R) \subseteq st(S)$

6. Suppose that R is a binary relation. Prove or disprove each of the following.

(a) If R is transitive, then $s(R)$ is transitive.

(b) $rst(R)$ is an equivalence relation.

(c) $str(R)$ is an equivalence relation.

(d) $rst(R) = srt(R)$

7. Suppose that R and S are binary relations on a set A. Prove each of the following.

(a) $r(R \cup S) = r(R) \cup r(S)$

(b) $s(R \cup S) = s(R) \cup s(S)$

(c) $t(R \cup S) \supseteq t(R) \cup t(S)$

(d) Give an example of relations where the inclusion in part (c) is a proper inclusion.

8. Suppose that R is a binary relation, and define two closure operations as follows: The *plus operation* is defined by $R^+ = t(R)$, and the *star operation* is defined by $R^* = tr(R)$. Prove each of the following.

(a) $(R^+)^+ = R^+$

(b) $(R^*)^* = R^*$

(c) $R \circ R^* = R^+ = R^* \circ R$

(d) R is a strict partial ordering if and only if $R \cap R^{-1} = \emptyset$ and $R = R^+$.

(e) R is a partial ordering if and only if $R \cap R^{-1} = \emptyset$ and $R = R^*$.

9. Suppose that R is a binary relation on the set $A = \{1, 2, \ldots, n\}$.

(a) Prove that $t(R) = \cup_{i=1}^{n} R^i$. The point here is that we need only take finitely many powers of R in the constructive description of $t(R)$, indeed, we need take only the first n powers if A has n elements.

(b) Give an example to show that $N = n$ is the least natural number so that $t(R) = \cup_{i=1}^{N} R^i$.

(*) 5.15 MEETS AND JOINS OF PARTITIONS

We have considered before the collection $\Pi(A)$ of all partitions of the set A, and we have seen in Theorem 5.13.4 that refinement $\leq$ of partitions is a partial order on $\Pi(A)$. We study here two binary operations, called the "meet" and "join" on $\Pi(A)$. Just as for the closure operations discussed in the previous section, the "meet" and the "join" can be defined directly or they can be defined by a collection of characterizing properties. For some further experience with the "elegant" point of view, we begin here with these characterizing properties.

Suppose that π_1, π_2 are partitions of the set $A \neq \emptyset$. We define the *meet* of π_1 and π_2 to be the partition $\pi_1 \wedge \pi_2$ of A so that

(1) $\pi_1 \wedge \pi_2$ refines both π_1 and π_2.

(2) If π is another partition of A which refines both π_1 and π_2, then π refines $\pi_1 \wedge \pi_2$.

In the same way, we define the *join* of π_1 and π_2 to be the partition $\pi_1 \vee \pi_2$ so that

(3) Both π_1 and π_2 refine $\pi_1 \vee \pi_2$.

(4) If π is another partition of A which both π_1 and π_2 refine, then $\pi_1 \vee \pi_2$ refines π.

Thus, the meet $\pi_1 \wedge \pi_2$ is the "largest or coarsest partition refining both π_1 and π_2", whereas the join $\pi_1 \vee \pi_2$ is the "smallest or finest partition refined by both π_1 and π_2". In other words, we have

$$\pi_1 \vee \pi_2 = lub\{\pi_1, \pi_2\},$$
$$\pi_1 \wedge \pi_2 = glb\{\pi_1, \pi_2\},$$

as the reader should check, where we take the *glb* and *lub* under refinement of partitions. The usage of the same symbols "$\wedge$" and "$\vee$" here as for conjunction and disjunction is unfortunate, but the notation is reasonably standard. In this context, one reads $\pi_1 \wedge \pi_2$ as "π_1 cap π_2", "π_1 meet π_2", or "π_1 wedge π_2", and one reads $\pi_1 \vee \pi_2$ as "π_1 cup π_2", "π_1 join π_2", or "π_1 vee π_2".

It is not difficult to prove that the characterizing properties uniquely determine the meet and the join. For instance, suppose that π is another partition on A which satisfies conditions (1)-(2) above. Since both π and $\pi_1 \wedge \pi_2$ refine π_1 and π_2, we conclude that $\pi \leq \pi_1 \wedge \pi_2$ by property (2) for $\pi_1 \wedge \pi_2$, and $\pi_1 \wedge \pi_2 \leq \pi$ by property (2) for π. We have seen that $\leq$ is a partial ordering on $\Pi(A)$, and in particular, antisymmetry of $\leq$ then implies that $\pi = \pi_1 \wedge \pi_2$. Thus, properties (1)-(2) do indeed uniquely determine the meet, and we leave the analogous argument for the join as an exercise.

In order to complete this "elegant" definition, we must prove that there is a relation $\pi_1 \wedge \pi_2$ satisfying properties (1)-(2) as well as another relation $\pi_1 \vee \pi_2$ satisfying properties (3)-(4).

5.15.1 THEOREM *Suppose that π_i is a partition of the set $A \neq \emptyset$, and let $R_i = R_{\pi_i}$, for $i = 1, 2$. Then $R_1 \cap R_2$ is an equivalence relation on A, and the corresponding partition is the meet $\pi_1 \wedge \pi_2$. Furthermore, the transitive closure $t(R_1 \cup R_2) = \cup_{n \geq 1}(R_1 \cup R_2)^n$ is an equivalence relation on A, and the corresponding partition is the join $\pi_1 \vee \pi_2$.*

Proof We have already observed that $R = R_1 \cap R_2$ is an equivalence relation and must verify properties (1)-(2) for the partition π corresponding to R. Since $R = R_1 \cap R_2$, we conclude that $R_1 \subseteq R$ and $R_2 \subseteq R$, so π refines both π_1 and π_2 by Theorem 5.13.3. As to condition (2), suppose that π' refines both π_1 and π_2, and let $R' = R_{\pi'}$ denote the corresponding equivalence relation on A. Since π' refines π_1 and π_2, we conclude that $R' \subseteq R_1$ and $R' \subseteq R_2$ again by Theorem 5.13.3. It follows immediately that $R_1 \cap R_2 \supseteq R'$, i.e., we have $R \supseteq R'$, so $\pi_1 \wedge \pi_2 = A/R$, as desired.

Turning to the second part, we must first show that $t(R_1 \cup R_2)$ is actually an equivalence relation. We have already seen that the union of reflexive or symmetric relations is again reflexive or symmetric, so $R_1 \cup R_2$ is both reflexive and symmetric. Thus, by Theorem 5.14.4, we have $R = t(R_1 \cup R_2) = tsr(R_1 \cup R_2)$ is the smallest equivalence relation containing both R_1 and R_2. Since $R_1 \subseteq R$ and $R_2 \subseteq R$, we conclude that both π_1 and π_2 refine A/R by Theorem 5.13.3. Moreover, any partition refined by both π_1 and π_2 induces an equivalence relation containing both R_1 and R_2 again by Theorem 5.13.3. Finally, since $R = t(R_1 \cup R_2)$ is the smallest such equivalence relation, it follows that A/R refines any such partition. Thus, conditions (3)-(4) hold, so $\pi_1 \vee \pi_2 = A/R$. *q.e.d.*

In keeping with our image of a partition as a method of decomposing a set, we next give diagrammatic representations of the meet and join. Given partitions π_1 and π_2 of a set $A \neq \emptyset$, the meet $\pi_1 \wedge \pi_2$ is obtained from the decomposition of A by π_1 by further decomposing each block of π_1 into its various intersections with the blocks of π_2. In other words, the blocks of $\pi_1 \wedge \pi_2$ are simply the non-empty sets $S \cap T$, where S is a block of π_1 and T is a block of π_2.

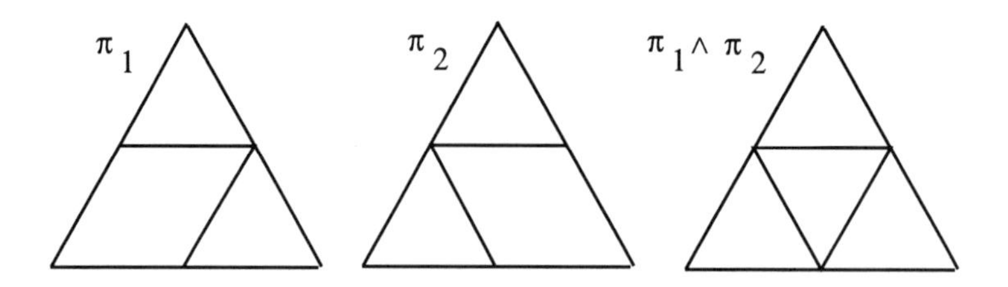

Turning to the join, observe that two elements $a, b \in A$ are in the same block of $\pi_1 \vee \pi_2$ if and only if there is a path from a to b in a digraph for the relation $R_1 \cup R_2$ on A. Thus, the blocks of $\pi_1 \vee \pi_2$ are "larger" than the blocks of either

π_1 or π_2 and are otherwise as small as possible.

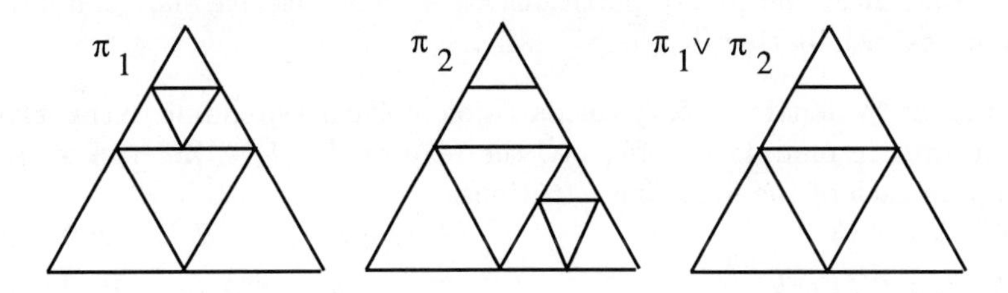

■ **Example 5.15.1** Take the set $A = \{1, 2, 3\}$ and define the following partitions of A:

$$a = \{\{1\}, \{2, 3\}\},$$
$$b = \{\{2\}, \{1, 3\}\},$$
$$c = \{\{3\}, \{1, 2\}\}.$$

Then we have

$$a \vee b = a \vee c = b \vee c = \{\{1, 2, 3\}\},$$
$$a \wedge b = a \wedge c = b \wedge c = \{\{1\}, \{2\}, \{3\}\}$$

for instance.

■ **Example 5.15.2** Take the set $A = \{1, 2, 3, 4, 5\}$ and define the following partitions of A:

$$a = \{\{1, 2, 3\}, \{4\}, \{5\}\},$$
$$b = \{\{1\}, \{2, 3\}, \{4, 5\}\},$$
$$c = \{\{1, 2\}, \{3, 4, 5\}\}.$$

Then we have

$$a \wedge b = \{\{1\}, \{2, 3\}, \{4\}, \{5\}\}, \qquad a \vee b = \{\{1, 2, 3\}, \{4, 5\}\},$$
$$a \wedge c = \{\{1, 2\}, \{3\}, \{4\}, \{5\}\}, \qquad a \vee c = \{\{1, 2, 3, 4, 5\}\},$$
$$b \wedge c = \{\{1\}, \{2\}, \{3\}, \{4, 5\}\}, \qquad b \vee c = \{\{1, 2, 3, 4, 5\}\},$$

for instance

A "lattice" is a poset $(X, \leq)$ so that any doubleton $\{a, b\} \subseteq X$ has both a least upper bound $a \vee b$ and a greatest lower bound $a \wedge b$, which are referred to as the "join" and "meet" respectively. In particular for any set A, the poset $\Pi(A)$ together with the meet and join operations is actually a lattice; this is the content of Theorem 5.15.1. There is a rich theory of lattices, and the lattices arising as partitions of sets turn out to be important examples of general lattices. We shall briefly discuss the basic theory of lattices in the next section.

EXERCISES

5.15 MEETS AND JOINS OF PARTITIONS

1. Prove that meet and join or partitions are each associative and commutative operations distributing over one another.

2. Recall that we denote by $\mathbb{Z}/p$ the partition of $\mathbb{Z}$ corresponding to the relation of equivalence modulo $p \in \mathbb{N}$, i.e., the relation $\{(a,b) \in \mathbb{Z}^2 : m \equiv n(p)\}$. Compute each of the following partitions.
   (a)$\mathbb{Z}/2 \vee \mathbb{Z}/3$ (b) $(\mathbb{Z}/2 \vee \mathbb{Z}/3) \wedge (\mathbb{Z}/2 \vee \mathbb{Z}3 \vee \mathbb{Z}/6)$
   (c)$\mathbb{Z}/2 \wedge \mathbb{Z}/4 \wedge \mathbb{Z}5 \wedge \mathbb{Z}/6$ (d) $(\mathbb{Z}/2 \wedge \mathbb{Z}/3 \wedge \mathbb{Z}/6) \vee \mathbb{Z}/4 \vee \mathbb{Z}/6$

3. Recall that $\mathbb{Z}/p$ refines $\mathbb{Z}/q$ if and only if p divides q. For each $p, q \in \mathbb{N}$, describe $\mathbb{Z}/p \wedge \mathbb{Z}/q$ and $\mathbb{Z}/p \vee \mathbb{Z}/q$.

4. Suppose that π_1 refines π_2. Prove that $\pi_1 \wedge \pi_2 = \pi_1$ and $\pi_1 \vee \pi_2 = \pi_2$ in $\Pi(A)$.

5. Fix a set A and consider the partial ordering of refinement on the set $\Pi(A)$ of all partitions of A.

   (a) Show that $\pi_1 \wedge \pi_2$ is the *glb* and $\pi_1 \vee \pi_2$ is the *lub* of $\{\pi_1, \pi_2\}$.

   (b) Prove that any finite non-empty collection of partitions has a *glb* and *lub*.

   (c) What about *lub*'s and *glb*'s of infinite collections of partitions? Prove your assertions.

(*) *5.16 LATTICES*

Suppose that $(A, \leq)$ is a poset. If $a, b \in A$, then define the *meet* of a and b, denoted $a \wedge b$, to be the *glb* of $\{a, b\}$ if it exists. Also define the *join* of a and b, denoted $a \vee b$ to be the *lub* of $\{a, b\}$ if it exists. The poset $(A, \leq)$ is said to be a *lattice* if both $a \wedge b$ and $a \vee b$ exist for each $a, b \in A$. Of course, it follows by induction that in a lattice any finite non-empty subset has both a greatest lower bound and a least upper bound.

■ **Example 5.16.1** For any set A, the poset $(\mathcal{P}(A), \subseteq)$ is a lattice, where the usual intersection and union of sets are the respective lattice operations of meet and join. A lattice of the form $(\mathcal{P}(A), \subseteq)$ for some set A is called a *Boolean lattice*.

■ **Example 5.16.2** Fix a natural number n, consider the set D_n consisting of all natural divisors of n, and observe that the relation $|$ of "divides" is a partial ordering on D_n. The join of any two elements of D_n is their *lcm*, their meet is their *gcd*, and $(D_n, \mid)$ is a lattice.

■ **Example 5.16.3** Fix a set A and consider the poset $\Pi(A)$ of all partitions of A under refinement as in Theorem 5.13.4. According to Theorem 5.15.1, the meet and the join are indeed defined, so the poset $\Pi(A)$ is actually a lattice.

If $(A, \leq)$ is a poset, then we define the *dual order* $\geq$ on A as follows: $a \geq b$ if and only if $b \leq a$ for each $a, b \in A$. Using that $(A, \leq)$ is a poset, one checks easily that $(A, \geq)$ is a poset as well. In particular, if $(A, \leq)$ is a lattice with meet and join denoted $\wedge$ and $\vee$ respectively, then $(A, \geq)$ is actually a lattice, called the *dual lattice*, whose meet and join are given by $\vee$ and $\wedge$ respectively; the reader may prove this fact directly from the definitions. For instance, the dual of the Boolean lattice $(\mathcal{P}(A), \subseteq)$ is the lattice $(\mathcal{P}(A), \supseteq)$, where the meet and join are $\cup$ and $\cap$ (rather than $\cap$ and $\cup$ as usual).

The next result is a list of algebraic properties enjoyed by all lattices. It is worth emphasizing that there are many other such identities, and we include these particular ones primarily for later application.

5.16.1 PROPOSITION *For any lattice $(A, \leq)$ and any $a, b, c \in A$, we have the following identities:*

(a) *$\wedge$ and $\vee$ are each idempotent, i.e., $a \wedge a = a = a \vee a$.*

(b) *$\wedge$ and $\vee$ are each commutative and associative.*

(c) $a \wedge (a \vee b) \ = \ a \ = \ a \vee (a \wedge b)$

(d) *If $a \leq b$, then $a \wedge c \ \leq \ b \wedge c$ and $a \vee c \ \leq \ b \vee c$.*

(e) $(a \wedge b) \vee (a \wedge c) \ \leq \ a \wedge (b \vee c)$

(f) $a \vee (b \wedge c) \ \leq \ (a \vee b) \wedge (a \vee c)$

Proof To see that $\wedge$ is idempotent in part (a), recall that the definition of $a \wedge a$ is as the *glb* of $\{a, a\} = \{a\}$, and this *glb* is a. Since the meet $\wedge$ is thus idempotent, then so too must $\vee$ be idempotent since $\vee$ is the meet of the dual lattice.

The remaining proofs are left as analogous routine manipulations of *glb*'s and *lub*'s. Furthermore, one again argues "by duality" as in part (a) to see that certain assertions above are consequences of others. *q.e.d.*

As a point of terminology, part (c) of the previous proposition is often called the "absorption law".

5.16.2 PROPOSITION *The following statements are equivalent for any lattice $(A, \leq)$.*

(1) *$(a \wedge b) \vee (a \wedge c) = a \wedge (b \vee c)$ for all $a, b, c \in A$.*

(2) *$(a \vee b) \wedge (a \vee c) = a \vee (b \wedge c)$ for all $a, b, c \in A$.*

(3) $[(a \vee b) \wedge c] \ \leq \ [a \vee (b \wedge c)]$ *for all* $a, b, c \in A$.

Proof We begin by proving that (1)⇒(2), and assume (1) to hold while substituting $a \vee b$ for a and a for b to get

$$[(a \vee b) \wedge a] \vee [(a \vee b) \wedge c] \quad = \quad (a \vee b) \wedge (a \vee c).$$

On the other hand, we compute the lefthand side of this equation

$$\begin{aligned}(a \vee b) \wedge (a \vee c) &= a \vee [(a \vee b) \wedge c] \\ &= a \vee [(a \wedge c) \vee (b \wedge c)] \\ &= a \vee (b \wedge c),\end{aligned}$$

as required, where the first and last equalities follow from Proposition 5.16.1b and 5.16.1c, and the second holds by hypothesis with a and c interchanged.

To prove that (2)⇒(3), observe that since $c \leq a \vee c$ by definition, Proposition 5.16.1d together with our assumption that (b) holds shows that

$$(a \vee b) \wedge c \leq (a \vee b) \wedge (a \vee c) = a \vee (b \wedge c),$$

as desired.

To complete the proof of the theorem, it remains to show that (3)⇒(1), or in other words that (3) implies the reverse of the inequality in Proposition 5.16.1f. To this end, two successive applications of the hypothesis (3) with suitable substitutions give

$$\begin{aligned}(a \vee b) \wedge (a \vee c) &\leq a \vee [b \wedge (a \vee c)] \\ &\leq a \vee [a \vee (b \wedge c)] \\ &= a \vee (b \wedge c),\end{aligned}$$

as required. *q.e.d.*

A lattice which satisfies any one of the equivalent properties in Proposition 5.16.2 is called a *distributive* lattice.

■ **Example 5.16.4** For any set A, the lattice $(\mathcal{P}(A), \subseteq)$ of Example 5.16.1 is distributive by Theorem 3.4.2. Thus, every Boolean lattice is distributive.

■ **Example 5.16.5** The lattice $(\Pi(A), \leq)$ of partitions of a set A considered in Example 5.16.3 is not usually distributive. For instance, we take the set $A = \{1, 2, 3\}$ and define the partitions a, b, c as in Example 5.15.1 to find

$$(a \wedge b) \vee (a \wedge c) = \{\{1\}, \{2\}, \{3\}\},$$

while

$$a \wedge (b \vee c) = \{\{1\}, \{2, 3\}\}.$$

Thus, the identity in Proposition 5.16.2a does not hold, so the lattice of partitions of the set $\{1,2,3\}$ is not distributive

Suppose now that A has a greatest element 1 and a least element 0 for $\leq$. For instance, in Example 5.16.1, we find $0 = \emptyset$, and $1 = A$, while in Example 5.16.3, we find that $0 \in \Pi(A)$ denotes the partition $\{\{a\} \in \mathcal{P}(A) : a \in A\}$, and $1 \in \Pi(A)$ denotes the partition $\{A\}$. Furthermore, if $(A, \leq)$ is any lattice whose underlying set A is finite, then A does indeed have a greatest and least element as we have observed above.

A lattice $(A, \subseteq)$ which has a greatest element 1 and least element 0 is said to be *complemented* if for each $a \in A$ there is some $a' \in A$ satisfying

$$a \wedge a' = 0 \quad \text{and} \quad a \vee a' = 1,$$

and we call $a' \in A$ a *complement* of $a \in A$ in this case.

Actually, if $(A, \leq)$ is distributive, then there can be at most one a' satisfying these conditions for each $a \in A$. For instance, if b and c were each complements of a, then by Proposition 5.16.2b and 5.16.2c, we have

$$b \;=\; b \wedge 1 \;=\; b \wedge (a \vee c) \;\leq\; (a \wedge b) \vee c \;=\; 0 \vee c.$$

Thus, we have $b \leq c$ and in the same way $c \leq b$, so $b = c$ by antisymmetry of $\leq$, as was claimed. It follows that in a distributive lattice, the complement a'' of the complement a' is again a itself.

■ **Example 5.16.6** In the Boolean lattice $(\mathcal{P}(A), \subseteq)$, the lattice complement of $X \in \mathcal{P}(A)$ is simply the set-theoretic complement $A - X \in \mathcal{P}(A)$. Thus, every Boolean lattice is complemented.

■ **Example 5.16.7** The lattice $(D_n, \mid\,)$ considered in Example 5.16.2 is not complemented in general. Taking $n = 18$ for instance, we find

$$D_{18} = \{1, 2, 3, 6, 9, 18\},$$

so 3 has no complement as the reader may exhaustively verify.

Part of our motivation for studying lattices here is to explain the following relationship with Boolean algebras, which were defined at the end of §3.6 and found to exhibit properties common to both propositional calculus and set theory.

5.16.3 THEOREM *Suppose that $(A, \leq)$ is a lattice which is both distributive and complemented. Then $(A, \vee, \wedge, {}', 0, 1)$ is a Boolean algebra.*

The proof is a routine verification of the properties of a Boolean algebra enumerated in §3.6 using some of our calculations here and is best left to the reader.

This is the sort of verification which a beginning mathematician or scientist should actually do once and for all in a lifetime.

The previous result shows that any complemented and distributive lattice gives rise to a corresponding Boolean algebra. On the other hand, given an arbitrary Boolean algebra $(A, +, \cdot, ^{-}, 0, 1)$, we may define a binary relation $\preceq$ on A by setting

$$a \preceq b \quad \Leftrightarrow \quad a + b = b \text{ for any } a, b \in A.$$

5.16.4 THEOREM *Given a Boolean algebra* $(A, +, \cdot, ^{-}, 0, 1)$, *define the relation* $\preceq$ *on* A *where* $a \preceq b$ *if and only if* $a + b = b$. *Then* $(A, \preceq)$ *is a complemented and distributive lattice.*

Again, the proof is a routine verification, which the reader should actually perform once in a lifetime.

As a mathematical structure, therefore, the concept of a Boolean algebra is entirely equivalent to the concept of a complemented and distributive lattice. Recall that we have said that a lattice is "Boolean" if it is of the form $(\mathcal{P}(A), \subseteq)$ for some set A and have verified that a Boolean lattice is necessarily complemented and distributive. Conversely, it is an important theorem (which is beyond the scope of our brief discussion here) that a complemented and distributive lattice is necessarily Boolean; that is, any complemented and distributive lattice is of the form $(\mathcal{P}(A), \subseteq)$ for some appropriate set A. Together with our previous remarks, this shows that Boolean algebras and Boolean lattices are entirely equivalent mathematical structures. Thus, one may think of arbitrary lattices as a certain generalization of Boolean lattices, that is, of Boolean algebras.

EXERCISES

5.16 LATTICES

1. Prove Proposition 5.16.1; prove Theorem 5.16.3; prove Theorem 5.16.4.

2. Suppose that L_1 and L_2 are lattices corresponding to underlying orderings on a common set A. Determine which of the following are again lattices and prove your assertions.

   (a) $L_1 \cap L_2$ (b) $(L_1 - L_2) \cup (L_2 - L_1)$

   (c) $L_1 \cup L_2$ (d) The restriction of L_1 to $B \subseteq A$

3. Suppose that $(L, \leq)$ is a distributive lattice and that $a, b \in L$ where $a \leq b$. Define the subset $W = \{x \in U : a \leq x \leq b\}$ and consider the restriction (also denoted $\leq$) of $\leq$ to W .

   (a) Prove that $(W, \leq)$ is also a distributive lattice.

   (b) Prove that for each $x \in W$, there is at most one $y \in U$ satisfying $x \wedge y = a$ and $x \vee y = b$.

(c) Show that if $(U, \leq)$ is complemented, then so too is $(W, \leq)$.

4. Suppose that $(L, \leq)$ is a complemented distributive lattice. If $x_1, x_2, \ldots, x_n \in U$, then we write $x_1 \oplus x_2 \oplus \cdots \oplus x_n$ for $x_1 \vee x_2 \cdots \vee x_n$ provided that $x_i \wedge x_j = 0$ for each $i \neq j$. Prove each of the the following for $x, y, z \in U$.

(a) $x \oplus y = 1$ if and only if x is the complement of y.

(b) $\oplus$ is commutative, i.e., we have $x \oplus y = y \oplus x$.

(c) $\oplus$ is associative, i.e., we have $(x \oplus y) \oplus z = x \oplus (y \oplus z)$.

Chapter 6

Functions

We have informally described a function f from a set A to a set B as a "black box" which returns an element $f(a)$ of B when given an element a of A. We made this precise in Chapter 3 by defining a function f from A to B to be a relation $f \subseteq A \times B$, so that for each $a \in A$, there is a unique $b \in B$ so that $(a, b) \in f$. We did not pursue this definition in Chapter 3 and now finally undertake a serious study of functions, which are again among the most basic of mathematical objects.

For an explicit example, the "function" $f(x) = x^2$ from $\mathbb{R}$ to $\mathbb{R}$ which returns $x^2 \in \mathbb{R}$ when given $x \in \mathbb{R}$ is to be formally regarded as the collection of ordered pairs

$$f = \{(x, x^2) \in \mathbb{R}^2 : x \in \mathbb{R}\} \subseteq \mathbb{R}^2,$$

which we may draw in the plane $\mathbb{R}^2$ as the "graph" of f:

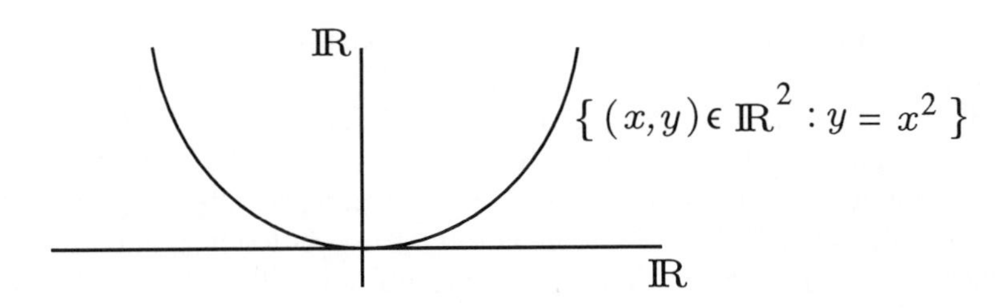

The formal definition of a function as a graph is especially useful for proving many important facts which we shall discuss below. In contrast, this formal definition is not terribly intuitive, and we shall give other graphical interpretations of functions which will hopefully enliven the reader's mathematical imagination about them; in a sense, if sets are the "nouns" of mathematics, then functions are the "verbs". In particular, the optional §6.7 treats certain elementary topics from the field of math called "dynamical systems", which illustrate certain important properties of functions. This material is further applied in §6.8 to the study of "permutations", which themselves nicely illustrate generalities about an important class of functions.

6.1 DEFINITIONS AND EXAMPLES

Let A and B be sets. A *function* f (sometimes called simply a *map*) from A to B is a binary relation $f \subseteq A \times B$ so that $\forall a \in A \exists ! b \in B[(a,b) \in f]$. In other words, $f \subseteq A \times B$ is a function from A to B if

(1) For each $a \in A$, there is some $(a,b) \in f$.

(2) For each $a \in A$, if $(a,b),(a,c) \in f$ for some $b, c \in B$, then $b = c$.

In case f is a function from A to B, we write $f : A \to B$, while one often writes $A \overset{f}{\to} B$ at the blackboard in this case. If $(a,b) \in f$ for $a \in A$, then we write $f(a) = b$, or we sometimes write $f : a \mapsto b$. In an expression such as $f(a) = b$, we refer to a as the *argument* of f and to b as the *image* of a under f or as the *value* of f at a.

Two functions $f \subseteq A \times B$ and $g \subseteq C \times D$ are regarded as identical if they are the same as binary relations, and according to our notion of identity of binary relations, if f and g are identical, then $A = C$, $B = D$, and $f = g$ as subsets of $A \times B = C \times D$. Thus, by definition, the function f uniquely determines the sets A and B where $f \subseteq A \times B$. The first factor A is called the *domain* of f, and the second factor B is called the *range* or the *codomain* of f. We sometimes say simply that "f is a function on A" if f is a function whose domain is the set A.

As usual, to get some feeling for these definitions, we begin with a number of examples.

■ **Example 6.1.1** Let f be the function from $\{1,2,3\}$ to $\{1,2\}$ given by $f = \{(1,1),(2,2),(3,2)\}$, so $f(1) = 1$ and $f(2) = 2 = f(3)$. Notice that whereas f is indeed a function, the binary relation $\{(1,1),(2,2),(3,2),(2,1)\}$ is not a function since condition (2) above is violated, and $\{(1,1),(2,2)\}$ is not a function since condition (1) above is violated. Furthermore, we might regard f as a subset of $\{1,2,3\}^2 \supseteq \{1,2,3\} \times \{1,2\}$ so as to obtain a function $g : \{1,2,3\} \to \{1,2,3\}$ where $g(x) = f(x)$ for each $x \in \{1,2,3\}$, and g is different from f (even though f and g have the same domains and take the same values) since these functions have different codomains.

■ **Example 6.1.2** Consider the empty set $\emptyset$ as a relation $\emptyset \subseteq A \times B$. In order that $\emptyset$ be a function according to the definition, we must have $A = \emptyset$ while B can be any set. The empty set is called the *empty function* when it is regarded as a function $\emptyset \to B$ in this way. Furthermore, if $B = \emptyset$ and $f : A \to B$ is a function from A to B, then f must be the empty function $f : \emptyset \to \emptyset$.

Given the function $f : A \to B$, we draw an associated digraph G which gives a vivid mental image of f as follows. Each element of A and B gives rise to a distinct vertex of G (and each element of $A \cap B$, if any, gives rise to two vertices of G), and we draw the vertices of G corresponding to A on the lefthand side of our figure and those corresponding to B on the righthand side. For each element $a \in A$, there is exactly one ordered pair $(a, b) \in f$ by definition, and we draw an oriented edge in G with initial point a and terminal point b. For instance, the function f in the Example 6.1.1 gives rise to the digraph

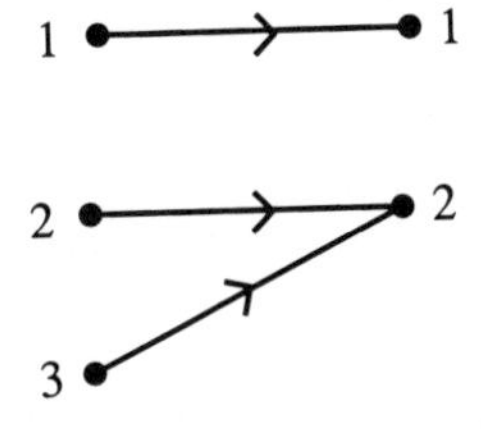

while the function g gives rise to the digraph

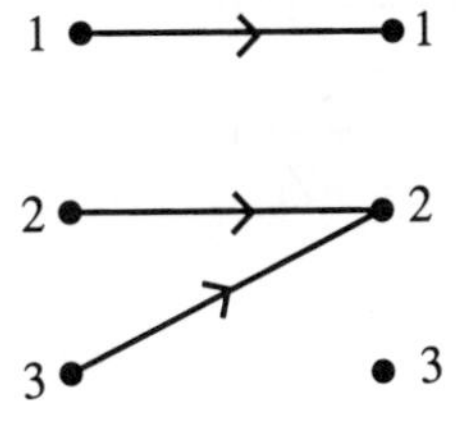

As in these examples, each vertex of G corresponding to an element of A has exactly one edge emanating from it, whereas a vertex of G corresponds to an element of B can have any number (including zero) of edges terminating at it.

We introduce this digraph G associated to the function f here because this is an instructive way to imagine a function in the mind's eye: A function is a collection of "arrows" (namely, edges of G) which begin at (vertices corresponding to) the elements of the domain and end at (vertices corresponding to) certain elements of the codomain. The condition that f is a function requires that for each vertex corresponding to an element of the domain, there is exactly one "arrow" emanating from it.

■ **Example 6.1.3** Consider the function

$$\begin{aligned} f : \mathbb{R}^2 &\to \mathbb{R}^2 \\ (x, y) &\mapsto (2x + y, x + y) \end{aligned}$$

where we have used the symbol "$\mapsto$" here in order to define the function $f \subseteq (\mathbb{R}^2)^2$. In other words, as a binary relation, we have

$$f = \{((x, y), (2x + y, x + y)) \in (\mathbb{R}^2)^2 : (x, y) \in \mathbb{R}^2\}.$$

Of course, it is not practical to try and draw the digraph associated with this function f since there are infinitely many arrows, yet the mental image of f as two copies of the plane $\mathbb{R}^2$ (one copy corresponding to the domain and the other to the codomain of f) together with arrows beginning at (x, y) and ending at $(2x + y, x + y)$ for each $(x, y) \in \mathbb{R}^2$ is still useful in the mind's eye.

■ **Example 6.1.4** Define the function

$$\begin{aligned} f : \mathbb{N} &\to \{0, 1\} \\ n &\mapsto \begin{cases} 0, & \text{if } n \text{ is even;} \\ 1, & \text{if } n \text{ is odd,} \end{cases} \end{aligned}$$

so that $f(n)$ is simply the remainder of n upon division by two. Again the associated digraph is awkward to draw

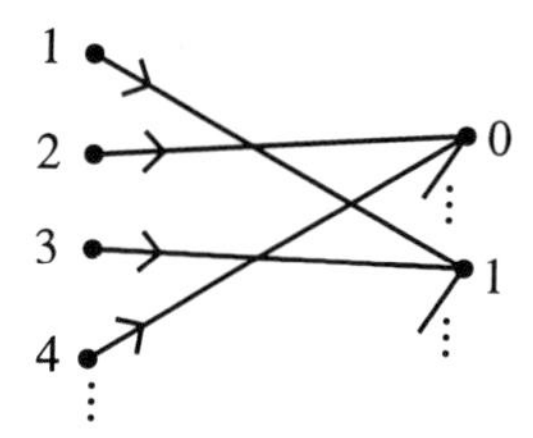

but useful to imagine.

The notation in the previous example for defining the values of a function depending on various cases of the argument is standard. Another instance of this notation, we have

■ **Example 6.1.5** Define the function

$$\begin{aligned} f : \mathbb{Z} &\to \mathbb{N} \cup \{0\} \\ n &\mapsto \begin{cases} +n, & \text{if } n \geq 0; \\ -n, & \text{if } n \leq 0, \end{cases} \end{aligned}$$

so $f(n)$ is simply the absolute value $|n|$ of $n \in \mathbb{Z}$.

Notice that in order for the previous kind of definition to make sense, any possible overlap in the various cases for the argument must lead to consistent values. For instance in the previous example, the only overlap between the cases $n \geq 0$ and $n \leq 0$ is the case in which $n = 0$, and in this case, we have $f(0) = 0 = -0$, as required. Furthermore, in order that the definition actually determines a function with the correct domain, the various cases must be "exhaustive" in the sense that at least one case holds for each element of the domain. For instance, in the previous example, any integer is either non-negative or non-positive, so the cases are indeed exhaustive.

Our next several examples give general constructions of families of functions under various circumstances.

■ **Example 6.1.6** Suppose that A is any set, and define the *identity map*

$$\begin{aligned} 1_A : A &\to A \\ a &\mapsto a, \end{aligned}$$

which is evidently a function. Indeed, the identity map on A is simply the binary relation of equality on A regarded as a function.

■ **Example 6.1.7** Suppose that A and $B \neq \emptyset$ are sets and choose some $b \in B$. Define the *constant map* $c_b : A \to B$ *with value* b, where $c_b(a) = b$ for each element $a \in A$, so as a binary relation, we have

$$c_b = \{(a, b) \in A \times B : a \in A\}.$$

Thus, c_b is the function which has value b for each value $a \in A$ of the argument. For instance, the digraph corresponding to the constant map $\{1, 2, 3\} \to \{2, 3\}$ with value 2 is

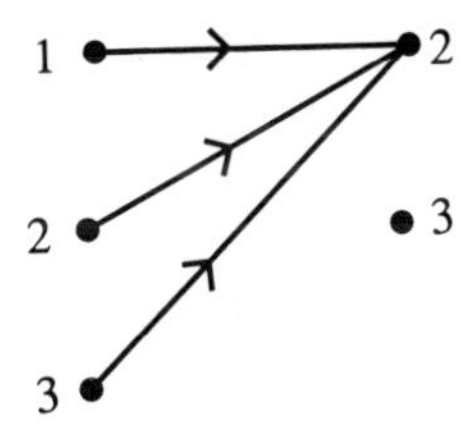

■ **Example 6.1.8** Suppose that $B \subseteq A$ are sets and define the *inclusion map*

$$\begin{aligned} i : B &\to A \\ b &\mapsto b. \end{aligned}$$

Of course, this makes sense since $B \subseteq A$ so $b \in B$ is actually also an element of A. The inclusion map $i : B \to A$ evidently just "includes" B into A hence the terminology. For instance, the digraph of the inclusion of $\{1, 2\}$ into $\{1, 2, 3\}$ is

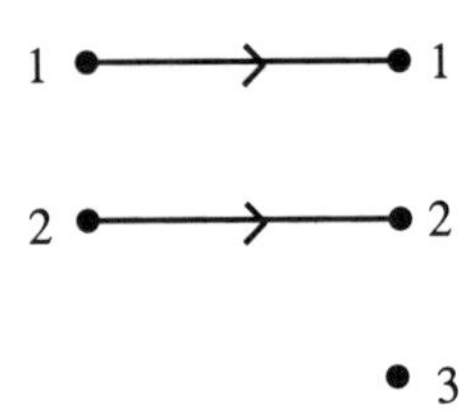

■ **Example 6.1.9** Suppose that A is a set and $\otimes$ is a binary operation on A, i.e., given two elements $a, a' \in A$, there is a well-defined element $a \otimes a' \in A$. In other words, a binary operation on a set A is simply a function

$$\begin{aligned} \otimes : A^2 &\to A \\ (a, a') &\mapsto a \otimes a'. \end{aligned}$$

In the same way, a ternary operation is simply a function $A^3 \to A$, and so on, and of course a unary operation on A is just a function $A \to A$. Thus, in a sense, we have been studying functions since the first chapters of this book.

■ **Example 6.1.10** Consider the Cartesian product $\times_{i=1}^n A_i$ of sets $A_1, \ldots, A_n$ for some $n \geq 1$ and choose an index $i \in \{1, \ldots, n\}$. Define the ith *projection map* or *projection onto the* i^{th} *factor*

$$\begin{aligned} \pi_i : \times_{i=1}^n A_i &\to A_i \\ (a_1, \ldots, a_n) &\mapsto a_i, \end{aligned}$$

which simply forgets all the entries of the n-tuple $(a_1, \ldots, a_n)$ except the ith. There are evidently n distinct projection maps on an n-fold Cartesian product. For instance, there are the two familiar projection maps

$$\pi_1(x, y) = x \quad \text{and} \quad \pi_2(x, y) = y$$

onto the coordinate axes from $\mathbb{R}^2$.

■ **Example 6.1.11** Let $\sim$ be some equivalence relation on a non-empty set A, and consider the quotient $A/\sim$. Define the *quotient map*

$$\begin{aligned} f : A &\to (A/\sim) \\ a &\mapsto [a], \end{aligned}$$

so to each element $a \in A$ is associated its equivalence class. For instance, if $\sim$ is the universal relation on A, then $A/\sim$ is a singleton, and the quotient map $A \to (A/\sim)$ is actually the constant map. If $\sim$ is the relation of equality on A for instance, then the quotient map is the assignment $a \mapsto \{a\}$, which associates the singleton $\{a\}$ to the element $a \in A$.

For a more interesting example of a quotient map, suppose that $\sim$ is equivalence modulo $p \in \mathbb{N}$ on $A = \mathbb{Z}$. As in Examples 5.12.4, 5.12.11, and 5.12.18, the quotient $\mathbb{Z}/\sim$ is identified with the finite set $\mathbb{Z}/p = \{[0], \ldots, [p-1]\}$, and the quotient map $\mathbb{Z} \to \mathbb{Z}/p$ assigns to $n \in \mathbb{Z}$ the equivalence class of the remainder of n on division by p. Of course, this is closely related to Example 6.1.4 above when $p = 2$.

For a final example of a quotient map, recall from Exercises 5.12.5, 5.12.12, and 5.12.19 the equivalence relation $\sim$ on $\mathbb{R}$ where $(x \sim y) \Leftrightarrow (x - y \in \mathbb{Z})$. The quotient $\mathbb{R}/\sim$ has been identified with the interval $\{t \in \mathbb{R} : 0 \leq t < 1\}$, and it is natural to imagine adjoining to this interval also its other endpoint 1, which must of course be identified with 0. One imagines the quotient therefore as the circle of circumference one, where the quotient map $\mathbb{R} \to (\mathbb{R}/\sim)$ "wraps" the real line infinitely often around the circle $\mathbb{R}/\sim$ as is indicated below

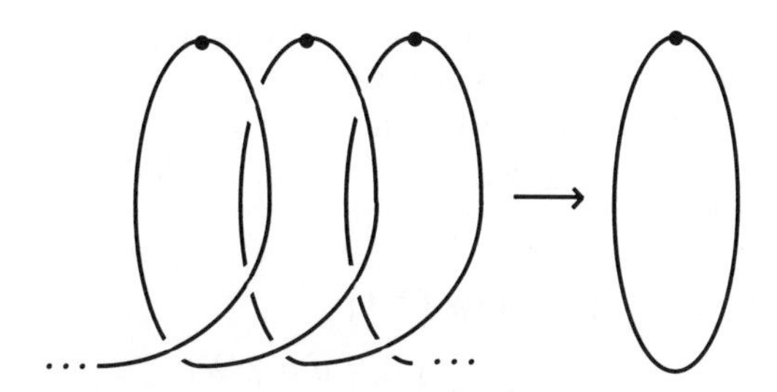

If A and B are sets, then we define A^B to be the set of all functions with domain B and codomain A. In order to better understand the set A^B, let us consider the case that A and B are finite sets, say A has $m \geq 1$ elements and B has $n \geq 1$ elements. We claim that A^B has m^n elements, and the idea of the proof is that each of the n elements of B has m possible values, and since there are n such elements, there are m^n possible functions. More formally, enumerate the elements of B as $B = \{b_1, b_2, \ldots, b_n\}$ once and for all. Given a function $f \in A^B$, we may then enumerate the values $(f(b_1), f(b_2), \ldots, f(b_n)) \in A^n$ taken by f in this order to determine a function

$$\begin{aligned} A^B &\to A^n \\ f &\mapsto (f(b_1), f(b_2), \ldots, f(b_n)) \end{aligned}$$

which identifies A^B with A^n. It follows from Theorem 3.5.4 that A^n has m^n elements, which proves our claim.

To close this section, we make several parenthetical but fundamental observations about the collection of all real-valued functions defined on some set A. Suppose that $f, g : A \to \mathbb{R}$ are functions. Since $\mathbb{R}$ is endowed with the usual arithmetic operations of addition, multiplication, and so forth, we may combine f and g using these operations in various ways to produce new real-valued functions on A. For instance, we might define a new function $h : A \to \mathbb{R}$ where $h(a) = f(a)+g(a)$; one usually denotes this function h simply as $h = f+g$, called the *pointwise sum* of f and g. For instance, the pointwise sum of the function $f : \mathbb{N} \to \mathbb{R}$ where $f(n) = n$ and the function $g : \mathbb{N} \to \mathbb{R}$ where $g(n) = -n$ is simply the constant function $f + g : \mathbb{N} \to \mathbb{R}$ taking the value zero. For another example, the pointwise sum of the previous function $f : \mathbb{N} \to \mathbb{R}$ with itself is simply the function $f + f : \mathbb{N} \to \mathbb{R}$ defined by $(f + f)(n) = 2n$. Similarly, given functions $f, g : A \to \mathbb{R}$, we might define a new function $h : A \to \mathbb{R}$, where $h(a) = f(a)g(a)$, which is called the *pointwise product* of f and g and usually denoted simply $h = fg$.

It is also convenient in this context to denote the constant function $A \to \mathbb{R}$ with value $t \in \mathbb{R}$ simply by $t : A \to \mathbb{R}$. Thus, if $f : A \to \mathbb{R}$ is a real-valued function on A, then for each $a \in A$, we have $(1 + f)(a) = 1 + f(a)$ and $(2f)(a) = 2f(a)$ for instance.

Clearly, there are many other natural ways to combine real-valued functions, and we simply wish to observe here that there is a rich structure on the set of real-valued functions defined on some fixed set. Still more generally, arbitrary binary operations on the common codomain of a family of functions can be used to construct new functions in this same way.

Having made these remarks, we finally discuss a very special class of real-valued functions defined on a given set A. Suppose that $X \subseteq A$, and define the *characteristic map of X*

$$\begin{aligned} \chi_X : A &\to \{0,1\} \\ a &\mapsto \begin{cases} 0, & \text{if } a \notin X; \\ 1, & \text{if } a \in X, \end{cases} \end{aligned}$$

so the value $\chi_X(a)$ of χ_X at $a \in A$ simply answers the question " Is $a \in X$?" by returning the value 0 if $a \notin X$ and the value 1 if $a \in X$. For instance, taking $A = \mathbb{N}$ and $X = \{n \in \mathbb{N} : n \equiv 0(2)\}$, the characteristic map χ_X takes at $n \in \mathbb{N}$ the value 0 if n is even and the value 1 if n is odd.

Not only do characteristic functions arise quite often in practice, but they are also a basic theoretical tool in mathematics since sums $\sum_{i=1}^{n} a_i\chi_i$ (where χ_i is a characteristic function and a_i is a real number) approximate arbitrary functions as n gets large. There are also lovely relationships between characteristic functions on A and the set-theoretic operations on $\mathcal{P}(A)$ as we indicate in our closing result.

6.1.1 Theorem *Suppose that $X, Y \subseteq A$ and let $\chi_Z : A \to \{0,1\}$ denote the characteristic map of any subset $Z \subseteq A$. Then we have*

(a) $\chi_{X\cap Y} = \chi_X\chi_Y$

(b) $\chi_{X-Y} = \chi_X[1 - \chi_Y]$

(c) $\chi_{X\cup Y} = \chi_X + \chi_Y - \chi_X\chi_Y$

Proof We prove only part (a) and leave the proofs of (b),(c) as analogous exercises. Part (a) asserts the equality of the functions $\chi_X\chi_Y$ and $\chi_{X\cap Y}$, where $\chi_X\chi_Y : A \to \mathbb{R}$ is the pointwise product and takes the value $(\chi_X\chi_Y)(a) = \chi_X(a)\chi_Y(a)$ for each $a \in A$ using the notation above. Observe first that these functions at least have the same domain (namely, the set A) as well as the same codomain (namely, the set $\mathbb{R}$), and it remains only to show that $\chi_X(a)\chi_Y(a) = \chi_{X\cap Y}(a)$ for each $a \in A$. To this end, notice that $\chi_X(a)\chi_Y(a)$ vanishes unless $a \in X$ and $a \in Y$ (since it is a multiple of zero), and in this case, $\chi_X(a) = \chi_Y(a) = 1$ so $\chi_X(a)\chi_Y(a) = 1$. Since these are exactly the values of $\chi_{X\cap Y}(a)$, the proof of part (a) is complete. *q.e.d.*

EXERCISES

6.1 DEFINITIONS AND EXAMPLES

1. Prove the following results from the text.

(a) Part (b) of Theorem 6.1.1.

(b) Part (c) of Theorem 6.1.1.

2. Which of the following digraphs represent functions from $A = \{a, b, c\}$ to $B = \{x, y, z\}$? For those that do not represent functions, explain why not.

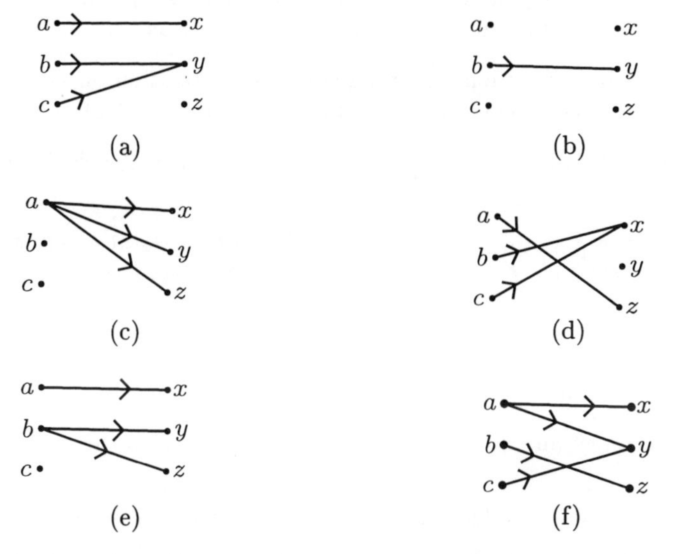

3. Which of the following relations are actually functions? For those that are not functions, explain why not.

(a) The relation $\{(1,1),(2,1),(1,2),(2,3)\}$ from $\{1,2\}$ to $\{1,2,3\}$.

(b) The relation $\{(1,2),(2,2),(3,2),(4,2)\}$ from $\{1,2,3,4\}$ to $\{1,2,3\}$.

(c) The relation $\{(x, y) \in \mathbb{R}^2 : y^2 = x\}$ from $\mathbb{R}$ to $\mathbb{R}$.

(d) The relation $\{(x, y) \in \mathbb{R}^2 : x^2 = y\}$ from $\mathbb{R}$ to $\mathbb{R}$.

4. Given a real number $x \in \mathbb{R}$, we define *floor* x to be the greatest integer $\lfloor x \rfloor \in \mathbb{Z}$ less than or equal to x and *ceiling* x to be the least integer $\lceil x \rceil \in \mathbb{Z}$ greater than or equal to x.

(a) Prove that floor and ceiling define function from $\mathbb{R}$ to $\mathbb{Z}$.

(b) Prove that for any $x \in \mathbb{R}$, we have the identity $\lfloor x \rfloor = -\lceil -x \rceil$

(c) Prove that for any $x \in \mathbb{R}$, we have the identity $\lceil x \rceil = -\lfloor -x \rfloor$

(d) Suppose that n and d are natural numbers. Prove that there are $\lfloor n/d \rfloor$ natural numbers less than or equal to n which are divisible by d.

5. Suppose that $\leq$ is a linear order on a set A. We say that a function $f : A \to A$ is *monotone* if $f(x) \leq f(y)$ whenever $x \leq y$. In the following questions, we consider the particular case $A = \mathbb{R}$ with the usual ordering $\leq$ of weak inequality.

(a) Prove that the function $f(x) = 2x$ is monotone.

(b) Prove that the function $f(x) = x^2$ is not monotone.

(c) Prove that if f and g are monotone, then the pointwise sum $f + g(x) = f(x) + g(x)$ is also monotone.

(d) Is it true that if f and g are monotone, then the pointwise product $fg(x) = f(x)g(x)$ is also monotone? Prove your assertion.

6. Let $f : A \to B$ be a function and S an equivalence relation on B. Define a binary relation R on A by setting aRb if and only if $f(a)Sf(b)$.

(a) Prove that R is an equivalence relation on A, called the "pull-back" of S by f.

(b) Consider the partitions A/R of A and B/S of B; we also call A/R the "pull-back" of B/S by f. Directly describe the blocks of A/R in terms of those of B/S.

7. Let $f : A \to \mathbb{R}$ be a real-valued function on A and define a binary relation R on A where aRb if and only if $f(a) \leq f(b)$.

(a) Find necessary and sufficient conditions on f for R to be a partial ordering.

(b) Find necessary and sufficient conditions for R to be a linear ordering.

(c) Find necessary and sufficient conditions for R to be a well ordering.

8. Suppose that I is a possibly infinite index set for an indexed family of sets A_i, $i \in I$. Define the *Cartesian product*

$$\times_{i \in I} A_i \;=\; \{\text{functions } a : I \to \cup\{A_i : i \in I\} \;:\; a(i) \in A_i, \text{ for each } i \in I\}.$$

Explain the sense in which this definition extends our previous definition of Cartesian product for finite sets.

9. Suppose that $(U, \leq)$ is a poset so that for each $x, y \in U$, the collection $\{u \in U : x \leq u < y\}$ is finite. The *Möbius function* of such a poset $(U, \leq)$ is the function $\mu : U \times U \to \mathbb{Z}$ defined recursively by setting $\mu(x, x) = 1$ and $\mu(x, y) = -\sum_{x \leq u < y} \mu(x, u)$ for all $x, y \in U$.

(a) Show that the Möbius function of the poset $(\mathcal{P}(X), \subseteq)$ for any finite set X is given by $\mu(S, T) = 0$ if S is not a subset of T, and otherwise, we have $\mu(S, T) = (-1)^{\epsilon}$, where ϵ is the number of elements of the symmetric difference $(S \cup T) - (S \cap T)$.

(b) Show that the Möbius function of the poset $\mathbb{N} \cup \{0\}$ with the partial ordering $|$ of divides is defined and has values $\mu(m, n) = (-1)^j$ if $m|n$ and n/m is the product of $j > 1$ distinct primes, $\mu(m, n) = 1$ if $m = n$, and $\mu(m, n) = 0$ otherwise.

(c) In particular for the poset $\mathbb{N} \cup \{0\}$ with the ordering $|$ of divides, there is the associated function $\mu(n) = \mu(1, n)$ of a single argument, which is also called the *Möbius function.* Prove that $\sum_{d|n} \mu(d) = 0$, for each $n > 1$.

(d) Let f_n, for $n > 0$, be a collection of numbers so that $f_n = 0$ for $n > N$. Using part (c), prove the following *Möbius inversion formulas.*

$$g_n = \sum_{k=1,3,5,\ldots}^{[N/n]} f_{kn} \quad \Rightarrow \quad f_n = \sum_{m=1,3,5,\ldots}^{[N/n]} \mu(m) g_{mn},$$

$$h_n = \sum_{k=1,3,5,\ldots}^{[N/n]} (-1)^{\frac{k-1}{2}} f_{kn} \quad \Rightarrow \quad f_n = \sum_{m=1,3,5,\ldots}^{[N/n]} (-1)^{\frac{m-1}{2}} \mu(m) h_{mn},$$

where $[x]$ denotes the integral part of x.

6.2 COMPOSITION

Suppose that $f : A \to B$ and $g : B \to C$ are functions where the codomain of f agrees with the domain of g as indicated, so the composition of f and g as relations is defined. In a departure from the notation for relations, we shall write this composition as

$$g \circ f = \Big\{(a, c) \in A \times C : \exists b \in B\big[(f(a) = b) \wedge (g(b) = c)\big]\Big\}$$

whereas the composition as relations would have been written in the other order as "$f \circ g$". The reason for this change of notation is that $g \circ f$ is actually a function (as we shall see presently) where

$$[(a, c) \in g \circ f] \Leftrightarrow [c = (g \circ f)(a)] \Leftrightarrow [c = g(f(a))],$$

and the notation $(g \circ f)(a) = g(f(a))$ is perhaps more natural than its competitor from this point of view. Of course, the order in which we choose to write compositions (as a relation or as a function) is just a point of notation, and in any case, the reader may see both notations in the literature. (In fact, in older works one sometimes finds the other order $(f \circ g)(a) = g(f(a))$ for composition of functions.) Here we shall always use the "functional notation" $g \circ f(a) = g(f(a))$ for composition which was introduced above.

6.2.1 Theorem *If $f : A \to B$ and $g : B \to C$ are functions, then their composition $g \circ f$ is a function from A to C and $g \circ f(a) = g(f(a))$ for each $a \in A$.*

Proof Of course $g \circ f$ is a relation from A to C, and we must first show that $g \circ f$ is a function, that is, for each $a \in A$, there is a unique $c \in C$ with $(a, c) \in g \circ f$. To this end, given $a \in A$, the ordered pair $(a, g(f(a)))$ lies in $g \circ f$ by definition. To see that there is a unique ordered pair in $g \circ f$ whose first entry is $a \in A$, suppose that $(a, c_1), (a, c_2) \in g \circ f$. By definition of composition, there are $b_1, b_2 \in B$ so that $f(a) = b_1, g(b_1) = c_1$ and $f(a) = b_2, g(b_2) = c_2$. Since f is a function and $(a, b_1), (a, b_2) \in f$, we must have $b_1 = b_2$. Since g is a function and $(b_1, c_1), (b_1, c_2) \in g$, we must have $c_1 = c_2$, as required.

Thus, $g \circ f$ is indeed a function with domain A and codomain C. For the final assertion, suppose that $(a, c) \in g \circ f$ so there is some $b \in B$ with $(a, b) \in f$ and $(b, c) \in g$, i.e., we have $f(a) = b$ and $g(b) = c$, so $g(f(a)) = g(b) = c = g \circ f(a)$, as was claimed. *q.e.d.*

In order to understand the composition $g \circ f : A \to C$ of functions $f : A \to B$ and $g : B \to C$, it is especially illuminating to consider the digraphs of f and g as functions. Thus, one imagines arranging the vertices corresponding to A on the extreme left, those corresponding to B in the middle, and those corresponding to C on the extreme right. Between the extreme left and the middle, we draw the arrows corresponding to elements of f, and between the middle and the extreme right, we draw the arrows corresponding to the elements of g. By combining the digraphs for f and g in this way, one can easily imagine the digraph for $g \circ f$. Namely, simply ignore the middle of the diagram, and include one arrow for each oriented edge-path of length two in our combined diagram.

For some examples, suppose that

$$f = \{(1,1), (2,2)\} : \{1,2\} \to \{1,2,3\}$$
$$g = \{(1,1), (2,2), (3,2)\} : \{1,2,3\} \to \{1,2\},$$

so f is just an inclusion map and g is the function considered in Example 6.1.1. The diagram combining the digraphs of f, g and the digraph corresponding to $g \circ f$ is

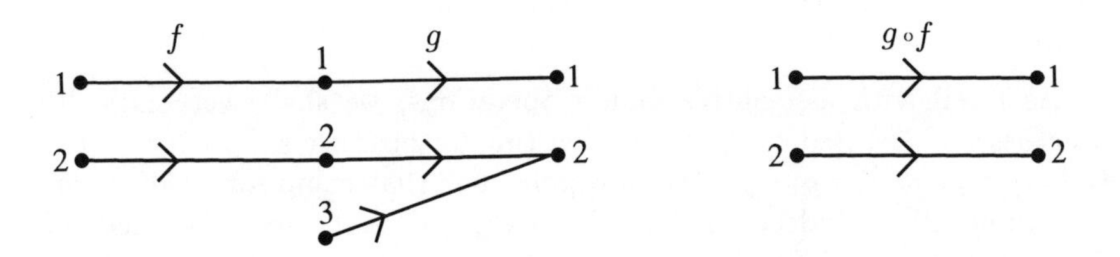

so in fact, $g \circ f$ is simply the identity map on $\{1, 2\}$. For another example, let g be as above and suppose that

$$h = \{(1,1), (2,3), (3,2), (4,3)\} \subseteq \{1,2,3,4\} \times \{1,2,3\},$$

so the diagram combining the digraphs of h, g and the digraph corresponding to $g \circ h$ is

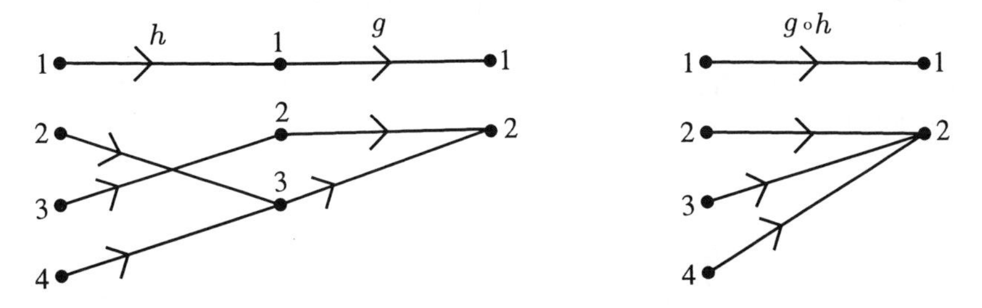

This mental image of a composition of two functions f, g as arising by combining the arrows of f with those of g as above is most useful as an imaginative tool in discovering and visualizing proofs.

6.2.2 Theorem *Suppose that $f : A \to B$ is a function, and let $1_A : A \to A$, $1_B : B \to B$ denote the identity maps on A and B. Then we have the equalities*

$$f \circ 1_A = f = 1_B \circ f$$

of functions.

Proof This follows immediately from Theorem 5.6.2, but we urge the reader to supply her or his own proof independently. *q.e.d.*

6.2.3 Theorem *Composition of functions is associative. That is, given functions $f : A \to B$, $g : B \to C$, and $h : C \to D$, we have*

$$h \circ (g \circ f) = (h \circ g) \circ f.$$

Proof This follows immediately from Theorem 5.6.1, but we again urge the reader to supply her or his own proof independently here. *q.e.d.*

As usual with associative binary operations, we shall systematically omit parentheses in iterated compositions, writing for instance simply $h \circ g \circ f$ instead of $h \circ (g \circ f) = (h \circ g) \circ f$. We also point out that composition of functions is *not* commutative: Indeed, if $f : A \to B$, $g : B \to C$ are functions, then the composition $g \circ f$ is defined, whereas $f \circ g$ is defined only if $A = C$. Furthermore, even if both compositions $g \circ f$ and $f \circ g$ are defined, they need not be equal, and we leave it to the reader to supply an example of this phenomenon.

In the special case that $f : A \to A$ is a function with the same domain and codomain, recursively define the *iterates* $f^n : A \to A$ by

Basis: Set $f^0 = 1_A$.

Induction: $f^{n+1} = f^n \circ f$ for each $n \geq 0$.

Arguing as in Example 3.8.8 (and as in the optional §5.7), one has

6.2.4 THEOREM *For any $m, n \geq 0$ and any function $f : A \to A$, we have the identities*

$$f^m \circ f^n = f^{m+n},$$
$$(f^m)^n = f^{mn}.$$

We next parenthetically discuss another type of digraph which one sometimes associates with families of functions, and the setting is the following. Let $\{A_1, \ldots, A_n\}$ be a collection of non-empty sets, and suppose that $\{f_1, \ldots, f_m\}$ is a collection of functions so that for each $k = 1, \ldots, m$, both the domain and the codomain of f_k are among the sets $A_1, \ldots, A_n$. Construct a digraph G from this data, where G has one vertex for each of the sets A_i, for $i = 1, \ldots, n$. We also "label" each vertex with its corresponding set, by which we mean simply that one writes A_i in the figure next to its corresponding vertex. Adjoin one oriented edge for each of the functions f_k, where the adjoined edge begins at the vertex corresponding to the domain of f_k and ends at the vertex corresponding to the codomain of f_k; we also associate the "label" f_k with this adjoined edge, for each $k = 1, \ldots, m$.

For instance, associated with the collection $\{A, B, C\}$ of sets and the collection

$$\{f, g, h, g \circ f, h \circ g, h \circ (g \circ f), (h \circ g) \circ f\}$$

of functions, where $f : A \to B$, $g : B \to C$, $h : C \to D$, we have the labeled digraph

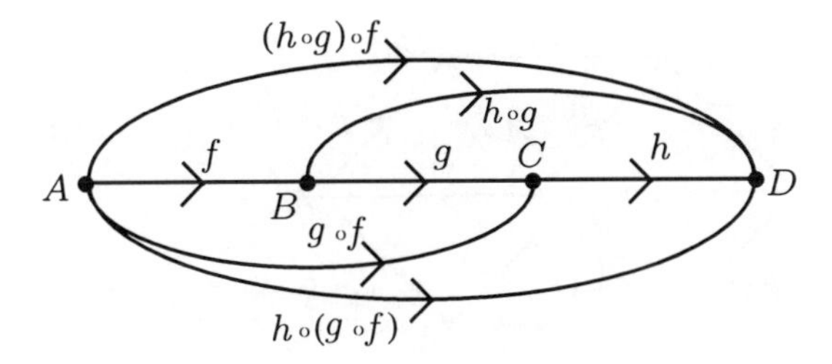

Actually, we shall only draw such a digraph G under certain circumstances as follows. Given an oriented edge-path on G, it travels along a certain sequence of edges of G in a certain order. Since these edges are labeled by functions, there is a corresponding composition of functions; for instance, in the previous example, the composition $h \circ g \circ f$ corresponds to the edge-path (A, B, C, D). We shall only draw such a digraph G under the following circumstances: Given two oriented edge-paths in the digraph G with the same initial and terminal points, the corresponding compositions of functions are required to be identical. For instance, the previous digraph does indeed satisfy this condition by Theorem 6.2.3. A digraph G arising as above from a collection of functions $\{f_1, \ldots, f_m\}$ satisfying this condition is called a *commutative diagram*, and commutative diagrams are evidently a convenient way to express potentially complicated relationships among families of functions.

For another example, the commutative diagram

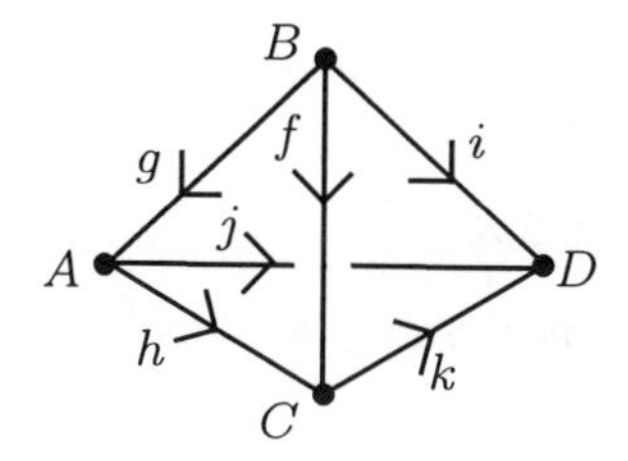

describes six functions satisfying the identities

$$h \circ g = f, k \circ f = i, k \circ h = j, \quad \text{and} \quad j \circ g = i.$$

Notice that we have drawn the previous commutative diagram as if it were a "stick figure" in space (drawing the edge labeled i "behind" the edge labeled f). One often draws commutative diagrams in this way as an aid to visualization.

As a final example of a commutative diagram, we have the following "commutative cube"

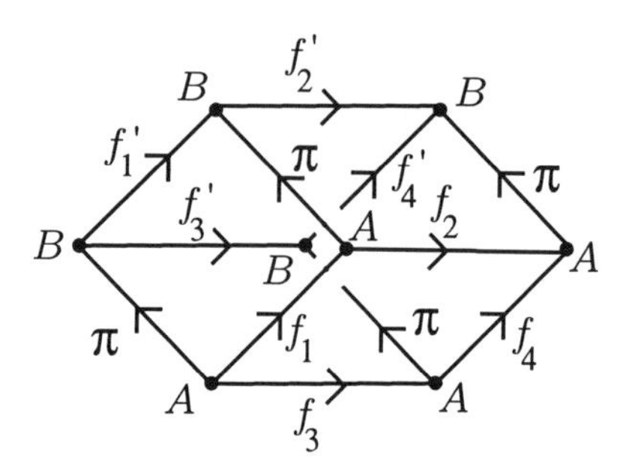

which summarizes the six equalities

$$f_2 \circ f_1 = f_4 \circ f_3, \quad f'_2 \circ f'_1 = f'_4 \circ f'_3, \quad \text{and } f'_i \circ \pi = \pi \circ f_i, \quad \text{for } i = 1, 2, 3, 4$$

among the various functions. Notice that we have included several copies of vertices labeled by A or B in this example as is typical in a situation where one has various functions with the same domain and codomain.

As indicated in these examples, a commutative diagram is just an intuitively appealing way to summarize families of equalities among functions. Though the concept of a commutative diagram is a relatively recent one (dating from the 1950-1960's), they arise rather often in practice, and this is why we have mentioned them briefly here. We shall not really rely on them in the sequel.

EXERCISES

6.2 COMPOSITION

1. Give a direct proof of Theorem 6.2.2.

2. Give a direct proof of Theorem 6.2.3.

3. Give an example of two functions f and g so that both $f \circ g$ and $g \circ f$ are defined and yet are not equal.

4. Define functions $f : \mathbb{R} \to \mathbb{R}$ by $f(t) = t^3 - 1$ and $g : \mathbb{R} \to \mathbb{R}$ by $g(t) = t^2 + 4$. Find the functions $f \circ g$ and $g \circ f$.

5. Define three functions f, g, and h from $\mathbb{Z}$ to $\mathbb{Z}$, where $f(n) = 2n$, $h(n) = n^3 + 2$, and $g(n)$ is defined to be 0 if n is even and 1 if n is odd. Find the following compositions.

   (a) $f \circ g$ (b) $g \circ f$ (c) $h \circ g$
   (d) $g \circ h$ (e) $f \circ h$ (f) $f \circ g \circ h$
   (g) $g \circ h \circ f$ (h) $h \circ g \circ f$ (i) $h \circ f \circ g$

6. Let A denote the set $\{a, b, c\}$. Find all functions $f \in A^A$ so that the following conditions hold.

(a) $f^2(x) = x$ (b) $f^2(x) = f(x)$ (c) $f^3(x) = f(x)$

7. Let A denote the set $\{a, b, c\}$, and let $f : A \to A$ denote the function where $f(a) = c$, $f(b) = a$, and $f(c) = b$. Find all functions $g \in A^A$ so that $g \circ f = f \circ g$.

8. Let A denote the set $\{a, b, c\}$, and let $f : A \to A$ denote the function where $f(a) = a$, $f(b) = a$, and $f(c) = a$. Find all functions $g \in A^A$ so that $g \circ f = f \circ g$.

9. Consider the functions $f(x) = x + 1$, $g(x) = 2x$, $h(x) = 2x - 2$, and $k(x) = x - 1$ from $\mathbb{R}$ to $\mathbb{R}$. Draw a commutative diagram reflecting identities among the various compositions of these functions.

10. Recall that a function $f : A \to A$ on a linearly ordered set $(A, \leq)$ is *monotone* if $f(x) \leq f(y)$ whenever $x \leq y$. Prove that the composition of two monotone functions is again monotone.

11. A function $f : \mathbb{R} \to \mathbb{R}$ is said to be *linear* if it is of the form $f(x) = ax + b$ for some $a, b \in \mathbb{R}$. Prove that the composition of two linear functions is again a linear function.

12. Suppose that is A_i an indexed family of sets for $i \in I$. Set $P = \times_{i \in I} A_i$ and let $\pi_i : P \to A_i$ denote projection onto the ith factor.

(a) Show that given any family $g_i : X \to A_I$, there exists a unique $h : X \to P$ so that $\pi_i \circ h = g_i$, for all $i \in I$. (Compare also Exercise 6.1.9.)

(b) Show that in fact the property of P described in part (a) uniquely characterizes the Cartesian product.

6.3 RESTRICTION AND EXTENSION

This short section is dedicated to various families of functions associated with a given function f, where we alter the domain or codomain of f to produce another closely related function. The material of this section thus allows a certain flexibility with regard to domains and codomains of functions.

Suppose that $f : A \to B$ is a function. If $A' \subseteq A$, then f induces a function from A' to B, namely f induces the function

$$\begin{aligned} f|_{A'} : A' &\to B \\ a &\mapsto f(a), \end{aligned}$$

so $f|_{A'}$ is none other than the function $f \circ i$, where $i : A' \to A$ is the inclusion of A' into A as the reader should carefully prove. The notation $f|_{A'}$ is entirely standard and is read in English as "f cut down to A'" or "f restricted to A'". We say that $f|_{A'}$ arises from f by *restricting the domain* to $A' \subseteq A$.

In a similar way, if $B \subseteq B'$, then there is an induced function

$$\begin{aligned} g : A &\to B' \\ a &\mapsto f(a), \end{aligned}$$

so g is none other than the function $j \circ f$, where $j : B \to B'$ is the inclusion of B into B' as the reader should carefully prove. There is no standard notation for this induced function g, but one says that g arises from f by *extending the codomain* to $B' \supseteq B$.

A single example of these simple constructions should suffice.

■ **Example 6.3.1** Let $f : \mathbb{Z} \to \mathbb{N} \cup \{0\}$ be the function $f(n) = |n|$ considered above which gives the absolute value $|n|$ of $n \in \mathbb{Z}$. Restricting the domain of f to the subset $\mathbb{N} \subseteq \mathbb{Z}$, we find the function

$$\begin{aligned} f|_{\mathbb{N}} : \mathbb{N} &\to \mathbb{N} \cup \{0\} \\ n &\mapsto |n|, \end{aligned}$$

which is evidently just the inclusion of $\mathbb{N}$ into $\mathbb{N} \cup \{0\}$.

We might also extend the range of this function $f : \mathbb{Z} \to \mathbb{N} \cup \{0\}$ to the set $\mathbb{Z} \supseteq \mathbb{N} \cup \{0\}$ to produce the function

$$\begin{aligned} g : \mathbb{Z} &\to \mathbb{Z} \\ n &\mapsto |n|, \end{aligned}$$

which takes only non-negative values.

We turn next to the question of restricting the codomain of a function $f : A \to B$. It may happen that there is some subset $B' \subseteq B$ so that $f(a) \in B'$ for all $a \in A$. In this case, there is an induced function

$$\begin{aligned} g : A &\to B' \\ a &\mapsto f(a) \end{aligned}$$

whose definition makes sense only because of our assumption on f. There is no standard notation for this induced function g, but we say that g arises from f by *restricting the codomain* to $B' \subseteq B$.

■ **Example 6.3.2** Let $f : \mathbb{N} \to \mathbb{N}$ be the function defined by $f(n) = 2n$, so all the values of f in fact lie in the subset $\mathcal{E} = \{n \in \mathbb{N} : n \equiv 0(2)\}$ of even natural numbers. We may restrict the codomain of f to produce a function $g : \mathbb{N} \to \mathcal{E}$ where $g(n) = 2n$.

We turn finally to perhaps the most interesting situation where $f : A \to B$ is a function whose domain is a subset of some other set $A' \supseteq A$. In this case, we seek a function $g : A' \to B$ so that for each $a \in A \subseteq A'$, we have $g(a) = f(a)$. In other words, we seek a function $g : A' \to B$ so that $g|_A = f$. We say in this case that g arises by *extending the domain* of f to $A' \supseteq A$, and the commutative diagram corresponding to this extension of domain is:

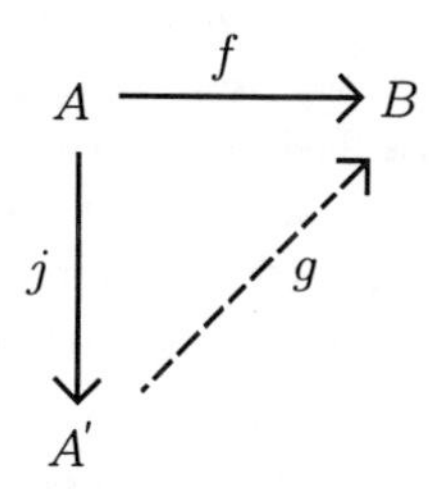

where j denotes the inclusion map $j : A \to A'$. As we shall see, there can be many different functions g extending the domain of f to the same superset A'. We mention that there are whole classes of problems and results in mathematics which amount to proving the existence of a function with special properties which extends the domain of a given function.

■ **Example 6.3.3** Let $f : \mathbb{N} \to \mathbb{Z}$ be the function defined by $f(n) = n+1$. We may easily extend the domain of f to $\mathbb{Z}$ by simply taking the function $g : \mathbb{Z} \to \mathbb{Z}$ given by the same formula $g(n) = n+1$, so of course, $g|_{\mathbb{N}} = f$. Another extension of domain to $\mathbb{Z}$ is given by the function

$$h : \mathbb{Z} \to \mathbb{Z}$$
$$n \mapsto \begin{cases} n+1, & \text{if } n \in \mathbb{N}\,; \\ 0, & \text{if } n \notin \mathbb{N}\,, \end{cases}$$

and again $h|_{\mathbb{N}} = f$ by definition.

Each of the four possibilities of extension or restriction of domain or codomain can be a useful technique in practice. As particular examples of extensions, we finally consider unions of functions.

6.3.1 Theorem *Let $f : A \to X$ and $g : B \to Y$ be functions and define $C = A \cap B$. Then $f \cup g : (A \cup B) \to (X \cup Y)$ is a function if and only if $f' = I_X \circ f|_C : C \to (X \cup Y)$ is identical with $g' = I_Y \circ g|_C : C \to (X \cup Y)$, where $I_X : X \to X \cup Y$ and $I_Y : Y \to X \cup Y$ denote the inclusions. In particular, if $C = \emptyset$, then $f \cup g$ is automatically a function.*

Proof Suppose first that $f \cup g$ is a function and choose $c \in C = A \cap B$. Since $C \subseteq A$ and $f : A \to X$ is a function, there is a unique ordered pair $(c, x) \in f$, and similarly, since $g : A \to Y$ is a function, there is a unique ordered pair (c, y).

Thus, $(c, x), (c, y) \in f \cup g$, and since $f \cup g$ is a function, we must conclude that $x = y$. Since $c \in A \cap B$ was arbitrary, we find $f' = g'$ as desired.

Conversely, we suppose that $f' = g'$ and must show that $f \cup g$ is a function from $A \cup B$ to $X \cup Y$. To this end if $t \in A \cup B$, then there are three cases depending on whether $t \in A \cap B$, $t \in A - (A \cap B)$, or $t \in B - (A \cap B)$. In the first case, there are two ordered pairs $(t, x) \in f$, $(t, y) \in g$, and we must have $x = y$ by our hypothesis that $f' = g'$. In this case, there is thus a unique ordered pair $(t, x) = (t, y) \in f \cup g$ with first entry t as desired. In the second case, there is an ordered pair $(t, x) \in f$ and since $t \notin B$, this is the only ordered pair in $f \cup g$ with first entry t. Thus, there is again a unique ordered pair $(t, x) \in f \cup g$. The third case $t \in B - (A \cap B)$ is entirely analogous to the second one. *q.e.d.*

■ **Example 6.3.4** Example 6.1.5 is simply a special case of the previous theorem, in which we have $A = \{n \in \mathbb{Z} : n \geq 0\}$, $B = \{n \in \mathbb{Z} : n \leq 0\}$, $f(n) = n$, and $g(n) = -n$.

EXERCISES

6.3 RESTRICTION AND EXTENSION

1. (a) Describe the restriction of the function $f : \mathbb{N} \to \mathbb{N}$ defined by $f(n) = 2n$ to the subset $\{1, 2, 3\} \subseteq \mathbb{N}$ of the domain.

(b) Describe the restriction of the function $f : \mathbb{Z} \to \mathbb{Z}$ defined by $f(n) = 3n - 2$ to the subset $\mathbb{N} \subseteq \mathbb{Z}$ of the domain.

2. (a) Describe an extension of the function $f : \{1, 2, 3\} \to \mathbb{N}$ given by $f = \{(1, 2), (2, 4), (3, 6)\}$ to the superset $\mathbb{N} \supseteq \{1, 2, 3\}$ of the domain.

(b) Describe an extension of the function $f : \mathbb{N} \to \{0, 1\}$ defined by $f(n) = 0$ if n is even and $f(n) = 1$ if n is odd to the superset $\mathbb{R} \supseteq \mathbb{N}$ of the domain.

3. Prove that if $A' \subseteq A$ is proper, then the restriction of the identity map $1_A : A \to A$ to A' is *not* the identity map $1_{A'}$.

4. Suppose that $A' \subseteq A$, let $i : A' \to A$ denote the inclusion map, and suppose that $f : A \to B$ is a function. Prove that $f|_{A'} = f \circ i$.

5. Suppose that $B \subseteq B'$, let $j : B \to B'$ denote the inclusion map, and suppose that $f : A \to B$ is a function. Prove that the function obtained from f by extending the codomain to B' is $j \circ f$.

6.4 INJECTIVITY, SURJECTIVITY, AND BIJECTIVITY

This section is dedicated to the study of certain special conditions on functions of truly basic importance, as follows: If $f : A \to B$ is a function, then we say that

f is *surjective* or *onto* if for each $b \in B$, there is some $a \in A$ so that $f(a) = b$. Dually, we say that f is *injective* or *one-to-one* or *monic* if whenever $a, a' \in A$ satisfy $a \neq a'$, then we have $f(a) \neq f(a')$. In other words, f is injective if and only if $\forall a, a' \in A[f(a) = f(a') \Rightarrow a = a']$. Furthermore, if f is both surjective and injective, then we say that f is *bijective*. A surjective function is called simply a *surjection* or an *epimorphism*, an injective function is called simply an *injection* or a *monomorphism*, and a bijective function is called simply a *bijection* or a *one-to-one correspondence*.

Let G be the digraph associated with the function $f : A \to B$ as before consisting of disjoint copies of vertices corresponding to elements of A and B together with one arrow starting at a and ending at b for each ordered pair $(a, b) \in f$. According to the definitions, f is surjective if for each element of B, there is some arrow of G terminating at b, and f is injective if no two distinct arrows of G terminate at a common vertex. The following digraphs of functions illustrate the various possibilities.

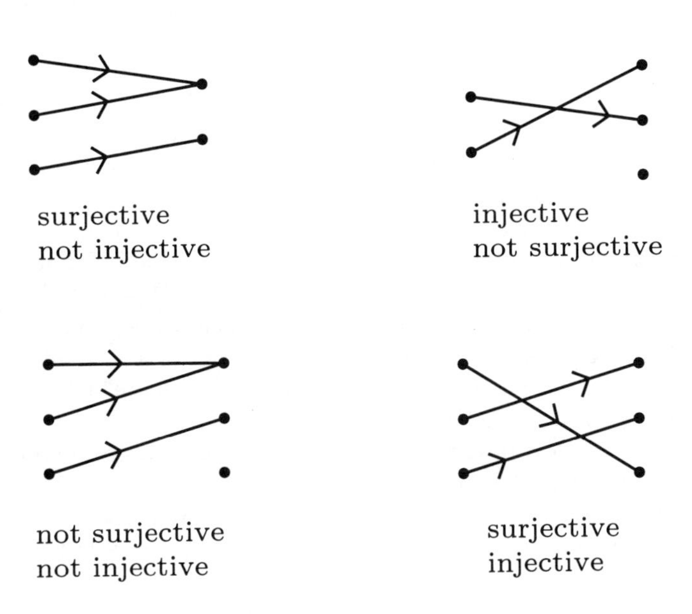

We next give several examples illustrating these definitions.

■ **Example 6.4.1** The function $f : \{1, 2, 3\} \to \{1, 2, 3\}$ given by

$$f = \{(1, 2), (2, 2), (3, 1)\}$$

is not surjective since f does not attain the value 3 for any argument, and it is not injective since $f(1) = 2 = f(2)$.

■ **Example 6.4.2** The function $f : \mathbb{N} \to \mathbb{N}$ defined by $f(n) = 3n$ is not surjective since for instance it does not attain the value 1 for any argument, but it is injective since if $f(n_1) = f(n_2)$, then $3n_1 = 3n_2$, so dividing by 3, we find $n_1 = n_2$.

■ **Example 6.4.3** Given $a < b \in \mathbb{R}$, let $[a, b]$ denote the closed interval

$$[a, b] = \{t \in \mathbb{R} : a \leq t \leq b\}$$

and define the function

$$\begin{aligned} f : [a, b] &\to [0, 1] \\ t &\mapsto \frac{t - a}{b - a}. \end{aligned}$$

The function f is surjective since if $x \in [0, 1]$, then $f(t) = x$ where $t = a + (b - a)x$ and $t \in [a, b]$ since $x \in [0, 1]$. The function f is also injective since if $f(t_1) = f(t_2)$, then $(t_1 - a)/(b - a) = (t_2 - a)/(b - a)$, so multiplying by $b - a$ and adding a, we find that $t_1 = t_2$. Since f is both injective and surjective, f is a bijection by definition.

It is worth saying explicitly here that on a certain level if there is a bijection $f : A \to B$, then one imagines the two sets A, B as being the same "size", and we shall make this comment precise in the next chapter. For finite sets A and B, if there is a bijection from A to B, then A and B clearly have the same number of elements. Indeed, one may think of a bijection from A to B as simply a "relabeling" of the the elements B by the elements in A. For instance, the bijection $f : \{1, 2, 3\} \to \{a, b, c\}$ defined by $f(1) = a$, $f(2) = b$, $f(3) = c$ simply "relabels" a, b, c by $1, 2, 3$, respectively. Indeed we shall define a set to be "finite" in the next chapter if it can be put in bijection with one of a particular collection of such sets. In fact, we shall also say a set is "countably infinite" if it can be put in bijection with the natural numbers. In a sense we shall make precise, a countably infinite set is the smallest possible infinite set, and we shall give infinitely many examples of strictly larger infinite sets.

Suppose that $f : A \to A'$ is a bijection, where $A' \subseteq A$, and A is finite. As finite sets in bijective correspondence, A and $A' \subseteq A$ must clearly have the same number of elements, so in fact, $A' = A$ in this case. This is not a complete proof (see Theorem 7.1.6 below for a complete proof), but the fact that "a subset of a finite set with the same number of elements as the set itself must be the entire set" is empirically obvious from everyday experience with finite sets.

To indicate that naive intuition about finite sets may fail, however, for infinite sets, we next give an example of a bijection between an infinite set and a proper subset of it.

■ **Example 6.4.4** Consider the subset $\mathbb{N} \subseteq \mathbb{Z}$ and the function $f : \mathbb{Z} \to \mathbb{N}$ defined by

$$f(n) = \begin{cases} 1, & \text{if } n = 0; \\ 2n, & \text{if } n > 0; \\ 1 - 2n, & \text{if } n < 0. \end{cases}$$

We are here simply defining

$$f(0) = 1, \quad f(1) = 2, \quad f(-1) = 3, \quad f(2) = 4, \quad f(-2) = 5, \quad f(3) = 6, \ldots,$$

and so on, so the digraph corresponding to f is

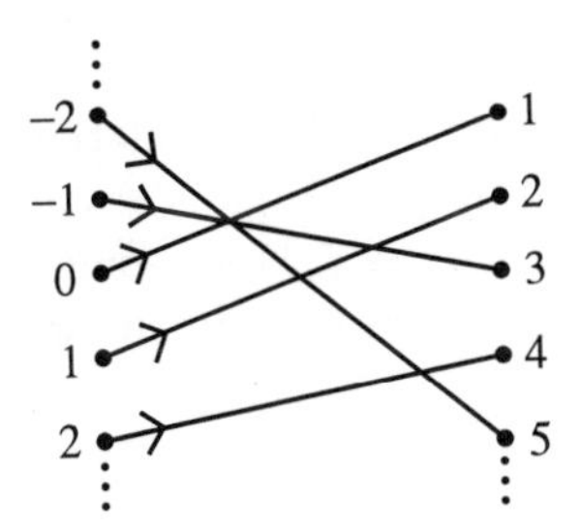

It is evident by inspection of the digraph that f is indeed a bijection, and we next give a complete proof of this.

First we show that f is injective, and to this end assume that $f(m) = f(n) = k$ for $m, n \in \mathbb{Z}$ and $k \in \mathbb{N}$. If $k = 1$, then the very definition of f implies that $m = n = 0$. We may suppose then that $k \neq 1$. According to the definition of f, if k is even, then both m and n are positive, while if k is odd, then both m and n are negative. In the former case, we compute that

$$n = \frac{f(n)}{2} = \frac{k}{2} = \frac{f(m)}{2} = m,$$

and in the latter case, we similarly compute

$$n = \frac{1 - f(n)}{2} = \frac{1 - k}{2} = \frac{1 - f(m)}{2} = m.$$

Thus, in any case we find $m = n$, so f is indeed injective.

To see that f is surjective, suppose that $k \in \mathbb{N}$. If $k = 1$, then $k = f(0)$. If k is even and non-zero, then we take $n = k/2$ satisfying $f(n) = k$, while if k is odd, then we take $n = (1 - k)/2$ satisfying $f(n) = k$. Thus, in any case we find that there is some $n \in \mathbb{N}$ so that $f(n) = k$, and f is indeed surjective.

■ **Example 6.4.5** Suppose that $A \subseteq B$ and let $i : A \to B$ denote the inclusion map. i is an injection since if $i(a_1) = i(a_2)$, then $a_1 = i(a_1) = i(a_2) = a_2$ for any $a_1, a_2 \in A$. On the other hand, i is not a surjection unless $A = B$ since if $b \in B - A$, then b is not the value of i for any choice of argument. In particular, the identity map 1_A on any set A is injective and moreover obviously surjective (since given $a \in A$, we have $1_A(a) = a$), so the identity map on any set is in fact a bijection.

■ **Example 6.4.6** Suppose that $f : A \to B$ is an injection and choose any subset $A' \subseteq A$. We may restrict the domain of f to A' to get a function $f|_{A'}$:

$A' \to B$ which is also an injection. Indeed, if $f(a_1) = f(a_2)$ for $a_1, a_2 \in A' \subseteq A$ and f is injective, then of course $a_1 = a_2$, so, indeed, $f|_{A'}$ must be injective. Furthermore, if we can restrict the codomain of f to get a function $g : A \to B'$ where $B' \subseteq B$, then a similar argument (that is, again just using the definitions) shows that g is also injective. Thus, the restriction of either domain or codomain of an injection again produces an injection.

■ **Example 6.4.7** Let R be an equivalence relation on a set $A \neq \emptyset$. The quotient map $f : A \to A/R$ defined by $f(a) = [a]$ is always a surjection, since any element $S \in A/R$ is of the form $S = [a] = f(a)$ for some $a \in A$. Thus, a quotient map is always a surjection, but it is not an injection unless $R = E_A$ is the relation of equality on A.

■ **Example 6.4.8** Let A be any set and consider the power set $\mathcal{P}(A)$ of A and the set $\{0,1\}^A$ of all functions with domain A and codomain $\{0,1\}$. Recall from §6.1 the characteristic function $\chi_B : A \to \{0,1\}$ of an arbitrary subset $B \subseteq A$ defined by

$$\chi_B(a) = \begin{cases} 0, & \text{if } a \notin B; \\ 1, & \text{if } a \in B. \end{cases}$$

The assignment of characteristic function to subset actually defines a function

$$(*) \qquad \begin{aligned} \mathcal{P}(A) &\to \{0,1\}^A \\ A &\mapsto \chi_A, \end{aligned}$$

and we shall prove here that this function is itself a bijection.

To prove surjectivity, suppose that $f \in \{0,1\}^A$ is a function and define the set

$$B = \{a \in A : f(a) = 1\}.$$

The characteristic function χ_B of B takes the same values as f, and surjectivity of the function $(*)$ is therefore established.

As to injectivity of $(*)$, suppose that $B_1, B_2 \subseteq A$ and $\chi_{B_1} = \chi_{B_2}$, i.e., the characteristic maps of B_1, B_2 agree as functions. Observe that by definition $a \in B_1$ if and only if $\chi_{B_1}(a) = 1$ with a similar comment for B_2. Equality of characteristic functions therefore implies equality of sets, whence $B_1 = B_2$, and injectivity is established.

Our next result asserts that the properties of surjectivity and injectivity are invariant under composition of functions.

6.4.1 THEOREM *Suppose that $f : A \to B$ and $g : B \to C$ are functions, so the composition $g \circ f : A \to C$ is defined. Then*

(a) *If both f and g are surjective, then so too is $g \circ f$.*

(b) *If both f and g are injective, then so too is $g \circ f$.*

(c) *If both f and g are bijective, then so too is $g \circ f$.*

Proof According to the definitions, part (c) follows immediately from parts (a) and (b). As to part (a), suppose that $c \in C$. We must show that there is some $a \in A$ so that $g \circ f(a) = c$. To this end, since g is surjective, there is some $b \in B$ so that $g(b) = c$, and since f is surjective, there is furthermore some $a \in A$ so that $f(a) = b$. Thus, $g \circ f(a) = g(f(a)) = g(b) = c$, as desired.

Turning to part (b), suppose that $a_1, a_2 \in A$ and $g \circ f(a_1) = g \circ f(a_2)$, i.e., we have $g(f(a_1)) = g(f(a_2))$ and must show that $a_1 = a_2$. Since g is injective, we conclude that $f(a_1) = f(a_2)$, and since f is injective, we finally conclude that indeed $a_1 = a_2$. *q.e.d.*

As a partial converse to the previous result, we have

6.4.2 THEOREM *Suppose that $f : A \to B$ and $g : B \to C$ are functions, so the composition $g \circ f : A \to C$ is defined. Then*

(a) *If $g \circ f$ is surjective, then so too is g.*

(b) *If $g \circ f$ injective, then so too is f*

(c) *If $g \circ f$ is bijective, then g is surjective and f is injective.*

Proof As before, part (c) follows directly from parts (a) and (b). For part (a), we suppose that $c \in C$ and must exhibit some $b \in B$ so that $g(b) = c$. To this end, since $g \circ f$ is surjective, there is some $a \in A$ so that $g \circ f(a) = g(f(a)) = c$, so $b = f(a) \in B$ satisfies the requirement that $g(b) = c$, as desired.

Turning to part (b), we suppose that $a_1, a_2 \in A$ satisfy $f(a_1) = f(a_2)$ and must show that $a_1 = a_2$. Indeed, we find

$$g \circ f(a_1) = g(f(a_1)) = g(f(a_2)) = g \circ f(a_2),$$

since $f(a_1) = f(a_2)$, so by injectivity of $g \circ f$, we conclude that in fact $a_1 = a_2$. *q.e.d.*

Thus, if $g \circ f$ is surjective (and injective respectively), then g is surjective (and f is injective) even though we *cannot* necessarily conclude that f is surjective (and g is injective). For an example where the latter condition fails, just recall a previous example where $f : \{1,2\} \to \{1,2,3\}$ is the inclusion map and $g : \{1,2,3\} \to \{1,2\}$ is the function defined by $g(1) = 1$ and $g(2) = 2 = g(3)$. The composition $g \circ f$ is the identity map on $\{1,2\}$ as we have seen, which is both surjective and injective. The inclusion map f is injective but not surjective, while the map g is surjective but not injective. Thus, the precise converse of Theorem 6.4.1 does not hold, and Theorem 6.4.2 is the most we can assert in this regard.

EXERCISES

6.4 INJECTIVITY, SURJECTIVITY, AND BIJECTIVITY

1. For each of the following functions, determine whether it is injective, surjective, bijective, or has none of these attributes. Justify your answer.

(a) $f : [0,1] \to [0,1]$
$t \mapsto \frac{t}{3} + \frac{1}{9}$

(b) $f : \mathbb{R} \to \mathbb{R}$
$t \mapsto 2^t$

(c) $f : \mathbb{N} \to \mathbb{N}^2$
$n \mapsto (n, n+2)$

(d) $f : \mathbb{R} \to \mathbb{R}$
$t \mapsto 8$

(e) $f : \mathbb{Z} \to \mathbb{Z}$
$n \mapsto 2n - 1$

(f) $f : (0,1) \to \{x \in \mathbb{R} : x \geq 0\}$
$t \mapsto \frac{1}{t}$

(g) $f : \mathbb{R}^2 \to \mathbb{R}$
$(x, y) \mapsto x^2 + y^2$

(h) $f : \{x \in \mathbb{R} : x \geq 0\} \to \mathbb{R}$
$t \mapsto \frac{1}{1+t}$

2. Give an example of a function $\mathbb{N} \to \mathbb{N}$ satisfying each of the properties below.

   (a) Neither one-to-one nor onto

   (b) One-to-one but not onto

   (c) Onto but not one-to-one

3. Construct a bijection from A to B in each case below.

   (a) $A = \{1, 2, 3\}$ and $B = \{a, b, c\}$

   (b) $A = \mathbb{Z}$ and $B = \mathbb{N}$

   (c) $A = (-1, 1)$ and $B = (0, 1)$

   (d) $A = (0, 1)$ and $B = \mathbb{R}$

   (e) $A = [0, 1)$ and $B = (\frac{1}{2}, 1]$

4. Suppose that $a, b \in \mathbb{R}$, where $a \neq 0$, and define $f : \mathbb{R} \to \mathbb{R}$ by $f(t) = at + b$.

   (a) Prove that f is surjective.

   (b) Prove that f is injective.

5. (a) Let $A = \{a, b\}$. Enumerate all the injections, surjections, and bijections in A^A.

   (a) Let $A = \{a, b, c\}$. Enumerate all the injections, surjections, and bijections in A^A.

6. Fix a set X and define a binary relation $\sim$ on $\mathcal{P}(X)$, where $A \sim B$ if and only if there is some bijection from A to B. Prove that $\sim$ is an equivalence relation on $\mathcal{P}(X)$.

7. Prove that for any set A there is an injection from A to $\mathcal{P}(A)$.

8. Fix a set A and suppose that $f \in A^A$. Prove that if f is injective (surjective

and bijective, respectively), then f^n is also injective (surjective and bijective) for each $n \in \mathbb{N}$.

9. Let $\sim$ be an equivalence relation on a set A. Show that that quotient map $A \to (A/\sim)$ is not injective unless $\sim$ is the relation of equality on A.

10. Let $f : A \to \mathbb{R}$ and $g : A \to \mathbb{R}$ be real-valued functions on a set A. If f and g are each injections, then is the pointwise sum $(f+g)(x) = f(x)+g(x)$ also an injection? Prove your assertion.

11. Give an example of a bijection between the interval $[0,1] \subseteq \mathbb{R}$ and a proper subset of it.

6.5 INVERSES

There are actually several notions of inverses which we shall discuss in this section. The simplest case occurs for a *bijection* $f : A \to B$. In this case, it turns out that there is another associated bijection $f^{-1} : B \to A$, where $f^{-1} \circ f = 1_A$ and $f \circ f^{-1} = 1_B$. When f is a surjection, then we can only conclude that there is some function $g : B \to A$ so that $f \circ g = 1_B$; g is called a "right inverse" to f since when you compose f with g on the righthand side, the resulting composition is the identity map. Dually, if f is an injection, then we can only conclude that there is some function $h : B \to A$ so that $h \circ f = 1_A$, and h is called a "left inverse" to f for reasons similar to the previous case. It is worth emphasizing that given a bijection f, there is a unique "two-sided" inverse f^{-1}, while if f is only an injection or surjection, then there can be many different "one-sided" (i.e., left or right respectively) inverses.

To begin, we assume that $f : A \to B$ is a bijection and define the *inverse* f^{-1} of f to be simply the inverse relation (in the sense of §5.5) of f; that is, we have

$$f^{-1} = \{(b,a) \in B \times A : (a,b) \in f\}.$$

For instance, given the bijection $f : \{1,2,3\} \to \{1,2,3\}$ defined by $f(1) = 2$, $f(2) = 3$, $f(3) = 1$, the inverse is given by $f^{-1}(2) = 1$, $f^{-1}(3) = 2$, $f^{-1}(1) = 3$ and is seen to be a bijection in this case. On the level of digraphs of functions, we have:

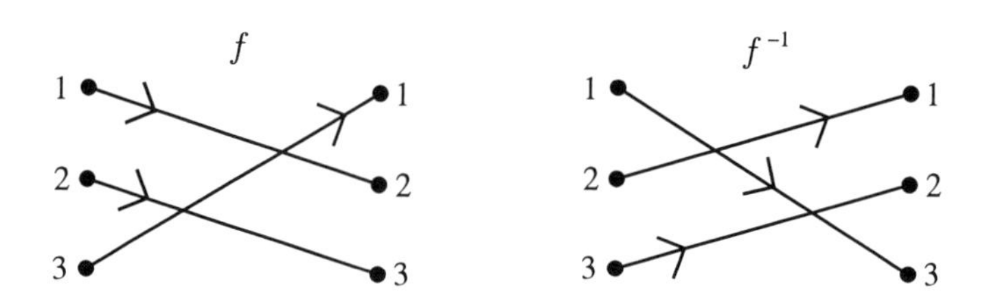

6.5.1 THEOREM *If $f : A \to B$ is a bijection, then $f^{-1} : B \to A$ is a bijection as well. Furthermore, we have*

$$f^{-1} \circ f = 1_A \text{ and } f \circ f^{-1} = 1_B.$$

Proof We first show that f^{-1} is actually a function, and to this end suppose that $b \in B$. We must prove that there is a unique ordered pair (b, a) in f^{-1}, or in other words by definition of inversion, given $b \in B$, there is a unique ordered pair $(a, b) \in f$. First of all, such an ordered pair must exist since f as a bijection is in particular surjective. Secondly, if (b, a') is another ordered pair in f^{-1}, then again by definition of inversion we have $(a', b), (a, b) \in f$. Since f is a bijection and in particular injective, we must conclude that $a = a'$, so f^{-1} is indeed a function.

Next we prove that $f^{-1} \circ f = 1_A$ and suppose that $(a, a') \in f^{-1} \circ f$. By definition of composition, there is some $b \in B$ so that $f(a) = b$ and $f^{-1}(b) = a'$, i.e., we have $(a, b) \in f$ and $(a', b) \in f$. Since f is injective, we must conclude that $a = a'$, so $f^{-1} \circ f \subseteq 1_A$. For the reverse inclusion, given $a \in A$, there is a unique $(a, b) \in f$ since f is a function, so $(b, a) \in f^{-1}$, and $(a, a) \in f^{-1} \circ f$, i.e., we indeed have $1_A \subseteq f^{-1} \circ f$.

Turning to the composition $f \circ f^{-1}$, suppose that $f \circ f^{-1}(b) = b'$ for some $b, b' \in B$. By definition of composition, there is some $a \in A$ so that $f^{-1}(b) = a$ and $f(a) = b'$'. Thus, we have $(a, b), (a, b') \in f$, and since f is a function, we must conclude that $b = b'$, so $f \circ f^{-1} \subseteq 1_B$. Again, the reverse inclusion follows easily and is left as an exercise.

Using these equalities $f^{-1} \circ f = 1_A$ and $f \circ f^{-1} = 1_B$, bijectivity of f^{-1} follows directly from Theorem 6.4.2, completing the proof. *q.e.d.*

As a consequence of the previous theorem, we conclude that if $f(a) = b$, then we may simply "apply f^{-1} to each side of the equality" to obtain $a = f^{-1}(f(a)) = f^{-1}(b)$. In the same way, if $f^{-1}(b) = a$, then we may "apply f to each side" to obtain $b = f(f^{-1}(b)) = f(a)$.

The inverse f^{-1} of a bijection f has been shown to be a bijection and hence has an inverse function $(f^{-1})^{-1}$.

6.5.2 THEOREM *If f is a bijection, then $(f^{-1})^{-1} = f$.*

Proof We proved this fact about inversion in Theorem 5.5.1a, and the reader should easily independently reprove this directly here. *q.e.d.*

Turning away now from the setting of bijections, we consider "one-sided" inverses of injections and surjections. Given a function $f : A \to B$, we say that $g : B \to A$ is a *right inverse* to f if $f \circ g = 1_B$. Dually, we say that $h : B \to A$ is a *left inverse* to f if $h \circ f = 1_A$. It is worth pointing out that simply remembering that the right inverse lies "on the right" in the composition $f \circ h$, one can figure out that in fact $f \circ h = 1_B$. Indeed, the identity $f \circ h = 1_A$ cannot hold since 1_A and $f \circ h$ have different domains (unless $A = B$).

■ **Example 6.5.1** Let f be the inclusion $f : \{1,2\} \to \{1,2,3\}$, so f is therefore an injection by Example 6.4.5; the digraph of the function f is

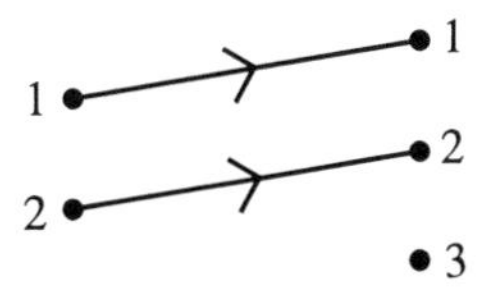

Let us consider "building" a left inverse $g : \{1,2,3\} \to \{1,2\}$ to f: Since $f(1) = 1$ and $f(2) = 2$ and we wish to arrange that $g \circ f = 1_A$, the function g we seek must satisfy $g(1) = 1$ and $g(2) = 2$; in order to complete the definition of a legitimate function g, it remains to define a value for g at 3. Of course, there are two choices for $g(3)$, namely, $g(3) = 1$ or $g(3) = 2$, and there are therefore two different left inverses of f. The following digraphs illustrate the two possible left inverses of f.

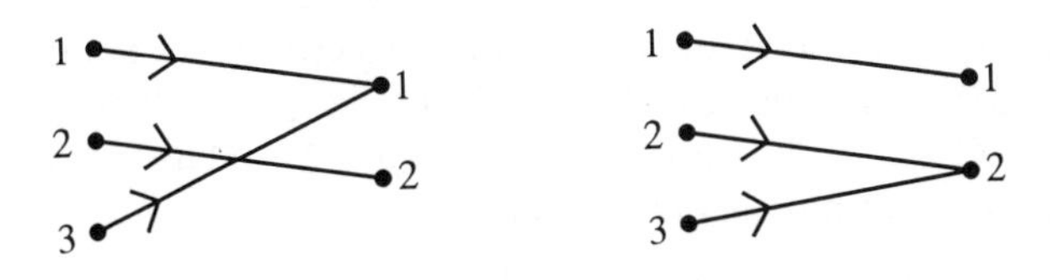

■ **Example 6.5.2** Let $f : \{1,2,3\} \to \{1,2\}$ be defined by $f(1) = 1 = f(2)$ and $f(3) = 2$, so f is evidently a surjection with corresponding digraph

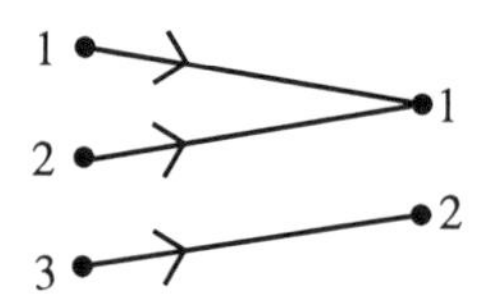

Again we consider "building" a right inverse $h : \{1,2\} \to \{1,2,3\}$. In order that $f \circ h = 1_B$, we must clearly have $h(2) = 3$. On the other hand, we have a choice for the value of h at 1: Indeed, this value might be either $h(1) = 1$ or $h(1) = 2$, so there are two different right inverses to f with corresponding digraphs

■ **Example 6.5.3** Let $f : \mathbb{N} \to \mathbb{Z}$ be the inclusion. In this case, f has infinitely many different left inverses. For instance, we may choose any $m \in \mathbb{N}$ and define

$$g_m : \mathbb{Z} \to \mathbb{N}$$
$$n \mapsto \begin{cases} n, & \text{if } n \in \mathbb{N}\,; \\ m, & \text{if } n \notin \mathbb{N}\,. \end{cases}$$

We leave it for the reader to check (for instance, by imagining the corresponding digraphs) that for each $m \in \mathbb{N}$, the function g_m is indeed a left inverse to f. Of course, there are many other left inverses to f as well in this example.

■ **Example 6.5.4** Consider equivalence modulo $p \in \mathbb{N}$ on $\mathbb{Z}$, and consider the quotient map $f : \mathbb{Z} \to \mathbb{Z}/p$, which is a surjection by Example 6.4.7. We have already identified the quotient $\mathbb{Z}/p$ with the set whose elements are the sets $[q] = \{q + kp \in \mathbb{Z} : k \in \mathbb{Z}\}$ where $0 \le q \le p-1$. Thus, we may choose a collection $k_0, \ldots, k_{p-1}$ of integers and define

$$h : \mathbb{Z}/p \to \mathbb{Z}$$
$$[q] \mapsto q + k_q p$$

which is a right inverse for each choice of integers $k_0, \ldots, k_{p-1}$, as the reader should verify.

There is a fundamental and elegant relationship between one-sided inverses, injectivity, and surjectivity as we next describe.

6.5.3 THEOREM *Suppose that $f : A \to B$ is a function and $A \neq \emptyset$. Then*

(a) *f is injective if and only if it has a left inverse.*

(b) *f is surjective if and only if it has a right inverse.*

(c) *f has both a left inverse g and right inverse h if and only if it is bijective. f has a two-sided inverse f^{-1}, g and h are the unique left and right inverses, respectively, and $g = f^{-1} = h$ in this case.*

Proof Let us remark at the outset that we have already seen elementary instances of portions of the proofs of parts (a) and (b) in Examples 6.5.1-4 above.

For part (a), if $g : B \to A$ is a left inverse to f, then $g \circ f = 1_A$. The identity map 1_A is bijective, hence f is injective by Theorem 6.4.2a. Conversely, if f is injective, then we may choose an element $a_0 \in A \neq \emptyset$ and define the function $g : B \to A$ defined by

$$g(b) = \begin{cases} a, & \text{if there is some } a \in A \text{ so that } f(a) = b; \\ a_0, & \text{if there is no } a \in A \text{ so that } f(a) = b. \end{cases}$$

Thus, if there is some ordered pair $(a, b) \in f$, then we simply define $g(b) = a$, and otherwise we stipulate that $g(b)$ is our chosen element a_0. This definition depends subtly on the fact that f is an injection: In the first case of the definition of g, if there is some $a \in A$ with $(a, b) \in f$, then there must be a unique such a since f is injective, so the specification $g(b) = a$ is uniquely determined in the first case, and g is indeed a function. To see that g is actually a left inverse to f, choose some $a \in A$, set $b = f(a) \in B$, and observe that $g \circ f(a) = g(b) = a$ by definition, that is, $g \circ f = 1_A$.

For part (b), suppose that $h : B \to A$ is a right inverse to f so $f \circ h = 1_B$. The identity map 1_B is bijective, hence f is surjective by Theorem 6.4.2b. Conversely, if f is surjective, then for each $b \in B$, there is some $a \in A$ so that $f(a) = b$. Thus, for each $b \in B$, the set $S_b = \{a \in A : f(a) = b\}$ is non-empty, so $X = \{S_b : b \in B\}$ is a set of non-empty sets. According to the Axiom of Choice, there is a choice function $F : X \to \cup X$ so that $F(S_b) \in S_b$ for each $b \in B$. Notice that since $F(S_b)$ is an element of S_b, we must have $f(F(S_b)) = b$ for any possible choice function. It follows that we may take the choice function $h = F$ itself as a right inverse, and the choice function F exists (i.e., the hypothesis of the Axiom of Choice that X is a set of *non-empty* sets is satisfied) because f is a surjection, completing the proof of part (b).

The first assertion in part (c) follows immediately. For the second assertion, let f^{-1} be the two-sided inverse of f and let g and h, respectively, denote the left and right inverse to f. We calculate

$$g = g \circ 1_B = g \circ (f \circ f^{-1}) = (g \circ f) \circ f^{-1} = 1_A \circ f^{-1} = f^{-1}$$

and similarly

$$h = 1_A \circ h = (f^{-1} \circ f) \circ h = f^{-1} \circ (f \circ h) = f^{-1} \circ 1_B = f^{-1},$$

where we have used associativity of composition and Theorem 6.2.2 in these various calculations. *q.e.d.*

For some practice with commutative diagrams (and hopefully also to illuminate the previous proof), we include commutative diagrams illustrating the relationships between the various functions in the previous two inset equations:

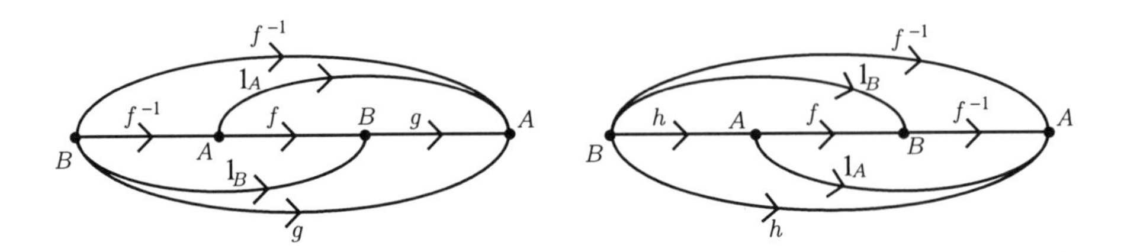

It is worth remarking that in the proof that surjectivity implies the existence of a right inverse, we have actually relied on the Axiom of Choice. In fact, the assertion we prove here is logically equivalent to the Axiom of Choice assuming the Zermelo-Frenkel Axioms. This is the first of several instances in this book where we really require the Axiom of Choice.

We should also mention that the previous result is a useful tool for proving that a given function f is injective, surjective, or bijective: One simply exhibits a left inverse, right inverse, or two-sided inverse, respectively, and applies the theorem. For instance, in Example 6.4.3, we considered the function

$$f : [a, b] \to [0, 1]$$
$$t \mapsto \frac{t-a}{b-a}.$$

To see that f is a bijection, we may simply define the function

$$g : [0, 1] \to [a, b]$$
$$x \mapsto a + (b-a)x$$

and check that $f \circ g = 1_{[0,1]}$, $g \circ f = 1_{[a,b]}$ to conclude that f is a bijection by Theorem 6.4.2c.

EXERCISES

6.5 INVERSES

1. Give a direct proof of Theorem 6.5.2.

2. Let $X = \mathbb{R} - \{2\}$ and define

$$\begin{aligned} f : X &\to X \\ t &\mapsto \frac{2t-1}{t-2}. \end{aligned}$$

   (a) Prove that f is a bijection.

   (b) Find f^{-1}.

3. Let $\mathbb{R}_+ = \{t \in \mathbb{R} : t\, 0\}$.

   (a) Let

$$\begin{aligned} f : \mathbb{R}_+ &\to \mathbb{R}_+ \\ t &\mapsto \frac{1}{t}. \end{aligned}$$

   Prove that f is a bijection and find f^{-1}.

   (b) Let

$$\begin{aligned} f : \mathbb{R}_+ &\to \mathbb{R}_+ \\ t &\mapsto \frac{1}{t^2}. \end{aligned}$$

   Prove that f is a bijection and find f^{-1}.

4. Let A and B be sets and find a bijection $A \times B \to B \times A$.

5. Show that

$$\begin{aligned} f : (\mathbb{N} \cup \{0\})^2 &\to \mathbb{N} \\ (a, b) &\mapsto 2^a(2b+1) \end{aligned}$$

   is a bijection and find f^{-1}.

6. Prove that for each $m \in \mathbb{N}$, the function

$$\begin{aligned} g_m : \mathbb{Z} &\to \mathbb{N} \\ n &\mapsto \begin{cases} n, & \text{if } n \in \mathbb{N}; \\ m, & \text{if } n \notin \mathbb{N}; \end{cases} \end{aligned}$$

   is a left inverse to the inclusion $\mathbb{N} \to \mathbb{Z}$.

7. Prove that for each $p \in \mathbb{N}$ and each choice of $k_0, k_1, \ldots, k_{p-1} \in \mathbb{Z}$, the function

$$\begin{aligned} h : (\mathbb{Z}/p) &\to \mathbb{Z} \\ [q] &\mapsto q + k_q p \end{aligned}$$

is a right inverse to the quotient map $\mathbb{Z} \to (\mathbb{Z}/p)$, where we write $\mathbb{Z}/p = \{[0], [1], \ldots, [p-1]\}$.

8. For each of the functions in Exercise 1 of Section 6.4, provide a left inverse, right inverse, two-sided inverse, or explain why one or another of these fails to exist.

9. Let $f : A \to B$ and $g : B \to C$ be functions.

(a) Is it true that if $g \circ f$ and f are each bijections, then so too is g? Prove your assertion.

(b) Is it true that if $g \circ f$ and g are each bijections, then so too is f? Prove your assertion.

10. (a) Suppose that f and g are bijections so that $f \circ g$ is defined. Prove that $(f \circ g)^{-1} = g^{-1} \circ f^{-1}$.

(b) Suppose that $f_1, f_2, \ldots, f_n$ are bijections so that $f_1 \circ f_2 \circ \cdots \circ f_n$ is defined for some $n \geq 2$. Prove that $(f_1 \circ f_2 \circ \cdots \circ f_n)^{-1} = f_n^{-1} \circ \cdots \circ f_2^{-1} \circ f_1^{-1}$.

6.6 IMAGES AND PRE-IMAGES

Given a function $f : A \to B$, we first define here an induced function $\mathcal{P}(A) \to \mathcal{P}(B)$. Namely, given $f : A \to B$ and a subset $A' \subseteq A$, we associate here a subset $f(A') = \{b \in B : \exists a \in A'(f(a) = b)\}$ called the "image" of A'. Thus, $f(A') \subseteq B$ is simply the collection of values that f takes as the argument a various over the subset $A' \subseteq A$, or in terms of the digraph of the function f, $f(A')$ is simply the collection of vertices $b \in B$ so that there is some arrow in f pointing from some $a \in A'$ towards b.

Dually, given $f : A \to B$, we also define an induced function $\mathcal{P}(B) \to \mathcal{P}(A)$. Namely, given $f : A \to B$ and a subset $B' \subseteq B$, we associate a subset $f^{-1}(B') = \{a \in A : f(a) \in B'\}$ called the "pre-image" of B'. In fact, given $B' \subseteq B$, the "pre-image" is simply the collection of arguments $a \in A$ whose values lie in the given subset $B' \subseteq B$, or in terms of the digraph of the function f, the pre-image $f^{-1}(B')$ is simply the collection of vertices $a \in A$ so that there is an arrow in f pointing from a to a vertex in B'.

This section is dedicated to analyzing certain set-theoretic aspects of images and pre-images.

Suppose that $f : A \to B$ is some function, and let $A' \subseteq A$ be a subset of the domain. Define the *image of A' under f* to be the subset

$$f(A') = \{b \in B : \exists a \in A'(f(a) = b)\} \subseteq B.$$

In particular, taking $A' = A$, we define the *image* of the function f itself to be $f(A) \subseteq B$.

■ **Example 6.6.1** Let $f : \{1,2,3\} \to \{1,2\}$ be the function defined by $f(1) = 1 = f(2)$ and $f(3) = 2$. We have

$$\begin{aligned} f(\{1\}) &= f(\{2\}) = f(\{1,2\}) = \{1\}, \\ f(\{3\}) &= \{2\}, \\ f(\{1,3\}) &= f(\{2,3\}) = f(\{1,2,3\}) = \{1,2\} \end{aligned}$$

■ **Example 6.6.2** Let $f : \mathbb{N} \to \mathbb{N}$ be defined by $f(n) = 2n$, so the image of the set $\{2,5,8\}$ is

$$f(\{2,5,8\}) = \{4,10,16\}$$

for instance. The image of the function f itself in this case is the set of all even positive integers.

In fact, taking images of subsets "respects" inclusions in the following sense.

6.6.1 THEOREM *Suppose that $f : A \to B$ is a function. If $A_1 \subseteq A_2 \subseteq A$, then $f(A_1) \subseteq f(A_2)$.*

Proof Suppose that $b = f(a_1) \in f(A_1)$ for some $a_1 \in A_1$. Since $A_1 \subseteq A_2$ by hypothesis, we conclude that $a_1 \in A_2$ as well, so $b = f(a_1) \in f(A_2)$. *q.e.d.*

It is worth pointing out that the converse of the previous result does not hold. For an example, simply take $f : \{1,2,3\} \to \{1,2\}$ where $f(1) = 1 = f(2)$ and $f(3) = 2$, so $f(\{1,2,3\}) \subseteq f(\{1,3\})$, while of course $\{1,2,3\} \not\subseteq \{1,3\}$.

Another basic result about images whose proof is essentially a reformulation of the definitions is

6.6.2 THEOREM *Suppose that $f : A \to B$ is a function. Then f is surjective if and only if the codomain B is the image of f.*

Proof By definition, the image $f(A)$ of f is the collection of values taken by f, and this collection of values is all of B exactly when f is surjective. *q.e.d.*

Our final basic result about images asserts that the restriction of codomain to the image of an injection is actually a bijection.

6.6.3 THEOREM *Suppose that $f : A \to B$ is an injection and consider $f(A) \subseteq B$. If $g : A \to f(A)$ arises by restricting the codomain of f to $f(A)$, then g is a bijection.*

Proof As the restriction of an injection, g is itself injective as in Example 6.4.6. Furthermore, g is a surjection by the very definitions, so g is indeed a bijection. *q.e.d.*

Dually, given a function $f : A \to B$ and a subset $B' \subseteq B$, define the *pre-image of B' under f* to be

$$f^{-1}(B') = \{a \in A : f(a) \in B'\} \subseteq A,$$

so on the level of digraphs, $f^{-1}(B')$ is the set of initial points of arrows of f which point towards elements of B'.

It is unfortunate but standard notation that the same symbol "f^{-1}" is used for pre-images and for two-sided inverses: The usage of "f^{-1}" in the expression "$f^{-1}(B')$" above does not refer to the two-sided inverse of f. Indeed, we are *not* assuming that f is a bijection in the definition of pre-image but simply that f is some function, so this two-sided inverse might not even exist. We feel obliged to explicitly point out this redundant use of the symbols "f^{-1}" since it can easily completely confuse the beginner.

■ **Example 6.6.3** If $f : \{1,2,3\} \to \{1,2\}$ is defined by $f(1) = 1 = f(2)$ and $f(3) = 2$, then we have

$$f^{-1}(\emptyset) = \emptyset, \quad f^{-1}(\{1\}) = \{1,2\}, \quad f^{-1}(\{2\}) = \{3\}, \text{ and } f^{-1}(\{1,2\}) = \{1,2,3\}.$$

■ **Example 6.6.4** Letting $i : \{1,2\} \to \{1,2,3\}$ denote inclusion, we find

$$i^{-1}(\{1\}) = \{1\}, \quad i^{-1}(\{1,2\}) = \{1,2\}, \quad i^{-1}(\{3\}) = \emptyset, \text{ and } i^{-1}(\{1,3\}) = \{1\}$$

for instance.

■ **Example 6.6.5** Consider the function $f : \mathbb{N} \to \mathbb{N}$ defined by $f(n) = 2n$, let $\mathcal{O} \subseteq \mathbb{N}$ and $\mathcal{E} \subseteq \mathbb{N}$, respectively, denote the set of odd and even natural numbers. Thus, we have

$$f^{-1}(\mathcal{E}) = \mathbb{N} \text{ and } f^{-1}(\mathcal{O}) = \emptyset.$$

Furthermore, letting $\mathcal{F} = f(\mathcal{E}) = \{n \in \mathbb{N} : 4 \text{ divides } n\}$, we find

$$f^{-1}(\mathcal{F}) = \mathcal{E}$$

as the reader should check.

Given a function $f : A \to B$ and a point $b \in B$ consider the pre-image $f^{-1}(\{b\})$ of the singleton $\{b\}$. Such a pre-image is called a *point pre-image* for the obvious reason and is called by some authors a *fiber of f*. If b is not in the image of f, then the point pre-image $f^{-1}(b)$ is empty, while if b does lie in the image of f, then there is at least one arrow in the digraph of f pointing towards b, and $f^{-1}(b)$ is simply the set of initial points of all such arrows.

■ **Example 6.6.6** For the function $f : \{1,2,3\} \to \{1,2,3,4\}$ defined by $f(1) = f(2) = 1$, $f(3) = 3$, the point pre-images are

$$f^{-1}(\{1\}) = \{1,2\}, \quad f^{-1}(\{2\}) = \emptyset, \quad f^{-1}(\{3\}) = \{3\}, \quad \text{and} \quad f^{-1}(\{4\}) = \emptyset.$$

As a side-note, we mention that given a function $f : A \to B$, the non-empty point pre-images of f actually form a partition of A, or in other words

6.6.4 THEOREM *Suppose that $f : A \to B$ is a function, and define a binary relation $\sim_f$ on A where $[a \sim_f a'] \Leftrightarrow [f(a) = f(a')]$. Then $\sim_f$ is an equivalence relation on A whose blocks are the non-empty point pre-images of f.*

Proof We must prove that $\sim_f$ is reflexive, symmetric, and transitive, and each property follows directly from the corresponding property of equality on B as the reader should carefully check. *q.e.d.*

■ **Example 6.6.7** The equivalence relation $\sim_f$ on $\{1,2,3\}$ for the function f in the previous example is

$$\sim_f \;=\; \{(1,1),(2,2),(3,3),(1,2),(2,1)\}.$$

■ **Example 6.6.8** Consider the function $f : \mathbb{Z} \to \mathbb{Z}$ defined by $f(n) = n^2$, so the point pre-image of $0 \in \mathbb{Z}$ is $f^{-1}(\{0\}) = \{0\}$, and the point pre-image of any integer which is not a square (for instance, any negative integer) is empty. Indeed, a point pre-image $f^{-1}(\{n\})$ for $n \neq 0$ is non-empty if and only if the positive square-root $\sqrt{n}$ is again a positive integer, and in this case the point pre-image is a doubleton

$$f^{-1}(n) = \{\sqrt{n}, -\sqrt{n}\}.$$

The corresponding equivalence relation $\sim_f$ on $\mathbb{Z}$ is therefore given by

$$(m \sim_f n) \Leftrightarrow (m = \pm n),$$

and the blocks of the corresponding partition of $\mathbb{Z}$ are simply the sets $\{\pm m\}$. Thus, the block containing 0 is a singleton and all the other blocks are doubletons. There is a natural one-to-one correspondence

$$\begin{aligned} g : (\mathbb{Z} / \sim_f) &\to \mathbb{N} \cup \{0\} \\ \{\pm m\} &\mapsto |m| \end{aligned}$$

between the quotient $\mathbb{Z} / \sim_f$ and the set of non-negative integers.

It is worth pointing out a pleasant detail of our notation as follows. Suppose that $f : A \to B$ is actually a bijection so the two-sided inverse $f^{-1} : B \to A$ is defined. If $B' \subseteq B$, then the collection of symbols "$f^{-1}(B')$" has two sensible interpretations according to our notation: This could either represent the pre-image of B' under f or it could represent the image of B' under the function f^{-1}. It is evident that if f is a bijection (which is the only circumstance under which f^{-1} is defined and such a double interpretation is possible), then these two sets are actually identical as the reader should carefully prove. Thus, our notation is entirely consistent, and the definition of the set $f^{-1}(B')$ makes sense whether interpreted as an image or a pre-image. Similarly if $b \in B$, then we have $f^{-1}(\{b\}) = \{f^{-1}(b)\}$, where the lefthand side of the equation denotes the point pre-image of the singleton $\{b\}$ under f and the righthand side denotes the singleton containing as element the value $f^{-1}(b)$ of f^{-1} at b.

Dual to Theorem 6.6.1, our next basic result shows that pre-images "respect" inclusion in the following sense.

6.6.5 THEOREM *Let $f : A \to B$ be a function. If $B_2 \subseteq B_1 \subseteq B$, then we have $f^{-1}(B_2) \subseteq f^{-1}(B_1)$.*

Proof Suppose that $a \in f^{-1}(B_2)$ or in other words that $f(a) \in B_2$. Insofar as $B_2 \subseteq B_1$ by hypothesis, we find that $f(a) \in B_1$ or in other words $a \in f^{-1}(B_1)$, as desired. *q.e.d*

■ **Example 6.6.9** Let $A = \{1, 2, 3, 4, 5\}$, $B = \{x, y, z\}$, and define $f : A \to B$ by setting

$$f(n) = \begin{cases} x, & \text{if } n \text{ is even;} \\ y, & \text{if } n \text{ is odd,} \end{cases}$$

so if $B' = \{y, z\}$, then $f^{-1}(B') = \{1, 3, 5\}$ while the image of $f^{-1}(B')$ under f is $f(f^{-1}(B')) = \{y\}$. Thus, in this example $f(f^{-1}(B')) \subseteq B'$ and $f(f^{-1}(B')) \neq B'$.

For a dual example, let $A' = \{3, 5\} \subseteq A$, so $f(A') = \{y\}$ and $f^{-1}(f(A')) = \{1, 3, 5\}$, so in this case $A' \subseteq f^{-1}(f(A'))$ and $A \neq f^{-1}(f(A'))$.

The previous example illustrates general phenomena, as we next see.

6.6.6 THEOREM *Suppose that $f : A \to B$ is a function and let $A' \subseteq A$ and $B' \subseteq B$ be subsets of the domain and codomain respectively.*

(a) *We have $f(f^{-1}(B')) \subseteq B'$ with equality $f(f^{-1}(B')) = B'$ if and only if $B' \subseteq f(A)$.*

(b) *We have $A' \subseteq f^{-1}(f(A'))$ with equality $A' = f^{-1}(f(A'))$ if f is injective.*

Proof For part (a), let $b \in f(f^{-1}(B'))$, so there is some $a \in f^{-1}(B')$ satisfying $f(a) = b$. Since $a \in f^{-1}(B')$, we must have $f(a) \in B'$, so $b = f(a) \in B'$, proving the first part (namely, $[b \in f(f^{-1}(B'))] \Rightarrow [b \in B']$).

For the second assertion in part (a), we must prove a bi-implication and begin by supposing that $B' \subseteq f(A)$. In light of the first part, it remains only to show that $B' \subseteq f(f^{-1}(B'))$, and to this end, we choose $b \in B'$. By our hypothesis $B' \subseteq f(A)$, there is some $a \in A$ with $f(a) = b$, and so $a \in f^{-1}(B')$ where $b = f(a) \in f(f^{-1}(B'))$, as desired. Conversely, suppose $f(f^{-1}(B')) = B'$ and let $b \in B'$, so there must be some $a \in f^{-1}(B') \subseteq A$ with $f(a) = b$. By definition of images, we must have $b \in f(A)$. Since $b \in B'$ was arbitrary, we conclude that indeed $B' \subseteq f(A)$.

Turning now to part (b), suppose that $a \in A'$ and let $b = f(a) \in f(A')$. By definition of pre-images, we conclude that $a \in f^{-1}(f(A'))$, so indeed $A' \subseteq f^{-1}(f(A'))$.

For the second part, we suppose that f is injective and must show that $f^{-1}(f(A')) \subseteq A'$. To this end, choose some $a \in f^{-1}(f(A'))$, so there is some $a' \in A'$ so that $f(a) = f(a')$ by definition of image and pre-image. Thus, $a = a' \in A'$ since f is injective as required. *q.e.d.*

In order to warn the reader of a common mistake, we observe that given the function $f : A \to B$, we have

$$[a \in f^{-1}(B')] \Leftrightarrow [f(a) \in B'],$$

and we leave the proof of this bi-implication as an elementary exercise for the reader. In contrast, only the implication

$$[a \in A'] \Rightarrow [f(a) \in f(A')]$$

holds for an arbitrary function; the reverse implication only holds if f is injective, that is,

$$\text{if } f \text{ is injective, then } [f(a) \in f(A')] \Rightarrow [a \in A'].$$

Again we leave the proofs as elementary exercises for the reader and recapitulate simply that some care is required in applying f^{-1} to equations of membership, as above.

Generalizing the implication $[a \in A'] \Rightarrow [f(a) \in f(A')]$ and indicating that care is also required when applying f or f^{-1} to equalities of sets, we have

6.6.7 THEOREM *Suppose that $f : A \to B$ is a function.*

(a) *If f is injective and $A_1, A_2 \subseteq A$, then $[f(A_1) = f(A_2)] \Rightarrow [A_1 = A_2]$.*

(b) *If f is surjective and $B_1, B_2 \subseteq B$, then $[f^{-1}(B_1) = f^{-1}(B_2)] \Rightarrow [B_1 = B_2]$.*

Proof For part (a), suppose that f is injective and $f(A_1) = f(A_2)$. We prove that $A_1 \subseteq A_2$, the reverse inclusion being entirely analogous. To this end, choose $a_1 \in A_1$ so we have $f(a_1) \in f(A_1) = f(A_2)$, whence there is some $a_2 \in A_2$ satisfying $f(a_1) = f(a_2)$. Since f is injective, we must conclude that $a_1 = a_2 \in A_2$, so indeed, $a_1 \in A_2$.

Dually in part (b), suppose that f is surjective and $f^{-1}(B_1) = f^{-1}(B_2)$. We prove that $B_1 \subseteq B_2$, the reverse inclusion again being entirely analogous. To this end, choose some $b \in B_1 \subseteq B$. Since f is surjective, there is some $a \in A$ so that $f(a) = b$, whence

$$a \in f^{-1}(\{b\}) \subseteq f^{-1}(B_1) = f^{-1}(B_2),$$

where the inclusion follows from Theorem 6.6.5. Thus, $b = f(a) \in B_2$ as required. *q.e.d.*

To close this section, we investigate images and pre-images of intersections and unions and show that "pre-images respect both unions and intersections" while "images respect only unions". Precisely, we have

6.6.8 THEOREM *Suppose that $f : A \to B$ is a function, and let $\{A_\alpha | \alpha \in \Gamma\} \subseteq A$ be an indexed family of subsets of A and $\{B_\beta | \beta \in \Delta\} \subseteq B$ an indexed family of subsets of B. Then we have*

(a) $f^{-1}(\cup_{\beta\in\Delta} B_\beta) = \cup_{\beta\in\Delta} f^{-1}(B_\beta)$.

(b) $f^{-1}(\cap_{\beta\in\Delta} B_\beta) = \cap_{\beta\in\Delta} f^{-1}(B_\beta)$.

(c) $f(\cup_{\alpha\in\Gamma} A_\alpha) = \cup_{\alpha\in\Gamma} f(A_\alpha)$.

(d) $f(\cap_{\alpha\in\Gamma} A_\alpha) \subseteq \cap_{\alpha\in\Gamma} f(A_\alpha)$.

Proof For part (a), suppose first that $a \in f^{-1}(\cup_{\beta\in\Delta} B_\beta)$, or in other words $f(a) \in \cup_{\beta\in\Delta} B_\beta$, so there is some $\beta \in \Delta$ with $f(a) \in B_\beta$. For this partiticular index β, we have $a \in f^{-1}(B_\beta)$, so

$$a \in f^{-1}(B_\beta) \subseteq f^{-1}(\cup_{\beta\in\Delta} B_\beta),$$

where the inclusion follows from Theorem 6.6.5. For the reverse inclusion, suppose that $a \in \cup_{\beta\in\Delta} f^{-1}(B_\beta)$, so there is some $\beta \in \Delta$ so that $a \in f^{-1}(B_\beta)$. Thus, we have

$$f(a) \in B_\beta \subseteq \cup_{\beta\in\Delta} B_\beta,$$

or in other words $a \in f^{-1}(\cup_{\beta\in\Delta} B_\beta)$ as required.

Similarly for part (b), suppose that $a \in f^{-1}(\cap_{\beta\in\Delta} B_\beta)$, so it follows that $f(a) \in \cap_{\beta\in\Delta} B_\beta$. In other words, for each $\beta \in \Delta$, we have $f(a) \in B_\beta$. Thus, for each index β, we have $a \in f^{-1}(B_\beta)$, so in fact $a \in \cap_{\beta\in\Delta} f^{-1}(B_\beta)$. For the reverse inclusion, suppose that $a \in \cap_{\beta\in\Delta} f^{-1}(B_\beta)$, so for each $\beta \in \Delta$, we have $a \in f^{-1}(B_\beta)$, or in other words $f(a) \in B_\beta$. Thus, $f(a) \in \cap_{\beta\in\Delta} B_\beta$, so $a \in f^{-1}(\cap_{\beta\in\Delta} B_\beta)$ completing the proof of part (b).

Part (c) follows directly from the logical equivalences

$$\begin{aligned} b \in \cup_{\alpha\in\Gamma} f(A_\alpha) &\Leftrightarrow \exists\alpha \in \Gamma[b \in f(A_\alpha)] \\ &\Leftrightarrow \exists\alpha \in \Gamma \exists a \in A_\alpha[b = f(a)] \\ &\Leftrightarrow b \in f\big(\cup_{\alpha\in\Gamma} A_\alpha\big). \end{aligned}$$

Finally for part (d), suppose that $b \in f(\cap_{\alpha\in\Gamma} A_\alpha)$, i.e., there is some $a \in \cap_{\alpha\in\Gamma} A_\alpha$ so that $b = f(a)$. Thus, for each $\alpha \in \Gamma$, we have $[a \in A_\alpha] \wedge [b = f(a)]$, or in other words, for each index $\alpha \in \Gamma$, we have $b \in f(A_\alpha)$. It follows then that $b \in \cap_{\alpha\in\Gamma} f(A_\alpha)$, proving the inclusion. *q.e.d.*

To see that the reverse inclusion in part (d) does not hold, consider

■ **Example 6.6.10** Define the function $f : \{1,2,3,4\} \to \{1,2,3\}$, where $f(1) = 1$, $f(2) = f(3) = 2$, and $f(4) = 3$, and let $A_1 = \{1,2\}$, $A_2 = \{3,4\}$. Thus, $A_1 \cap A_2 = \emptyset$, so $f(A_1 \cap A_2) = \emptyset$. On the other hand, we have $f(A_1) = \{1,2\}$, $f(A_2) = \{2,3\}$, so $f(A_1) \cap f(A_2) = \{2\}$.

EXERCISES

6.6 IMAGES AND PRE-IMAGES

1. Prove Theorem 6.6.4.

2. For each of the following digraphs representing functions, describe the corresponding equivalence relation determined by the point pre-images and specify the associated quotient map.

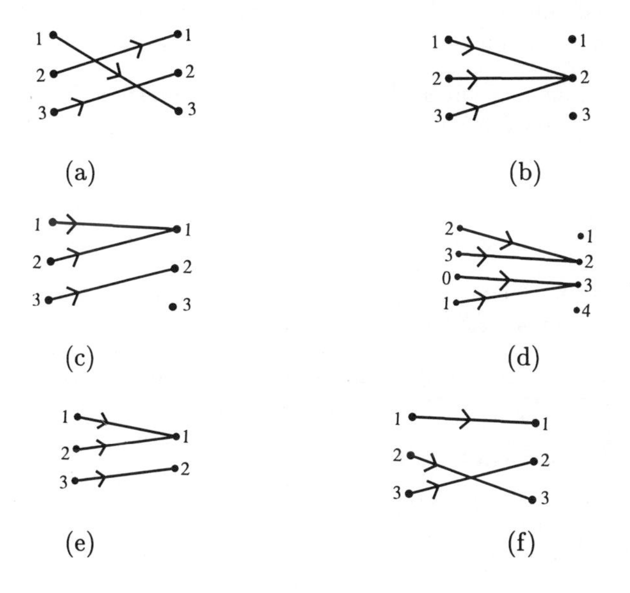

3. Let $f : \mathbb{R} \to \mathbb{R}$ be the function determined by $f(t) = 2t + 1$. Find the images $f(\mathbb{N})$, $f(\mathbb{Z})$, $f(\mathbb{R})$ and $f(\mathcal{E})$, where $\mathcal{E}$ denotes the set of all even integers.

4. Let $f : \mathbb{R} \to \mathbb{R}$ be the function determined by $f(t) = 2t^2 + 1$. Find the following images and pre-images.

(a) $f([2,3])$
(b) $f(\{1,2,3\})$
(c) $f^{-1}([6,28])$
(d) $f^{-1}(\{1,2,3\})$
(e) $f^{-1}([0,1])$
(f) $f^{-1}(f(\{1,2,3\}))$
(g) $f^{-1}(f([0,1]))$
(h) $f(f^{-1}([0,2]))$

5. Let $\mathcal{F} = \{n \in \mathbb{N} : 4 \text{ divides } n\}$ and define the function $f : \mathbb{N} \to \mathbb{N}$ by setting $f(n) = 2n$. Prove that $f^{-1}(\mathcal{F})$ is the set of all even natural numbers.

6. Suppose that $f : A \to B$ is a bijection, let $B' \subseteq B$, let $A_1 \subseteq A$ denote the pre-image of B' under f, and let $A_2 \subseteq A$ denote the image of B' under the inverse f^{-1}. Prove that $A_1 = A_2$.

7. Suppose that $f : X \to Y$ is an injection and $A \subseteq X$. Prove that $f(a) \in f(A)$ if and only if $a \in A$.

8. Suppose that $f : A \to B$ is a function, $X \subseteq A$, and $Y \subseteq B$.

(a) Prove that $f(A) - f(X) \subseteq f(A - X)$.

(b) Under what conditions do we have equality $f(A) - f(X) = f(A - X)$ in part (a)? Prove your assertion.

(c) Under what conditions do we have the equality $f^{-1}(B - Y) = A - f^{-1}(Y)$? Prove your assertion.

(d) Under what conditions do we have the equality $f(X \cap f^{-1}(Y)) = f(X) \cap Y$? Prove your assertion.

9. For each function in Exercise 1 of Section 6.4, describe the image of the function and the equivalence relation on the domain determined by the point pre-images.

10. Suppose that $f : A \to B$ is a function and π is a partition of B. Show that $\{f^{-1}(S) : S \in \pi\}$ is a partition of A; it is called the *pull-back* of π under f.

11. Let $\pi_1 : A \times B \to A$ denote projection onto the first factor and suppose that $X \subseteq A$. Prove that $\pi_1^{-1}(X) = X \times B$.

12. Let $\pi_1 : A \times B \to A$ denote the projection onto the first factor, let $Z_1, Z_2 \subseteq A \times B$, and let $X_1, X_2 \subseteq A$. Prove or disprove each of the following equalities.

(a) $\pi_1(Z_1 \cup Z_2) = \pi_1(Z_1) \cup \pi_1(Z_2)$

(b) $\pi_1(Z_1 \cap Z_2) = \pi_1(Z_1) \cap \pi_1(Z_2)$

(c) $\pi_1^{-1}(X_1 \cup X_2) = \pi_1^{-1}(X_1) \cup \pi_1^{-1}(X_2)$

(d) $\pi_1^{-1}(X_1 \cap X_2) = \pi_1^{-1}(X_1) \cap \pi_1^{-1}(X_2)$

13. Define functions $f_i : \mathbb{Z} \to \mathbb{Z}$ for $i = 1, 2, 3, 4$ by setting

$$f_1(n) = \begin{cases} +1, & \text{if } n \geq 0; \\ -1, & \text{if } n < 0, \end{cases} \qquad f_3(n) = n,$$

$$f_2(n) = \begin{cases} -1, & \text{if 3 divides } n; \\ +1, & \text{if 3 does not divide } n, \end{cases} \qquad f_4(n) = 5,$$

and let E_i denote the corresponding equivalence relation induced on the domain for $i = 1, 2, 3, 4$.

(a) For each $i = 1, 2, 3, 4$, describe the quotient $\mathbb{Z}/E_i$.

(b) For each $i = 1, 2, 3, 4$, find the image of $+1$ and -1 under the quotient map $\mathbb{Z} \to (\mathbb{Z}/E_i)$.

(c) Draw the digraph of the poset $\{\mathbb{Z}/E_1, \mathbb{Z}/E_2, \mathbb{Z}/E_3, \mathbb{Z}/E_4\}$ under refinement of partitions.

14. Let $f : A \to B$ be a function and define the induced function

$$\begin{aligned} F : \mathcal{P}(B) &\to \mathcal{P}(A) \\ X &\mapsto f^{-1}(X). \end{aligned}$$

Show that F is injective if and only if f is injective.

15. Prove the equalities and inclusions of sets in Theorem 6.6.8

(a) Directly when the index set Γ is a doubleton.

(b) Extend part (a) by induction to the situation where Γ is finite.

16. Let $(A, \leq)$ be a linearly ordered set, and $f : A \to A$ a monotone function, so $f(x) \leq f(y)$ if $x \leq y$. Prove or disprove each of the following.

(a) If $I \subseteq A$ is an interval, then $f(I) \subseteq A$ is also an interval.

(b) If $I \subseteq A$ is an interval, then $f^{-1}(I) \subseteq A$ is an interval.

(*) 6.7 *ELEMENTS OF DYNAMICAL SYSTEMS*

One of the central interests in the field of mathematics called "dynamical systems" is the study of the iterates

$$f^n = \underbrace{f \circ \cdots \circ f}_{n \text{ times}} : A \to A$$

of a given function $f : A \to A$ from A to itself. This optional section is dedicated in part to various elementary aspects of this study. In fact, iterates are of substantial interest because many important problems in pure mathematics and in the physical and computer sciences can be solved using iterates of functions. Another rationale for undertaking a discussion of dynamical systems at this point is that we thereby hopefully enrich the intuition of the reader regarding functions.

The mental image of a function $f : A \to A$ which we wish to develop is as follows: Imagine each point $a \in A$ as a "particle", so A itself is a collection of these particles. Applying the function f to $a \in A$, we think of the particle a as "jumping" to its image, which is again a particle $f(a) \in A$. Of course, one should *not* think of A a particle at a time, but rather one should think of A as a collection of particles, each of which jumps to its image when applying f. Thus, before applying f, one imagines the set A as a collection of particles, and then at some instant in time, one applies f, and each particle $a \in A$ simultaneously jumps to its image particle $f(a) \in A$.

The mental image of the iterate $f^n : A \to A$ is then as follows: One again imagines A as a collection of particles, and at some some instant in time, each particle $a \in A$ jumps to its image $f(a) \in A$. At a later instant in time, each particle a again jumps to it image, so the net effect so far on a point $a \in A$ is to jump twice to $f(f(a)) = f^2(a)$. At a later instant in time, each particle again jumps to its image, so the net effect so far on a point $a \in A$ is to jump thrice to $f(f^2(a)) = f^3(a)$. After n such applications of f at n successive instants in time, each particle a has jumped to $f^n(a)$, and this is our mental image of the iterate $f^n : A \to A$.

■ **Example 6.7.1** Define the function $f : \{1,2,3,4\} \to \{1,2,3,4\}$, where $f(1) = 1$, $f(2) = 4$, $f(3) = 2$, and $f(4) = 3$ and consider the iterate f^3 of f. With the first application of f, the particle 1 simply sits where it is, 2 jumps to 4, 3 jumps to 2, and 4 jumps to 3. With the second application of f, the particle 1 again sits where it is at 1, while the particle 2 has jumped to $f(4) = 3$, 3 has jumped to $f(2) = 4$, and 4 has jumped to $f(3) = 2$. With the third and final application of f, the particle 1 continues to sit at 1, 2 has jumped to $f(3) = 2$, 3 has jumped to $f(4) = 3$, and 4 has jumped to $f(2) = 4$. Thus, under f^3 each particle returns to its starting point, or in other words, f^3 is simply the identity map on $\{1,2,3,4\}$. Of course, f^0 is also the identity map on $\{1,2,3,4\}$ by definition.

Suppose that $f : A \to A$ is a function and define the *forward orbit* of a point $a \in A$ to be the set

$$\mathcal{O}_f^+(a) = \{b \in A : b = f^n(a) \text{ for some } n \geq 0\}.$$

Thus, in terms of our view of particles jumping around under the application of f, the forward orbit $\mathcal{O}_f^+(a)$ of $a \in A$ is simply the collection of particles $b \in A$ so that the particle a jumps to b under some iterate of f; that is, a "visits" b under some repeated application of f. Define the *backward orbit* of a point $a \in A$ to be the set

$$\mathcal{O}_f^-(a) = \{b \in A : a = f^n(b) \text{ for some } n \geq 0\},$$

so in terms of our view of particles jumping around, the backward orbit of $a \in A$ is simply the collection of particles $b \in A$ so that b "visits" a under some repeated application of f. In particular, notice that for each $a \in A$, we have $a \in \mathcal{O}_f^+(a) \cap \mathcal{O}_f^-(a)$ by definition of forward and backward orbits.

■ **Example 6.7.2** In the previous example, the forward and backward orbits are

$$\mathcal{O}_f^+(1) = \{1\} = \mathcal{O}_f^-(1),$$

and

$$\mathcal{O}_f^+(2) = \mathcal{O}_f^+(3) = \mathcal{O}_f^+(4) = \{2,3,4\} = \mathcal{O}_f^-(2) = \mathcal{O}_f^-(3) = \mathcal{O}_f^-(4).$$

■ **Example 6.7.3** Define $g : \{1,2,3\} \to \{1,2,3\}$, by $g(1) = g(2) = 3$ and $g(3) = 1$. The forward orbits are

$$\mathcal{O}_+(1) = \{1,3\} = \mathcal{O}_+(3) \text{ and } \mathcal{O}_+(2) = \{1,2,3\}$$

while the backward orbits are

$$\mathcal{O}_-(1) = \{1,2,3\} = \mathcal{O}_-(3) \text{ and } \mathcal{O}_-(2) = \{2\}.$$

Given a function $f : A \to A$, we say that $a \in A$ is a *periodic point* if there is some $n \in \mathbb{N}$ so that $f^n(a) = a$. In terms of our view of particles jumping around, $a \in A$ is a periodic point if it eventually jumps back to itself. The *period* of a periodic point a is defined to be the least $n \in \mathbb{N}$ so that $f^n(a) = a$, or in other words, the least n so that a jumps back to itself under n iterations of f. In particular, a periodic point $a \in A$ of period one is called simply a *fixed point* of f, so a is a fixed point of f if $a = f(a)$ does not move under f, i.e., it jumps to itself. Furthermore, notice that if $a \in A$ is a common fixed point of $f, g : A \to A$, then a is a fixed point of $f \circ g$ as well.

■ **Example 6.7.4** In Example 6.7.1 above, 1 is the unique fixed point of f, while each of 2,3,4 are periodic points of period three.

In fact, if a is a periodic point of f with period n, then a is a fixed point of f^n, as the reader should carefully prove. More generally, we have

6.7.1 LEMMA *Suppose that $f : A \to A$ is a function and $a = f^n(a)$ is a periodic point of period n. Then $a = f^{nk}(a)$ is a fixed point of f^{nk} for each $k \geq 0$.*

Proof The proof is by induction on $k \geq 0$, and the basis $f^{n\cdot 0}(a) = f^0(a) = a$ follows from the definition. For the induction step, observe that

$$f^{n(k+1)}(a) = f^{nk} \circ f^n(a) = f^{nk}(f^n(a)) = f^{nk}(a) = a$$

where we have used associativity of $\circ$ in the first and the induction hypothesis in the last equality. *q.e.d.*

We say that $a \in A$ is an *eventually periodic point* of $f : A \to A$ if the forward orbit $\mathcal{O}_f^+(a)$ is finite. In particular, a periodic point is automatically eventually periodic.

■ **Example 6.7.5** In Example 6.5.1 above, there are no fixed points, 1 and 3 are periodic points of period two, and 2 is an eventually periodic point.

■ **Example 6.7.6** Consider the function $f : \mathbb{Z} \to \mathbb{Z}$ defined by $f(n) = n^2$, which has the two fixed points $1 = 1^2$ and $0 = 0^2$. The point -1 is eventually periodic since its forward orbit is $\mathcal{O}_f^+(-1) = \{-1, 1\}$, while the forward orbit of each point other than $0, \pm 1$ is infinite. The backward orbit of 0 is $\mathcal{O}_f^-(0) = \{0\}$, and the backward orbit of 1 is $\mathcal{O}_f^-(1) = \{\pm 1\}$. The backward orbit of any negative integer is a singleton, whereas the backward orbit of a natural number of the form $n = k^{2^m}$ for some $k, m \in \mathbb{N}$, for instance, is $\mathcal{O}_f^-(n) = \{\pm k^{2^j} \in \mathbb{Z} : 1 \leq j < m\} \cup \{k^{2^m}\}$.

6.7.2 LEMMA *Suppose that $f : A \to A$ is a function and $a \in A$ is an eventually periodic point. Then there is $\ell \in \mathbb{N} \cup \{0\}$ so that $f^\ell(a)$ is a periodic point.*

Proof Consider the forward orbit $\mathcal{O}_f^+(a)$ of the eventually periodic point $a \in A$. Since a is eventually periodic, the set $\mathcal{O}_f^+(a)$ is finite, yet it contains infinitely many iterates $f^n(a)$ for $n \in \mathbb{N} \cup \{0\}$. There must therefore be distinct elements $k, \ell \in \mathbb{N} \cup \{0\}$ so that $f^k(a) = f^\ell(a)$ and we may assume without loss of generality that $\ell < k$. Thus, we have

$$f^\ell(a) = f^k(a) = f^{k-\ell} \circ f^\ell(a) = f^{k-\ell}(f^\ell(a)),$$

so $f^\ell(a)$ is indeed a periodic point of period at most $k - \ell$. *q.e.d.*

Thus, an eventually periodic point $a \in A$ of $f : A \to A$ jumps around a bit under the iterates of f, but it eventually lands on a periodic orbit. Of course, this explains the terminology "eventually periodic". Furthermore, in the case that f is an injection, we have

6.7.3 THEOREM *Suppose that $f : A \to A$ is an injection. Then $a \in A$ is an eventually periodic point if and only if it is a periodic point.*

Proof As we observed above, periodic implies eventually periodic. Turning to the converse, suppose that $a \in A$ is eventually periodic, so by the previous lemma, there is some $\ell \in \mathbb{N} \cup \{0\}$ so that $f^\ell(a)$ is periodic. The proof that a

is periodic is by induction on $\ell \geq 0$, and for the basis step $\ell = 0$, we find that $f^0(a) = a$ is indeed a periodic point of f. For the induction step, suppose that $\ell \geq 1$ and consider $a' = f(a)$. Of course, a' is an eventually periodic point of f since $a' = f(a)$ implies that $\mathcal{O}_f^+(a') \subseteq \mathcal{O}_f^+(a)$, and

$$f^{\ell-1}(a') = f^{\ell-1}(f(a)) = f^{\ell}(a)$$

is a periodic point of f. The induction hypothesis therefore shows that a' is actually a periodic point. Thus, $f^n(a') = a'$ for some $n \in \mathbb{N}$, or in other words

$$f(f^{n-1}(a')) = f^n(a') = a' = f(a).$$

Since f is injective, we must have $a = f^{n-1}(a') = f^{n-1}(f(a)) = f^n(a)$, so a is then also periodic of period at most n. *q.e.d.*

In the special case that $f : A \to A$ is an injection and A is a *finite* set, f is necessarily a bijection and the forward and backward orbits of a point actually coincide, as we next see.

6.7.4 THEOREM *Suppose that A is a finite set and $f : A \to A$ is an injection. Then f is actually a bijection, and there is some $m \in \mathbb{N} \cup \{0\}$ so that $f^m = f^{-1}$. Furthermore, for each $a \in A$, we have $\mathcal{O}_f^+(a) = \mathcal{O}_f^-(a)$.*

Proof For each $a \in A$, we have $\mathcal{O}_f^+(a) \subseteq A$ by definition, so $\mathcal{O}_f^+(a)$ is finite, and a is therefore eventually periodic. By the previous theorem, a is actually periodic, and we let n_a denote the period of a, so $f^{n_a}(a) = a$ for each $a \in A$. Letting $n = \prod_{a \in A} n_a$ denote the (finite) product of the periods of the elements of A, it follows from Lemma 6.7.1 that $f^n(a) = a$ for each $a \in A$. Thus, f^n is actually the identity map on A, or in other words, we have $f^{n-1} \circ f = 1_A = f \circ f^{n-1}$, so f is bijective and $f^{-1} = f^{n-1}$ by Theorem 6.5.3c. Finally assuming $f^m = f^{-1}$, we find that $b = f^n(a) \Leftrightarrow a = (f^{-1})^n(b) \Leftrightarrow a = f^{mn}(b)$, proving the equality of forward and backward orbits. *q.e.d.*

For any function $f : A \to A$ and any point $a \in A$, we finally define the *full orbit* or simply the *orbit* of a under f to be the set

$$\mathcal{O}_f(a) = \mathcal{O}_f^+(a) \cup \mathcal{O}_f^-(a),$$

so in case f is an injection and A is finite, we find $\mathcal{O}_f(a) = \mathcal{O}_f^+(a) = \mathcal{O}_f^-(a)$ for each $a \in A$ by the previous result. In fact, if $f : A \to A$ is an injection (even if A is not necessarily finite), we have

6.7.5 THEOREM *Given an injection $f : A \to A$, the set $\{\mathcal{O}_f(a) \in \mathcal{P}(A) : a \in A\}$ of full orbits is a partition of A.*

Proof Consider the binary relation $\equiv_f$ on A defined by $a \equiv_f b$ if and only if $a \in \mathcal{O}_f(b)$. In other words $a \equiv_f b$ if and only if there is some $n \in \mathbb{N} \cup \{0\}$ so that $[a = f^n(b)] \vee [b = f^n(a)]$ The binary relation $\equiv_f$ is evidently reflexive since $a = f^0(a)$ for each $a \in A$, and is moreover symmetric by commutativity of disjunction. As to transitivity, suppose that $a \equiv_f b$ and $b \equiv_f c$, so there are $m, n \in \mathbb{N} \cup \{0\}$ so that

$$[(a = f^n(b)) \vee (b = f^n(a))] \wedge [(b = f^m(c)) \vee (c = f^m(b))].$$

There are thus four cases, as follows:

- If $[a = f^n(b)] \wedge [b = f^m(c)]$, then we have $a = f^{n+m}(c)$.
- If $[b = f^n(a)] \wedge [c = f^m(b)]$, then we have $c = f^{n+m}(a)$.
- Suppose that $[a = f^n(b)] \wedge [c = f^m(b)]$. If $n \geq m$, then we have $f^{n-m}(c) = f^{n-m}(f^m(b)) = f^n(b) = a$, whereas if $m \geq n$, then we similarly have $f^{m-n}(a) = c$.
- Suppose that $f^n(a) = b = f^m(c)$ and proceed by induction on the minimum p of m and n. For the basis step $p = 0$, if $m = 0$, then $f^n(a) = b = f^0(c) = c$, so $c = f^n(a)$; a similar remark holds if $n = 0$. For the induction step with $p \geq 1$, observe that $f(f^{n-1}(a)) = f^n(a) = b = f^m(c) = f^(f^{m-1}(c))$, so injectivity of f guarantees that $f^{n-1}(a) = f^{m-1}(c)$. It therefore follows from the inductive hypothesis that indeed $c \in \mathcal{O}_f(a)$, as desired.

In any case, we find $a \equiv_f c$, so $\equiv_f$ is transitive as well and hence is an equivalence relation.

By construction, the equivalence class of a under $\equiv_f$ is exactly the full orbit $\mathcal{O}(a)$, so the collection of all full orbits is a partition of A by Theorem 5.13.1, as desired. *q.e.d.*

■ **Example 6.7.7** In Example 6.5.1 above, there are two blocks in the partition of $\{1, 2, 3, 4\}$ corresponding to f, namely $\{1\}$ and $\{2, 3, 4\}$.

EXERCISES

6.7 ELEMENTS OF DYNAMICAL SYSTEMS

1. Prove that if $a \in A$ is a periodic point of the function $f : A \to A$ of period n, then a is a fixed point of f^n.

2. Let $f : \mathbb{R} \to \mathbb{R}$ be defined by $f(t) = t^3$. Determine the forward and backward orbits of $-1, 0, 1$, and 2.

3. Let $A = \{1, 2, \ldots, 10\}$. For each of the following bijections $f : A \to A$, determine the forward and backward orbits of each element of A. We describe

the function f in each case by specifying the 10-tuple $(f(1), f(2), \ldots, f(10))$ of values taken by f on $1, 2, \ldots, 10$.

(a) $(f(1), f(2), \ldots, f(10)) = (3, 9, 5, 10, 1, 8, 2, 4, 7, 6)$

(b) $(f(1), f(2), \ldots, f(10)) = (2, 3, 4, 5, 6, 7, 8, 9, 10, 1)$

(c) $(f(1), f(2), \ldots, f(10)) = (10, 9, 8, 7, 6, 5, 4, 3, 2, 1)$

(d) $(f(1), f(2), \ldots, f(10)) = (4, 6, 5, 2, 3, 1, 7, 10, 9, 8)$

4. Let $f : \mathbb{N} \to \mathbb{N}$ be determined by $f(n) = 2n$. Determine the forward and backward orbits of each $n \in \mathbb{N}$.

5. Let $f : \mathbb{Z} \to \mathbb{Z}$ be determined by $f(n) = 3n$. Determine the forward and backward orbits of each $n \in \mathbb{Z}$.

6. Define the function

$$f : \mathbb{Z} \to \mathbb{Z}$$
$$n \mapsto \begin{cases} -n, & \text{if } n \leq 0; \\ n+1, & \text{if } n > 0 \text{ and } n \neq 10; \\ n, & \text{if } n = 10. \end{cases}$$

Find the forward and backward orbits of each $n \in \mathbb{Z}$.

7. Suppose that $f : A \to B$ is a bijection and $g : A \to A$ has a periodic point $a \in A$. Prove that $f \circ g \circ f^{-1}$ has $f(a)$ as periodic point.

(*) *6.8 PERMUTATIONS*

A bijection from a finite set A to itself is called a *permutation* of A, and we observe that Theorem 6.7.4 guarantees that an injection from A to A is actually a permutation. This optional section contains a reasonably thorough discussion of permutations.

6.8.1 THEOREM *Suppose that A is a finite set with n elements. Then there are $n!$ distinct permutations of A.*

Proof The assertion is trivial for $A = \emptyset$, and we let $a_1, \ldots, a_n$ denote the elements of $A \neq \emptyset$. A function $f : A \to A$ is determined by specifying its values $f(a_1), \ldots, f(a_n)$, and f is a bijection if and only if each of these values is distinct, i.e., if and only if f is an injection. Beginning with a_1, the value $f(a_1)$ can be any of the n elements of A. Having specified the value $f(a_1)$, the value $f(a_2)$ can be any of the $n-1$ elements of $A - \{f(a_1)\}$. In general, having specified the values $f(a_1), \ldots, f(a_i)$, the value $f(a_{i+1})$ can be any of the $n-i$ elements of $A - \{f(a_1), \ldots, f(a_i)\}$. There are thus $n! = n(n-1)\cdots 3 \cdot 2 \cdot 1$ possible assignments of values of f in order that f be an injection, that is, in order that f

be a permutation. Thus, the $n!$ distinct functions described here are exactly the various permutations of A. *q.e.d.*

In order to be explicit, we fix the finite set

$$\overline{n} = \{1, 2, \ldots, n\}$$

consisting of the first n natural numbers, for each $n \geq 1$, defining also $\overline{0} = \emptyset$. For each $n \in \mathbb{N} \cup \{0\}$, we furthermore define

$$\Sigma_n = \{\sigma \in \overline{n}^{\overline{n}} : \sigma \text{ is a bijection}\},$$

which is called the *symmetric group* on n elements. An element of Σ_n is called simply a *permutation on* n. An explanation of the terminology "symmetric group" would force a digression from our main thread of discussion and will not be taken up here. We mention however that these symmetric groups are one of the basic objects in all of mathematics, and they arise in many different contexts and guises. Σ_n is also the prototypical example of a fundamental mathematical object, called a "group".

Our discussion here centers on dynamical properties of elements of Σ_n, that is, periodic points, orbits, and so on, and we let $\sigma : \overline{n} \to \overline{n}$ be some permutation $\sigma \in \Sigma_n$ for $n \geq 0$. There is a simple notation for describing σ as follows: Simply list in order the values $f(1), \cdots, f(n)$ of f enclosed in parentheses, and write $\sigma = (f(1) \cdots f(n))$. For instance, if $\sigma \in \Sigma_4$ is the permutation defined by $\sigma(1) = 2$, $\sigma(2) = 4$, $\sigma(3) = 1$, and $\sigma(4) = 3$, then we write simply $\sigma = (2\ 4\ 1\ 3)$ to represent σ as a "table of values". For another example, if $\sigma \in \Sigma_5$ satisfies $\sigma(j) = j + 1$ for $j \neq 5$, then the corresponding table of values is $\sigma = (2\ 3\ 4\ 5\ 1)$. In the extreme case $\sigma \in \Sigma_0$ where $\sigma : \emptyset \to \emptyset$ is the empty function, we write simply $\sigma = ()$ for the corresponding empty table of values.

Using this notation, it is easy to compute the table of values for a composition of σ and τ in terms of the table of values of $\sigma, \tau \in \Sigma_n$. For instance, if $\sigma = (2\ 4\ 1\ 3), \tau = (2\ 3\ 4\ 1) \in \Sigma_4$, then $\sigma \circ \tau = (4\ 1\ 3\ 2)$ and $\tau \circ \sigma = (3\ 1\ 2\ 4)$, as the reader may check. In particular, $\sigma \circ \tau$ and $\tau \circ \sigma$ are apparently different functions, so composition of permutations is not necessarily commutative.

Now, fix some $\sigma \in \Sigma_n$, choose some $i \in \overline{n}$, and consider the orbit

$$\mathcal{O}_\sigma(i) = \{\sigma^j(a) \in \overline{n} : j \geq 0\}$$

of i under σ. According to Theorem 6.7.3, the element $i \in \overline{n}$ is actually a periodic point of σ, say of period $m \geq 0$, and the orbit of i is simply the set

$$\mathcal{O}_\sigma(i) = \{\sigma^0(i), \sigma(i), \ldots, \sigma^{m-1}(i)\}.$$

Furthermore, the elements $\sigma^0(i), \sigma^1(i), \ldots, \sigma^{m-1}(i)$ are all distinct since if $\sigma^p(i) = \sigma^q(i)$ where $0 \leq p, q < m$, then $\sigma^{|p-q|}(i) = i$; on the other hand, we have $|p - q| < m$ and therefore contradict that m is the period of i unless $p = q$.

Notice that if $j \in \mathcal{O}_\sigma(i)$, then $j = \sigma^t(i)$ for some t, so $\sigma(j) = \sigma^{t+1}(i)$ again lies in $\mathcal{O}_\sigma(i)$. Thus, we have $\sigma(\mathcal{O}_\sigma(i)) \subseteq \mathcal{O}_\sigma(i)$, so the restriction

$$\sigma|_{\mathcal{O}_\sigma(i)} : \mathcal{O}_\sigma(i) \to \mathcal{O}_\sigma(i)$$

is defined, where we have restricted both the domain and the codomain of σ. This restriction is an injection as the restriction of such and is therefore a bijection by Theorem 6.7.4. Furthermore, this restriction is quite easy to understand: $\mathcal{O}_\sigma(i)$ consists of the distinct elements $\sigma^0(i), \ldots, \sigma^{m-1}(i)$, the image of $\sigma^t(i)$ under $\sigma|_{\mathcal{O}_\sigma(i)}$ is simply $\sigma^{t+1}(i)$ provided $0 \le t < m-1$ while the image of $\sigma^{m-1}(i)$ under $\sigma|_{\mathcal{O}_\sigma(i)}$ is $\sigma^0(i)$.

■ **Example 6.8.1** Consider the permutation $\sigma = (2\ 4\ 9\ 6\ 10\ 1\ 3\ 5\ 7\ 8) \in \Sigma_{10}$. The various orbits of σ are

$$\mathcal{O}_\sigma(1) = \{1, 2, 4, 6\}, \quad \mathcal{O}_\sigma(3) = \{3, 9, 7\}, \quad \text{and } \mathcal{O}_\sigma(5) = \{5, 10, 8\}.$$

The effect of $\sigma|_{\mathcal{O}_\sigma(1)}$ is

$$1 \mapsto 2 \mapsto 4 \mapsto 6 \mapsto 1,$$

the effect of $\sigma|_{\mathcal{O}_\sigma(3)}$ is

$$3 \mapsto 9 \mapsto 7 \mapsto 3,$$

and the effect of $\sigma|_{\mathcal{O}_\sigma(5)}$ is

$$5 \mapsto 10 \mapsto 8 \mapsto 5,$$

using the obvious notation.

In the previous example, the restriction of σ to each orbit is easy to understand, and the total effect of σ itself is determined by its separate effects on each of the orbits. In fact, this behavior is typical of permutations, and the remainder of this section is dedicated to making this precise.

Given a collection $i_1, i_2, \ldots, i_m$ of distinct natural numbers where $i_j \le n$ for each $j = 1, \ldots, m$, define the permutation

$$(i_1, i_2, \ldots, i_m) : \overline{n} \to \overline{n}$$
$$j \mapsto \begin{cases} i_1, & \text{if } j = i_m; \\ i_{k+1}, & \text{if } j = i_k \text{ and } 0 < k < m; \\ j, & \text{if } j \notin \{i_1, \ldots, i_m\}\ , \end{cases}$$

which is called an *m-cycle* in Σ_n. When the natural number m is fixed or is not important, we may refer to such a permutation simply as a *cycle* or as a *cyclic permutation*. The effect of the m-cycle $(i_1, i_2, \ldots, i_m)$ is to "cyclically permute" $i_1, i_2, \ldots, i_m$ in the sense that

$$i_1 \mapsto i_2 \mapsto i_3 \quad \cdots \quad i_{m-1} \mapsto i_m \mapsto i_1$$

while leaving fixed $j \notin \{i_1, i_2, \ldots, i_m\}$.

Our notation $(i_1, i_2, \ldots, i_m)$ is distinguished from the table of values considered before in that we include commas between the entries in our current "cycle notation". Furthermore, the reader should notice that the cycle notation is in a sense incomplete since the domain and codomain of the function $(i_1, i_2, \ldots, i_m)$ is intentionally suppressed. For instance, the cycle $(2, 3, 4)$ represents the permutation with table (1 3 4 2) when regarded as an element of Σ_4, while it represents the permutation with table (1 3 4 2 5) when regarded as an element of Σ_5. This incompleteness of the notation is completely standard and is in fact convenient, but we must be careful to specify domains when using the cycle notation. Notice finally that a 1-cycle in Σ_n is simply the identity map on $\overline{n}$.

It is easy to calculate the table of values corresponding to the composition of two cycles, as we next indicate in examples.

■ **Example 6.8.2** Consider the cycles $\sigma_1 = (4, 1, 2, 3)$ and $\sigma_2 = (5, 6, 4, 1, 8)$ as elements of Σ_9. To compute the value of $1 \in \overline{9}$ for instance under the composition $\sigma = \sigma_1 \circ \sigma_2$, we first compute that $\sigma_2(1) = 8$ and then that $\sigma_1(8) = 8$, so $\sigma(1) = 8$. Continuing in this way, one finds that that table of values corresponding to σ is (8 3 4 2 6 1 7 5 9). Thus, the various orbits of σ are

$$\mathcal{O}_\sigma(1) = \{1, 8, 5, 6\}, \quad \mathcal{O}_\sigma(2) = \{2, 3, 4\}, \quad \mathcal{O}_\sigma(7) = \{7\}, \quad \text{and} \quad \mathcal{O}_\sigma(9) = \{9\}.$$

In particular, the product σ of the two cycles σ_1, σ_2 is not itself a cycle.

For another example, consider the composition $\sigma_2 \circ \sigma_1$ in the other order, whose table of values is given by (2 3 1 8 6 4 7 5 9).

If $\sigma = (i_1, i_2, \ldots, i_p)$ and $\tau = (j_1, j_2, \ldots, j_q)$ are cycles in Σ_n, then we say that σ and τ are *disjoint cycles* provided

$$\{i_1, i_2, \ldots, i_p\} \cap \{j_1, j_2, \ldots, j_q\} = \emptyset.$$

The basic fact about disjoint cycles is that they commute:

6.8.2 PROPOSITION *If $\sigma, \tau \in \Sigma_n$ are disjoint cycles, then $\sigma \circ \tau = \tau \circ \sigma$.*

Proof Let $\sigma = (i_1, i_2, \ldots, i_p)$ and $\tau = (j_1, j_2, \ldots, j_q)$ as above. We must show that $\sigma \circ \tau$ and $\tau \circ \sigma$ take the same values, and to this end, we choose some $k \in \overline{n}$. There are three cases for the chosen k: $k \in \{i_1, i_2, \ldots, i_p\}$, $k \in \{j_1, j_2, \ldots, j_q\}$, or $k \in \overline{n} - (\{i_1, i_2, \ldots, i_p\} \cup \{j_1, j_2, \ldots, j_q\})$, and exactly one of these possibilities holds by our assumption that σ and τ are disjoint cycles. In the first case, both k and $\sigma(k)$ are elements of $\{i_1, i_2, \ldots, i_p\}$, and τ fixes each element of this set, so we find

$$\sigma \circ \tau(k) = \sigma(\tau(k)) = \sigma(k) = \tau(\sigma(k)) = \tau \circ \sigma(k).$$

A similar argument handles the second case. In the third case, we find that k is a fixed point of both σ and τ, and hence

$$\sigma \circ \tau(k) = \sigma(\tau(k)) = \sigma(k) = k = \tau(k) = \tau(\sigma(k)) = \tau \circ \sigma(k).$$

q.e.d.

We are now in a position to make precise our remark above that Example 6.8.1 is typical of permutations. Let $\sigma \in \Sigma_n$ be a fixed permutation and consider the collection of orbits of σ. According to Theorem 6.7.5, two orbits of σ are either identical or disjoint, and we let $\mathcal{O}_1, \mathcal{O}_2, \ldots, \mathcal{O}_T$ denote the various distinct orbits of σ. The restriction $\sigma|_{\mathcal{O}_t}$ is simply the restriction of a cycle $\sigma_t \in \Sigma_n$ to $\mathcal{O}_t \subseteq \overline{n}$ for each $t = 1, \ldots, T$ by previous comments, and any two of these cycles are disjoint (as they are blocks of the corresponding partition). We next show that $\sigma = \sigma_1 \circ \cdots \circ \sigma_T$ is exactly the composition of these cycles.

6.8.3 THEOREM *Suppose that $\sigma \in \Sigma_n$ is a permutation, let $\mathcal{O}_1$, $\mathcal{O}_2$, ..., $\mathcal{O}_T$ denote the distinct orbits of σ, and let $\sigma_t \in \Sigma_n$ be the cycle whose restriction to $\mathcal{O}_t$ agrees with the restriction of σ to $\mathcal{O}_t$ for each $t = 1, \ldots, T$. Then we have $\sigma = \sigma_1 \circ \cdots \circ \sigma_T$.*

Proof We must prove that σ and $\sigma_1 \circ \cdots \circ \sigma_T$ take the same values, and to this end, we choose some $k \in \overline{n}$. Since $\{\mathcal{O}_1, \mathcal{O}_2, \ldots, \mathcal{O}_T\}$ is a partition of $\overline{n}$, there is a unique $t \in \{1, \ldots, T\}$ so that $k \in \mathcal{O}_t$, and of course this orbit $\mathcal{O}_t$ also contains $\sigma(k)$. Since the cycles $\sigma_1, \ldots, \sigma_T$ are disjoint, we conclude that k and $\sigma_t(k) = \sigma(k)$ are fixed points of σ_s for each $s \neq t$, where $s = 1, \ldots, T$. We find

$$\begin{aligned}
\sigma_1 \circ \cdots \circ \sigma_T(k) &= [\sigma_1 \circ \cdots \circ \sigma_{t-1}] \circ \sigma_t \circ [\sigma_{t+1} \circ \cdots \circ \sigma_T](k) \\
&= [\sigma_1 \circ \cdots \circ \sigma_{t-1}](\sigma_t([\sigma_{t+1} \circ \cdots \circ \sigma_T](k))) \\
&= [\sigma_1 \circ \cdots \circ \sigma_{t-1}](\sigma_t(k)) \\
&= \sigma_t(k) \\
&= \sigma(k),
\end{aligned}$$

where we have used that k is a fixed point of $\sigma_{t+1} \circ \cdots \circ \sigma_T$ in the third equality and that $\sigma_t(k) = \sigma(k)$ is a fixed point of $\sigma_1 \circ \cdots \circ \sigma_{t-1}$ in the fourth equality. Since $k \in \overline{n}$ was arbitrary, we conclude that σ and $\sigma_1 \circ \cdots \circ \sigma_T$ do indeed take the same values. *q.e.d.*

To recapitulate, the dynamics of a permutation $\sigma \in \Sigma_n$ is easy to describe: The domain $\{1, 2, \ldots, n\}$ breaks up into disjoint sets, namely the orbits of σ, and the effect of σ on each such set is simply that of cyclic permutation. In this way, one regards cyclic permutations as the "building blocks" of general permutations.

In fact, an arbitrary cycle $(i_1, i_2, \ldots, i_n)$ of length $n \geq 2$ can be written as the composition $(i_1, i_2, \ldots, i_n) = (i_1, i_2) \circ (i_2, i_3) \circ \cdots \circ (i_{n-1}, i_n)$ of 2-cycles (which are *not* disjoint) as the reader should check. Thus, 2-cycles, which are also called *transpositions*, may be regarded as still more basic "building blocks" of a cyclic or general permutation. A basic difference, though, is that whereas the expression of a general permutation as a product of disjoint cycles is essentially unique (up to the order of composition of the disjoint cycles), there are many truly different ways to write a given permutation as a product of transpositions.

EXERCISES

6.8 PERMUTATIONS

1. Prove that the n-cycle $(i_1, i_2, \ldots, i_n)$ may be written as the composition

$$(i_1, i_2, \ldots, i_n) = (i_1, i_2) \circ (i_2, i_3) \circ \cdots \circ (i_{n-1}, i_n)$$

of transpositions.

2. Consider the following permutations in Σ_{10} given as tables of values

$$\sigma_1 = (3\ 9\ 5\ 10\ 1\ 8\ 2\ 4\ 7\ 6), \qquad \sigma_2 = (2\ 3\ 4\ 5\ 6\ 7\ 8\ 9\ 10\ 1),$$
$$\sigma_3 = (10\ 9\ 8\ 7\ 6\ 5\ 4\ 3\ 2\ 1), \qquad \sigma_4 = (4\ 6\ 5\ 2\ 3\ 1\ 7\ 10\ 9\ 8),$$

whose dynamics were analyzed in Exercise 3 of Section 6.7.

(a) Express each σ_i as a composition of disjoint cycles.

(b) Express the composition $\sigma_1 \circ \sigma_2$ as a composition of disjoint cycles and as a composition of transpositions.

3. Express each permutation in Σ_{12} given below as a table of values as a composition of disjoint cycles.

(a) (4 10 12 8 6 2 3 1 7 5 9 11)

(b) (2 8 11 7 12 6 3 1 5 10 4 9)

(c) (4 6 11 1 3 2 5 9 8 12 7 10)

(d) (5 9 11 4 10 2 1 8 6 3 7 12)

4. For each of the permutations $\sigma \in \Sigma_{12}$ in Exercise 3 above, find the inverse $\sigma^{-1} \in \Sigma_{12}$.

5. (a) Enumerate all of the permutations in Σ_3.

(b) Enumerate all of the permutations in Σ_4.

6. Characterize the permutations $\sigma \in \Sigma_n$ so that $\sigma^2 = \sigma$. Prove your assertion.

7. Let σ denote the composition

$$\sigma = (i^1_1, i^1_2, \ldots, i^1_{n_1}) \circ (i^2_1, i^2_2, \ldots, i^2_{n_2}) \circ \cdots \circ (i^k_1, i^k_2, \ldots, i^k_{n_k})$$

of k disjoint cycles for some $k \in \mathbb{N}$. How does one express the inverse σ^{-1} as a composition of disjoint cycles? Prove your assertion.

8. Prove that if $\sigma \in \Sigma_n$, then $\sigma^{n!}$ is the identity.

9. A *group* is a set G together with an associative binary operation $\cdot$, a distinguished element $1 \in G$ (called the *unit*) where $1 \cdot g = g = g \cdot 1$ for all $g \in G$, and a unary operation $g \mapsto \overline{g} \in G$ so that $g \cdot \overline{g} = 1 = \overline{g} \cdot g$ for all $g \in G$. Prove that Σ_n with the binary operation of composition is a group with identity map as unit.

Chapter 7

Cardinality

This chapter is dedicated to the study of "sizes" of sets. Of course, the reader has ample everyday experience with sizes of finite sets, and we begin with a precise treatment and formalization of "empirically obvious" results in this setting. We extend this material to infinite sets in the remaining sections, and there are many surprises and thought-provoking insights in this extension. For instance, we shall find that in a precise sense there are infinitely many different sizes of infinite sets. Much of the foundational material on infinite sets which we describe here was pioneered by Georg Cantor over a century ago, and there was great resistance in the late 1800's towards Cantor's remarkable discoveries. Indeed, Cantor's career and life were seriously impaired by persecution as a mathematician and as a Jew, and his life ended in an insane asylum. Cantor's ideas may have seemed so "far out" in the 19th century that intellectual resistance is perhaps understandable, yet the theory we describe is thought-provoking and of compelling cogency. We hasten to emphasize that Cantor's work on cardinalities of infinite sets is uncontroversial nowadays and is fully accepted by most mathematicians.

7.1 FINITE SETS

We need a "standard model" of each of the various finite sets and to this end define

$$\overline{n} = \{1, 2, \ldots, n\} \subseteq \mathbb{N},$$

so $\overline{n}$ has n elements, and we set $\overline{0} = \emptyset$ by convention. We say a set A is *finite* if there is some $n \in \mathbb{N} \cup \{0\}$ and a bijection $\overline{n} \to A$; thus a set A is finite if it is in bijective correspondence with one of the standard finite sets $\overline{0}, \overline{1}, \ldots$.

Our first result is *the* fundamental tool in the study of finite sets, and is presumably "obvious" to the reader from everyday experience.

7.1.1 LEMMA *If there is an injection $f : \overline{m} \to \overline{n}$, then we must have $m \leq n$.*

This lemma (or sometimes its consequence Theorem 7.1.3 below) is called the *Pigeonhole Principle* because of the following interpretation of (the contrapositive of) the result: Suppose that $m > n$ and imagine m pigeons numbered 1 through m alighting in n pigeonholes numbered 1 through n; there clearly must be some pigeonhole containing at least two pigeons since there are more pigeons than pigeonholes, so the natural function from the set of pigeons to the set of pigeonholes could not be injective. While this "real" interpretation of Lemma 7.1.1 is indeed a compelling argument from everyday experience with finite sets, we turn next to the proof.

Proof We argue by induction on $n \in \mathbb{N} \cup \{0\}$, and for the basis step $\overline{n} = \emptyset$, the existence of a function (let alone an injection) $f : \overline{m} \to \overline{n}$ necessarily implies that $\overline{m} = \emptyset$ as well. In other words, $n = 0$ implies $m = 0$, as desired.

For the induction step, suppose that $f : \overline{m} \to \overline{n}$ is an injection where $n \geq 1$. We must show that $m \leq n$ and first consider the case that $n \in \overline{n}$ does not lie in the image of f. In this case, we may restrict the codomain of f to produce a function $\overline{m} \to \overline{n} - \{n\} = \overline{(n-1)}$ which is again an injection as the restriction of such. By the inductive hypothesis, we conclude that $m \leq n - 1 < n$, as was asserted.

It remains to consider the possibility that n does lie in the image of f, so by injectivity of f, there is a unique $j \in \overline{m}$ so that $f(j) = n$. We may thus restrict the domain to define a function $f' : \overline{m} - \{j\} \to \overline{(n-1)}$ which is again an injection as the restriction of such.

There is another function we shall also require in this proof, namely

$$g : (\overline{m} - \{j\}) \to \overline{(m-1)}$$
$$k \mapsto \begin{cases} k, & \text{if } k < j; \\ k - 1, & \text{if } k > j. \end{cases}$$

We prove that g is a bijection using Theorem 6.5.3 by writing down a two-sided inverse

$$h : \overline{(m-1)} \to \overline{m} - \{j\}$$
$$\ell \mapsto \begin{cases} \ell, & \text{if } \ell < j; \\ \ell + 1, & \text{if } \ell \geq j, \end{cases}$$

leaving it for the reader to check that the compositions $h \circ g$ and $g \circ h$ are indeed the respective identity maps.

Finally, the composition

$$f' \circ h : \overline{(m-1)} \to \overline{(n-1)}$$

is injective by Theorem 6.4.1b as the composition of the injection f' and the bijection h, so by the inductive hypothesis, we conclude that $m - 1 \leq n - 1$ so that $m \leq n$, as required. *q.e.d.*

Summarizing part of the previous argument, namely, the existence of the bijection g, for later use, we have

7.1.2 LEMMA *For any $j \in \overline{m}$, there is a bijection $(\overline{m} - \{j\}) \to \overline{(m-1)}$.*

Suppose now that A is a finite set, so A is in bijective correspondence $f : \overline{n} \to A$ with some $\overline{n}$. If A were also in bijective correspondence $g : \overline{m} \to A$ with $\overline{m}$, then there would be induced bijections $f \circ g^{-1}$ and $g \circ f^{-1}$, so that $m \leq n$ and $n \leq m$ by the Pigeonhole Principle, or in other words, $m = n$ by antisymmetry of $\leq$.

Armed with this consequence of the Pigeonhole Principle, we may define the *cardinality* of a finite set A to be the *unique* $n \in \mathbb{N} \cup \{0\}$ so that $\overline{n}$ is in bijective correspondence with A. Introducing some notation, we shall denote the cardinality n of a set A by the symbol $n = |A|$, so for instance $|\overline{n}| = n$. We say that two finite sets A and B are *equinumerous* or *equipotent* if $|A| = |B|$, or in other words, if A and B have the same number of elements.

Our next result is essentially a reformulation of the previous lemma and is itself sometimes referred to as the "Pigeonhole Principle".

7.1.3 THEOREM *Suppose that A and B are finite sets with $|A| = m$ and $|B| = n$, where $m > n$. Then there is no injection $A \to B$.*

Proof By definition there are bijections $f : \overline{m} \to A$ and $g : \overline{n} \to B$. An injection $h : A \to B$ induces a corresponding injection

$$g^{-1} \circ h \circ f : \overline{m} \to \overline{n}.$$

Thus, we have $m \leq n$ by Lemma 7.1.1, contradicting our supposition that $m > n$. *q.e.d.*

7.1.4 COROLLARY *Suppose that A and B are finite sets. Then A and B are equinumerous if and only if there is some bijection $A \to B$.*

Proof If A and B are equinumerous, say $|A| = n = |B|$, then there are bijections $f : \overline{n} \to A$ and $g : \overline{n} \to B$, so the composition $g^{-1} \circ f$ is a bijection from A to B.

Conversely, if $h : A \to B$ is a bijection, then in particular it is an injection, so $|A| \leq |B|$ by the previous theorem. Of course, h^{-1} is likewise an injection, so $|B| \leq |A|$ again by the previous theorem, whence A and B are equinumerous. *q.e.d.*

We close with a final application of the Pigeonhole Principle to the effect that a subset of a finite set is necessarily finite and has cardinality strictly less than that of the superset, a fact which is again presumably "empirically obvious" to the reader. To prove this, we require the following

7.1.5 LEMMA *Suppose $b \in B$ and B is a finite set. Then $|B - \{b\}| = |B| - 1$.*

Proof Since B is finite and non-empty, there is some bijection $f : \overline{n} \to B$ and $b = f(j)$ for some $j \in \overline{n}$. Define the function

$$\begin{aligned} g : (\overline{n} - \{j\}) &\to (B - \{b\}) \\ k &\mapsto f(k), \end{aligned}$$

which is an injection as the restriction of such and a bijection by construction.

Thus, g is a bijection, and Lemma 7.1.2 guarantees the existence of another bijection $h : \overline{(n-1)} \to (\overline{n} - \{j\})$, so the composition $g \circ h : \overline{(n-1)} \to B$ provides the required bijection. *q.e.d.*

Armed with the previous lemma, we turn to the final major result of this section regarding subsets of finite sets.

7.1.6 THEOREM *Suppose that A and B are sets where $A \subseteq B$. If B is finite, then so too is A. Furthermore, if A is a proper subset of B, then $|A| < |B|$.*

Proof Beginning with the proof that A is finite if B is, we proceed by induction on $|B| = n$. For the basis step $n = 0$, we have $B = \emptyset$, so $A \subseteq B$ implies that $A = \emptyset$ as well, and A is indeed finite.

For the induction step, we may choose some $b \in B \neq \emptyset$, so of course $A - \{b\} \subseteq B - \{b\}$, and $|B - \{b\}| = |B| - 1 = n - 1$ by Lemma 7.1.5.

By the inductive hypothesis, we conclude that $A - \{b\}$ is finite, so there is some $m \in \mathbb{N} \cup \{0\}$ and a bijection $g : \overline{m} \to (A - \{b\})$. In particular, if $b \notin A$, then $A = A - \{b\}$ is itself finite, as desired. On the other hand, if $b \in A$, then we define the function

$$h : \overline{(m+1)} \to A$$
$$k \mapsto \begin{cases} g(k), & \text{if } k \le m; \\ b, & \text{if } k = m+1, \end{cases}$$

which is evidently a bijection. This completes the induction step and therefore the proof that A is finite.

To complete the proof of the theorem, we suppose that A is a proper subset of B. In this case, we may choose some $b \in B - A$ and compute

$$|A| = |A - \{b\}| \le |B - \{b\}| = |B| - 1,$$

where the inequality follows from the Pigeonhole Principle and the final equality holds by Lemma 7.1.5. Thus, $|A| \le |B| - 1$, so in fact $|A| < |B|$, as desired. *q.e.d.*

We have observed before that if $f : A \to B$ is an injection, then restricting the codomain produces a function $F : A \to f(A)$ which is a bijection. This is a generally useful technique, which give us here the following two consequences of the previous result.

7.1.7 COROLLARY *If B is finite, then there is no injection $f : B \to B$ with $f(B)$ a proper subset of B.*

Proof Suppose to get a contradiction that $f(B)$ is indeed a proper subset of B, so $|f(B)| < |B|$ by Theorem 7.1.6. As above, f induces a bijection between B and $f(B)$, so we have $|B| = |f(B)| < |B|$, which is absurd by the Pigeonhole Principle. *q.e.d.*

We mention parenthetically that the converse of Corollary 7.1.7 holds as well, i.e., a set B is "infinite" (that is, not finite) if and only if there is some injection $f : B \to B$ where $f(B)$ is a proper subset of B. We shall prove this converse of Corollary 7.1.7 in Theorem 7.2.2 below.

7.1.8 COROLLARY *Suppose that $f : A \to B$ is an injection, where A and B are finite sets so that $|A| = |B|$. Then f is surjective as well, so f is in fact a bijection.*

Proof Again as above, f induces a bijection between A and $f(A) \subseteq B$, so $|A| = |f(A)|$. If $f(A)$ is a proper subset of B, then $|f(A)| < |B| = |A| = |f(A)|$, where the inequality follows from Theorem 7.1.6, and this is absurd again by the Pigeonhole Principle. *q.e.d.*

We mention parenthetically that the special case of Corollary 7.1.8 in which $A = B$ was proved using dynamical systems techniques in Theorem 6.7.4.

Given finite sets A, B, say with $|A| = m, |B| = n$, it is natural to try to compute the cardinalities of the various sets constructed from A, B using the usual set-theoretic operations of union, Cartesian product, and so on. One can calculate that in fact

$$\begin{aligned} |A \cup B| &= |A| + |B| - |A \cap B| \\ |A \times B| &= |A||B| \\ |A^B| &= |A|^{|B|} \\ |\mathcal{P}(A)| &= 2^{|A|} \end{aligned}$$

but we shall not actually require these identities in this chapter and include these formulas (to be taken up in §8.1) here just for completeness.

It is worth saying explicitly that the reader may well feel that this section is so much bluster because it gives painfully detailed proofs of entirely obvious facts about finite sets. The value of our discussion is that it sets the stage for the treatment of infinite sets in the next sections, where there are surprising deviations from the natural intuition starting from finite sets.

EXERCISES

7.1 FINITE SETS

1. Given finite sets A and B, for each of the sets $A \cup B$, $A \times B$, A^B, and $\mathcal{P}(A)$
   (a) Directly prove finiteness.
   (b) Calculate the cardinality in terms of $|A|$, $|B|$, and $|A \cap B|$.
2. Prove the following converse of Corollary 7.1.8: If A and B are finite equinumerous sets and $f : A \to B$ is a surjective funtion, then f is also injective.
3. Suppose there are at least two people at a party. Show that at least two people know the same number of people at the party (where we make the egalitarian assumption that if x knows y, then also y knows x).
4. Prove the following generalization of the Pigeonhole Principle: Suppose that A and B are finite sets, where $|B| = m \geq 1$ and $|A| \geq km + 1$ with $k \geq 1$, and let $f : A \to B$ be any function. Then there exists $b \in B$ so that $|f^{-1}(b)| \geq k + 1$.

7.2 INFINITE SETS

We consider two definitions of "infinite sets" only to find that the two definitions are entirely equivalent. One of these formulations is especially natural while the other is useful in practice, so each formulation is important. We also discuss and prove various facts about infinite sets analogous to certain previous results for finite sets.

Recall that a set A is finite if there is some $n \in \mathbb{N} \cup \{0\}$ and a bijection $\overline{n} \to A$, where $\overline{n} = \{1, 2 \ldots, n\}$. Our first most natural definition is that a set A is said to be *infinite* if it is not finite, i.e., if there is no bijection $\overline{n} \to A$ for any $n \in \mathbb{N} \cup \{0\}$. This is an awkward definition to use: To show that a given set is infinite, we must prove the non-existence of a collection of bijections, and proving such a non-existence result directly can be difficult.

We say a set A is *Dirichlet infinite* if there is an injection $f : A \to A$ so that $f(A)$ is a proper subset of A. For instance the function $f : \mathbb{N} \to \mathbb{N}$ given by $f(n) = n + 1$ is an injection with proper image, so $\mathbb{N}$ is Dirichlet infinite. Since a finite set is not Dirichlet infinite by Corollary 7.1.7, we conclude that if A is Dirichlet infinite, then A is also infinite.

In fact, if A is infinite then A is also Dirichlet infinite, as we shall prove below, so the two notions of "infinite" are found to be entirely equivalent. To prove that a set is Dirichlet infinite, hence infinite, we need therefore simply exhibit a suitable injection, and this can often be done directly, as in the case of $\mathbb{N}$ above.

In order to prove the equivalence of our two notions of infinite, we use the Axiom of Choice to prove

7.2.1 LEMMA *If A is infinite, then there exists an injection $f : \mathbb{N} \to A$.*

Proof By the Zermelo Well-Ordering Principle formulation of the Axiom of Choice there is some well-ordering $\leq$ on the infinite set A. We shall define an injection $F : \mathbb{N} \to A$ by inductively specifying its values $F(1), F(2), \ldots$ as follows. For the basis step, let $F(1)$ denote the least element of A under $\leq$ (using that $A \neq \emptyset$ since A is infinite). Inductively assuming that $F(0), \ldots, F(n-1) \in A$ are distinct elements of A, the set $A - \{F(i) | 1 \leq i \leq n - 1\}$ cannot be empty, for otherwise $A \subseteq \{F(i) : 1 \leq i \leq n - 1\}$ is finite by Theorem 7.1.6, and this contradicts our assumption that A is infinite. Thus, the non-empty set $A - \{F(i) : 1 \leq i \leq n - 1\}$ has a least element for $\leq$, and we define $F(n)$ to be this least element.

The function $F : \mathbb{N} \to A$ thereby defined is the desired injection. *q.e.d.*

7.2.2 THEOREM *A set is infinite if and only if it is Dirichlet infinite.*

Proof As we have noted above, Dirichlet infinite implies infinite by Corollary 7.1.7. Conversely, suppose that A is infinite, so by the previous lemma, there is some injection $F : \mathbb{N} \to A$, and set $a_n = F(n) \in A$. Define another function

$$f : A \to A$$
$$a \mapsto \begin{cases} a_{n+1}, & \text{if } a = a_n \text{ for some } n; \\ a, & \text{otherwise,} \end{cases}$$

which is evidently an injection (as the reader should carefully check). Furthermore, we have $a_1 \notin f(A)$ by construction, so $f : A \to A$ is an injection with proper image, whence A is Dirichlet infinite. *q.e.d.*

In light of the previous result, we shall typically refer to a Dirichlet infinite set simply as an "infinite set" with no loss of precision.

As was observed before, the set $\mathbb{N}$ is infinite since there is an injection $n \mapsto n+1$ from $\mathbb{N}$ to $\mathbb{N}$ with proper image, and we next give a few more examples of infinite sets. The function

$$f : \mathbb{Z} \to \mathbb{Z}$$
$$n \mapsto 2n$$

is an injection (as the reader should check) whose image (namely, the set of even integers) is a proper subset of $\mathbb{Z}$. For another example, consider the function

$$g : \mathbb{Q} \to \mathbb{Q}$$
$$t \mapsto \begin{cases} t+1, & \text{if } t \geq 0; \\ t, & \text{if } t < 0; \end{cases}$$

which is again an injection with proper image (as the reader should check).

A superset of an infinite set is necessarily infinite, as we next see.

7.2.3 LEMMA *If $A' \subseteq A$ and A' is infinite, then A is infinite.*

Proof This is just the contrapositive of Theorem 7.1.6 using the definition that infinite means not finite. The reader might also give a direct proof here using the definition that infinite means Dirichlet infinite. *q.e.d.*

7.2.4 THEOREM *Suppose that $f : A \to B$ is an injection. If A is infinite, then B is infinite.*

Proof Suppose that A is infinite and consider the image $f(A) \subseteq B$. The injection f induces a bijection $A \to f(A)$, so $f(A)$ is infinite. Thus, $B \supseteq f(A)$ is infinite by the previous lemma, as desired. *q.e.d.*

7.2.5 COROLLARY *A set A is infinite if and only if there is some injection $\mathbb{N} \to A$.*

Proof If there is an injection $\mathbb{N} \to A$, then A is infinite by the previous theorem. The converse is simply a restatment of Lemma 7.2.1. *q.e.d.*

To close this section, we prove that various set-theoretic constructions involving infinite sets produce infinite sets.

7.2.6 COROLLARY *Suppose that A is an infinite set, and let B be any non-empty set. Then each of*

$$\mathcal{P}(A), \quad A \cup B, \quad A \times B, \quad \text{and} \quad A^B$$

is an infinite set.

Proof In each case, we exhibit an injection whose domain is the infinite set A and apply Theorem 7.2.4, and we begin with the power set $\mathcal{P}(A)$. The function

$$\begin{aligned} A &\to \mathcal{P}(A) \\ a &\mapsto \{a\} \end{aligned}$$

is the required injection, so $\mathcal{P}(A)$ is indeed infinite.

For the union $A \cup B$, observe that the function

$$\begin{aligned} A &\to (A \cup B) \\ a &\mapsto a \end{aligned}$$

provides the required injection, so $A \cup B$ is infinite.

Since $B \neq \emptyset$, we may choose some $b_0 \in B$ and define

$$\begin{aligned} A &\to A \times B \\ a &\mapsto (a, b_0) \end{aligned}$$

to show that $A \times B$ is also infinite.

Finally, for each $a \in A$, let $c_a : B \to A$ denote the constant function with value a (so that $c_a(b) = a$ for each $b \in B \neq \emptyset$), and define

$$\begin{aligned} A &\to A^B \\ a &\mapsto c_a. \end{aligned}$$

The reader should carefully check that this function is an injection (which actually follows from the definitions), so A^B is again an infinite set. *q.e.d.*

EXERCISES

7.2 INFINITE SETS

1. Prove the following analogue of the Pigeonhole Principle: If an infinite set is partitioned into finitely many blocks, then at least one block of the partition must be infinite.

2. Prove that if A is finite and B is infinite, then $B - A$ is infinite.

3. Prove directly that if A is finite and B is Dirichlet infinite, then $B - A$ is Dirichlet infinite.

4. Show that a set A is infinite if and only if for every $n \in \mathbb{N}$, there is a subset of A with n elements.

7.3 COUNTABLE SETS

As a first distinction about sizes of infinite sets, we study here "countably infinite" sets, which are simply sets which can be put into bijective correspondence with the set $\mathbb{N}$. Techniques for recognizing and constructing countably infinite sets are described and examples are given.

We say that a set A is *countably infinite* if there is a bijection $f : \mathbb{N} \to A$. If A is countably infinite, then we write $|A| = \aleph_0$ and say that the set A "has cardinality aleph zero". The symbol "$\aleph$" is the Hebrew letter "aleph", and this notation which is due to Cantor is entirely standard. The appearance of the subscript zero suggests the fact to be discussed later that there are infinitely many different cardinalities $\aleph_0 < \aleph_1 < \aleph_2 < \cdots$ of infinite sets.

Of course, $|\mathbb{N}| = \aleph_0$ since the identity map is a bijection $\mathbb{N} \to \mathbb{N}$. Furthermore, the function

$$g : \mathbb{N} \to \mathbb{Z}$$
$$n \mapsto \begin{cases} n/2, & \text{if } n \text{ is even;} \\ -(n+1)/2, & \text{if } n \text{ is odd;} \end{cases}$$

is a bijection as the reader should prove (cf. Example 6.4.4), so we also have $|\mathbb{Z}| = \aleph_0$. Thus, $\mathbb{N}$ is a proper subset of $\mathbb{Z}$ with cardinality $|\mathbb{N}| = \aleph_0 = |\mathbb{Z}|$, and of course this behavior for finite sets is prohibited by the Pigeonhole Principle.

As an immediate application of Theorem 7.2.1, we have

7.3.1 THEOREM *An infinite set necessarily contains a countably infinite one.*

It is in the sense of this result that countably infinite sets are the "smallest" infinite sets

A set A is said to be *countable* or *denumerable* if either A is finite or A is countably infinite, and a set which is not countable is said to be *uncountable.* We shall give examples of uncountable sets in the next section and dedicate the rest of this section to the study of countable sets.

In fact, given a countable set A, it can be difficult to actually exhibit a particular bijection $\mathbb{N} \to A$, and we first discuss another condition which is equivalent to countability and is easier to check in practice. To describe this alternative formulation of countability, we require some definitions, as follows. An *initial segment* I of $\mathbb{N}$ is either a finite subset $I = \overline{n} \subseteq \mathbb{N}$ for some $n \in \mathbb{N} \cup 0$, or it is $I = \mathbb{N}$ itself. Given a set A, an *enumeration* of A is a surjection $f : I \to A$, where I is an initial segment of $\mathbb{N}$, and if A admits an enumeration, then we say that A is enumerable. The enumeration f is said to be an *enumeration without repetitions* if f is also injective, so f is a bijection. If $f : I \to A$ is an enumeration of A, then we shall specify f as a (possibly finite) sequence

$$f(0), f(1), f(2), \ldots,$$

where we list the values of f in the natural way using the linear ordering on $I \subseteq \mathbb{N}$. Let us next give some examples of enumerations.

■ **Example 7.3.1** Consider the set $A = \{a, b, c, d\}$, and define the enumeration $f : \overline{5} \to A$ defined by $f(1) = a = f(2)$, $f(3) = b$, $f(4) = c$, $f(5) = d$, or in other words, determined by the sequence a, a, b, c, d of values. Another example of an enumeration $\overline{6} \to A$ has corresponding sequence of values a, d, c, c, a, b. Of course, the function $\overline{6} \to A$ determined by a, b, c, a, b, c is not surjective and is therefore not an enumeration. For a final example, the infinite sequence $a, b, c, c, d, a, a, \ldots$ describes an enumeration $g : \mathbb{N} \to A$ where $g(n) = a$ whenever $n \geq 5$.

■ **Example 7.3.2** We prove here that the set $\mathbb{Q}_+ = \{q \in \mathbb{Q} : q > 0\}$ of positive rationals is enumerable, and we first consider the collection $\mathbb{N} \times \mathbb{N}$. There is an obvious surjection

$$\begin{aligned} f : \mathbb{N} \times \mathbb{N} &\to \mathbb{Q}_+ \\ (p, q) &\mapsto \frac{p}{q}, \end{aligned}$$

which is not an injection (since $1/2 = f(1,2) = f(2,4) = f(3,6) = \cdots$ for instance). We shall exhibit below a bijection $g : \mathbb{N} \to \mathbb{N} \times \mathbb{N}$, so the composition $f \circ g$ is an enumeration of $\mathbb{Q}_+$, as was asserted.

To describe the bijection $g : \mathbb{N} \to \mathbb{N} \times \mathbb{N}$ informally, the reader should imagine $\mathbb{N} \times \mathbb{N}$ as a subset of the plane $\mathbb{R}^2$ in the usual way, as in

$$
\begin{array}{ccccccccc}
\vdots & & \vdots & & \vdots & & \vdots & & \iddots \\
& \nwarrow & & \searrow & & \nwarrow & & & \\
(1,4) & & (2,4) & & (3,4) & & (4,4) & & \cdots \\
\uparrow & \searrow & & \nwarrow & & \searrow & & \nwarrow & \\
(1,3) & & (2,3) & & (3,3) & & (4,3) & & \cdots \\
& \nwarrow & & \searrow & & \nwarrow & & \searrow & \\
(1,2) & & (2,2) & & (3,2) & & (4,2) & & \cdots \\
\uparrow & \searrow & & \nwarrow & & \searrow & & \nwarrow & \\
(1,1) & & (2,1) & \to & (3,1) & & (4,1) & \to & \cdots
\end{array}
$$

We illustrate also an infinite directed path in the figure composed of oriented edges running from $g(n)$ to $g(n+1)$ for each $n \in \mathbb{N}$. This path begins at $(1,1)$ and passes through each element of $\mathbb{N}^2$ exactly once and provides a vivid mental image of the bijection g.

More specifically, we describe g by inductively specifying the values of g at $1, 2, \ldots$, and we set

$$
\begin{aligned}
&g(1) = (1,1), \quad g(2) = (1,2), \quad g(3) = (2,1), \\
&g(4) = (3,1), \quad g(5) = (2,2), \quad g(6) = (1,3)
\end{aligned}
$$

as is indicated with arrows in the previous figure. The enumeration continues in this manner "traveling" along lines of slope -1 in the plane as well as along the x and y axes. More formally, we may proceed inductively and define $g(1) = (1,1)$ and $g(2) = (1,2)$. Assuming inductively that g is defined for each natural number at most $n \geq 2$ and letting $g(n) = (p,q)$, we define

$$
g(n+1) = \begin{cases} (p, q+1), & \text{if } p = 1; \\ (p+1, q), & \text{if } q = 1; \\ (p-1, q+1), & \text{if } p \neq 1 \text{ and } p+q \text{ is even}; \\ (p+1, q-1), & \text{if } q \neq 1 \text{ and } p+q \text{ is odd}; \end{cases}
$$

and leave it to the exercises to check that this accurately reflects the informal discussion above and is indeed a bijection.

Thus, $|\mathbb{N} \times \mathbb{N}| = \aleph_0$ since g is a bijection, and $\mathbb{Q}_+$ is enumerable since $f \circ g$ is surjective.

Our purpose in presenting this example here is two-fold: This "trick" of enumerating $\mathbb{N} \times \mathbb{N}$ is one of the basic tools for countable sets (cf. Theorem 7.3.4 below), and $\mathbb{Q}_+$ is an example of an enumerable set which is not obviously in one-to-one correspondence with $\mathbb{N}$. The next result shows for instance that $\mathbb{Q}_+$ must in fact be countably infinite.

7.3.2 THEOREM *A set is enumerable if and only if it is countable.*

Proof One implication here is trivial, for if A is countable then either A is finite or A is countably infinite, and in either case, there is an enumeration (indeed there is a bijection unless $A = \emptyset$) $f : I \to A$ with I an initial segment of $\mathbb{N}$ by definition.

Conversely, suppose that A is enumerable. If A is finite, then A is countable by definition, so we suppose that A is an infinite enumerable set and must prove that A is countably infinite.

Let $f : I \to A$ be an enumeration and suppose first that $I \subseteq \mathbb{N}$ is finite. Since f is surjective, it has a right inverse $g : A \to I$ by Theorem 6.5.3b which is an injection by Theorem 6.5.3a since f is a left inverse. Thus, A is in one-to-one correspondence with the subset $g(A) \subseteq I$, which is finite by Theorem 7.1.6 since I is finite. Since A is infinite by assumption, this contradiction guarantees that the initial segment is not finite, i.e., $I = \mathbb{N}$.

We inductively construct from the enumeration $f : \mathbb{N} \to A$ a bijection $h : \mathbb{N} \to A$ as follows. To begin, we define $h(1) = f(1)$. To define $h(2)$, observe that $A - \{h(1)\} \neq \emptyset$, for otherwise $A \subseteq \{h(1)\}$ is finite. Moreover, since f is surjective, there is some $k \geq 1$ so that $f(k) \neq f(1)$. There is therefore a least k satisfying $f(k) \neq f(1)$ by Theorem 3.7.3, and we let k_2 denote this least such k finally defining $h(2) = f(k_2)$. Thus, $h(2) = f(k_2)$, where k_2 is the least element of $\mathbb{N}$ satisfying $f(k_2) \notin \{f(1)\}$. Continue inductively in this way, defining $h(n+1) = f(k_{n+1})$, where k_{n+1} is the least element of $\mathbb{N}$ so that $f(k_{n+1}) \notin \{f(1), f(k_2), \ldots, f(k_n)\}$.

The function $h : \mathbb{N} \to A$ just constructed using f is an injection by construction (as the reader might prove inductively), and we claim that h is furthermore a surjection. Indeed, surjectivity of h follows from surjectivity of f as follows: Given $a \in A$, we have $f^{-1}(\{a\}) \neq \emptyset$ (since f is assumed to be surjective) and let $k \in \mathbb{N}$ denote the least element of $f^{-1}(\{a\})$. Of course $f(k) \notin \{f(j) : 1 \leq j < k\}$ since k is least, but on the other hand $f(k) \in \{h(j) : 1 \leq j \leq k\}$ by construction (as the reader might again prove inductively). Thus, for some $j = 1, \ldots, k$, we find $h(j) = f(k) = a$ and conclude that h is indeed surjective.

We have shown that the function $h : \mathbb{N} \to A$ constructed above is a bijection, so A is indeed countable. *q.e.d.*

7.3.3 COROLLARY *A subset of a countable set is itself countable.*

Proof Suppose that $A \subseteq B$ where B is countable and let $f : \mathbb{N} \to B$ be a bijection. Choose $a \in A$ (unless $A = \emptyset$ and is hence countable), and define a function $h : \mathbb{N} \to A$ by setting

$$h(n) = \begin{cases} f(n), & \text{if } f(n) \in A; \\ a, & \text{otherwise.} \end{cases}$$

Surjectivity of f implies surjectivity of h, so A is enumerable, hence countable by the previous theorem. *q.e.d.*

A fundamental result about countable sets is

7.3.4 THEOREM *A countable union of countable sets is itself countable.*

Proof Let $I \subseteq \mathbb{N}$ be an initial segment of $\mathbb{N}$ and re-index the sets in the countable family of sets to be $\{A_i : i \in I\}$. Thus, we must show that $A = \cup_{i \in I} A_i$ is countable, where A_i is countable for each $i \in I$.

If $I = \emptyset$ or if each $A_i = \emptyset$, then $A = \emptyset$ so A is countable. Furthermore, if $A_j = \emptyset$ for some j, then of course $A = \cup_{i \in I - \{j\}} A_i$, so we may simply omit j from I. We may therefore assume that $I \neq \emptyset$ and each A_i is countable and non-empty.

Choose an enumeration $f_i : \mathbb{N} \to A_i$ for each $i \in I$ and list the values of f_i as $a_{i1} = f_i(1), a_{i2} = f_i(2), \ldots$ for each $i \in I$. Arranging the elements of A in an "array" in the plane $\mathbb{R}^2$ in the natural way (where the ith row in the array is the ordered tuple of values of f_i), we find

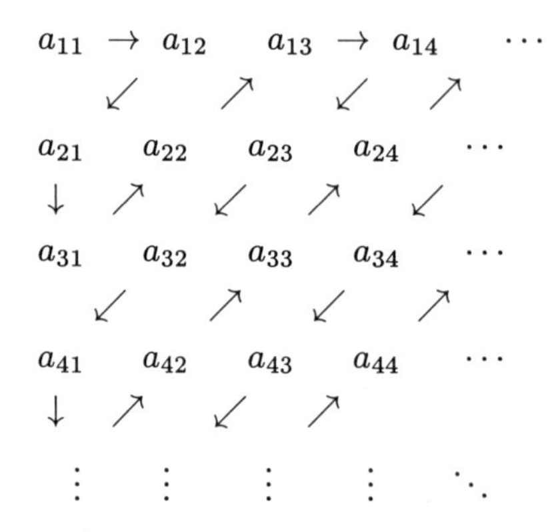

It may happen that I itself is finite, so the array above would have only finitely many rows. In this case, let us simply infinitely often repeat the last row in the array so as to produce a "bi-infinite" array so that the figure above is accurate.

Indicated in the figure is an infinite directed path, which describes an enumeration of A as in Example 7.3.2. Thus, A is enumerable hence countable. *q.e.d.*

We can easily conclude at this point that $\mathbb{Q}$ itself is countable, for $\mathbb{Q}_+$ is enumerable by Example 7.3.2 hence countable. Since $\mathbb{Q}_- = \{q \in \mathbb{Q} : q < 0\}$ is in obvious bijective correspondence with $\mathbb{Q}_+$, the set $\mathbb{Q}_-$ is also countable, so by the previous result, we conclude that $\mathbb{Q} = \mathbb{Q}_+ \cup \{0\} \cup \mathbb{Q}_-$ is also countable.

Another consequence of the previous theorem, we next show that "removing a small set from a large one leaves a large set". Explicitly, we have

7.3.5 THEOREM (a) *If A is countably infinite and $B \subseteq A$ is finite, then $A - B$ is countably infinite.*

(b) *If A is uncountable and B is countable, then $A - B$ is uncountable.*

Proof For part (a), since $A - B \subseteq A$ and A is countable, we conclude that $A - B$ is countable by Corollary 7.3.3. To get a contradiction, suppose that $A - B$ is finite, so $A = (A - B) \cup B$ is a finite union of finite sets, and therefore itself finite as the reader should carefully prove (or see Theorem 8.1.2 for a complete proof). This contradicts our assumption that A is countably infinite.

Similarly for part (b), if $A - B$ were countable, then $A = (A - B) \cup B$ would be a finite union of countable sets. A is therefore itself countable by Theorem 7.3.4, and this is contrary to hypothesis. *q.e.d.*

Another basic result, we have

7.3.6 THEOREM *If A and B are countable sets, then the Cartesian product $A \times B$ is also countable.*

Proof We may write the product

$$A \times B = \cup_{b \in B} A \times \{b\}$$

as a countable union of countable sets and apply the previous theorem. *q.e.d.*

For some further examples of countably infinite sets, we may apply the previous results and remarks to conclude that the following are countable sets:

■ **Example 7.3.3** $\mathbb{Z}^n$ and $\mathbb{Q}^n$ are countable for any $n \in \mathbb{N} \cup \{0\}$.

■ **Example 7.3.4** Consider the collection $P_n(x)$ of all polynomials in x of degree at most n with integral coefficients, that is $P_n(x)$ is the set of all expressions of the form $a_n x^n + a_{n-1} x^{n-1} + \cdots + a_1 x + a_0$, where x is an indeterminate and $a_i \in \mathbb{Z}$ for each $i = 1, \ldots, n$. The set $P_n(x)$ is countable for each $n \geq 0$.

■ **Example 7.3.5** The collection of all polynomials (of any degree) with rational coefficients is countable.

We shall give examples of uncountable sets in the next section. Indeed, we shall find that if A is a set with at least two elements, then $A^{\mathbb{N}}$ is uncountable. In contrast to this, we have

7.3.7 THEOREM *If B is countable and A is finite, then B^A is countable.*

Proof If $A = \emptyset$, then B^A is the singleton containing the empty function, while if $B = \emptyset$, then $B^A = \emptyset$, so the theorem holds in these trivial cases. Furthermore, if B is finite as well as A, then B^A is finite as the reader should carefully prove (or see Theorem 8.1.5 for a complete proof).

We thus suppose that $A, B \neq \emptyset$ with B countably infinite, $|A| = n \geq 1$, and choose a bijection $g : \mathbb{N} \to B$. For each $k \in \mathbb{N}$, define

$$F_k = \{f \in B^A : f(A) \subseteq g(\{1, 2, \ldots, k\})\},$$

so each F_k is finite with cardinality $|F_k| = |\{1, 2, \ldots, k\}^A|$. Since A is finite, the image $f(A) \subseteq B$ is also finite for each $f \in B^A$. For each fixed $f \in F_k$, there is therefore some $m \in \mathbb{N}$ so that for all $k > m$, we have $f \in F_k$. It follows that

$$B^A = \cup_{k \in \mathbb{N}} F_k$$

is a countable union of finite sets, and is therefore countable by Theorem 7.3.4.
q.e.d.

EXERCISES

7.3 COUNTABLE SETS

1. Show that $f(m, n) = 2^{m-1}(2n - 1)$ defines a bijection $f : \mathbb{N} \times \mathbb{N} \to \mathbb{N}$.
2. Show that if $X \subseteq \mathbb{R}$ is well ordered, then X is countable. [HINT: For any $x \in X$, choose a rational between x and the next element of X if any.]
3. Find a set $\mathcal{A}$ of open intervals in $\mathbb{R}$ so that each element of $\mathbb{Q}$ lies in at least one interval yet $\cup \mathrm{A} \neq \mathbb{R}$. [HINT: Limit the sum of the lengths of the intervals.]
4. Given a set A, define a *finite sequence* in A to be a function $f : \bar{n} \to A$ for some $n \in \mathbb{N}$. Prove that if $A \neq \emptyset$ is countable, then the set of all finite sequences in A is also countable.
5. Be more precise in the proof of Theorem 7.3.2 and construct h from f with a careful application of the Axiom of Choice.

7.4 UNCOUNTABLE SETS

We turn to a discussion of uncountable sets. The basic technique for proving that a set is uncountable is informally called "Cantor's diagonal argument", and our discussion begins with a description of this technique.

7.4.1 THEOREM *Suppose that A is finite set with cardinality at least two. Then $A^{\mathbb{N}}$ is uncountable.*

Proof Suppose that $|A| = n$ and write $A = \{a_1, \ldots, a_n\}$, where $n \geq 2$. We may specify $f \in A^{\mathbb{N}}$ by listing the ordered tuple of values taken by f, namely, given $f \in A^{\mathbb{N}}$, we may write $f(1), f(2), f(3), \ldots$ in this order.

Suppose to get a contradiction that $A^{\mathbb{N}}$ is actually countable, and suppose that $f_1, f_2, \ldots$ is an enumeration of $A^{\mathbb{N}}$. Form an infinite two-dimensional array whose i^{th} row consists of the values $a_{i1} = f_i(1), a_{i2} = f_i(2), \ldots$. In this way, we produce an infinite two-dimensional array

$$\begin{array}{cccc} a_{11} & a_{12} & a_{13} & a_{14}\cdots \\ a_{21} & a_{22} & a_{23} & a_{24}\cdots \\ a_{31} & a_{32} & a_{33} & a_{34}\cdots \\ a_{41} & a_{42} & a_{43} & a_{44}\cdots \\ & \vdots & & \ddots \end{array}$$

whose respective rows are the values taken by the functions $f_1, f_2, \ldots$.

The idea of the proof is to define a function $g \in A^{\mathbb{N}}$ which cannot occur as a row in this array. We construct this function g using the entries occurring on the "diagonal" $a_{11}, a_{22}, a_{33}, \ldots$ of this array by defining

$$\begin{aligned} g : \mathbb{N} &\to A \\ n &\mapsto \begin{cases} a_1 & \text{, if } a_{nn} \neq a_1; \\ a_2 & \text{, if } a_{nn} = a_1, \end{cases} \end{aligned}$$

so we require $n \geq 2$ in order to make this definition.

Having defined this function $g : \mathbb{N} \to A$, we claim that g is not among the functions $f_1, f_2, \ldots$. Indeed, if g did occur among these functions, then $g = f_i$ for some $i \geq 1$, and we consider $f_i(i)$. By construction, if $f_i(i) = a_1$, then $g(i) = a_2$, while if $f_i(i) \neq a_1$, then $g(i) = a_1$, so in either case, f_i and g take different values at $i \in \mathbb{N}$. Thus, $g \neq f_i$ and since i was arbitrary, we conclude that $g \in A^{\mathbb{N}}$ is not among the functions $f_1, f_2, \ldots$. This contradicts that $f_1, f_2, \ldots$ is an enumeration of $A^{\mathbb{N}}$, so $A^{\mathbb{N}}$ is uncountable, as was asserted. *q.e.d.*

The technique above of showing that a set is uncountable by producing an element of the set which cannot occur in a given enumeration is due to Cantor. This technique occurs in many different guises in the theory of computability.

To extend the utility of the previous result, we require

7.4.2 LEMMA *if A is uncountable and $f : A \to B$ is an injection, then B is also uncountable.*

Proof Suppose to the contrary that B is countable, and let $g : \mathbb{N} \to B$ be an enumeration. Since f is an injection, it has a left inverse $h : B \to A$ which is itself a surjection by Theorem 6.5.3b. The composition $h \circ g : \mathbb{N} \to A$ is then a surjection as the composition of such by Theorem 6.4.1a, so A is enumerable hence countable, which is contrary to hypothesis. *q.e.d.*

7.4.3 COROLLARY *Suppose that B is infinite and A is either infinite or is finite with at least two elements. Then A^B is uncountable.*

Proof Since B is infinite, it contains a countably infinite set $B_1 \subseteq B$, and by our hypotheses on A, it contains a set $A_1 \subseteq A$ with two elements. We choose an element $a \in A$ and define a function

$$F : A_1^{B_1} \to A^B$$
$$f \mapsto g$$

where we define

$$g : B \to A$$
$$b \mapsto \begin{cases} f(b) \in A_1 \subseteq A, & \text{if } b \in B_1; \\ a, & \text{if } b \in B - B_1, \end{cases}$$

so g arises from f by extending the codomain from A_1 to A and the domain from B_1 to B, where $g(b) = a$ for each $b \in B - B_1$.

Of course $A_1^{B_1}$ is uncountable by Theorem 7.4.1, so by Lemma 7.4.2, it remains to show that $F : A_1^{B_1} \to A^B$ is an injection. To this end, suppose that $f_1, f_2 \in A_1^{B_1}$ satisfy $F(f_1) = F(f_2) \in A^B$. In particular, f_1 and f_2 take the same values on B_1, so indeed $f_1 = f_2$, as required. *q.e.d.*

To close this section, we apply Cantor's diagonal argument to show that the interval $[0,1] = \{t \in \mathbb{R} : 0 \le t \le 1\}$ is uncountable. This was actually Cantor's original application of this technique. Since we have not rigorously defined the set $[0,1]$ here (or the set $\mathbb{R}$ for that matter), our treatment is necessarily a bit imprecise (but nevertheless communicates the main points).

7.4.5 THEOREM [Cantor] *The interval* $[0,1]$ *is uncountable.*

Proof Each $t \in [0,1]$ admits an infinite decimal expansion $t = .t_0t_1t_2\ldots$, where each $t_i \in \{0,1,\ldots,9\}$ is a decimal digit. This decimal expansion is not unique; for instance, the rational number $1/5$ admits the decimal expansions $.2000\ldots$ and $.1999\ldots$. Conversely, this indeterminacy (arising from an infinite string of consecutive digits nine, i.e., an infinite string of consecutive digits zero, i.e., corresponding to a rational number) is the only indeterminacy in the infinite decimal expansion of a real number. We take these facts about decimal expansions of reals for granted here.

Suppose to get a contradiction that there is actually an enumeration $f : \mathbb{N} \to [0,1]$, and write $f(i)$ in a decimal expansion as $f(i) = .a_{i1}a_{i2}a_{i3}\ldots$ for each $i \in \mathbb{N}$ (where we may choose one of the two possible decimal expansions if necessary). As before, arrange this data in an infinite two-dimensional array listing the decimal digits $a_{i1}a_{i2}a_{i3}\ldots$ in the ith row, and consider the infinite decimal expansion $.a_1a_2a_3\ldots$ defined by setting

$$a_i = \begin{cases} 1, & \text{if } a_{ii} \ne 1; \\ 2, & \text{if } a_{ii} = 1. \end{cases}$$

There corresponds a unique real number t to this decimal expansion (because there are no zeroes or nines in the decimal expansion above), and its decimal expansion is not among $\{f(i) : i \in \mathbb{N}\}$ as before. This contradicts that f is an enumeration, so $[0,1]$ is uncountable.

q.e.d.

A set A is said to have cardinality $\mathbf{c}$ and we write $|A| = \mathbf{c}$ if there is a bijection $A \to [0,1]$. The symbol $\mathbf{c}$ stands for the word "continuum", and the set $[0,1]$ itself (or any set in bijective correspondence with $[0,1]$) is called a *continuum.* Another standard bit of terminology, one might say that A "has the power of the continuum" if $|A| = \mathbf{c}$. Of course, the previous result shows that any continuum is uncountable. We close this section with several other examples of continua. We also remark here that we shall show in the next section that $|2^{\mathbb{N}}| = \mathbf{c}$ as well.

■ **Example 7.4.1** Let $[a,b] = \{t \in \mathbb{R} : a \leq t \leq b\}$. The map

$$\begin{aligned} [0,1] &\to [a,b] \\ t &\mapsto (b-a)t + a \end{aligned}$$

is a bijection as in Example 6.4.3 (and the reader should check this independently here), so any "closed interval" $[a,b]$ has cardinality $\mathbf{c}$, i.e., $[a,b]$ is a continuum.

■ **Example 7.4.2** Let $(a,b) = \{t \in \mathbb{R} : a < t < b\}$ be any open interval. As in the previous example, any open interval (a,b) is in one-to-one correspondence with the interval $(0,1)$, so it suffices to exhibit a bijection $[0,1] \to (0,1)$. To this end, we let

$$A = \{0\} \cup \{\frac{1}{n} : n \in \mathbb{N}\} = \{0, 1, \frac{1}{2}, \frac{1}{3}, \ldots\}$$

and define

$$\begin{aligned} f : [0,1] &\to (0,1) \\ t &\mapsto \begin{cases} \frac{1}{2}, & \text{if } t = 0; \\ \frac{1}{n+2}, & \text{if } t = \frac{1}{n} \text{ for } n \in \mathbb{N}; \\ t, & \text{if } t \in [0,1] - A. \end{cases} \end{aligned}$$

Thus, f restricts to the identity on the complement of A, while its restriction to A simply advances each element to the successor of its successor in the natural linear ordering on $A \subseteq \mathbb{R}$. We leave it for the reader to check that f is actually a bijection.

■ **Example 7.4.3** The set $\mathbb{R}$ itself is a continuum, for the map

$$\begin{aligned} (0,1) &\to \mathbb{R} \\ t &\mapsto \frac{\frac{1}{2} - 1}{x(1-x)} \end{aligned}$$

is a bijection, as we leave it for the reader to verify using Theorem 6.5.3c.

■ **Example 7.4.4** The collection $\mathbb{R} - \mathbb{Q}$ of all irrational numbers is the complement of a countable set (by Example 7.3.2) in an uncountable set (by Theorem 7.4.4 and Lemma 7.4.2) and is therefore itself uncountable by Theorem 7.3.5b.

EXERCISES

7.4 UNCOUNTABLE SETS

1. A real number is *algebraic* if it is a root of some polynomial with real coefficients, and a number is *transcendental* if it is not algebraic. Show that there are uncountably many transcendental numbers.

2. Show that the set of all partitions of $\mathbb{N}$ has the power of the continuum.

3. Let $X \subseteq \mathcal{P}(\mathbb{N})$ have the property that for any subset S of $\mathbb{N}$, either $S \in X$ or $\overline{S} \in X$ but not both. Show that X has the power of the continuum.

4. (a) Show that a collection of pairwise disjoint circular disks in the plane must be countable.

   (b) Must a collection of pairwise disjoint circles in the plane be countable? Prove your assertion.

5. Let $R_1, R_2, \ldots$ be a sequence of equivalence relations on $\mathbb{N}$. Build an equivalence relation R on $\mathbb{N}$ by diagonalization so that $R \neq R_i$ for any $i \geq 1$. [HINT: Consider the corresponding partitions.]

7.5 COMPARING CARDINALITIES

This section is dedicated to the comparison of cardinalities of infinite sets. Our first major result here is the Cantor-Schroeder-Bernstein Theorem, which is a basic tool for proving the existence of bijections between sets. We then give an elegant argument of Cantor which implies in particular the existence of infinitely many different cardinalities of infinite sets. As an application of the Axiom of Choice in its incarnation as Zorn's Lemma, we finally prove a major result called Zermelo's Trichotomy Theorem.

If A and B are sets, then we write $|A| \leq |B|$ if there is some injection $A \to B$. We also write $|A| = |B|$ if there is some bijection $A \to B$, and we say in this case that A and B are *equipotent*. Finally, we write $|A| < |B|$ if $|A| \leq |B|$ and $|A| \neq |B|$, or in other words, there is an injection $A \to B$ but no bijection $A \to B$. If $|A| < |B|$, then we say that "A has smaller cardinality than B", and we think of A as being of a strictly smaller "size" than B. By the Pigeonhole Principle, this generalizes our usual notion of the "size" of a finite set.

To begin, we make the following basic observation, which was proved already in Example 6.4.8.

7.5.1 THEOREM *For any set A, we have $|(\{0,1\})^A| = |\mathcal{P}(A)|$.*

Clearly, we have $|A| = |A|$, and one is tempted to regard equipotence as a "reflexive relation on the collection of all sets", but there is no set of all sets by Russell's Paradox (cf. §3.3). Thus, to regard $=$ as a relation, we must seemingly depart from ZFC. There are actually several ways to make sense of equipotence as an equivalence relation using the ZFC axioms, but we shall not take this up here. Rather, we shall abuse terminology slightly and say simply that "$=$ is reflexive", for instance, since $|A| = |A|$ for any set A. This terminology is quite natural, for instance in

7.5.2 THEOREM *Equipotence of sets is an equivalence relation.*

Proof Reflexivity and symmetry of equipotence are obvious, and transitivity follows directly from the fact Theorem 6.4.1c that a composition of bijections is again a bijection. *q.e.d.*

In the same spirit, $\leq$ is obviously reflexive, and transitivity of $\leq$ follows from the fact Theorem 6.4.1a that a composition of injections is again an injection. In fact, anti-symmetry of $\leq$ is a serious result, the Cantor-Schroeder-Bernstein Theorem, whose proof will be given presently, but first let us observe that assuming this result, we have shown that $\leq$ is a partial ordering. Furthermore, $<$ is just the strict partial ordering corresponding to the partial ordering $\leq$.

7.5.3 THEOREM [Cantor-Schroeder-Bernstein] *$\leq$ is anti-symmetric. That is, if $|A| \leq |B|$ and $|B| \leq |A|$, then we have $|A| = |B|$.*

In other words, to prove that there is a bijection between the sets A and B, it suffices to exhibit an injection $A \to B$ and another injection $B \to A$. This provides a powerful method of proving that two sets are equipotent as we shall see in examples later.

Proof Suppose that

$$f : A \to B \quad \text{and} \quad g : B \to A$$

are injections, and let $A_* = f(A) \subseteq B$ and $B_* = g(B) \subseteq A$ denote the images. There are therefore induced bijections

$$f_* : A \to B_* \quad \text{and} \quad g_* : B \to A_*$$

with inverses

$$f_*^{-1} : B_* \to A \text{ and } g_*^{-1} : A_* \to B.$$

Given some $a \in A$, it may be that $a \in A_*$, and in this case, we may apply g_*^{-1} to get $g_*^{-1}(a) \in B$, which we shall call the "first forbear" of a. It may happen that then $g_*^{-1}(a) \in B_*$, in which case we can apply f_*^{-1} to get $f_*^{-1}(g_*^{-1}(a)) \in A$, which we shall call the "second forbear" of a. If $f_*^{-1}(g_*^{-1}(a))$ happens to lie in A_*, then we can continue to obtain the "third forbear" $g_*^{-1}(f_*^{-1}(g_*^{-1}(a)))$, and so on.

Each point of A thus gives rise to its set of forbears and by definition a point $a \in A - A_*$ has zero forbears. On the other hand, if A has only finitely many forbears, then $g(f(a))$ has two more forbears than a; furthermore, a has an even number of forbears if and only if its last forbear lies in A.

Define the sets

$$\begin{aligned} A_\infty &= \{a \in A : a \text{ has infinitely many forbears}\} \\ A_e &= \{a \in A : a \text{ has a finite even number of forbears}\} \\ A_o &= \{a \in A : a \text{ has a finite odd number of forbears}\} \end{aligned}$$

so evidently $A = A_o \cup A_e \cup A_\infty$, where the sets A_o, A_e, A_∞ are pairwise disjoint, and $A - A_* \subseteq A_e$ by definition. Define subsets B_o, B_e, B_∞ in the analogous way using the same functions f and g.

By construction (and as the reader should check), f induces bijections

$$f_\infty : A_\infty \to B_\infty \text{ and } f_e : A_e \to B_o,$$

and g_*^{-1} induces a bijection

$$g_o^{-1} : A_o \to B_e.$$

Finally, define

$$h : A \to B$$

$$a \mapsto \begin{cases} f(a) \in B_\infty, & \text{if } a \in A_\infty; \\ f(a) \in B_o, & \text{if } a \in A_e; \\ g^{-1}(a) \in B_e, & \text{if } a \in A_o. \end{cases}$$

We leave it as an exercise for the reader to check that h is indeed a bijection (using that each of f_∞, f_e and g_o^{-1} are bijections and that each of A_o, A_e, A_∞ and B_o, B_e, B_∞ are pairwise disjoint), and this completes the proof. *q.e.d.*

As was observed before, $\leq$ is also reflexive and transitive, so we have the immediate

7.5.4 COROLLARY *$\leq$ is a partial ordering.*

We also mentioned above that the Cantor-Schroeder-Bernstein Theorem gives a practical method for showing that two sets are equipotent, and we next give some examples of this kind of application.

■ **Example 7.5.1** Let us first give a quick proof of the fact that $|(0,1)| = |[0,1]|$ as was already discussed in Example 7.4.2. The inclusion map $(0,1) \to [0,1]$ is of course an injection, and we define

$$f : [0,1] \to (0,1)$$
$$t \mapsto \frac{n}{3} + \frac{1}{2}$$

leaving it for the reader to check by exhibiting a left inverse that f is also an injection. Having exhibited the required injections, we conclude from Theorem 7.5.3 that there is indeed a one-to-one correspondence between $(0,1)$ and $[0,1]$.

■ **Example 7.5.2** We have already described a surjection $\overline{2}^{\mathbb{N}} \to [0,1]$ using infinite decimal expansions in the proof of Theorem 7.4.4, and the right inverse of this map is an injection $[0,1] \to \overline{2}^{\mathbb{N}}$, so $|[0,1]| \le |2^{\mathbb{N}}|$. On the other hand, we define a function $F : \overline{2}^{\mathbb{N}} \to [0,1]$, where we specify the decimal digit $F(f) = .f(1)\ 0\ f(2)\ 0\ f(3)\ 0\ \cdots \in [0,1]$ which alternates the values of $f \in \overline{2}^{\mathbb{N}}$ with zeroes. We leave it to the reader to check that F is actually an injection again by exhibiting a left inverse. It follows from the Cantor-Schroeder-Bernstein Theorem, that $|\overline{2}^{\mathbb{N}}| = |[0,1]| = \mathbf{c}$, or in other words $\overline{2}^{\mathbb{N}}$ is a continuum. Of course, it follows from Theorem 5.4.1 that $\mathcal{P}(\mathbb{N})$ then also has the power of the continuum.

Thus, the Cantor-Schroeder-Bernstein Theorem can be used to prove that sets are equipotent, but it is of no use in proving that two sets are not equipotent. Indeed, it can be tricky to show that two sets are not equipotent as it requires proving the non-existence of a bijection (or of one of the two injections in Theorem 7.5.3).

Our next result which is again due to Cantor and closely parallels Russell's Paradox proves a non-equipotence result which is used to produce many interesting examples of non-equipotent infinite sets.

7.5.5 THEOREM [Cantor] *For any set A, we have $|A| < |\mathcal{P}(A)|$.*

Proof To see that $|A| \le |\mathcal{P}(A)|$, simply observe that the function

$$f : A \to \mathcal{P}(A)$$
$$a \mapsto \{a\}$$

is an injection. To show that $|A| \ne |\mathcal{P}(A)|$, suppose that $g : A \to \mathcal{P}(A)$ is an arbitrary function. We shall show that g cannot be surjective, so g cannot be bijective. Since g has domain A and codomain $\mathcal{P}(A)$, it assigns to each element $a \in A$ a subset $g(a) \subseteq A$, and a may or may not be in its corresponding subset $g(a)$. Define the set

$$S = \{a \in A : a \notin g(a)\} \subseteq A.$$

If g were surjective, then there would be some $a \in A$ so that $g(a) = S$, and in this case

$$a \in S \quad \Leftrightarrow \quad a \notin g(a) \quad \Leftrightarrow \quad a \notin S.$$

This contradiction shows that S is not in the image of g, as desired. *q.e.d.*

Thus, we can construct a countably infinite family of infinite sets, each of which has larger cardinality than its predecessor:

$$|\mathbb{N}| < |\mathcal{P}(\mathbb{N})| < |\mathcal{P}(\mathcal{P}(\mathbb{N}))| < \cdots.$$

Furthermore, it is not difficult to show that $<$ is transitive (and we leave the proof as an exercise using Theorem 7.5.3), so we have actually produced infinitely many infinite sets of different cardinalities!

We digress briefly to discuss another axiom of set theory which is sometimes added to ZFC. This new axiom is called the "Continuum Hypothesis", which is usually abbreviated simply "CH", and it says, roughly, that a continuum has the "smallest uncountable cardinality". More precisely, the *Continuum Hypothesis* asserts that if A is a set so that $\aleph_0 \leq |A| \leq c$, then either $|A| = \aleph_0$ or $|A| = \mathbf{c}$. In fact, Paul Cohen showed in the 1960's that CH is "undecidable" using ZFC, that is, it can be neither proven nor refuted in ZFC set theory (actually, Gödel had already shown that CH could not be disproved from ZFC), and the acceptance of CH is actually a somewhat controversial matter in mathematics today. As a historical remark, we mention that Cantor published an incorrect "proof" of the Continuum Hypothesis in 1884 though the mistake was quickly discovered, and he spent part of his life in vain trying to prove an undecidable result! In fact, Hilbert also published a faulty proof of the Continuum Hypothesis early in the century.

For a further perspective on CH and to explain the notation "$\aleph_0$", given a set A, we shall write $|A| = \aleph_{i+1}$ if $\aleph_i < |A|$ and for each set B, if $\aleph_i \leq |B| \leq |A|$, then either $|B| = \aleph_i$ or $|B| = |A|$. Thus, $\aleph_{i+1}$ is the "next cardinality after $\aleph_i$". According to Cantor's Theorem 7.5.5, we have

$$\begin{aligned}
\aleph_0 &= |\mathbb{N}| \\
\aleph_1 &\leq |\mathcal{P}(\mathbb{N})| = \mathbf{c} \\
\aleph_2 &\leq |\mathcal{P}(\mathcal{P}(\mathbb{N}))| \\
&\vdots
\end{aligned}$$

and the Continuum Hypothesis is simply the equation $\aleph_1 = \mathbf{c}$.

We close this section with a result of Zermelo's to the effect that given two sets A and B, either there is an injection from A to B or there is an injection from B to A (or both, in which case A and B are equipotent by the Cantor-Schroeder-Bernstein Theorem).

7.5.6 THEOREM [Zermelo's Trichotomy] *Given two sets A and B, exactly one of the following possibilities holds: $|A| < |B|$, $|A| = |B|$, or $|B| < |A|$.*

Proof As was mentioned in the introduction to this section, the proof will rely on the Zorn Lemma formulation of the Axiom of Choice, which says (we recall from the optional §5.11) that given a non-empty poset so that each linearly ordered subset has an upper bound in the poset, there is a maximal element in the poset.

The poset to which we shall apply Zorn's Lemma is defined as follows. The underlying set X on which we shall define a partial ordering is the set X of all injections $f' : A' \to B$, where $A' \subseteq A$, so X is evidently non-empty. We shall write $f_1 \leq f_2$ for two injections $f_1 : A_1 \to B$ and $f_2 : A_2 \to B$ if $A_1 \subseteq A_2$ and $f_1 = f_2|_{A_1}$; in other words, regarding the functions $f_1 \subseteq A_1 \times B \subseteq A \times B$ and $f_2 \subseteq A_2 \times B \subseteq A \times B$ formally as relations, we have $f_1 = f_1 \cap f_2$.

The binary relation $\leq$ on the set X is reflexive, anti-symmetric, and transitive (as we leave it for the reader to check), so $\leq$ is in fact a partial ordering on X, as required.

Next we must check the hypotheses of Zorn's Lemma, namely, we must show that every linearly ordered subset of X has an upper bound in X. To this end, suppose that $L \subseteq X$ is a linearly ordered subset. To illuminate the upcoming argument, we shall regard L as a family

$$L = \{f_\ell \in B^{A_\ell} : \ell \in \mathcal{L}\}$$

of functions indexed by some linearly ordered set $\mathcal{L}$, where $f_k, f_\ell \in L$ satisfy $f_k \leq f_\ell$ in X if and only if $k \leq \ell$ in $\mathcal{L}$. Our passage here from L to $\mathcal{L}$ is simply a point of notation.

If $\mathcal{L} = \emptyset$, then any element of X is an upper bound for L, so we assume that $\mathcal{L} \neq \emptyset$ and choose some $\ell_0 \in \mathcal{L}$. Define the subset

$$A_* = \cup\{A_\ell \subseteq A : \ell \in \mathcal{L} \text{ and } \ell_0 \leq \ell\} \subseteq A$$

and the corresponding relation

$$f = \{(a,b) \in A \times B : a \in A_* \text{ and } f_\ell(a) = b \text{ for some } \ell \in \mathcal{L} \text{ where } \ell_0 \leq \ell\}.$$

We first claim that $f \subseteq A \times B$ is actually a function, and to this end choose some $a \in A_*$, so $a \in A_\ell$ for some $\ell \in \mathcal{L}$ with $\ell_0 \leq \ell$. Since f_ℓ is a function, there is some ordered pair $(a,b) \in f_\ell$, and there is therefore some ordered pair $(a,b) \in f$. If (a,b') is another ordered pair in f, then there is some $\ell' \in \mathcal{L}$ so that $(a,b') \in f_{\ell'}$, and we may assume that in fact $\ell \leq \ell'$ in $\mathcal{L}$. Since $f_\ell = f_{\ell'}|_{A_\ell}$ and $A_\ell \subseteq A_{\ell'}$ by hypothesis (namely, L is a linearly ordered subset), we must have $b = b'$, so $f : A_* \to B$ is indeed a function.

We furthermore claim that f is an injection, and in the contrary case, there are $a, a' \in A_*$ so that $f(a) = f(a')$. Arguing as in the previous paragraph, there are $\ell, \ell' \in \mathcal{L}$ so that $f_\ell(a) = f_{\ell'}(a')$, and we may assume that in fact $\ell \leq \ell'$ in $\mathcal{L}$. Again using that L is linearly ordered, we have $f_\ell = f_{\ell'}|_{A_\ell}$, so $f_{\ell'}(a') = f_{\ell'}(a')$. This contradicts that $f_{\ell'}$ is an injection, so $f : A_* \to B$ is indeed an injection.

Thus, f is actually an element of X and is by construction an upper bound to L. Since the linearly ordered subset L was arbitrary, we conclude that the hypotheses of Zorn's Lemma are indeed satisfied.

Applying Zorn's Lemma, we conclude that X has a maximal element and let $f_m : A_m \to B$ denote some maximal element, where $A_m \subseteq A$. Thus, f_m is an injection, and if $A_m = A$, then we have produced the required injection from A to B. If $f_m : A_m \to B$ is a surjection, then it has a right inverse $g : B \to A_m$ which is an injection, and we may compose with the inclusion $i : A_m \to A$ to produce a function $i \circ g : B \to A$ which is the required injection since it is the composition of injections.

In order to get a contradiction, assume that f_m is not surjective and A_m is a proper subset of A. We may therefore choose $a \in A - A_m$ and $b \in B$ which does not lie in the image of f_m. Define $F = f_m \cup \{(a, b)\}$ so that $F : A \cup \{a\} \to B$ is an injection and $f_m = F|_{A_m}$ by construction. Thus, $F \in X$ and $f_m \leq F$, which contradicts that f_m is a maximal element of X.

We conclude that the maximal element f_m is either surjective or $A_m = A$, so as above, there is indeed either an injection $A \to B$ or an injection $B \to A$. If both injections exist, then A and B are equipotent by the Cantor-Schroeder-Bernstein Theorem, completing the proof. *q.e.d.*

An immediate consequence of the previous result, we have

7.5.7 COROLLARY *$\leq$ is a linear ordering.*

The application of Zorn's Lemma here is a good paradigm for many such applications, and the reader will observe three basic steps above:

(1) One concocts a suitable poset and must be careful to check that the poset is non-empty.

(2) One checks the condition that linearly ordered subsets have upper bounds; this is usually the key hypothesis of Zorn's Lemma and its verification the key technical step in applications.

(3) One uses the existence of a maximal element as in the conclusion of Zorn's Lemma to finally complete the proof.

This kind of application of the Axiom of Choice as Zorn's Lemma is not uncommon in algebra, for instance, and is a basic proof technique. A first clue of the viability of this proof technique in practice is that the desired result follows from a suitably maximal element. That is, one often recognizes an opportunity for the application of Zorn's Lemma by first discovering the proof of step (3) above.

EXERCISES

7.4 COMPARING CARDINALITIES

1. Prove that if $A \subseteq B \subseteq C$ and $|A| = |C|$, then we must have $|A| = |B| = |C|$.

2. For any set X, let $\mathcal{F}(X) \subseteq \mathcal{P}(X)$ denote the collection of all finite subsets of X. Prove that X is infinite if and only if $|X| = |\mathcal{F}(X)|$.

3. Prove that if $|A| \leq |B|$ and $|B| < |C|$, then $|A| < |C|$.

4. Does there exist a set X so that $|\mathcal{P}(X)| = \aleph_0$? What about $|\mathcal{P}(X)| = c$? Prove your assertions.

5. Suppose that A and B are disjoint sets with $|A| = a$ and $|B| = b$, and define the following arithmetic operations $a+b = |A \cup B|$, $a \cdot b = |A \times B|$, $a^b = |A^B|$ of *cardinal arithmetic.*

(a) Show that $2^{\aleph_0} \cdot \aleph_0 = 2^{\aleph_0}$ and $n + \aleph_0 = \aleph_0$ for any finite cardinality n.

(b) Prove analogues for cardinal arithmetic of various standard properties of arithmetic for natural numbers.

Chapter 8

Elements of Combinatorics

In this chapter, we discuss various aspects of the field of mathematics known as "combinatorics". In essence, this field addresses the basic problem of "counting" which encompasses many different problems and results. Here we begin by calculating the sizes of various finite sets constructed using the usual set-theoretic operations of union, Cartesian product, and so on. We turn next to the consideration of more sophisticated but still basic counting problems, and we give several illustrations of a method of proof, called "combinatorial proof", in this setting. In the remainder of this chapter, we turn to a fundamental construction in combinatorics, called "recursive sequences", which provide a method of defining functions $\mathbb{N} \to A$ for any set A. Many diverse examples of this construction are given, and we finally turn our attention to several important techniques for calculating them.

8.1 FINITE CARDINALITIES

This section introduces the precise notion of the "size" of a finite set, which is called the "cardinality" of the set. We then calculate the cardinalities of various finite sets.

We say that a set A is *finite* if either $A = \emptyset$ or if there is some $n \in \mathbb{N}$ and a bijection $\{1, 2, \ldots, n\} \to A$, and we define the sets $\overline{n} = \{1, 2, \ldots, n\}$ for $n \in \mathbb{N}$ setting $\overline{0} = \emptyset$ for convenience. Thus, a set is finite if and only if there is a bijection $\overline{n} \to A$ for some $n \in \mathbb{N} \cup \{0\}$.

A basic fact about finite sets is the

8.1.1 PIGEONHOLE PRINCIPLE *Suppose that A and B are finite sets, say $\overline{m} \to A$ and $\overline{n} \to B$ are bijections, where $m, n \in \mathbb{N} \cup \{0\}$. If there is an injection $A \to B$, then we must have $m \leq n$.*

The reader may prove this independently here by induction on n (or see Lemma 7.1.1 for a complete proof). The reason for the terminology "Pigeonhole Principle" is as follows: If m pigeons alight at n pigeonholes and each pigeon lands in a different hole, then we must have $m \leq n$. Of course, this fact is "empirically obvious" from everyday experience with finite sets and pigeons.

It follows from the Pigeonhole Principle that if $f : \overline{m} \to A$ and $g : \overline{n} \to A$ are bijections, then we must have $m = n$ since $g^{-1} \circ f$ and $f^{-1} \circ g$ are bijections as the composition of such. Thus, the Pigeonhole Principle guarantees that $m \leq n$ and $n \leq m$, so in fact $m = n$. We may therefore define the *cardinality* of a finite set A to be the *unique* $n \in \mathbb{N} \cup \{0\}$ so that there is some bijection $\overline{n} \to A$, and we write $|A| = n$ in this case, so $|\overline{n}| = n$ for each $n \in \mathbb{N} \cup \{0\}$ for instance. Thus, the cardinality of a finite set is simply the number of elements in the set.

Given finite sets A and B, recall that $A \cup B$ denotes the union of A and B, $A \times B$ denotes the Cartesian product of A and B, A^B denotes the set of all functions from B to A, and $\mathcal{P}(A)$ denotes the power set of A. This section is dedicated to the calculation of the cardinalities of these various sets, and summarizing the results to be proved here, we have

$$
\begin{aligned}
|A \cup B| &= |A| + |B| - |A \cap B| \\
|A \times B| &= |A||B| \\
|A^B| &= |A|^{|B|} \\
\mathcal{P}(A) &= 2^{|A|}
\end{aligned}
$$

To begin, we prove the formula above for $|A \cup B|$, which is sometimes called the *Rule of Sum*.

8.1.2 RULE OF SUM *If A and B are finite sets, then we have $|A \cup B| = |A| + |B| - |A \cap B|$.*

Proof We first consider the special case in which $A \cap B = \emptyset$. Let $f : \overline{m} \to A$, $g : \overline{n} \to B$ be bijections and define

$$F : A \cup B \to \overline{(m+n)}$$
$$x \mapsto \begin{cases} f(x), & \text{if } x \in A; \\ g(x) + n, & \text{if } x \in B. \end{cases}$$

We leave it to the reader to produce a two-sided inverse to F and thereby conclude using Theorem 6.5.3c that F is a bijection, so that

$$|A \cup B| = |\overline{(n+m)}| = n + m = |A| + |B| = |A| + |B| - |A \cap B|,$$

as desired, where we have used $A \cap B = \emptyset$, i.e., $|A \cap B| = 0$, in the last equality.

Turning to the general case in which $A \cap B$ may be non-empty, observe that $A \cup B = A \cup (B - A)$ and $A \cap (B - A) = \emptyset$, so the previous argument applies to give

$$|A \cup B| = |A| + |B - A|.$$

Furthermore, observe that $B = (B - A) \cup (B \cap A)$ and $(B - A) \cap (B \cap A) = \emptyset$, so the previous argument again applies to give

$$|B| = |B - A| + |B \cap A|,$$

i.e., we have $|B - A| = |B| - |B \cap A|$. Substituting this expression for $|B - A|$ into the equation $|A \cup B| = |A| + |B - A|$ derived above gives the desired result. *q.e.d.*

In particular, if $A \cap B = \emptyset$, then it follows that $|A \cup B| = |A| + |B|$. This result extends by induction (and this extension is left as an exercise) to

8.1.3 COROLLARY *Suppose that A_i is a set, where $i = 1, \ldots, n$. If these sets are pairwise disjoint, i.e., if $A_i \cap A_i = \emptyset$ unless $i = j$, then we have*

$$|A_1 \cup A_2 \cup \ldots \cup A_n| \quad = \quad |A_1| + |A_2| + \cdots + |A_n|.$$

In fact, there is a result called the "Inclusion/Exclusion Principle" which gives the natural extension of the Rule of Sum itself; we shall discuss this result in Theorem 8.2.8 below.

Turning next to the cardinality of Cartesian products, we prove the formula above for $|A \times B|$, which is sometimes called the *Rule of Product.*

8.1.4 RULE OF PRODUCT *If A and B are finite sets, then we have $|A \times B| = |A||B|$.*

Proof If either A or B is empty, then so too is $A \times B$ and the asserted equality holds in this case. In general, suppose that $|A| = m$ and $|B| = n$ and let $f : (\{0\} \cup \overline{(m-1)}) \to A$ and $g : (\{0\} \cup \overline{(n-1)}) \to B$ be bijections. (It is more convenient here to use these bijections rather than the usual bijections $\overline{m} \to A$ and $\overline{n} \to B$.) Given any $j \in \{0\} \cup \overline{(nm-1)}$, we may apply the Division Algorithm 4.2.1 to write j uniquely as $j = q_j m + r_j$, where $0 \leq r_j < m$, and $0 \leq q_j < n$ since $0 \leq j < nm$ by hypothesis. We may therefore define a function

$$F : (\{0\} \cup \overline{(nm-1)}) \to A \times B$$
$$j \mapsto (f(q_j), g(r_j)),$$

and we leave it to the reader to exhibit a two-sided inverse to F. Thus, F is a bijection, so $|A \times B| = |\{0\} \cup \overline{(nm-1)}| = 1 + (nm-1) = nm = |A||B|$ where the second equality follows from the Rule of Sum. *q.e.d.*

As an application of the previous result, we finally consider the cardinality of A^B for finite sets A and B.

8.1.5 THEOREM *If A and B are finite sets, then we have $|A^B| = |A|^{|B|}$.*

Proof First notice that if $B = \emptyset$, then there is a unique function with domain B and codomain A, namely the empty function $\emptyset \subseteq A^B$, and we find $|A|^{|B|} = |A|^0 = 1$. For the other extreme case, there is no function from B to A if $A = \emptyset$ unless also $B = \emptyset$, and we have $|A|^{|B|} = 0^{|B|} = 0 = |A^B|$ if $A = \emptyset$ and $B \neq \emptyset$. Thus, if either A or B is empty, then the asserted equality does indeed hold.

In the more general case when $A, B \neq \emptyset$, let us choose bijections $f : \overline{m} \to A$ and $g : \overline{n} \to B$. Given a function $F : B \to A$, we consider the induced function

$$F' = g^{-1} \circ F \circ f : \overline{m} \to \overline{m}.$$

We may list in order the values taken by F' as in

$$T(F) = (F'(0), F'(1), \ldots, F'(n-1)) \in (\overline{n})^m = \underbrace{\overline{n} \times \cdots \times \overline{n}}_{m \text{ times}}$$

to produce a function

$$T : A^B \to (\overline{n})^m$$
$$F \mapsto T(F),$$

and we again leave it to the reader to exhibit a two-sided inverse to T. Thus, $F : A^B \to (\overline{n})^m$ is a bijection, so $|A^B| = |(\overline{n})^m|$. Finally, $|(\overline{n})^m| = n^m$ by the previous theorem, as desired. *q.e.d.*

To finish up our promised list, we next calculate the cardinality of the power set $\mathcal{P}(A)$ of a finite set A.

8.1.6 THEOREM *If A is a finite set, then we have $|\mathcal{P}(A)| = 2^{|A|}$.*

Proof For each subset $A' \subseteq A$ of A, let

$$\begin{aligned} \chi_{A'} &: A \to \{0,1\} \\ a &\mapsto \begin{cases} 0, & \text{if } a \notin A'; \\ 1, & \text{if } a \in A'; \end{cases} \end{aligned}$$

denote the characteristic map of A' and consider the function

$$\begin{aligned} I &: \mathcal{P}(A) \to (\{0,1\})^A \\ A' &\mapsto \chi_{A'}. \end{aligned}$$

Two characteristic functions χ_{A_1} and χ_{A_2} are equal as functions if and only if A_1 and A_2 are equal as sets by definition, so the mapping I is injective. Furthermore, given $F \in (\{0,1\})^A$, we define the set $A' = \{a \in A : F(a) = 1\}$ so that $F = \chi_{A'}$, whence I is also surjective. Thus $I : \mathcal{P}(A) \to (\{0,1\})^A$ is a bijection, and $|(\{0,1\})^A| = 2^{|A|}$ by the previous theorem, so $|\mathcal{P}(A)| = 2^{|A|}$. *q.e.d.*

EXERCISES

8.1 FINITE CARDINALITIES

1. Prove the following strong version of the Pigeonhole Principle: Suppose that $q_1, \ldots, q_n$ are natural numbers, where $n \geq 1$. If $q_1 + \cdots + q_n - n + 1$ objects are put into n boxes, then box i must contain at least q_i elements for some $i = 1, \ldots, n$.

2. Let $\mathcal{A}$ be a collection of distinct pairwise disjoint subsets of $\overline{n}$. Prove that then $|\mathcal{A}| \leq 2^{n-1}$. Furthermore, prove that there is some collection $\mathcal{B} \supseteq \mathcal{A}$ with the given properties where $|\mathcal{B}| = 2^{n-1}$.

3. Prove that if $n+1$ distinct integers are chosen from among $\{1, 2, \ldots, 2n\}$, then some pair must be relatively prime.

4. Let $A \in \mathcal{P}(X)$. How many functions f in X^X fix A "setwise" (i.e. $f(A) = A$)? How many functions f in X^X fix A "pointwise" (i.e., $f(a) = a$ for each $a \in A$)? Prove your assertions.

5. If $A \in \mathcal{P}(X)$ and $B \in \mathcal{P}(Y)$, then how many functions in Y^X map A into B (i.e., $f(A) \subseteq B$)? Prove your assertions.

8.2 COMBINATORIAL PROOFS

We begin with certain standard counting problems associated with a fixed finite set. Specifically, suppose that r is some integer satisfying $0 \leq r \leq n$, and define the sets

$$\begin{aligned} P_r(n) &= \{(a_1, a_2, \ldots, a_r) \in (\overline{n})^r : a_i \neq a_j \text{ unless } i = j\} \\ C_r(n) &= \{A \subseteq \overline{n} \ : \ |A| = r\}. \end{aligned}$$

Thus, $C_r(n)$ is the set whose elements are the subsets of $\overline{n}$ which have cardinality exactly r, and $P_r(n)$ is the set of all r-tuples of distinct elements of $\overline{n}$. Furthermore, there is a natural one-to-one correspondence between the set $P_r(n)$ and the set of all injections $\overline{r} \to \overline{n}$, as we leave it for the reader to check. Our primary goal for the remainder of this section is to compute the cardinalities of $P_r(n)$ and $C_r(n)$.

One may think of $C_r(n)$ and $P_r(n)$ as arising from an explicit procedure as follows. Imagine the finite set $\overline{n} = \{1, 2, \ldots, n\}$ as a "bag" containing its elements $1, 2, \ldots, n$. We might choose an element a_1 from this bag, remove a_1 from the bag, and put a_1 in a cup (which is not in the bag) labeled "1". Having performed this choice, removal and placement in a labeled cup, we might again choose an element a_2 from the bag (which at this point contains $\overline{n} - \{a_1\}$), remove it, and place it in a cup labeled "2". Continuing in this way, we successively make r such choices, removals, and placements, so as to produce an element $(a_1, \ldots, a_r)$ of $P_r(n)$. Thus, $P_r(n)$ is simply the collection of possible outcomes of our procedure, and an element of $P_r(n)$ is called a *permutation of n objects taken r at a time.*

For another such procedure, we might begin with two bags, one of which contains $1, 2, \ldots, n$ and the other of which is empty. Choosing a_1 from the first bag, we deposit it in the second bag. We may then successively make r such choices, removals and deposits in the second bag so as to produce a subset $A \subseteq \overline{n}$ where $|A| = r$ (namely, the elements of A are the elements of the second bag). Thus, $C_r(n)$ is simply the collection of possible outcomes of this second procedure, and an element of $C_r(n)$ is called a *combination of n objects taken r at a time.*

We mention parenthetically that each of the procedures above is called a *selection without replacement* for the obvious reason (that we do not replace an element into the original bag after choosing it). In contrast, there are *selections with replacement*; in the corresponding procedure for permutations, for instance, we simply r times select an element of $\overline{n}$, and the various possible outcomes are evidently in one-to-one correspondence with the set $(\overline{n})^{\overline{r}}$, which has cardinality n^r.

Our interest here is in the cardinalities of the sets $P_r(n)$ and $C_r(n)$, and we define

$$\begin{bmatrix} n \\ r \end{bmatrix} = |\ P_r(n)\ | \quad \text{and} \quad \binom{n}{r} = |\ C_r(n)\ |.$$

The notation $\binom{n}{r}$ is entirely standard in mathematics while our notation $\left[\begin{smallmatrix} n \\ r \end{smallmatrix}\right]$ is not. The collection of symbols "$\binom{n}{r}$" is read in English as "n choose r" and is called a *binomial symbol* or a *binomial coefficient* for reasons which will become clear.

8.2.1 THEOREM *The number of permutations of n objects taken r at a time is given by*

$$\begin{bmatrix} n \\ r \end{bmatrix} = \frac{n!}{(n-r)!}.$$

Proof In the trivial case $r = 0$, we have $\left[\begin{smallmatrix} n \\ r \end{smallmatrix}\right] = 1$, as required, since there is only one empty sequence. In case $r > 0$, there are n possible choices for the selection of the first element a_1 of r elements of $\overline{n}$. Since the selection is without replacement, the second choice can be any element of the set $\overline{n} - \{a_1\}$, which has cardinality $n - 1$ as the reader could prove here by induction on n (or see Lemma 7.1.5 for a complete proof). Similarly (that is, by induction) there are $n - i + 1$ possible values for the selection of the ith object, and so

$$\begin{bmatrix} n \\ r \end{bmatrix} = n(n-1)\cdots(n-r+1) = \frac{n!}{(n-r)!}$$

by the Rule of Product. *q.e.d.*

8.2.2 COROLLARY *There are exactly $n!$ permutations of $\overline{n}$, that is, bijections from $\overline{n}$ to $\overline{n}$.*

Proof We have already remarked that there is a natural one-to-one correspondence between $P_n(n)$ and the collection of all injections from $\overline{n} \to \overline{n}$. Since an injection $\overline{n} \to \overline{n}$ is necessarily a bijection as the reader may prove here using the Pigeonhole Principle (or see Corollary 7.1.8 for a complete proof), there are exactly $|P_n(n)|$ such bijections, and $|P_n(n)| = n!$ by the previous theorem. *q.e.d.*

8.2.3 THEOREM *The binomial coefficients are given by*

$$\binom{n}{r} = \frac{n!}{r!(n-r)!}.$$

Proof An ordered list of r elements of $\overline{n}$ can be produced by first choosing r unordered elements, i.e., by choosing a subset $A \subseteq \overline{n}$ of cardinality r, and then choosing an ordering on the set A. It follows from the Rule of Product that

$$\begin{bmatrix} n \\ r \end{bmatrix} = \binom{n}{r} \cdot r!$$

since there are $r!$ orderings of A (that is, bijections of a set of cardinality r with itself). Thus,

$$\binom{n}{r} = \frac{1}{r!}\begin{bmatrix} n \\ r \end{bmatrix} = \frac{n!}{r!(n-r)!},$$

where the second equality follows from the previous theorem. *q.e.d.*

The explicit interpretation of $C_r(n)$ and $P_r(n)$ as outcomes of certain procedures allows an elegant and powerful proof technique which we finally discuss. For a simple example of this proof technique, which is called a "combinatorial proof", consider the identity

$$\binom{n}{r} = \binom{n}{n-r}$$

which expresses the simple fact that a subset of $\overline{n}$ of cardinality r is uniquely determined by its complement, which has cardinality $n - r$. In light of the interpretation of combinations discussed above, this observation gives an immediate proof of the previous identity. Of course, one might also prove the result via explicit calculation using Theorem 8.2.3, and we leave this as an exercise.

A proof such as this, in which a fact or observation about counting proves for instance a numerical identity, is called a "combinatorial proof" of the identity.

For another example of a combinatorial proof, we have

8.2.4 THEOREM *For every integer $n \geq 0$, we have*

$$\sum_{r=0}^{n} \binom{n}{r} = 2^n.$$

Proof Notice that an arbitrary subset A of $\overline{n}$ has cardinality $r = |A|$ for some unique $r = 0, 1, \ldots, n$, so $\mathcal{P}(A)$ breaks up into the disjoint union of sets $C_r(n)$ for $0 \leq r \leq n$. The set $C_r(n)$ has cardinality $\binom{n}{r}$, so $\mathcal{P}(A)$ therefore has cardinality $\sum_{r=0}^{n} \binom{n}{r}$ by Theorem 8.2.3. On the other hand, $\mathcal{P}(A)$ has cardinality 2^n by Theorem 8.1.6, and equating these two expressions proves the identity. *q.e.d.*

Our next example of a combinatorial proof relies on the elementary dichotomy that given a subset and an element of a set, the subset either does or does not contain the element. Such elementary observations can often lead to non-trivial numerical facts.

8.2.5 THEOREM *For any integers $n \geq 0$ and $r \in \{0, 1, \ldots, n\}$, we have the identity*

$$\binom{n+1}{r} = \binom{n}{r-1} + \binom{n}{r}.$$

Proof Consider the element $(n+1) \in \overline{(n+1)}$. Any subset of $\overline{(n+1)}$ either contains $n + 1$ or does not, so $C_r(n + 1)$ breaks up into the disjoint union of

$$S_1 = \{A \subseteq \overline{(n+1)} : (n+1) \in A\} \text{ and } S_2 = \{A \subseteq \overline{(n+1)} : (n+1) \notin A\}.$$

Thus, $|C_r(n + 1)| = |S_1| + |S_2|$ by the Rule of Sum. If $A \in S_1$, then $A - \{n + 1\}$ is a subset of $\overline{n}$ of cardinality $r - 1$, and S_1 is in bijective correspondence with $C_{r-1}(n)$, which has cardinality $\binom{n}{r-1}$ by Theorem 8.2.3. If $A \in S_2$, then A is in fact a subset of $\overline{n}$ of cardinality r and corresponds to an element of $C_r(n)$, which has cardinality $\binom{n}{r}$ again by Theorem 8.2.3. Thus, we have $\binom{n+1}{r} = |C_r(n+1)| = |S_1| + |S_2| = \binom{n}{r-1} + \binom{n}{r}$, as desired. *q.e.d.*

Our next example of a combinatorial proof is the famous *Binomial Theorem*:

8.2.6 BINOMIAL THEOREM *If $x, y \in \mathbb{R}$ and $n \in \mathbb{N} \cup \{0\}$, then we have*

$$(x+y)^n = \sum_{r=0}^{n} \binom{n}{r} x^r y^{n-r}.$$

Proof Write $(x+y)^n$ as the product

$$(x+y)^n = \underbrace{(x+y)(x+y)\cdots(x+y)}_{n \text{ times}},$$

and imagine multiplying out this product. Each factor $x+y$ contributes either a factor x or a factor y to each term, and we may therefore write

$$(*) \qquad (x+y)^n = \sum_{\epsilon \in (\{0,1\})^{(\{1,2,\ldots,n\})}} \prod_{i=1}^{n} x^{\epsilon(i)} y^{1-\epsilon(i)},$$

where the sum is over all functions $\epsilon : \{1,2,\ldots,n\} \to \{0,1\}$, and $\epsilon(i) = 1$ if and only if the ith factor $x+y$ contributes a factor x. Given a term in this sum corresponding to $\epsilon \in (\{0,1\})^{(\{1,2,\ldots,n\})}$, observe that

$$\prod_{i=1}^{n} x^{\epsilon(i)} y^{1-\epsilon(i)} = x^r y^{n-r},$$

where $r = \epsilon(1) + \epsilon(2) + \cdots + \epsilon(n)$.

We finally consider how many terms in the sum can give rise to a term $x^r y^{n-r}$ and claim that there are $\binom{n}{r}$ such terms. Indeed, we may choose r elements of $\{1,2,\ldots,n\}$ on which ϵ takes the value one and set the other values of ϵ to zero in order to uniquely determine a corresponding term in the sum. Collecting like terms in equation $(*)$ therefore yields the asserted formula. *q.e.d.*

Owing to the importance of the Binomial Theorem, we urge the reader to give a direct inductive proof using Theorem 8.2.3.

Another formulation of the Binomial Theorem (which arises immediately from this result by setting $x = 1$) is as follows.

8.2.7 COROLLARY *If $y \in \mathbb{R}$ and $n \in \mathbb{N} \cup \{0\}$, then we have*

$$(1+y)^n = \sum_{r=0}^{n} \binom{n}{r} y^r.$$

We turn finally to the promised generalization of the Rule of Sum and suppose that $A_1, A_2, \ldots, A_n$ is a collection of sets where $n \geq 1$. For each $r = 1, 2, \ldots, n$, let

μ_r denote the sum of the cardinalities of all possible intersections of r sets chosen without replacement from among $A_1, A_2, \ldots, A_n$. For instance, when $n = 2$, we have

$$\mu_1 = |A_1| + |A_2|,$$
$$\mu_2 = |A_1 \cap A_2|,$$

and if $n = 3$, then we have

$$\begin{aligned}\mu_1 &= |A_1| + |A_2| + |A_3|,\\ \mu_2 &= |A_1 \cap A_2| + |A_1 \cap A_3| + |A_2 \cap A_3|,\\ \mu_3 &= |A_1 \cap A_2 \cap A_3|.\end{aligned}$$

These numbers μ_r which depend on the collection of sets are clearly germane to an extension of the Rule of Sum, which reads in this notation simply

$$|A_1 \cup A_2| = \mu_1 - \mu_2.$$

In the next case of three sets, we find

$$\begin{aligned}|A_1 \cup A_2 \cup A_3| &= |A_1 \cup (A_2 \cup A_3)|\\ &= |A_1| + |A_2 \cup A_3| - |A_1 \cap (A_2 \cup A_3)|\\ &= |A_1| + |A_2| + |A_3| - |A_2 \cap A_3|\\ &\quad - |(A_1 \cap A_2) \cup (A_1 \cap A_3)|\\ &= |A_1| + |A_2| + |A_3| - |A_2 \cap A_3|\\ &\quad - |A_1 \cap A_2| - |A_1 \cap A_3|\\ &\quad + |(A_1 \cap A_2) \cap (A_1 \cap A_3)|\\ &= |A_1| + |A_2| + |A_3|\\ &\quad - |A_2 \cap A_3| - |A_1 \cap A_2| - |A_1 \cap A_3|\\ &\quad + |A_1 \cap A_2 \cap A_3|\\ &= \mu_1 - \mu_2 + \mu_3.\end{aligned}$$

Turning to the general case, we have

8.2.8 INCLUSION/EXCLUSION PRINCIPLE *Given a collection A_1, A_2, $\ldots A_n$ of finite sets, for each $r = 1, 2, \ldots, n$, let μ_r denote the sum of the cardinalities of all possible intersections of r sets chosen without replacement from among $A_1, A_2, \ldots, A_n$. Then we have*

$$(*) \qquad |A_1 \cup A_2 \cup \cdots \cup A_n| = \sum_{r=1}^{n} (-1)^{r+1} \mu_r.$$

Proof Let a be an element of $\cup_{r=1}^{n} A_r$ and suppose that a belongs to k of the sets $A_1, A_2, \ldots, A_n$. Define the set $B = \{j \in \mathbb{N} : a \in A_j\}$, so B is the collection of subscripts of sets which contain a. By our assumption on a, B contains exactly $k = \binom{k}{1}$ elements, and the first term μ_1 "counts" the element a exactly k times. The second term $-\mu_2$ "counts" a with a negative sign exactly once for each doubleton in B, and this number of doubletons is $\binom{k}{2}$. The third term adds back $\binom{k}{3}$, the fourth term subtracts $\binom{k}{4}$, and so forth. Thus, the number of times the element a is counted in $\sum_{r=1}^{n}(-1)^{r+1}\mu_r$ is

$$\binom{k}{1} - \binom{k}{2} + \binom{k}{3} - \cdots + (-1)^{k+1}\binom{k}{k},$$

and it remains only to check that this sum has value 1, for a is "counted" exactly once as an element of $\cup_{r=1}^{n} A_i$. To prove that the sum has value 1, we apply the Binomial Theorem as follows:

$$\begin{aligned}
0 &= (1 + (-1))^k \\
&= \binom{k}{0}(1)^k(-1)^0 + \binom{k}{1}(1)^{k-1}(-1)^1 + \cdots + \binom{k}{k}(1)^0(-1)^k \\
&= 1 - \binom{k}{1} + \binom{k}{2} - \cdots + (-1)^k\binom{k}{k}.
\end{aligned}$$

Subtracting 1 from both sides and multiplying through by -1, we find

$$1 = \binom{k}{1} - \binom{k}{2} + \cdots + (-1)^{k+1}\binom{k}{k},$$

as desired. *q.e.d.*

■ **Example 8.2.1** We compute here how many natural number less than or equal to 10,000 are divisible by 7,13, or 17. To this end, let A, B, and C denote the collection of all natural numbers less than or equal to 10,000 which are divisible by 7,13, and 17 respectively. Thus, we seek here the cardinality of $A \cup B \cup C$. By the Inclusion/Exclusion Principle, we find

$$\begin{aligned}
|A \cup B \cup C| = |A| + |B| + |C| - |A \cap B| - |A \cap C| \\
- |B \cap C| + |A \cap B \cap C|.
\end{aligned}$$

Of course, we have

$$|A| = [\frac{10,000}{7}] = 1,428, \quad |B| = [\frac{10,000}{13}] = 769, \quad |C| = [\frac{10,000}{17}] = 588,$$

where we are using the notation $[x]$ for the greatest integer less than or equal to x. Since any two of 7,13, and 17 are relatively prime, we conclude

$$|A \cap B| = [\frac{10,000}{91}] = 109, \quad |A \cap C| = [\frac{10,000}{119}] = 84, \quad |B \cap C| = [\frac{10,000}{21}] = 45.$$

91 and 17 are likewise relatively prime, and so

$$|A \cap B \cap C| = [\frac{10,000}{1,547}] = 6,$$

and we find

$$|A \cup B \cup C| = 1,428 + 769 + 588 - 109 - 84 - 45 + 6 = 2,553.$$

EXERCISES

8.2 COMBINATORIAL PROOFS

1. Describe the seclection procedure for combinations with replacement and enumerate the various outcomes.

2. Define the *multinomial coefficient* $\binom{n}{k_1 \cdots k_m}$ to be the number of ways of arranging n objects of m different sorts into a column, where there are k_i objects of type $i = 1, \ldots, m$, for $n \geq k_i \geq 0$.

(a) Prove that if $n = k_1 + \cdots + k_m$, then $\binom{n}{k_1 \cdots k_m} = \frac{n!}{k_1! \cdots k_m!}$.

(b) Prove that $\binom{n}{k_1 \cdots k_m}$ is the coefficient of $x_1^{k_1} \cdots x_m^{k_m}$ in $(x_1 + \cdots x_m)^n$.

(c) Show that if $p = \sum_{i=1}^{m} k_i$ is prime, then p divides $\binom{n}{k_1 \cdots k_m}$.

3. Give combinatorial proofs of the following identities.

(a) For any $n \geq 1$, we have $\binom{n}{0} + \binom{n}{2} + \cdots + \binom{n}{2[\frac{n}{2}]} = 2^{n-1}$, where $[\cdot]$ denotes integral part.

(b) For any $r, n \geq 0$, we have $\sum_{i=r}^{n} \binom{i}{r} = \binom{n+1}{r+1}$.

4. Prove that $\sum_{k=0}^{n} \binom{m_1}{k}\binom{m_2}{n-k} = \binom{m_1+m_2}{n}$ for any $m_1, m_2, n \in \mathbb{N}$.

5. (a) Prove the following *Theorem of Bollobás*: Let $A_1, \ldots, A_N$, $B_1, \ldots, B_N$ be subsets of $\overline{n}$ so that $A_i \cap B_j = \emptyset$ if and only if $i = j$. If $a_i = |A_i|$ and $b_i = |B_i|$, then

$$\sum_{i=1}^{N} \frac{1}{\binom{a_i+b_i}{a_i}} \leq 1.$$

[HINT: If π is a permutation of $\overline{n}$, we say that that π "contains A before B" if all elements of A occur in π before all elements of B, where we specify π by enumerating the values $\pi(1), \pi(2), \ldots, \pi(n)$ in this order. Show that there are

$$\binom{n}{a_i + b_i} a_i! b_i! (n - a_i - b_i)! = n! / \binom{a_i + b_i}{a_i}$$

permutations of $\overline{n}$ containing A before B, and to complete the proof, finally sum on i.]

(b) Define an *anti-chain* or a *clutter* to be a collection $\mathcal{A} = \{A_i\}$ of sets so that A_i is not a subset of A_j whenever $i \neq j$. Derive from Bollobás' Theorem above *Sperner's Theorem* that if $\mathcal{A} \subseteq \mathcal{P}(\overline{n})$ is a clutter then $|\mathcal{A}| \leq \binom{n}{[n/2]}$, where $[\cdot]$ denotes the integral part.

8.3 *RECURSIVELY DEFINED SEQUENCES*

A "sequence" is simply a function whose domain is $\mathbb{N}$, and a method of defining sequences called "recursion" is described here. Numerous examples of recursive definitions are discussed, and inductive arguments turn out to be a natural technique for proving basic results about recursively defined sequences.

If A is a set, then a *sequence in* A is simply a function $x : \mathbb{N} \to A$. As a standard notational convention, we shall typically write the values of x as $x_n = x(n) \in A$ for $n \in \mathbb{N}$, and one thereby thinks of a sequence as a collection of elements $x_n \in A$ indexed by $n \in \mathbb{N}$. Actually, it is convenient to allow some flexibility in the index set, regarding a function $\mathbb{N} \cup \{0\} \to A$, $\mathbb{Z} \to A$, or even $\mathbb{N}^2 \to A$ also as sequences.

For instance, the function

$$x : \mathbb{N} \to \mathbb{N}$$
$$n \mapsto n^2$$

is a sequence in $\mathbb{N}$, where $x_n = x(n) = n^2$ for each $n \in \mathbb{N}$. For another example, let $\mathbb{Z}[t]$ denote the collection of all polynomials in a variable t with coefficients in $\mathbb{Z}$, so for instance, we have $2, t+2, 3t^2 - 2t - 1 \in \mathbb{Z}[t]$. Define a sequence of polynomials

$$x : \mathbb{N} \to \mathbb{Z}[t]$$
$$n \mapsto (1+t)^n$$

so that $x_n = x(n) = (1+t)^n$.

There is a powerful method of defining sequences $x : B \to A$ (where $B = \mathbb{N}$ or $B = \mathbb{N}^2$ for instance) which depends on an inductive definition of the function $x \subseteq B \times A$ as a relation. As usual for an inductive definition, this method depends upon two pieces of data, as follows:

Basis Step: The specification of a subset $B_0 \subseteq B$ and a relation on it

$$R_0 = \{(b, a_b) \in B \times A : b \in B_0\} \subseteq B \times A.$$

Induction Step: A method for each $i \in \mathbb{N}$ of constructing a subset $B_{i+1} \subseteq B$ from the subset $B_i \subseteq B$ together with a method of constructing a collection

$$R_{i+1} = \{(b, a_b) \in B \times A : b \in B_{i+1}\} \subseteq B \times A$$

of ordered pairs whose first entries lie in B_{i+1}.

Actually, in the current setting of defining binary relations and functions from B to A, we modify our previous terminology and refer to such an inductive definition as a *recursive definition* or as a *recurrence system.* We also refer to the basis clause as the *initial condition* and to the induction clause as the *recurrence relation* or simply as the *recursion* of the recursive definition.

As usual, this inductive definition determines a binary relation

$$R = \cup_{i \geq 0} R_i \subseteq B \times A$$

from B to A, but the relation R will not necessarily be a *function* from B to A (and hence a sequence) without some further requirements. Namely, we must have

$$\forall b \in B[\exists! a \in A((b, a) \in R)]$$

by definition in order that $R \subseteq B \times A$ be a function. In particular, we must have $B = \cup_{i \geq 0} B_i$ in order that the putative function is defined on all of B.

In any given example of a recursive definition of a sequence, one can usually check directly by induction that the defined relation is indeed a function, and we next give several illustrations of this.

■ **Example 8.3.1** Recursively define the relation $p = \cup_{i \geq 0} p_i \subseteq (\mathbb{N} \cup \{0\}) \times \mathbb{N}$ as follows

Initial Condition: Define $B_0 = \{0\} \subseteq \mathbb{N}$ and set $p_0 = \{(0, 1)\}$.

Recurrence Relation: For each $n \in \mathbb{N}$, set $B_n = B_{n-1} \cup \{n\}$ and define

$$p_n = p_{n-1} \cup \{(n, nb_{n-1}) \in (\mathbb{N} \cup \{0\}) \times \mathbb{N} : (n-1, b_{n-1}) \in p_{n-1} \}.$$

In fact, the syntax above of presenting first the initial condition and then the recurrence relation is completely standard, and one typically omits the explicit labels "Initial Condition" and "Recurrence Relation" in practice.

Let us prove directly that the relation p defined recursively here is actually a function $p : (\mathbb{N} \cup \{0\}) \to \mathbb{N}$. First observe that $B = \cup_{i \geq 0} B_i$ is indeed all of $\mathbb{N} \cup \{0\}$ as one proves easily by induction over $\mathbb{N} \cup \{0\}$ (indeed, we find that $n \in B_n$ for each n). Next, we claim that given $n \in (\mathbb{N} \cup \{0\})$, there is exactly one ordered pair in p whose first entry is n, that is, $p_n - p_{n-1}$ is a singleton, as one again proves easily by induction over $n \in (\mathbb{N} \cup \{0\})$. It follows that $p : (\mathbb{N} \cup \{0\}) \to \mathbb{N}$ is indeed a function (that is, a sequence in $\mathbb{N}$), and of course $p(n) = n!$ as is again easily proved by induction.

Thus, elementary inductive proofs are evidently useful in understanding recursive definitions, for instance, in showing that a recursively defined relation is actually a function. Such inductive proofs are "standard" from a sufficiently enlightened point of view, and are often omitted from the recursive definition of a sequence in practice. For instance, a suitable recursive definition of the sequence p_n described above would be simply

$$p_0 = 1,$$

$$p_n = np_{n-1} \text{ for } n \geq 1,$$

and it is tacitly left for the reader to check by induction that this is a legitimate definition of a sequence. This somewhat cavalier treatment of a recursive definition of a function can be problematic as we next see.

■ **Example 8.3.2** Consider the following recursive "definition" of a function $f : \mathbb{N} \to \mathbb{N}$

Define $f(1) = 1$.

For each $m, n \in \mathbb{N}$, define $f(m+n) = f(m) + 2f(n)$.

Assuming this definition is valid (which it is not!), the recursion relation gives $f(2) = f(1) + 2f(1) = 1 + 2 = 3$. Again applying the recursion relation, we find

$$\begin{aligned} f(3) &= f(1+2) = f(1) + 2f(2) = 1 + 6 = 7 \\ &\neq 5 = 3 + 2 = f(2) + 2f(1) = f(2+1) = f(3), \end{aligned}$$

so there can be no function $f : \mathbb{N} \to \mathbb{N}$ satisfying our cavalier recursive definition.

In spite of the fact that one must apparently exercise care with cavalier recursive definitions of sequences in light of the previous example, the conventional way of recursively defining a sequence is actually with just such a cavalier definition! We shall conform to this practice in the sequel, but we first digress briefly to recall a general theorem which can often be applied to prove that a given recursive definition legitimately defines a sequence.

Recall the Recursion Theorem 3.7.4, which states that given a function $f : A \to A$ and an element $a \in A$, there is a unique function $x : \mathbb{N} \to A$ so that $x_1 = a$ and $x_{n+1} = f(x_n)$ for all $n \in \mathbb{N}$. We proved the Recursion Theorem in the optional §3.7, and the reader might either recall the proof or supply her or his own independently here. The Recursion Theorem can be used as a tool for recursively constructing sequences, where the choice of $a \in A$ corresponds to the initial condition and the specification of the function $f : A \to A$ corresponds to the recursion relation. For instance, take $A = \mathbb{R}$ and define $f(t) = t^2$ for each $t \in \mathbb{R}$. Choosing $a \in \mathbb{R}$, we may construct the sequence $x_n = a^{2^n}$. For another example, we again take $a \in A = \mathbb{R}$ this time defining $f(t) = bt$, where $b \in \mathbb{R}$ is some fixed real number. The Recursion Theorem thus produces the sequence $x_n = ab^{n-1}$.

With some chicanery, one can often use the Recursion Theorem to prove that a cavalier recursive definition of a function is legitimate, and we next give an example of this.

■ **Example 8.3.3** We wish to recursively define the sequence $p_n = n!$ for $n \in \mathbb{N}$ using the Recursion Theorem and to this end define the function

$$\begin{aligned} f : \mathbb{N}^2 &\to \mathbb{N}^2 \\ (n,p) &\mapsto (n+1, np) \end{aligned}$$

taking $(1,1) \in \mathbb{N}^2$ as initial condition. Thus, for instance, we have

$$f(1,1) = (2,1), \quad f(2,1) = (3,2), \quad f(3,2) = (4,6), \quad \ldots,$$

and by the Recursion Theorem, there is a sequence

$$\begin{aligned} x : \mathbb{N} &\to \mathbb{N}^2 \\ n &\mapsto (n+1, n!). \end{aligned}$$

The composition $p = \pi_2 \circ x : \mathbb{N} \to \mathbb{N}$ of x with the projection $\pi_2 : \mathbb{N}^2 \to \mathbb{N}$ onto the second factor, gives the desired sequence $p_n = n!$ for $n \geq 1$ as the reader should verify by induction, and we finally set $p_0 = 1$.

In fact, the Recursion Theorem is just the tip of an iceberg, called "Recursion Theory" in mathematical logic, in the sense that there are many more general results which guarantee that a given recursive definition is a legitimate definition of a function. We shall not undertake a serious discussion of such results here. Rather, as was mentioned above, we shall liberally give cavalier recursive definitions and leave it tacitly for the reader to check by induction (or check by applying the Recursion Theorem with some chicanery) that a given recursive definition of a sequence is legitimate.

We turn next to some typical examples of recursive definitions of sequences.

■ **Example 8.3.4** We recursively define here a function $\mathbb{N} \to \mathbb{N}$ called the *double factorial function*, and the value $n!! \in \mathbb{N}$ of $n \in \mathbb{N}$ under this function is given by the recursion

$$1!! = 1 \text{ and } 2!! = 2$$

$$n!! = n[(n-2)!!] \text{ for } n \geq 3$$

We have, for instance

$$\begin{aligned}3!! &= 3 \cdot 1 = 3\\ 4!! &= 4 \cdot 2 = 8\\ 5!! &= 5 \cdot 3 \cdot 1 = 15\\ 6!! &= 6 \cdot 4 \cdot 2 = 48\\ 7!! &= 7 \cdot 5 \cdot 3 \cdot 1 = 245\\ &\vdots\end{aligned}$$

so the double factorial is closely related to the factorial.

■ **Example 8.3.5** Define the *Fibonacci numbers* $f_n \in \mathbb{N} \cup \{0\}$ for $n \geq 0$ using the recursion relation $f_n = f_{n-1} + f_{n-2}$ and the initial conditions $f_0 = 0, f_1 = 1$. We have discussed the Fibonacci sequence before in Example 4.7.3. Notice that the initial conditions here require the two specifications $f_0 = 0$ and $f_1 = 1$. To emphasize that the initial condition itself is an important part of a recursive definition, we define also the *Lucas numbers* $\ell_n \in \mathbb{N}$ for $n \in \mathbb{N} \cup \{0\}$ using the same recursion relation $\ell_n = \ell_{n-1} + \ell_{n-2}$ with the alternate initial conditions $\ell_0 = 2$ and $\ell_1 = 1$. For instance, the first several Fibonacci numbers are

$$0, 1, 1, 2, 3, 5, 8, 13, 21, 34, \cdots,$$

and the first several Lucas numbers are

$$2, 1, 3, 4, 7, 11, 18, 29, 47, 76, \cdots.$$

■ **Example 8.3.6** Specify numbers $a, b \in \mathbb{R}$, and recursively define the *geometric progression* $g_n \in \mathbb{R}$ for $n \in \mathbb{N} \cup \{0\}$ using the recursion relation $g_{n+1} = g_n + b$ with the initial condition $g_0 = a$. Thus, $g_n = a + nb$ as the reader can verify by induction.

■ **Example 8.3.7** Recursively define the *harmonic numbers* $h_n \in \mathbb{Q}$ for $n \in \mathbb{N} \cup \{0\}$ by setting $h_0 = 0$ and $h_{n+1} = h_n + 1/(n+1)$ for $n \geq 1$. We have for instance

$$h_1 = 1, \quad h_2 = \frac{3}{2}, \quad h_3 = \frac{11}{6}, \quad h_4 = \frac{50}{24}, \ldots,$$

and in general

$$h_n = 1 + \frac{1}{2} + \frac{1}{3} + \cdots + \frac{1}{n} = \sum_{1 \leq k \leq n} \frac{1}{k}, \quad \text{for } n \geq 0,$$

as the reader may easily prove by induction.

In all of our examples so far, the sequences have had domain $\mathbb{N}$ or $\mathbb{N} \cup \{0\}$, and we next extend our horizons with some examples of sequences $x : \mathbb{N}^2 \to A$ for various sets A. Adopting notation as before, we shall often write $x_{mn} = x(m,n) \in A$ for $m,n \in \mathbb{N}$ and regard the sequence x as a collection of elements of A indexed by $\mathbb{N}^2$. One uses induction or the Recursion Theorem and generalizations of it as before to define x by specifying initial conditions (that is, some finite collection of values of x) and a recursion relation (that is, an expression for values of x in terms of previously defined values) as we see in the next two examples.

■ **Example 8.3.8** We have discussed the binomial coefficients $x_{nk} = \binom{n}{k}$ in §8.2. Pascal gave the following recursive definition:

$$\binom{n}{0} = 1 = \binom{n}{n} \text{ for each } n \geq 0$$

$$\binom{n}{k} = \binom{n-1}{k-1} + \binom{n-1}{k} \text{ for } n \geq 1 \text{ and } 0 < k < n$$

where the recursion relation was proved before as Theorem 8.2.5.

If A and B are finite sets with $|A| = m \leq |B| = n$, then by Theorem 8.2.1, there are $n!/(n-m)!$ injections from A to B, and in particular there are $n!$ bijections. We next turn to the related question of counting the number of surjections between fixed finite sets and give a recursive solution.

■ **Example 8.3.9** Let A and B be finite non-empty sets with $|A| = m$ and $|B| = n$, where $m \geq n$, and recursively define the sequence s_{mn} by setting

$$s_{m1} = 1,$$

$$s_{mn} = n^m - \sum_{r=1}^{n-1} \binom{n}{r} s_{mr}, \text{ for } m \geq n > 1.$$

We claim that this gives a recursive definition for the function

$$s_{mn} = |\{f \in B^A : f \text{ is surjective}\}|,$$

and the initial condition $s_{m1} = 1$ simply reflects the fact that there is only one surjection from A to $B = \{b\}$, namely the constant function with value b.

To verify that the recursion relation is obeyed by the number of surjections, suppose that $n > 1$. The number of surjections from A to B agrees with the total number of functions from A to B (and there are n^m such functions by Theorem 8.1.5) minus the number of functions from A to B whose images are proper subsets of B. Furthermore, if $B_1 \subseteq B$ is a fixed proper subset, say with $r = |B_1| < n$ elements, then there are s_{mr} functions from A to B whose image is B_1, as follows by induction. According to Theorem 8.2.3, there are $\binom{n}{r}$ distinct subsets of B of cardinality r, so there are $\binom{n}{r} s_{mr}$ different functions from A to B whose images have cardinality r. Summing over the various possibilities for the cardinality of a proper subset, we find that there are $\sum_{r=1}^{n-1} \binom{n}{r} s_{mr}$ functions from

A to B which are not surjections, so the number of surjections does indeed satisfy the recursion relation.

■ **Example 8.3.10** Suppose that $m, n \in \mathbb{N}$ and let $\left\{ {n \atop m} \right\}$ denote the number of ways to partition a set with exactly n elements into m nonempty disjoint subsets. The number $\left\{ {n \atop m} \right\}$ is called a *Sterling number of the second kind*, and the notation given here is standard. In fact, these Sterling numbers are related to the numbers s_{mn} of surjections considered in the previous example by the formula $s_{mn} = m! \left\{ {n \atop m} \right\}$, as we leave it for the reader to verify. A recursion for the Sterling numbers therefore follows directly from Example 8.3.9, and we calculate them explicitly in Exercise 8.4.4.

The previous two examples are prototypical in the sense that many naturally occuring sequences in mathematics admit simple recursive definitions.

In order to emphasize that a sequence $x : \mathbb{N} \to A$ may have an arbitrary set A as its codomain, we turn to a final family of important examples.

■ **Example 8.3.11** Consider the set $A = \mathbb{R}[x]$ of all polynomials in a variable x with real coefficients, and define a sequence $H_n(x) \in A$ for $n \in \mathbb{N} \cup \{0\}$ with recursion

$$H_n(x) = 2xH_{n-1}(x) - 2(n-1)H_{n-2}(x) \quad \text{for } n \geq 2$$

where $H_0(x) = 1$ and $H_1(x) = 2x$. The polynomials $H_n(x) \in \mathbb{R}[x]$ are called the *Hermite polynomials*, and we have

$$H_0(x) = 1,\ H_1(x) = 2x,\ H_2(x) = 4x^2 - 2,$$
$$H_3(x) = 8x^3 - 12x,\ H_4(x) = 16x^4 - 48x^2 + 12, \cdots,$$

for instance. The Hermite polynomials are the tip of an iceberg in mathematics called sequences of "orthogonal polynomials" $P_n(x) \in \mathbb{R}[x]$, which are defined by a "three-term recursion"

$$P_n(x) = (A_n x + B_n)P_{n-1}(x) - C_n P_{n-2}(x), \quad \text{for } n \geq 2,$$

subject to some initial condition specifying $P_0(x)$ and $P_1(x)$, where A_n, B_n, C_n are specified functions of n but are independent of x. For instance, we take $A_n = 2, B_n = 0, C_n = 2(n-1)$ above to define the Hermite polynomials.

For some other famous examples of orthogonal polynomials, we have the *Laguerre* polynomials $L_n^{(\alpha)}(x)$ for $n \in \mathbb{N} \cup \{0\}$ and $\alpha > -1$, which have recursion

$$nL_n^{(\alpha)}(x) = (-x + 2n + \alpha - 1)L_{n-1}^{(\alpha)}(x) - (n + \alpha - 1)L_{n-2}^{(\alpha)}(x), \quad \text{for } n \geq 2,$$

with initial conditions $L_0^{(\alpha)}(x) = 1,\ L_1^{(\alpha)}(x) = -x + \alpha + 1$. For a final famous example, define the *ultra spherical* or *Gegenbauer polynomials* $C_n^{\nu}(x)$, for $n \geq 0$ and $\nu > -1/2$ with recursion

$$(n+1)C_{n+1}^{\nu}(x) = (2\nu + 2n)xC_n^{\nu}(x) - (2\nu - 1 + n)C_{n-1}^{\nu}(x), \quad \text{for } n \geq 1,$$

with initial conditions $C_0^\nu(x) = 1$, $C_1^\nu(x) = 2\nu x$.

The real import of orthogonal polynomials involves differential calculus, that is, techniques of differentiation and integration, and is partly developed in the exercises.

EXERCISES

8.3 RECURSIVELY DEFINED SEQUENCES

1. Fix a finite sequence a_i of natural numbers, for $i = 1, \ldots, n \geq 1$ and write $a_1 + a_2 + \cdots + a_n$. Consider all possible legal ways to insert $n - 2$ pairs of parentheses into this expression to get a fully parenthesized expression; for instance, in case $n = 3$, we insert one pair of parentheses to get either $(a_1 + a_2) + a_3$ or $a_1 + (a_2 + a_3)$. Let $g(n)$, called the nth *Catalan number*, denote the number of distinct such expressions, so $g(4) = 3$ for instance. Prove that the Catalan numbers satisfy each of the recursions

$$g(n) = \sum_{k=1}^{n-1} g(k)\, g(n-k), \text{ for } n \geq 3,$$

$$g(n) = \frac{4n-6}{n}\, g(n-1), \text{ for } n \geq 2,$$

where $g(1) = g(2) = 1$ in either case.

2. Define a *derangement* of $\overline{n}$ to be a permutation $f : \overline{n} \to \overline{n}$ which has no fixed point, i.e., $f(k) \neq k$ for any $k \in \overline{n}$. Prove that numbers D_n of derangements of $\overline{n}$ satisfy each of the recursions

$$D_n = (n-1)(D_{n-2} + D_{n-1}), \text{ for } n \geq 3,$$

$$D_n = nD_{n-1} + (-1)^n, \text{ for } n \geq 2,$$

where we take $D_1 = 0$ and $D_2 = 1$ in either case.

3. This problem depends upon a knowledge of differential calculus. Suppose that $\alpha, \beta \in \mathbb{R}$ and $w(x)$ is a suitable real-valued function defined on the interval (α, β). We say that a family $\phi_n(x)$, $n \geq 0$, of polynomials is a set of *orthogonal polynomials* for $w(x)$ if $\phi_n(x)$ is of degree n for each n and for any p, q, we have $\int_\alpha^\beta \phi_p(x)\phi_q(x)\, w(x)dx = 0$ unless $p = q$.

(a) Show that if $\phi_n(x)$ is a set of orthogonal polynomials for $w(x)$, then they must satisfy a three-term recursion as described.

(b) Check that the Hermite polynomials are orthogonal for $w(x) = e^{-x^2}$ on the real line $(\alpha, \beta) = (-\infty, \infty)$.

(c) Check that the Laguerre polynomials are orthogonal for $w(x) = x^\alpha e^{-x}$ on the ray $(\alpha, \beta) = (0, \infty)$.

(d) Check that the ultra spherical polynomials are orthogonal for $w(x) = (1-x^2)^{\nu-1/2}$ on the interval $(\alpha, \beta) = (-1, 1)$.

8.4 *INDUCTIVE PROOFS AND SOLVING RECURSIONS*

To begin, we give examples of applying inductive techniques to prove facts about recursively defined sequences. We then turn our attention to the problem of "solving" a recursion, by which we mean the explicit calculation of the values of a sequence which is defined recursively.

As was discussed in the previous section, proofs by induction are a fundamental tool for the very definition of a sequence by recursion. In fact, inductive proofs are also of basic utility for establishing general facts about recursively defined sequences, and we begin with some examples of such inductive proofs.

■ **Example 8.4.1** We shall prove by induction on $m \geq 1$ that if $n = 2m \in \mathbb{N}$, then $n!! = 2^m\,[m!]$, and the basis step follows from the definition $2!! = 2 = 2^1 \cdot 1!$. For the induction step, we suppose that $n = 2(m+1)$ and compute that

$$\begin{aligned} n!! &= (2m+2)[(2m)!!] \\ &= 2(m+1)\left[2^m[m!]\right] \\ &= 2^{m+1}[(m+1)!] \end{aligned}$$

where the first equality follows from the recurrence relation for double factorials and the second relies on the inductive hypothesis.

■ **Example 8.4.2** We shall prove that the Fibonacci numbers satisfy the identity

$$f_{n+1} - f_n f_{n+2} = (-1)^n \quad \text{for } n \geq 0$$

by induction on n, and the basis step $f_1 - f_0 f_2 = 2 - 1 \cdot 1 = 1 = 1^0$ is checked directly. For the induction step, suppose that $f_{n+1}^2 - f_n f_{n+2} = (-1)^n$, and compute that

$$\begin{aligned} f_{n+2}^2 - f_{n+1} f_{n+3} &= f_{n+2}^2 - f_{n+1} f_{n+2} - f_{n+1}^2 \\ &= (f_{n+1} + f_n) f_{n+2} - f_{n+1}^2 - f_{n+1} f_{n+2} \\ &= f_{n+1} f_{n+2} + f_n f_{n+2} - f_{n+1}^2 - f_{n+1} f_{n+2} \\ &= -(f_{n+1}^2 - f_n f_{n+2}) \\ &= -(-1)^n \\ &= (-1)^{n+1} \end{aligned}$$

as desired, where we have used the recurrence relation of the Fibonacci numbers in the first two equalities and the induction hypothesis in the next-to-last equality.

■ **Example 8.4.3** From the definition of Fibonacci numbers, we find immediately that

$$\begin{aligned}
f_{n+3} &= f_{n+2} + f_{n+1} = 2f_{n+1} + f_n \\
f_{n+4} &= f_{n+3} + f_{n+2} = (2f_{n+1} + f_n) + (f_{n+1} + f_n) \\
f_{n+5} &= f_{n+4} + f_{n+3} = (3f_{n+1} + 2f_n) + (2f_{n+1} + f_n) = 5f + n + 1 + 3f_n \\
&\vdots \qquad\qquad\qquad \vdots
\end{aligned}$$

and, in general, we find by induction on m (as the reader should carefully prove) that

$$f_{n+m} = f_m f_{n+1} + f_{m-1} f_n$$

for any $n \in \mathbb{N} \cup \{0\}$ and $m \in \mathbb{N}$.

■ **Example 8.4.4** It may seem that the sequence h_n of harmonic numbers in Example 8.3.7 above does not get too large as n gets large since we are always adding smaller and smaller values. In contrast to this intuition, we prove here by induction that

$$h_{2^m} \geq 1 + \frac{m}{2} \text{ for } m > 0$$

so in fact the sequence h_n takes arbitrarily large values. The basis step $m = 1$ follows from the equality $h_{2^1} = h_2 = 3/2 = 1 + 1/2$. For the inductive step, notice that

$$\begin{aligned}
h_{2^{m+1}} &= h_{2^m} + \frac{1}{2^m + 1} + \frac{1}{2^m + 2} + \cdots + \frac{1}{2^{m+1}} \\
&> h_{2^m} + \underbrace{\frac{1}{2^{m+1}} + \cdots + \frac{1}{2^{m+1}}}_{2^m \text{ times}} \\
&= h_{2^m} + \frac{1}{2} \\
&\geq (1 + \frac{m}{2}) + \frac{1}{2} \\
&= 1 + \frac{m+1}{2},
\end{aligned}$$

where we have used the induction hypothesis in the next-to-last line.

■ **Example 8.4.5** Recall the Hermite polynomials $H_n(x)$, for $n \geq 0$, considered in Example 8.3.11, which are defined by the recursion $H_n(x) = 2xH_{n-1}(x) - 2(n-1)H_{n-2}(x)$ with initial condition $H_0(x) = 1$, $H_1(x) = 2x$. We shall first prove inductively that

$$H_{2n-1}(0) = 0, \text{ for each } n \geq 1,$$

and for the basis step $n = 1$, we observe that $H_1(x) = 2x$ and hence indeed $H_1(0) = 0$. For the induction step, we evaluate the recursion relation at $x = 0$ to find

$$H_{2(n+1)-1}(0) \quad = \quad -2\left[2(n+1) - 2\right] H_{2n-1}(0) = 0$$

where the second equality holds by the inductive hypothesis.

For a more interesting calculation, we shall prove inductively that

$$H_{2n}(0) = (-1)^{n+1}2^n\ (2n-1)!! \quad \text{for } n \geq 1,$$

where we are using here the double factorial function

$$(2n-1)!! = (2n-1) \cdots (2n-3) \cdot \cdots \cdot 5 \cdot 3 \cdot 1$$

discussed in Example 8.3.4. The basis step $n = 1$ amounts to $H_2(0) = -2 = (-1)^1\ 2^1\ 1!!$. For the induction step, evaluate the recursion relation at $x = 0$ to find

$$\begin{aligned} H_{2n}(0) &= -2(2n-1)H_{2n-2}(0) \\ &= -2(2n-1)[(-1)^n 2^{n-1}(2n-3)!!] \\ &= (-1)^{n+1}2^n(2n-1)!! \end{aligned}$$

where the second equality follows from the induction hypothesis.

Given a recursively defined sequence x_n in A, there is a basic associated problem: to compute the value of x_n as an explicit function of n. A solution to this basic associated problem is called simply a *solution of the recursion.* For instance, we gave the recursive definition of p_n in Example 8.3.1 and computed that its solution is $p_n = n!$, for $n \in \mathbb{N} \cup \{0\}$. For another example, the binomial coefficients $\binom{n}{k}$ were defined recursively in Example 8.3.8, and the solution of this recursion is given in Theorem 8.2.3 as $\binom{n}{k} = n!/k!(n-k)!$. In fact, it is not always possible to solve a recursion in closed form as in these examples.

On the other hand, there are various techniques available for solving recursions under suitable hypotheses, and the rest of this section as well as the next one are dedicated to an exposition of such techniques.

The most basic technique for solving a recursion defining a sequence $x : \mathbb{N} \to A$ amounts to simply listing the first several values $x_1, x_2, x_3, \ldots, x_m$ of x (where in practice m is perhaps five or ten), "guessing" the general form of x_n on the basis of this data, and finally proving by induction that the guess is correct. Proofs by induction are the most basic tool for solving recursions, and in spite of the guessing, provide a perfectly rigorous solution to a recursion.

The current discussion thus follows naturally our previous treatment of proofs by induction for recursively defined sequences, and we continue with two examples of solving a recursion in this way.

■ **Example 8.4.6** Define a sequence $x_n \in \mathbb{N}$ for $n \in \mathbb{N}$ recursively by setting

$$x_{n+2} = x_{n+1} + x_n + x_{n+1}x_n \quad \text{for} \quad n \in \mathbb{N}$$

subject to the initial conditions

$$x_1 = x_2 = 1.$$

Listing the next several values, we find

$$x_3 = 3, \quad x_4 = 7, \quad x_5 = 31, \quad x_6 = 255$$

and observe empirically that each term in the sequence is one less than a power of two. We are led to guess that x_n is of the form $2^{m_n} - 1$ for some m_n which depends on n. It is not immediately clear what these powers m_n should be, and we are led to list the corresponding powers

$$m_1 = 1, \quad m_2 = 1, \quad m_3 = 2, \quad m_4 = 3, \quad m_5 = 5, \quad m_6 = 8$$

and observe the appearance of the Fibonacci numbers.

This leads us to conjecture that $x_n = 2^{f_n} - 1$, where f_n denotes the nth Fibonacci number for $n \in \mathbb{N}$. To verify our conjecture (and the basis step of our inductive proof has already been verified), we compute

$$\begin{aligned} x_{n+1} &= x_n + x_{n-1} + x_n x_{n-1} \\ &= (2^{f_n} - 1) + (2^{f_{n-1}} - 1) + (2^{f_n} - 1)(2^{f_{n-1}} - 1) \\ &= 2^{f_n} + 2^{f_{n-1}} + 2^{f_n + f_{n-1}} - 2^{f_n} - 2^{f_{n-1}} - 1 \\ &= 2^{f_{n+1}} - 1. \end{aligned}$$

An inductive argument therefore proves our conjecture, and we have successfully solved the recursion.

■ **Example 8.4.7** A puzzle called the Tower of Hanoi, invented by the mathematician Edouard Lucas, consists of a board with three upright pegs and seven disks of different diameters which fit upon the the pegs. The game begins with all seven disks on one peg in a "pyramid" with the largest disk on the bottom, the next-largest upon it, and so on:

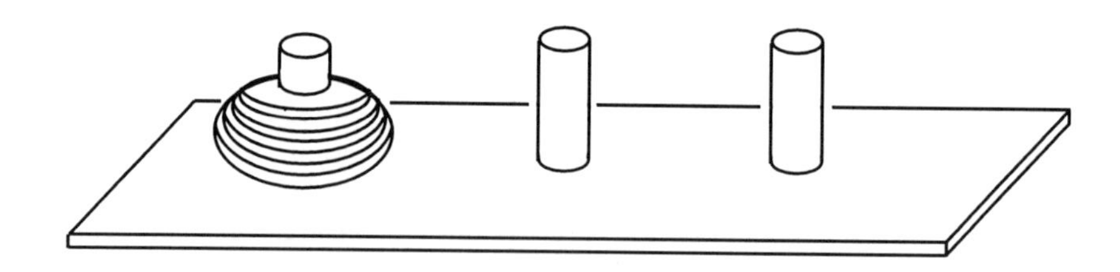

The object of the puzzle is to transfer all of the disks, one at a time and in the least number of moves, to the third peg to produce an identical pyramid. During the moves, one is not allowed to place a larger disk on top of a smaller one. We describe here an inductive solution to the problem and compute the minimum number of moves required to transfer the disks.

Assuming for the moment the existence of solutions, for each natural number n, let s_n denote the number of moves required to transfer n disks in this way from the first peg to the third peg, so Lucas' original game is the case $n = 7$. If we have $n + 1$ disks, then we can move the top n disks from the first to the second peg in s_n moves by the inductive hypothesis. We may then move the largest disk from the first to the third peg with a single move, and finally move the pyramid of remaining disks from the second to the third peg in s_n moves again by the inductive hypothesis. This describes an inductive method of solving the puzzle (where in the basis step $n = 1$, we simply move the unique disk from the first to the third peg), and we leave it to the reader to verify that this solution indeed requires the least number of moves. From this discussion, it follows that the numbers s_n satisfy the recursion

$$s_1 = 1$$

$$s_{n+1} = s_n + 1 + s_n = 2s_n + 1$$

Computing from this recursion, we find that the first several values of s_n are

$$s_1, s_2, s_3, \ldots \quad = \quad 1, 3, 7, 15, 31, 63, \ldots,$$

and we are led to guess from this data that $s_n = 2^n - 1$.

We prove this guess by induction, and the basis step $n = 1$ has already been verified. For the induction, we compute

$$s_{n+1} = 2s_n + 1 = 2(2^n - 1) + 1 = (2^{n+1} - 2) + 1 = 2^{n+1} - 1,$$

where the second inequality follows from the induction hypothesis.

This proves, for instance, that $s_7 = 2^7 - 1 = 127$ is the minimum number of moves required to solve Lucas' original Tower of Hanoi puzzle with 7 disks.

Just these two examples of this technique of solving a recursion by guessing and then inductively proving the verity of the guess should suffice here. Indeed, we have seen examples of just this procedure already in our inductive calculations of finite series in §A.5 of Chapter 1.

Turning then to a more systematic technique of solving recursions, we suppose that the recursion is of the special form

$$\begin{aligned} x_0 &= b_0 \\ x_n &= c_n x_{n-1} + b_n, \end{aligned}$$

where the "coefficients" c_n and b_n may be some functions of n. Such a system is called a "linear first-order inhomogeneous equation" for reasons we shall explain in the next section. To determine the nth term in the sequence, consider the following sequence of equations, where each equation arises from the previous one by applying the recurrence relation.

$$\begin{aligned}
x_n &= c_n x_{n-1} + b_n \\
c_n x_{n-1} &= c_n c_{n-1} x_{n-2} + c_n b_{n-1} \\
c_n c_{n-1} x_{n-2} &= c_n c_{n-1} c_{n-2} x_{n-3} + c_n c_{n-1} b_{n-2} \\
&\vdots \\
\prod_{i=0}^{n-2} c_{n-i} x_1 &= \prod_{i=0}^{n-1} c_{n-i} x_0 + \prod_{i=0}^{n-2} c_{n-i} b_1 \\
\prod_{i=0}^{n-1} c_{n-i} x_0 &= \prod_{i=0}^{n-1} c_{n-i} b_0
\end{aligned}$$

Separately adding the lefthand and righthand sides of these equations and cancelling like terms produces the equation

$$x_n = b_n + c_n b_{n-1} + c_n c_{n-1} b_{n-2} + \cdots + \prod_{i=0}^{n-2} c_{n-i} b_1 + \prod_{i=0}^{n-1} c_{n-i} b_0,$$

and in many cases one can calculate the righthand side using standard summation identities to give a solution to the recursion.

For an example of this approach, we have

■ **Example 8.4.8** Fix $b, c \in \mathbb{R}$ and recursively define the sequence $x_n \in \mathbb{R}$ for $n \in \mathbb{N} \cup \{0\}$ by

$$\begin{aligned}
x_0 &= b \\
x_n &= c x_{n-1} + b.
\end{aligned}$$

As prescribed in the previous paragraph, we form the set

$$\begin{aligned}
x_n &= c x_{n-1} + b \\
c x_{n-1} &= c^2 x_{n-2} + cb \\
c^2 x_{n-2} &= c^3 x_{n-3} + c^2 b \\
&\vdots \\
c^{n-1} x_1 &= c^n x_0 + c^{n-1} b \\
c^n x_0 &= c^n b
\end{aligned}$$

and separately sum the lefthand and righthand sides cancelling like terms to get

$$\begin{aligned}
x_n &= b + cb + c^2 b + \cdots + c^{n-1} b \\
&= b \sum_{i=0}^{n} c^i.
\end{aligned}$$

Finally applying the formula for the sum of a geometric series from Example B.5.3 of Chapter 1, we find

$$x_n = \begin{cases} (n+1)b, & \text{if } c = 1; \\ \frac{b(1-c^{n+1})}{(1-c)}, & \text{if } c \neq 1, \end{cases}$$

solving the recursion.

We admit that the techniques discussed in this section for solving recursions are somewhat ad hoc insofar as the first technique requires guessing and the second depends on applying known summation formulas. In fact, there is a whole grab-bag of special techniques for solving suitably special recursions, and we do not undertake a study of this grab-bag here. Two such basic techniques, each of whose study in contrast constitutes a beautiful and coherent theory, are described in the next two sections.

EXERCISES

8.4 INDUCTIVE PROOFS AND SOLVING RECURSIONS

1. Prove that the nth Fibonacci number satisfies $f_n = \binom{n}{0} + \binom{n-1}{1} + \cdots + \binom{n-k}{k}$.

2. Solve the recursion of the previous section for the Catalan numbers to conclude that $ng(n) = \binom{2n-2}{n-1}$.

3. Solve the recursion of the previous section for the derangement numbers to conclude that

$$\frac{1}{n!}D_n = 1 - \frac{1}{1!} + \frac{1}{2!} - \frac{1}{3!} + \cdots + (-1)^n \frac{1}{n!}.$$

4. Use inclusion/exclusion to prove that the $s_{m,n}$ of Example 8.3.9 are given by

$$s_{m,n} = \sum_{i=0}^{n} (-1)^i \binom{n}{i} (n-i)^m$$

and conclude that the Sterling numbers themselves are given by

$$\left\{ {n \atop m} \right\} = \frac{1}{n!} \sum_{i=0}^{n} (-1)^i \binom{n}{i} (n-i)^m.$$

5. The following problems on orthogonal polynomials require the differential calculus.

(a) Prove that the Hermite polynomials satisfy

$$e^{-x^2} H_n(x) = (-1)^n \left(\frac{d}{dx}\right)^n e^{-x^2}.$$

(b) Prove that the Laguerre polynomials satisfy

$$n!e^{-x}x^{\alpha}L_n^{(\alpha)}(x) = \left(\frac{d}{dx}\right)^n (e^{-x}x^{n+\alpha}).$$

(c) Prove that the ultra spherical polynomials satisfy

$$(1-x^2)^{v-1/2}C_n^{\nu}(x) = \frac{(-1)^n}{2^n n!}\left(\frac{d}{dx}\right)^n (1-x^2)^{n+\nu-1/2}.$$

(†) 8.5 CHARACTERISTIC EQUATIONS

Consider the recursive definition of a sequence of real numbers x_n whose recurrence relation is of the form $x_{n+2} = bx_{n+1} + cx_n$, where b and c are real numbers. We describe the "method of characteristic equations" for solving suitable such recurrence systems, give several examples, and briefly discuss more general recurrence systems to which this method of solution applies.

Before discussing this powerful method of solving suitable recursions, we introduce some terminology in order to define this suitable class. Consider a recurrence relation for a sequence x_n of real numbers of the form

$$b_0x_n + b_1ax_{n-1} + b_2x_{n-2} + \cdots + b_kx_{n-k} = a(n).$$

where $b_0, b_1 \ldots b_k$ are a fixed collection of real numbers and $a(n)$ is some fixed real-valued function of n. Such a recurrence relation is called a *linear recurrence relation with constant coefficients*, and it is said to have *order* k provided that both b_0 and b_k are non-zero. For example, $6a_r - 2a_{r-1} + 2a_{r-2} = r^3 + 2$ is a second order linear recurrence relation with constant coefficients. For another example, the recurrence relation $x_n - bx_{n-1} = c$ considered in Example 8.4.8 is a first order linear recurrence relation with constant coefficients.

Given k consecutive values $x_{m-k}, x_{m-k+1}, \ldots, x_{m-1}$, the recurrence relation allows the calculation of x_m, namely, we have

$$x_m = -\frac{1}{b_0}\left[b_1x_{m-1} + b_2x_{m-2} + \cdots + b_kx_{m-k} - a(m)\right].$$

It follows that a suitable initial condition for using such a recurrence relation to define a sequence $x : \mathbb{N} \to \mathbb{R}$ is the specification of the first k values of x, namely, $x_0, x_1, \ldots, x_{k-1}$.

To complete our introduction of new terminology, we shall say that a linear recurrence relation with constant coefficients is *homogeneous* if the function $a(n)$ above is the constant function with value zero, and we say the recurrence relation is *inhomogeneous* if it is not homogeneous. For instance, the first order linear recurrence relation in Example 8.4.8 is inhomogeneous, whereas the recurrence relation $f_{n+2} - f_{n+1} - f_n = 0$ of the Fibonacci sequence is homogeneous of order two.

There is a general theory of solving linear recursions, both homogeneous and inhomogeneous, with constant coefficients, but we shall not take up this general theory here. Instead, we illustrate aspects of this general theory by giving a fairly complete treatment for the case of second order linear homogeneous recurrence relations with constant coefficients, and comment specifically on the general theory later. Thus, we shall study here a recurrence system of the form

$$(*) \qquad \begin{cases} \text{Specify the values of } x_0, x_1 \in \mathbb{R} \\ \\ x_{n+2} = bx_{n+1} + cx_n \text{ for } n \geq 0 \end{cases}$$

where $b, c \in \mathbb{R}$. We shall fix the notation for this recurrence system $(*)$ of interest for the rest of this section, and we shall solve $(*)$ below under the additional numerical assumption that $b^2 + 4c > 0$. For instance, the recurrence relation for the Fibonacci numbers or the Lucas numbers satisfies this condition. We make this additional assumption here just for simplicity and remark that the more general case of unrestricted coefficients b, c is solved similarly (cf. Theorems 8.5.2 and 8.6.2 below).

It is convenient for now to consider the sequences defined by various specifications of initial condition with fixed recursion. More explicitly, we ignore the initial conditions for now and concentrate on solving just the recurrence relation. In the spirit of our ad hoc techniques in the previous section, we shall "guess" a solution of the recurrence relation above of the form $x_n = ar^n$, where $a, r \in \mathbb{R}$ are to be determined. That is, let us guess that this is a solution to the recurrence system and discover what restrictions are thereby imposed on a and r. To this end, we compute from the recurrence relation that

$$ar^n = x_n = bar^{n-1} + car^{n-2}.$$

Supposing that $a \neq 0$ and $r \neq 0$ for now, we may divide the previous equation by ar^{n-2} to get

$$r^2 - br - c = 0$$

which is called the *characteristic equation* of the recursion $(*)$. For instance, we take $b = c = 1$ to define the Fibnoacci or Lucas numbers, and have the characteristic equation $r^2 - r - 1 = 0$ in either case.

Thus, in order that $x_n = ar^n$ is a solution to the recurrence system, we must have that r is a root of the characteristic equation. By the quadratic formula, the characteristic equation has the roots

$$r_{\pm} = \frac{b \pm \sqrt{b^2 + 4c}}{2},$$

and by our assumption above that $b^2 + 4c > 0$, we find that the roots r_+, r_- are real and distinct. It is to guarantee that the roots of the characteristic equation are real and distinct that we make the assumption $b^2 + 4c > 0$ in the first place.

Having discovered this necessary condition that $r = r_+$ or $r = r_-$ in order that ar^n be a solution to the recurrence relation, one is led to see if these are actually solutions to the recurrence relation. We leave this verification to the reader since it follows from the next lemma; in the trivial case $a = 0$, we may still conclude that $ar^n = 0r^n = 0$ solves the recurrence relation with initial conditions $x_0 = x_1 = 0$. Having thereby found the two families $a_+r_+^n$ and $a_-r_-^n$ of solutions, one wonders if one might combine them in some way to produce more general solutions, and in fact one can:

8.5.1 LEMMA *For any choice of* $a_+, a_- \in \mathbb{R}$, *the sequence*

$$x_n = a_+r_+^n + a_-r_-^n$$

solves the recurrence system

$$\begin{aligned} x_0 &= a_+ - a_-, \quad x_1 = a_-r_- + a_+r_+ \\ x_{n+2} &= bx_{n+1} + cx_n \end{aligned}$$

where r_+, r_- *are the roots of the characteristic equation.*

Proof Fix a_+, a_- and define

$$x_n = a_+r_+^n + a_-r_-^n,$$

so the initial conditions above are obviously satisfied. To check the recurrence relation, we compute

$$\begin{aligned} x_n &= a_+r_+^n + a_-r_-^n \\ &= b(a_+r_+^{n-1} + a_-r_-^{n-1}) \ + \ c(a_+r_+^{n-2} + a_-r_-^{n-2}), \end{aligned}$$

so factoring out r_+^{n-2} and r_-^{n-2}, we find

$$0 = a_+r_+^{n-2}(r_+^2 - br_+ - c) + a_-r_-^{n-2}(r_-^2 - br_- - c),$$

as desired. *q.e.d.*

Thus, we have found the family $a_+r_+^n + a_-r_-^n$ of solutions to the recurrence relation, where the specification of a_+, a_- determines the initial condition for the recursion from the equations

$$x_0 = a_+ + a_- \quad \text{and} \quad x_1 = a_+r_+ + a_-r_-.$$

In fact, these equations for the initial condition can be inverted (solving for a_+ and a_- as a function of x_0, x_1), to get

$$a_+ = \frac{r_-x_0 - x_1}{r_- - r_+} \quad \text{and} \quad a_- = \frac{r_+x_0 - x_1}{r_+ - r_-}$$

as we leave it for the reader to verify by direct calculation.

This leads to our main result in this section about second order homogeneous linear recurrence relations with constant coefficients.

8.5.2 THEOREM *Consider the recursion relation*

Specify x_0 and x_1

$x_{n+2} = bx_{n+1} + cx_n$ for $n \geq 0$

(a) *Suppose that* $b^2 + 4c > 0$ *so the roots* r_+, r_- *of the characteristic equation are distinct real numbers. Then the sequence*

$$x_n = a_+ r_+^n + a_- r_-^n$$

solves the recursion, where

$$a_+ = \frac{r_- x_0 - x_1}{r_- - r_+} \quad \text{and} \quad a_- = \frac{r_+ x_0 - x_1}{r_+ - r_-}.$$

(b) *Suppose that* $b^2 + 4c = 0$, *so there is a repeated real root* r_* *of the characteristic equation* $(r - r_*)^2 = 0$. *Then the sequence*

$$x_n = \alpha \cdot r_*^n + \beta \cdot n \cdot r_*^n$$

solves the recursion, where

$$\alpha = x_0 \quad \text{and} \quad \beta = \frac{x_1}{r} - x_0.$$

Proof For part (a), the previous lemma shows that x_n satisfies the recurrence relation. Checking the initial condition is simply a question of observing that our expressions above for a_+, a_- solve the equations $x_0 = a_+ + a_-$ and $x_1 = a_+ r_+ + a_- r_-$ as we asked the reader to verify above.

The proof of part (b) is analogous to the previous argument and is left as an exercise (requiring the analogue of Lemma 8.5.1 in case $b^2 + 4c = 0$). *q.e.d.*

This theorem gives an example of the "method of characteristic equations" for solving recursions, and as we mentioned at the beginning of this section, there is an elaborate theory here for more general linear recurrence relations with constant coefficients. Before briefly discussing aspects of the general theory, we apply the previous theorem to several examples including the calculation of Fibonacci and Lucas numbers.

It is worth saying that one need *not* memorize the explicit formulas above for a_+, a_- in part (a) (or those for α, β in part (b)), for the expressions $x_0 = a_+ + a_-$,

$x_1 = a_+r_+ + a_-r_-$ follow directly from the definition of $x_n = a_+r_+^n + a_-r_-^n$ and can be easily solved directly in any given example.

■ **8.5.1 Example** Recall the Fibonacci numbers f_n for $n \geq 0$ with recursion

$$f_0 = 0 \text{ and } f_1 = 1$$

$$f_{n+2} = f_{n+1} + f_n \quad \text{for } n \geq 0$$

In the notation of the previous theorem, we have $b = c = 1$ so $b^2 + 4c = 5$ and part (a) of the previous theorem therefore applies. As was noted above, the characteristic equation is $r^2 - r - 1 = 0$, which has roots

$$r_\pm = \frac{1 \pm \sqrt{5}}{2}.$$

Furthermore, taking initial conditions $f_0 = 0, f_1 = 1$, we find the equations

$$0 = f_0 = a_+ + a_-,$$
$$1 = f_1 = a_+r_+ + a_-r_-,$$

which have the solution

$$f_n = \frac{1}{\sqrt{5}}\left(\frac{1+\sqrt{5}}{2}\right)^n - \frac{1}{\sqrt{5}}\left(\frac{1-\sqrt{5}}{2}\right)^n.$$

■ **Example 8.5.2** The Lucas numbers ℓ_n for $n \geq 1$ are defined by the same recurrence relation as in the previous example but with the initial condition $\ell_0 = 2, \ell_1 = 1$. We have therefore the same characteristic equation as in the previous example and therefore the same roots $r_\pm$, but the formulas for $a_\pm$ in the previous theorem give

$$a_+ = \frac{2r_- - 1}{r_- - r_+} = 1 \text{ and } a_- = \frac{2r_+ - 1}{r_+ - r_-} = 1,$$

so the solution of the sequence of Lucas numbers is theorefore

$$\ell_n = \left(\frac{1+\sqrt{5}}{2}\right)^n + \left(\frac{1-\sqrt{5}}{2}\right)^n.$$

■ **Example 8.5.3** Consider the recurrence system

$$x_0 = 1 \text{ and } x_1 = 3$$

$$x_{n+2} = 3x_{n+1} - 2x_n \quad \text{for } n \geq 0$$

The characteristic equation is $r^2 - 3r + 2 = 0$ which has roots $r_+ = 2$ and $r_- = 1$, so the sequence $x_n = a_+ 2^n + a_- 1^n = a_+ 2^n + a_-$ solves the recursion where we compute from Theorem 8.5.2 that $a_+ = 1 = -a_-$. The solution of the recursion is thus

$$x_n = 2^n - 1.$$

■ **Example 8.5.4** Consider the recurrence system

$$x_0 = 1 \text{ and } x_1 = 1$$

$$x_{n+2} = -4x_{n+1} + 4x_n \quad \text{for } n \geq 0$$

so the characteristic equation $r^2 - 4r + 4 = (r-2)^2 = 0$ has the repeated real root $r_* = 2$. According to Theorem 8.5.2, the sequence

$$x_n = \alpha \cdot r_*^n + \beta \cdot n \cdot r_*^n$$

solves the recursion, where

$$\alpha = x_0 = 1 \quad \text{and} \quad \beta = \frac{x_1}{r_*} - x_0 = \frac{1}{2} - 1 = -\frac{1}{2}.$$

We therefore get

$$x_n = 2^n - \frac{n}{2} 2^n = 2^n (1 - \frac{n}{2})$$

as the solution to the recursion.

To close this section, we comment briefly on the extension of the method of characteristic equations to more general homogeneous linear recurrence relations with constant coefficients. The case of a second order recursion relation whose characteristic equation has complex roots can actually be handled by analogous techniques.

For another extension of the method of characteristic equations, we might consider a homogeneous recursion relation of order higher than two, say the recurrence relation is

$$b_0 x_n + b_1 a x_{n-1} + b_2 x_{n-2} + \cdots + b_k x_{n-k} = 0.$$

There is an associated *characteristic equation*

$$b_0 r^n + b_1 r^{n-1} + b_2 r^{n-2} + \cdots + b_k r^{n-k} = 0,$$

which plays an analogous role here. Indeed, the reader may prove that if r_* is a real root of this characteristic equation, then $a r_*^n$ satisfies the recurrence relation. These solutions are then combined as in Lemma 8.5.1 and Theorem 8.5.2 to produce general solutions of a recurrence system. We emphasize that there is more here than perhaps meets the eye for instance since one cannot always find the roots of the characteristic equation in practice.

We have discussed here only the existence of solutions to recurrence systems; there is also a general theory about the uniqueness of solutions, and the solutions provided by Theorem 8.5.2 are the unique solutions. Furthermore, one can also sometimes solve inhomogeneous linear recurrence relations with constant coefficients using characteristic equations.

There is thus an elaborate theory of solving linear recurrence relations with constant coefficients using the method of characteristic equations, and we leave the pursuit of such applications to the untiring reader.

(†) 8.6 GENERATING FUNCTIONS

This section discusses another elegant method of solving a recurrence system by computing a certain expression, called the "generating function" associated to the corresponding sequence. In fact, techniques from differential calculus (namely integration and differentiation) are crucial tools in the application of generating functions, and since we do not assume differential calculus here, our discussion is necessarily somewhat limited. Nevertheless, the basic theory of generating functions is uncovered here.

We must begin with some preliminaries and define a *(formal) power series in the variable* z to be an expression of the form

$$\phi(z) = x_0 + x_1 z + x_2 z^2 + \cdots + x_n z^n + \cdots = \sum_{j\geq 0} x_j z^j,$$

where $x_i \in \mathbb{R}$ is a sequence of real numbers. Two power series $\phi(z) = \sum_{j\geq 0} x_j z^j$ and $\psi(x) = \sum_{j\geq 0} y_j z^j$ are the same if and only if $x_j = y_j$ for each $j \geq 0$, and we refer to x_j as the sequence of *coefficients* of the power series, for $j \geq 0$. For instance, the power series with all coefficients equal to one is the expression $1 + z + z^2 + z^3 + \cdots = \sum_{j\geq 0} z^j$. In fact, the simplest examples of power series are just the polynomial functions $\sum_{j=0}^{n} x_j z^j$ for $n \geq 0$ with real coefficients; these examples are characterized by the condition that the coefficients x_j are "eventually zero" in the sense that $\exists n \forall m[(m \geq n) \Rightarrow (x_m = 0)]$.

It is thus reasonable to imagine a general power series in z as a "polynomial function in the variable z which has terms of arbitrarily high degree". As with polynomials in z, two power series in z may be added in the natural way:

$$\sum_{j\geq 0} x_j z^j \ + \ \sum_{j\geq 0} y_j z^j \ = \ \sum_{j\geq 0} (x_j + y_j) z^j$$

by adding their coefficients. Similarly, one multiplies power series according to the rule

$$\left(\sum_{j\geq 0} x_j z^j\right) \cdot \left(\sum_{j\geq 0} y_j z^j\right) \ = \ \sum_{j\geq 0} u_j z^j,$$

where

$$u_m = \sum_{j=0}^{m} x_j y_{m-j} \quad \text{for } m \geq 0.$$

For instance, we find

$$u_0 = x_0 y_0, \; u_1 = x_0 y_1 + x_1 y_0, \; u_2 = x_0 y_2 + x_1 y_1 + x_2 y_0,$$
$$u_3 = x_0 y_3 + x_1 y_2 + x_2 y_1 + x_3 y_0.$$

This prescription again specializes to the correct operation of multiplication of polynomials in the variable z, as we leave it for the reader to check (by induction on m).

Let us define $\mathbb{R}[[z]]$ to be the collection of all formal power series in the variable z with real coefficients, so $\mathbb{R}[[z]]$ is endowed with the natural binary operations of addition and multiplication as above. Furthermore, the collection $\mathbb{R}[z]$ of all polynomials in the variable z with real coefficients is a subset $\mathbb{R}[z] \subseteq \mathbb{R}[[z]]$ of $\mathbb{R}[[z]]$ as above, and the usual addition and multiplication of polynomials are simply the restrictions of addition and multiplication of power series.

The connection between power series and sequences is this: Given a sequence x_j of real numbers for $j \in \mathbb{N} \cup \{0\}$, there is an associated power series

$$G(z) = x_0 + x_1 z + x_2 z^2 + \cdots = \sum_{j \geq 0} x_j z^j \in \mathbb{R}[[z]]$$

called the *generating function* of the sequence x_j. From the very definition of when two generating functions are equal, it follows that the generating function $G(z)$ uniquely determines the corresponding sequence x_i and conversely.

■ **Example 8.6.1** Consider the sequence $x_i = i$ for all $i \in \mathbb{N} \cup \{0\}$. The corresponding generating function $G(z)$ is simply

$$G(z) = 0 + z + 2z^2 + \cdots = \sum_{j \geq 0} j z^j.$$

■ **Example 8.6.2** Let $F(z)$ be the generating function of the Fibonacci numbers, so

$$F(z) = f_0 + f_1 z + f_2 z^2 + \cdots \in \mathbb{R}[[z]],$$

where

$$f_n = \frac{1}{\sqrt{5}}\left(\frac{1+\sqrt{5}}{2}\right)^n - \frac{1}{\sqrt{5}}\left(\frac{1-\sqrt{5}}{2}\right)^n$$

according to Example 8.5.1.

■ **Example 8.6.3** Let us fix $n \in \mathbb{N} \cup \{0\}$ and consider

$$x_0 = \binom{n}{0}, \quad x_1 = \binom{n}{1}, \quad \ldots, x_n = \binom{n}{n}, \quad x_{n+1} = 0, \quad x_{n+2} = 0, \quad \ldots$$

as a sequence, where we set $x_N = 0$ for all $N > n$. According to the statement of the Binomial Theorem in Corollary 8.2.7, we have

$$(1+z)^n = \sum_{r+0}^{n} \binom{n}{r} z^r,$$

so the generating function $G(z)$ for the sequence x_i is in this case simply the polynomial

$$G(z) = (1+z)^n.$$

When considering a sequence $x_{mn} \in \mathbb{R}$, where $m, n \in \mathbb{N} \cup \{0\}$ with *two* subscripts for instance, one might similarly construct a "generating function in two variables" w and z by setting

$$\begin{aligned} G(w,z) &= \sum_{i,j \geq 0} x_{ij} w^i z^j \\ &= x_{00} + x_{01}z + x_{10}w + x_{11}wz + \cdots + x_{ij}w^i z^j + \cdots. \end{aligned}$$

Thus, one considers "power series in two variables" in this setting adding and multiplying such power series as before. We shall not undertake a serious study of such more complicated power series here. Rather, we just mention parenthetically that one does sometimes consider power series in several variables and give the

■ **Example 8.6.4** Fix some $n \in \mathbb{N} \cup \{0\}$ and consider the sequence $x_{mn} \in \mathbb{R}$ for $m \in \mathbb{N} \cup \{0\}$ given by defining

$$x_{mn} = \binom{n}{m} \quad \text{for} \quad 0 \leq m \leq n$$

and setting $x_{mn} = 0$ for $m > n$. Applying Theorem 8.2.6, we find that

$$(w+z)^n = \sum_{m=0}^{n} \binom{n}{m} w^m z^{n-m},$$

so the generating function $G(w,z)$ in this example is simply

$$G(w,z) = (w+z)^n.$$

Before proceeding with some further examples of generating functions, we turn our attention briefly to certain generalities. In order to discuss these generalities, we must make the following definition: Suppose that x_i and y_i for $i \in \mathbb{N} \cup \{0\}$ are sequences of real numbers and define the sequence

$$u_m = \sum_{j=0}^{m} x_j y_{m-j} \quad \text{for } m \geq 0.$$

This sequence u_m is called the *convolution* of x and y and is written $u = x * y$. For instance, let x denote the constant sequence with value 1, so the generating function of x is $F(z) = 1 + z + z^2 + \cdots$, and let y denote an arbitrary sequence. The convolution $x * y$ is then given by

$$(x * y)_m = \sum_{i=0}^{m} y_i.$$

8.6.1 THEOREM *Suppose that*

$$F(z) = x_0 + x_1 z + x_2 + \cdots = \sum_{i \geq 0} x_i z^i$$

$$G(z) = y_0 + y_1 z + y_2 + \cdots = \sum_{i \geq 0} y_i z^i$$

are generating functions for the sequences x_i and y_i respectively. Then we have

(a) *The generating function for the sequence $u_i = x_i + y_i$ for $i \in \mathbb{N} \cup \{0\}$ is given by $H(z) = F(z) + G(z)$.*

(b) *The generating function for the convolution $u = x * y$ of x and y is given by $H(z) = F(z)G(z)$.*

Proof Part (a) follows immediately from the definition of the sum of two power series, and part (b) follows immediately from the definition of the product of power series and the definition of convolution. *q.e.d.*

Returning to some more illuminating examples of generating functions, we have

■ **Example 8.6.5** Define the sequence $x : \mathbb{N} \cup \{0\} \to \mathbb{N}$ by setting x_i to be the number of ways to select i objects from six objects with replacement, where one of the objects is labeled "X" and can be chosen at most twice in the selection, another of the objects is labeled "Y" and can be chosen at most three times in the selection, and the remaining four objects can be chosen at most once. We claim that the generating function for x_i is simply the polynomial

$$G(z) = (1 + z + z^2)(1 + z + z^2 + z^3)(1 + z)^4$$

To prove this, notice that the coefficient of z^r in $G(z)$ is exactly the number of ways to make up the term z^r from the factors $1 + z + z^2$, $1 + z + z^2 + z^3$ and the four factors $1 + z$. The contribution from the factor $1 + z + z^2$ can be 1, z, or z^2, which corresponds to selecting the object X zero, one, or two times.

Similarly, the contribution from the factor $1+z+z^2+z^3$ can be 1, z, z^2, or z^3, which corresponds to selecting the object Y zero, one, two, or three times. The contribution from the remaining four factors $1+z$ can be 1 or z, which corresponds to the the remaining four selections of objects either zero or one time.

This proves that $G(z)$ given above is indeed the required generating function. To use this fact for an explicit expression of the numbers x_i, we must expand this polynomial and compute

$$G(z) = 1 + 6z + 16z^2 + 26z^3 + 31z^4 + 31z^5 + 26z^6 + 16z^7 + 6z^8 + z^9,$$

so, for instance, there are 31 ways to select 4 such objects, and 16 ways to select two such objects.

Several comments are in order regarding the previous example, and we first remark that the generating function for the number of such selections is similarly given in general by an easily computed polynomial. A more general remark is that the previous example nicely illustrates the calculational utility of a generating function: We first abstractly recognize a certain known function (in our case, the polynomial $G(z)$) as the generating function of given sequence; we then use existing information about the polynomial (in our case, the explicit calculation of the coefficients of $G(z)$) to finally compute the sequence x_i. This is a fairly typical application of generating functions.

In our next example, an abstract argument again establishes an explicit expression for the generating function. The example is more sophisticated than the previous ones in that the sequence in this case is a sequence of polynomials $a_m(y)$ rather than just a sequence of real numbers as before. Thus, the generating function here is an expression of the form

$$\sum_{m\geq 0} a_m(y)z^m,$$

so there are *two* "variables" here, the variable z in the power series and the variable y as the argument of the coefficients $a_m(y)$. We pursue this example here just to indicate the fact that generating functions are also a fundamental tool in handling sequences more general than just real-valued ones.

■ **Example 8.6.6** For each $n \in \mathbb{N}$ and $m \in \mathbb{N} \cup \{0\}$ where $0 \leq m \leq n$, define the mth *elementary symmetric function* in the n variables $y_1, \ldots, y_n$ to be the polynomial

$$a_m(y_1, \ldots, y_n) = \sum_{1\leq j_1 < j_2 < \cdots < j_n \leq n} y_{j_1} y_{j_2} \cdots y_{j_n},$$

taking $a_0(y_1, \ldots, y_n) = 1$ by definition. For instance, we have

$$\begin{aligned}
a_0 &= 1 \\
a_1 &= y_1 + \cdots y_n \\
a_2 &= y_1(y_2 + y_3 + \cdots + y_n) + y_2(y_3 + y_4 + \cdots + y_n) \\
&\quad + \cdots + y_{n-2}(y_{n-1} + y_n) + y_{n-1}(y_n) \\
&\vdots \\
a_n &= y_1 y_2 \cdots y_n.
\end{aligned}$$

The crucial property of the elementary symmetric functions is that for any permutation σ of $\{1, 2, \ldots, n\}$ and any $m \in \mathbb{N}$, we have an identity

$$a_m(y_1, y_2, \ldots, y_n) = a_m(y_{\sigma(1)}, y_{\sigma(2)}, \ldots, y_{\sigma(n)}),$$

but we will not need this fact here and leave to the proof to the reader. We claim that the generating function for the elementary symmetric functions is given by

$$\begin{aligned}
G(z) &= 1 + a_1(y_1, \ldots, y_n)z + a_2(y_1, \ldots, y_n)z^2 + \cdots + a_n(y_1, \ldots, y_n)z^n \\
&= (1 + y_1 z)(1 + y_2 z) \cdots (1 + y_n z).
\end{aligned}$$

The proof is by induction on $n \geq 1$ and can safely be left to the reader.

To emphasize the point of the previous example, generating functions are a basic tool for sequences more general than just sequences of real numbers, for instance, for sequences of polynomials as in the example.

We turn our attention finally to the application of generating functions to the solution of second order linear homogeneous recurrence relations with constant coefficients. It turns out that one can actually easily use generating functions to solve such a recursion as we shall see, but before turning to the general theorem, we pursue an example.

■ **Example 8.6.7** Consider the generating function $F(z)$ (discussed in Example 8.6.2) for the Fibonacci numbers f_n, which satisfy the recursion $f_r = f_{r-1} + f_{r-2}$ with initial conditions $f_0 = 0, f_1 = 1$. Multiplying both sides of $f_r = f_{r-1} + f_{r-2}$ by z^r and summing over all natural numbers $r \geq 3$, we have

$$\sum_{r \geq 3} f_r z^r = \sum_{r \geq 3} f_{r-1} z^r + \sum_{r \geq 3} f_{r-2} z^r,$$

That is, we find

$$F(z) - f_2 z^2 - f_1 z - f_0 = z\big[F(z) - f_1 z - f_0\big] \; + \; z^2\big[F(z) - f_0\big],$$

and collecting like terms, we find

$$(1 - z - z^2)F(z) = z^2(f_2 - f_1 - f_0) + z(f_1 - f_0).$$

Substituting in the values $f_0 = 0, f_1 = 1, f_2 = 1$, we conclude that

$$(1 - z - z^2)F(z) = z.$$

It is left to the reader to generalize this example and prove the following result, which similarly gives an explicit equation, called a "functional equation", for the generating function of a second-order homogeneous linear recursion with constant coefficients.

8.6.2 THEOREM *Suppose that $x : \mathbb{N} \cup \{0\} \to \mathbb{R}$ is a recursively defined sequence determined by the recurrence system*

$$\text{Specify } x_0 \in \mathbb{R} \text{ and } x_1 \in \mathbb{R}$$

$$x_{n+2} = -bx_{n+1} + cx_n \quad \text{for } n \geq 0$$

The generating function $G(z) = \sum_{j \geq 0} x_j z^j$ of the sequence x satisfies the functional equation

$$(1 - bz - cz)\, G(z) \quad = \quad z^2(x_2 - bx_1 - cx_0) + z(x_1 - bx_0).$$

This result brings us to our final and fundamental remarks on generating functions. It is tempting to "solve" the functional equation in the previous theorem for the generating function. For instance, we are tempted to write the generating function in the previous example for the Fibonacci numbers as $F(z) = z/(z^2 - z - 1)$, but the complication is that we have here no explicit interpretation here of $z/(z^2 - z - 1)$ as a power series in z. One achievement of differential calculus is to give such an interpretation of a function, as a "Taylor series". By calculating these Taylor series, one calculates the corresponding generating function and thereby solves the recursion.

We shall not take up Taylor series here but remark that Theorem 8.6.2 together with these techniques from differential calculus indeed allow one to solve the functional equation in Theorem 8.6.2, and thereby compute the generating function

$$G(z) = \frac{z^2(x_2 - bx_1 - cx_0) + z(x_1 - bx_0)}{(1 - bz - cz)}$$

for an arbitrary second-order linear homogeneous recurrence equation.

The utility of differential calculus in manipulating generating functions does not end here for it is a basic tool in the application and calculation of generating functions in general.

EXERCISES

8.6 GENERATING FUNCTIONS

1. Calculate a generating function for the Sterling numbers of the second kind.

2. A *partition* of a natural number $n \in \mathbb{N}$ is a representation of n as a sum of natural numbers, where the order of the summands is immaterial, and we define $p(n)$ to be the number of partitions of n where $p(0) = 1$ by convention.

(a) Show that $p(n)$ is the number of distinct solutions to the equation $n = 1 \cdot x_1 + 2 \cdot x_2 + \cdots + nx_n$ in the variables $x_i \in \mathbb{N} \cup \{0\}$.

(b) Show that the generating function for $p(n)$ is given by $\prod_{i=1}^{\infty}(1 - x^k)^{-1}$.

(c) Fix distinct $t_i \in \mathbb{N}$, where $i = 1, \ldots, n$ and let $q(n)$ denote the number of partitions of n so that each summand in the partition is among $t_1, t_2, \ldots, t_n$. Show that the generating function for $q(n)$ is $\prod_{k=1}^{n}(1 - x^{t_k})^{-1}$.

3. The following problems on orthogonal polynomials again require the differential calculus.

(a) Derive the generating function

$$e^{2xt-t^2} = \sum_{n=0}^{\infty} \frac{H_n(x)}{n!} t^n$$

for the Hermite polynomials, and finally solve to calculate

$$\frac{H_n(x)}{n!} = \sum_{k=0}^{[n/2]} \frac{(-1)^k (2x)^{n-2k}}{(n-2k)!k!},$$

where $[\cdot]$ denotes integral part.

(b) Derive the generating function

$$(1-t)^{-1-\alpha} \exp\{\frac{-xt}{1-t}\} = \sum_{n=0}^{\infty} L_n^{\alpha}(x)\ t^n$$

for the Laguerre polynomials, and finally solve to calculate

$$\frac{L_n^{(\alpha)}(x)}{n!} = \sum_{k=0}^{n} \frac{(1+\alpha)_n}{(n-k)!(1+\alpha)_k} \frac{(-x)^k}{k!},$$

where $(a)_n = a(a+1)(a+2)\cdots(a+n-1)$ for $n \geq 1$ and any $a \in \mathbb{R}$, where also $(a)_0 = 1$ by convention provided $a \neq 0$.

(c) Derive the generating function

$$(1 - 2xt + t^2)^{-\nu} = \sum_{n=0}^{\infty} C_n^{\nu}(x)\ t^n$$

for the ultra spherical polynomials, and finally solve to calculate

$$C_n^\nu(x) = \sum_{k=0}^{[n/2]} \frac{(-1)^k (\nu)_{n-k} (2x)^{n-2k}}{(n-2k)!k!},$$

where $[\cdot]$ denotes integral part and $(a)_n$ is defined as above.

Chapter 9

Languages and Finite State Automata

This chapter begins with the definitions of "alphabet" and "language" in the precise sense of computer science, and roughly an alphabet is a finite collection of letters such as $\{a, b, \ldots, z\}$, and a language is some collection of words in these letters. Actually, as a model of English, Swedish, Chinese, Swahili, or any other such "natural language", our precise definitions are necessarily simplistic, but the languages we study here are actually *excellent* paradigms for compilers and interpretors. We define various operations on the set of all languages over a given alphabet, discuss the basic properties, and then use these operations to define a special class of languages called "regular languages" using certain formal "regular expressions". Next, we turn to a study of "finite state automata", which in a precise sense are "machines" which return the answer "Yes" when given a sequence of letters in the alphabet if and only if the sequence is actually a word in the language. Thus, a finite state automaton determines a language, namely, the collection of sequences of letters for which it answers "Yes". It is a famous result known as "Kleene's Theorem" that a language is regular, that is, defined by a regular expression, if and only if it is so determined by a finite state automaton, so Kleene's Theorem unites all the previous material. To prove Kleene's Theorem, we briefly discuss "non-deterministic finite state automata", which generalize the automata described before. Finally, we introduce the basic mathematical objects called "semigroups" and "monoids", study some of their basic theory, and review parts of this chapter from this more sophisticated viewpoint.

9.1 ALPHABETS, WORDS, AND CONCATENATION

An *alphabet* is simply a finite set A, and its elements are called *letters*. Actually, we shall define the symbols "ε", "$*$", "$\vee$", "(", and ")" to have special significance below, and we must also require here that none of these special symbols are letters in the alphabet, i.e., we assume that $\varepsilon, *, \vee, (,) \notin A$. Define a *(non-trivial) word* or a *(non-trivial) string* (*over* A) to be a function $w : \{1, 2, \ldots, n\} \to A$ for some $n \in \mathbb{N}$, and define $||w|| = n \in \mathbb{N}$ to be the *length* of the word w. By way of notation, it is usual to simply list the ordered sequence $w(1)\ w(2)\ \ldots\ w(n)$ of letters to describe the word w, and we shall adopt this practice here. We are thus tacitly identifying the set of words of length $n \geq 1$ with the Cartesian product A^n and further simplifying the notation by omitting the commas separating the entries as well as the initial and final parentheses.

We shall also consider the *trivial word* or *empty word* or *null word* ε (*over* A) which may similarly be regarded as the empty ordered sequence of letters (i.e., as the function $\varepsilon : \emptyset \to A$) and is defined to have length $||\varepsilon|| = 0$.

Define the set

$$A^* = \{\varepsilon\} \cup \{a_1\ a_2\ \ldots\ a_n : n \in \mathbb{N} \text{ and } a_i \in A \text{ for all } i = 1, \ldots, n \geq 1\}$$

so A^* consists of all possible words of all finite lengths. Notice that our tacit identification of the set of words of length n with the Cartesian product A^n identifies A^* with $\{\varepsilon\} \cup (\cup_{n \geq 1} A^n)$.

■ **Example 9.1.1** Taking the alphabet $A = \{0, 1\}$, we find:

- There is exactly one word in A^* of length zero, namely, ε.
- There are exactly two words in A^* of length one, namely, 0 and 1.
- There are exactly four words in A^* of length two, namely, 00, 01, 10, and 11.

In general, there are 2^n possible distinct words of length $n \geq 0$ over the alphabet $\{0, 1\}$ as the reader may check.

■ **Example 9.1.2** Over the usual English alphabet $\{a, b, \ldots, z\}$, the string of letters "automata" is a word of length eight as is the string "pppqqqqq".

Suppose that words $u, v \in A^*$ are given by $u = a_1 a_2 \cdots a_m$ and $v = b_1 b_1 \cdots b_n$, where $a_i \in A$ and $b_j \in A$ for $i = 1, 2, \ldots, m$ and $j = 1, 2, \ldots, n$. We may construct from u and v a new word $w \in A^*$ called the *concatenation* or *catenation*

$$w = uv = a_1 a_2 \cdots a_m b_1 b_2 \cdots b_n,$$

which evidently has length $m + n$. In particular, concatenation with the empty word ε is given by $\varepsilon v = v = v\varepsilon$ for each $v \in A^*$. One might thus give the following inductive definition for the set A^*

Basis Step: $\varepsilon \in A^*$

Induction Step: If $a \in A$ and $u \in A^*$, then $au \in A^*$.

■ **Example 9.1.3** Take $A = \{0, 1\}$ as in Example 9.1.1 and consider the words $u = 0010$ and $v = 110$, whose possible concatenations are $uv = 0010110$ and $vu = 1100010$. We might also take the concatenations of ε with $v = 110$ to get $\varepsilon v = 110 = v\varepsilon$. Furthermore, we have the concatenation

$$u\varepsilon v\varepsilon u = uvu = 00101100010.$$

■ **Example 9.1.4** Over the usual English alphabet, among the 6 possible concatenations of the words "tra", "vel" and "ing", only "traveling" is actually a natural word in the English language.

Thus, one concatenates words by simply "adjoining" the corresponding sequences of letters. As a binary operation $A^* \times A^* \to A^*$ on the set of words, concatenation is apparently not commutative. Concatenation of words is however evidently associative i.e., $(uv)w{=}u(vw)$ for any words $u, v, w \in A^*$, and we shall therefore usually just omit parentheses when concatenating words though we shall also include parentheses sometimes for clarity.

As usual when presented with an associative binary operation, we shall think of concatenation as a "multiplication" on A^*. For any $w \in A^*$ and $n \in \mathbb{N} \cup \{0\}$ we inductively define the *powers* w^n of w by

Basis Step: $w^0 = \varepsilon$

Induction Step: $w^{n+1} = w^n w$

The powers of a word are therefore simply iterated concatenations of the word with itself, and in particular $w^0 = \varepsilon$ and $w^1 = w$ for any word $w \in A^*$. Furthermore, arguing by induction as in Example 3.8.8, we have $w^m w^n = w^{m+n}$ and $(w^m)^n = w^{mn}$ for each $m, n \in \mathbb{N} \cup \{0\}$.

If $w = uv \in A^*$ is a concatenation of $u, v \in A^*$, then we say that u is a *prefix* and that v is a *suffix* of w. If w may be written as $w = uxv$ for $u, v, x \in A^*$, then we say x is a *subword* or *substring* of w. In particular, a prefix or suffix is necessarily a substring, and ε is both a prefix $w = \varepsilon w$, and a suffix $w = w\varepsilon$ of each word $w \in A^*$. Furthermore, a word w is both a prefix and a suffix of each power w^n for $n \geq 1$.

■ **Example 9.1.5** The word $w = 0101$ has possible prefixes 0, 01, 010, 0101, as well as ε, $\varepsilon 0$, 0ε, $0\varepsilon 1\varepsilon$, for instance, and possible suffixes 1, 01, 101, 0101, as well as others involving ε. For instance, 10 is a substring of $w = 0(10)01$, and

1 is a substring of $w = 0(1)01 = 010(1) = 010(1)$ in several different ways. By definition, we have

$$w^0 = \varepsilon,\ w^1 = 0101,\ w^2 = 01010101, \text{ and } w^3 = 010101010101$$

for instance.

■ **Example 9.1.6** Consider the words "tin" and "rin" over the English alphabet. We have

$$\text{rin}(\text{tin})^2 = \text{rintintin} \text{ and } ((\text{tin})^2\text{rin})^2(\text{tin})^2 = \text{tintinrintintinrintintin}$$

for instance.

We finally discuss orderings on A^* induced by a chosen linear ordering $\leq$ on the finite set A, and there are actually *two* different natural linear orderings $\leq_\ell$ and $\leq_s$ induced on A^*. Only $\leq_s$ is a well ordering on A^*, and this well ordering is of substantial utility for studying languages. On the other hand, $\leq_\ell$ is presumably already known to the reader as the usual "lexicographic" ordering in an English dictionary, and so both $\leq_\ell$ and $\leq_s$ are discussed here. We next define each of these linear orderings and begin with $\leq_\ell$.

Given the linear ordering $\leq$ on A, we define the *lexicographic ordering* $\leq_\ell$ on A^* by setting $u \leq_\ell v$ for $u, v \in A^*$ if either of the following conditions hold:

(i) u is a prefix of v

(ii) $u = wx$ and $v = wy$ where $w \in A^*$ is the longest common prefix of u and v, and the first letter of x (if any) precedes the first letter of y in the ordering $\leq$ on A. In particular, a word is preceded by each of its prefixes.

The lexicographic ordering of A^* corresponds to the usual "alphabetic ordering" of English words that one finds in dictionaries using the usual ordering $a \leq b \leq \cdots \leq z$ on the letters. In the ordering $\leq_\ell$, every element of A^* has an immediate successor, but if A has more than one element, then elements of A^* do not necessarily have an immediate successor.

■ **Example 9.1.7** Let $A = \{a, b\}$, and let $a \leq b$ describe the linear ordering on A. If $w \in A^*$, then the immediate successor of w is evidently wa, but there is no immediate predecessor of the string wb. Furthermore the set $\{b, ab, aab, \ldots\} = \{a^n b : n \geq 0\}$ has no least element, since $a^m b \leq_\ell a^n b$ if $m \geq n$.

Thus, the lexicographic ordering is not a well ordering in this example. The lexicographic ordering on A^* is however a linear ordering in general, as we leave it for the reader to check. Furthermore, restricting attention to the collection of words of some fixed length $n \geq 1$, the lexicographic ordering $\leq_\ell$ evidently corresponds to the usual lexicographic ordering considered in §5.9 on the Cartesian product A^n induced by $\leq$ on A, and the lexicographic ordering on A^n is a well ordering by Theorem 5.9.2 and Example 5.9.6. On the other hand, its extension $\leq_\ell$ to A^* where words have a variable length is not a well ordering.

We turn next to defining the other ordering $\leq_s$ on A^* induced by the chosen linear ordering $\leq$ on A: Define the *standard ordering* $\leq_s$ on A^* by setting $u \leq_s v$ for $u, v \in A^*$ if either of the following conditions hold:

(i) $||u|| < ||v||$

(ii) $||u|| = ||v||$, and u precedes v in the usual lexicographic ordering on the Cartesian product $A^{||u||}$, i.e., $||u|| = ||v||$ and $u \leq_\ell v$ in the lexicographic ordering on A^*.

■ **Example 9.1.8** Let $A = \{a, b, c\}$ be an alphabet with the linear ordering $a \leq b \leq c$. If $w \in A^*$, then the immediate successors under $\leq_s$ of wa, wb, and wc, respectively, are wb, wc, and va, where v is the immediate successor of w. The immediate predecessors under $\leq_s$ of wb and wc, respectively, are wa and wb. If $w \neq \varepsilon$, then the immediate predecessor of wa is vc, where v is the immediate predecessor of w. Furthermore, b is the least element of $\{\text{words } a^n b : n \geq 0\}$.

As before, it is not difficult to check that the standard ordering $\leq_s$ is a linear ordering on A^*, and this is again left as an exercise. Furthermore, given any non-empty set $X \subseteq A^*$, the least element of X for $\leq_s$ is simply the shortest element of X, i.e., the one of least length, which is least in the lexicographic ordering. Thus, any non-empty set has a least element, so $\leq_s$ is actually a well ordering.

To close, we summarize the previous discussion about orderings on A^* in

9.1.1 THEOREM *Fix a linear ordering $\leq$ on the alphabet A, and let $\leq_\ell$ and $\leq_s$ denote the corresponding lexicographic and standard orderings, respectively, which are induced by $\leq$ on A^*. Then $\leq_\ell$ is a linear ordering and $\leq_s$ is a well ordering on A^*.*

EXERCISES

9.1 ALPHABETS, WORDS, AND CONCATENATIONS

1. Given $w \in A^*$, define the *primitive root* $\rho(w)$ of w to be the shortest $u \in A^*$ so that $w = u^n$ for some $n \geq 1$.

   (a) Show that $uw = wu$ if and only if $\rho(u) = \rho(w)$.

   (b) Prove that if $u^m v^n = w^p$ for $m, n, p \geq 2$, then $\rho(u) = \rho(v) = \rho(w)$.

2. How many subwords are there of a given word w if there are no two occur-

rences of the same letter in w? Give upper bounds in various cases where not all the letters of w are distinct.

3. This exercise sketches some *Theorems of Thue* on repetitions. Fix a two letter alphabet $A = \{a, b\}$, and recursively define a word w_i of length 2^i for $i \geq 0$ as follows: $w_0 = a$, and w_{i+1} arises from w_i upon replacing each occurrence of a in w_i by ab (so for instance $w_1 = ab$) and each appearance of b in w_i by ba (so for instance $w_2 = abba$).

(a) Prove that $w_{i+1} = w_i w_i^{\#}$, where $w_i^{\#}$ arises from w_i upon replacing each occurrence of a in w_i by b and each occurrence of b in w_i by a.

(b) Show that none of $a^3, b^3, ababa, babab$ occurs as a subword of any w_i.

(c) Show that if a^2 or b^2 occurs as a subword of some w_i, then it starts with the jth letter of w_i, where j is necessarily even.

(d) Prove that each w_i is "strongly cube free" in the sense that it contains no subword $u^2 a$ where u is not empty and a is the first letter of u; in particular, no subword of any w_i is a cube.

9.2 LANGUAGES

A "language" is defined to be simply a set of words over a specified alphabet A. The usual set-theoretic operations such as union, intersection, and so on thus induce corresponding binary operations on the collection $\mathcal{L}(A)$ of all languages over A, and concatenation likewise induces a binary operation called "set product" on $\mathcal{L}(A)$. There is furthermore a unary operation on $\mathcal{L}(A)$ called the "Kleene closure" which is important in the sequel. These various operations give a rich structure to the collection of all languages over a fixed alphabet, and aspects of this structure are described.

We shall fix once and for all for this section an alphabet A. By definition, a *language* (*over* A) is simply a subset of A^*, that is, a collection of words over A. Let $\mathcal{L}(A)$ denote the set of all languages over A, so $\mathcal{L}(A)$ is simply the power set $\mathcal{P}(A^*)$ of A^*.

We mention parenthetically that the precise determination of a language $L \in \mathcal{L}(A)$ requires the specification of the set $L \subseteq A^*$ *together with* the specification of the alphabet A. For instance, the empty language $L = \emptyset$ represents different languages over different alphabets.

Given two languages $L, M \in \mathcal{L}(A)$ over A, we may evidently use the usual set-theoretic operations on $\mathcal{L}(A) = \mathcal{P}(A^*)$ to construct, for instance, the new languages $L \cup M$ and $L \cap M$ over A. Thus, $\mathcal{L}(A)$ comes equipped with the two commutative and associative binary operations

$$\begin{aligned} \mathcal{L}(A) \times \mathcal{L}(A) &\to \mathcal{L}(A) \\ (L, M) &\mapsto L \cup M \end{aligned}$$

and

$$\mathcal{L}(A) \times \mathcal{L}(A) \to \mathcal{L}(A)$$
$$(L, M) \mapsto L \cap M$$

Actually, unions of languages will turn out to be more interesting for us than intersections.

There is also another important binary operation on $\mathcal{L}(A)$ induced by concatenation which is defined as follows. If $L, M \in \mathcal{L}(A)$, then we define the *set product* or simply the *product* of L and M to be the language

$$LM = \{uv \in A^* : u \in L \text{ and } v \in M\} \subseteq A^*,$$

so in particular, we have $L\emptyset = \emptyset = \emptyset L$ and $L\{\varepsilon\} = L = \{\varepsilon\}L$ for any language L. Thus, an element of LM is simply the concatenation uv of words $u \in L$ and $v \in M$. Set product therefore provides yet another binary operation

$$\mathcal{L}(A) \times \mathcal{L}(A) \to \mathcal{L}(A)$$
$$(L, M) \mapsto LM$$

■ **Example 9.2.1** Over the alphabet $A = \{a, b\}$, define the languages $L = \{a, b, ab\}$ and $M = \{a, b\}$ to get

$$L \cup M = \{a, b, ab\}, \quad L \cap M = \{a, b\},$$

$$LM = \{a^2, ab, ba, b^2, aba, ab^2\}, \; ML = \{a^2, ab, a^2b, ba, b^2, bab\},$$

$$\text{and} \quad (L \cup M)(L \cap M) = \{a^2, ab, ba, b^2, aba, ab^2\}.$$

For a final explicit example of set products, we have

$$L(\{\varepsilon\} \cup M) = \{a, b, a^2, ab, ba, b^2, aba, ab^2\} = LM \cup L.$$

As was just illustrated, set product is not a commutative binary operation on $\mathcal{L}(A)$. On the other hand, we have

9.2.1 THEOREM *Set product is associative in the sense that for any languages $L, M, N \in \mathcal{L}(A)$, we have $(LM)N = L(MN)$.*

This result follows directly from associativity of concatenation of words itself and allows us to omit parentheses when convenient from multiple set products of languages. As usual under such circumstances, given the language $L \in \mathcal{L}(A)$, we make the inductive definition

Basis Step: $L^0 = \{\varepsilon\}$

Induction Step: $L^{n+1} = L^n L$ for $n \geq 1$

and prove the following result (left as an exercise) by induction.

9.2.2 THEOREM *For any language L over A and any $m, n \in \mathbb{N} \cup \{0\}$, we have*

$$L^m L^n = L^{m+n} \quad \text{and} \quad (L^m)^n = L^{mn}.$$

We next describe some relationships between the set-theoretic operations and the set product on $\mathcal{L}(A)$ and begin with the observation that "set products respect inclusions" in the following sense.

9.2.3 LEMMA *If $L_1 \subseteq L_2$ and $M_1 \subseteq M_2$ are languages over a common alphabet, then we have $L_1 M_1 \subseteq L_2 M_2$.*

Proof Given $w \in L_1 M_1$, we may write $w = uv$, where $u \in L_1 \subseteq L_2$ and $v \in M_1 \subseteq M_2$. Thus, in fact $w = uv \in L_2 M_2$, as desired. *q.e.d.*

Next, we prove that "set product distributes across unions but not across intersections" in the following sense.

9.2.4 THEOREM *Suppose that $M_\alpha \in \mathcal{L}(A)$ is a collection of languages over an alphabet A, where α lies in some index set Ω, and let $L \in \mathcal{L}(A)$. Then we have*

$$\begin{aligned} L(\cup_{\alpha \in \Omega} M_\alpha) &= \cup_{\alpha \in \Omega}(L M_\alpha), \\ (\cup_{\alpha \in \Omega} M_\alpha) L &= \cup_{\alpha \in \Omega}(M_\alpha L). \end{aligned}$$

On the other hand for intersections, we have only

$$\begin{aligned} L(\cap_{\alpha \in \Omega} M_\alpha) &\subseteq \cap_{\alpha \in \Omega}(L M_\alpha), \\ (\cap_{\alpha \in \Omega} M_\alpha) L &\subseteq \cap_{\alpha \in \Omega}(M_\alpha L). \end{aligned}$$

Proof We first prove the inclusion $L(\cup_{\alpha \in \Omega} M_\alpha) \subseteq \cup_{\alpha \in \Omega}(L M_\alpha)$, and to this end choose some $w \in L(\cup_{\alpha \in \Omega} M_\alpha)$, so $w = uv$ for some $u \in L$ and $w \in \cup_{\alpha \in \Omega} M_\alpha$. Thus, $w = uv$ where $u \in L$ and $v \in M_{\alpha_0}$ for some $\alpha_0 \in \Omega$, so in fact $w = uv \in L M_{\alpha_0} \subseteq \cup_{\alpha \in \Omega}(L M_\alpha)$.

For the reverse inclusion, suppose that $w \in \cup_{\alpha \in \Omega} L M_\alpha$, so that $w \in L M_{\alpha_0}$ for some $\alpha_0 \in \Omega$. Thus, $w = uv$ for some $u \in L$ and $v \in M_{\alpha_0} \subseteq \cup_{\alpha \in \Omega} M_\alpha$, so $w \in L(\cup_{\alpha \in \Omega} M_\alpha)$ using the previous lemma. This proves the first equality for unions, and the other one follows similarly.

Turning to the inclusion $L(\cap_{\alpha \in \Omega} M_\alpha) \subseteq \cap_{\alpha \in \Omega}(L M_\alpha)$, choose some $w \in L(\cap_{\alpha \in \Omega} M_\alpha)$, so $w = uv$, where $u \in L$ and $v \in \cap_{\alpha \in \Omega} M_\alpha$. Thus, $w = uv$ where $u \in L$ and $v \in M_\alpha$ for all $\alpha \in \Omega$, so $w = uv \in L M_\alpha$ for each $\alpha \in \Omega$ again by the previous lemma, and we conclude that $w \in \cap_{\alpha\in\Omega}(LM_\alpha)$, as desired. This proves the first inclusion for intersections, and again the other one follows similarly. *q.e.d.*

In particular, given languages $L \in \mathcal{L}(A)$ and $M_i \in \mathcal{L}(A)$ for $i \in \mathbb{N}$, we have

$$\begin{aligned} L(M_1 \cup M_2) &= LM_1 \cup LM_2 \text{ and } (M_1 \cup M_2)L = M_1L \cup M_2L, \\ L(\cup_{i\geq n} M_i) &= \cup_{i\geq n}(LM_i) \text{ and } (\cup_{i\geq n} M_i)L = \cup_{i\geq n}(M_iL), \end{aligned}$$

and

$$\begin{aligned} L(M_1 \cap M_2) &\subseteq LM_1 \cap LM_2 \text{ and } (M_1 \cap M_2)L \subseteq M_1L \cap M_2L, \\ L(\cap_{i\geq n} M_i) &\subseteq \cap_{i\geq n}(LM_i) \text{ and } (\cap_{i\geq n} M_i)L \subseteq \cap_{i\geq n}(M_iL). \end{aligned}$$

Let us give a simple example to prove that the reverse inclusions in the second part of the previous theorem do not hold.

■ **Example 9.2.2** Over the alphabet $A = \{a, b\}$, define the languages

$$L = \{b, ba\}, \quad M = \{a\}, \quad \text{and} \quad N = \{a^2\},$$

so $M \cap N = \emptyset$. We compute

$$LM = \{ba, ba^2\} \text{ and } LN = \{ba^2, ba^3\}$$

to find that $LM \cap LN = \{ba^2\}$ is not a subset of $L(M \cap N) = L(\emptyset) = \emptyset$.

Of basic importance to us below are the binary operations on languages of set product and union. There is also a basic unary operation $L \mapsto L^*$ called the *Kleene closure* or *star closure* on $\mathcal{L}(A)$ defined by

$$\begin{aligned} L^* &= \cup_{n\geq 0} L^n \\ &= L^0 \cup L^1 \cup L^2 \cup \cdots \\ &= \{\varepsilon\} \cup L \cup L^2 \cup \cdots. \end{aligned}$$

In particular, for any language L, we find $\varepsilon \in L^*$ and $L \subseteq L^*$.

■ **Example 9.2.3** Take the alphabet $A = \{0, 1\}$ and the language $L = \{01, 11\}$, so the Kleene closure L^* can be defined inductively by

Basis Step: $\varepsilon \in L^*$

Induction Step: If $w \in L^*$, then we have $w01, w11, 01w, 11w \in L^*$

In other words, the Kleene closure is simply the set whose elements are either ε or some word over the alphabet $A_L = \{01, 11\}$, that is, we have

$$L^* = A_L^* = \{\varepsilon, 01, 11, 0101, 0111, 1101, 1111, \ldots\}.$$

Among other things, the previous example illustrates the fact that our notation A^* for the words over an alphabet A and our notation L^* for the Kleene closure of a language $L \in \mathcal{L}(A)$ are consistent in the sense that if we regard the alphabet A itself as a language over A, then its Kleene closure as a language is none other than the set of all words over A.

■ **Example 9.2.4** Let $L \in \mathcal{L}(A)$ be any language. For any $n \geq 0$, we have

$$\begin{aligned} L^n L^* &= L^n \big(\cup_{i\geq 0} L^i\big) \\ &= \cup_{i\geq 0}\ L^n L^i \\ &= \cup_{i\geq 0}\ L^{n+i} \\ &= \cup_{i\geq 0}\ L^i L^n \\ &= \big(\cup_{i\geq 0} L^i\big) L^n \\ &= L^* L^n, \end{aligned}$$

where we have used distributivity of set product over unions Theorem 9.2.4 in the second and sixth equalities and Theorem 9.2.2 in several of the others.

By the previous example, for any language $L \in \mathcal{L}(A)$ and $n \geq 0$, we have $L^n L^* = L^n = L^* L^n$, so the Kleene closure L^* of L "commutes" with any power L^n of L. In fact, distributivity of set product over unions together with the definition of Kleene closure similarly lead to a whole class of identities, and we next give another such example.

■ **Example 9.2.5** Suppose that L, M are languages over a common alphabet. We compute

$$\begin{aligned} ML^* &= M\big(\cup_{i\geq 0} L^i\big) \\ &= M\big(L^0 \cup \cup_{i\geq 1} L^i\big) \\ &= ML^0 \cup M\big(\cup_{i\geq 1} L^i\big) \\ &= M \cup M\big(\cup_{i\geq 0} LL^i\big) \\ &= M \cup ML\big(\cup_{i\geq 0} L^i\big). \\ &= M \cup MLL^* \end{aligned}$$

Thus, for any languages L, M over a common alphabet, we find $ML^* = M \cup MLL^*$, and there is a similar formula $L^*M = M \cup LL^*M$. Further basic properties of Kleene closure are described in the following

9.2.5 THEOREM *Fix an alphabet A and let $L, M \in \mathcal{L}(A)$ be languages over A. Then*

(a) *If $L \subseteq M$, then $L^* \subseteq M^*$.*

(b) *We have* $L \subseteq LM^*$ *and* $L \subseteq M^*L$. *More generally, if* N *is any language with* $\varepsilon \in N$, *then* $L \subseteq NL$ *and* $L \subseteq LN$.

(c) *We have* $L^*L^* = L^* = (L^*)^*$.

(d) *We have* $(L^*M^*)^* = (L \cup M)^*$.

Proof For the first part, suppose that $w \in L^*$, so by definition $w \in L^n$ for some $n \geq 0$. Insofar as $L \subseteq M$, we conclude that $L^n \subseteq M^n$ by Lemma 9.2.3 using induction on n, whence $w \in L^n \subseteq M^n \subseteq M^*$, so $L^* \subseteq M^*$.

For part (b), observe that if $\varepsilon \in N$, then any word $w \in L$ can be written as $w = w\varepsilon \in LN$, so indeed $L \subseteq LN$. In particular, $\varepsilon \in M^*$ for any language M, so $L \subseteq LM^*$. One proves that $L \subseteq LN$ and hence $L \subseteq LM^*$ in the same way.

Turning to part (c), let us first prove the equality $L^*L^* = L^*$ and choose some $w \in L^*L^*$, so that w may be written in the form $w = uv$ for $u, v \in L^*$, that is $u \in L^m$ and $v \in L^n$ for some $m, n \geq 0$. Thus, $w = uv \in L^mL^n = L^{m+n} \subseteq L^*$, so we have $L^*L^* \subseteq L^*$. The reverse inequality follows from part (b) since $\varepsilon \in L^*$.

Continuing with part (c), we turn to the equality $(L^*)^* = L^*$, and $L^* = (L^*)^1 \subseteq (L^*)^*$ by definition. For the reverse inclusion, we choose some $w \in (L^*)^*$, so by definition $w \in (L^*)^n$ for some $n \geq 0$, i.e., $w = u_1u_2 \cdots u_n$, where each $u_i \in L^*$. Again by definition, each $u_i \in L^*$ lies in some L^{m_i} for some $m_i \geq 0$, so w is a concatenation of $\mu = \sum_{i=1}^n m_i$ words in L. Thus, $w \in L^\mu \subseteq L^*$, so indeed $(L^*)^* \subseteq L^*$, as desired.

Turning to part (d), we first concentrate on the inclusion $(L^*M^*)^* \subseteq (L \cup M)^*$. Observe that $L \subseteq L \cup M$ shows that $L^* \subseteq (L \cup M)^*$ by part (a). We similarly deduce that $M^* \subseteq (L \cup M)^*$, so the set product satisfies

$$\begin{aligned} L^*M^* &\subseteq (L \cup M)^*(L \cup M)^* \\ &= (L \cup M)^* \end{aligned}$$

where the equality follows from part (c) and the inclusion from Lemma 9.2.3. Applying parts (a) and (c) once again, we conclude that

$$(L^*M^*)^* \subseteq \left((L \cup M)^*\right)^* = (L \cup M)^*,$$

completing the proof of the first inclusion.

For the final inclusion $(L \cup M)^* \subseteq (L^*M^*)^*$ in part (d), observe that $L^* \subseteq L^*M^*$ by part (b), so also $L \subseteq L^* \subseteq L^*M^*$. In the same way, one deduces that $M \subseteq L^*M^*$, and so $L \cup M \subseteq L^*M^*$. Again applying part (a), we derive the required inclusion. *q.e.d.*

Part (b) of the previous result illustrates the fact that various general properties of a language L depend upon whether $\varepsilon \in L$, and indeed this is a basic distinction among languages.

Recall that if A is an alphabet with a linear ordering $\leq$, then there is an induced well ordering $\leq_s$ on A^* which was discussed at the end of the previous section. This well ordering restricts to a well ordering on any language $L \subseteq A^*$, so a linear ordering on an alphabet A determines a natural well ordering on any language L over A.

Strong induction over a language L using this well ordering can be a useful proof technique especially when the language L is defined inductively. As a particular case, one can often simply induct on the length of words in L, and we next pursue an example of this.

■ **Example 9.2.6** Take the alphabet $A = \{[,]\}$ with the linear ordering $[\ \leq\ ]$, and define the language $B \subseteq A^*$ (where the "B" stands for "bracket") according to the following inductive definition:

Basis Step: $\varepsilon \in B$

Induction Step: If $b \in B$, then we have $[\ ]b \in B$, $b[\ ] \in B$, and $[b] \in B$.

Given an element $b \in B$, we may define $left(b) \in \mathbb{N}$ and $right(b) \in \mathbb{N}$, respectively, to be the number of letters in b which are left brackets "[" and right brackets "]". We claim that for each $b \in B$, we have $left(b) = right(b)$, and the proof is by induction on the length $n \geq 0$ of a word $w \in B$. If $n = 0$, then $w = \varepsilon$, and indeed $left(w) = 0 = right(w)$. For the induction step, given a word w of length $n + 1$, suppose that $left(u) = right(u)$ for each $u \in B$ so that $u \leq_s w$ and $u \neq w$. (This is the induction hypothesis for strong induction). In particular, if $||u|| < ||w||$, then $u \leq_s w$ so $left(u) = right(u)$. Writing $w = [\ ]u$, $w = u[\ ]$, or $w = [u]$, we find that $left(w) = left(u) + 1 = right(u) + 1 = right(w)$, as desired.

For a more interesting example of such an inductive proof, suppose that $L, M \in \mathcal{L}(A)$ are languages over an alphabet A, and consider the "equation"

$$(*) \qquad X = LX \cup M$$

for a language $X \in \mathcal{L}(A)$. That is, given $L, M \in \mathcal{L}(A)$, we seek $X \in \mathcal{L}(A)$ so that $X = LX \cup M$ and observe that if $X = L^*M$, then indeed equation $(*)$ holds by Example 9.2.5. Thus, our equation $(*)$ has the solution $X = L^*M$ in any case.

■ **Example 9.2.7** Fix the alphabet $A = \{0, 1, a\}$ and define the languages $L = \{0, 01\}$, $M = \{aa\}$. The equation $X = LX \cup M$ has solution $X = \{0, 01\}^*\{aa\}$.

■ **Example 9.2.8** Suppose that $\varepsilon \in L$, let $N \supseteq M$ be any superset of M, and define $X = L^*M$, so the equation $(*)$ in this case reads $L^*N = LL^*N \cup M$. Since we are assuming that $\varepsilon \in L$, we have $LL^*N = L^*N$ and $M \subseteq N \subseteq L^*N$, so the equation $(*)$ holds for any choice of superset $N \supseteq M$. Thus, if $\varepsilon \in L$, the equation $(*)$ is solved by L^*N for any language $N \supseteq M$. We mention here that we shall solve $(*)$ in case $\varepsilon \notin L$ presently.

It is instructive to consider the conditions imposed on a language X which solves equation $(*)$, and of course $M \subseteq X$ by definition. Since $LX \subseteq X$ also by definition, we conclude that $LM \subseteq X$. Repeatedly applying this observation, i.e., as the reader should prove by induction, we conclude that $L^n M \subseteq X$ for each $n \geq 0$, so in fact $L^*M = \cup_{i\geq 0} L^i M \subseteq X$ for any solution X to $(*)$.

Conversely, if $w \in X = LX \cup M$, then either $w \in M$ or else $w \in LX$ in which case $w = uv$ has a prefix $u \in L$. If we assume moreover that $\varepsilon \notin L$, then this prefix is a non-trivial word, and hence $w \in X$ determines another word $v \in X$ of strictly smaller length; indeed, there is one such word $v \in X$ for each prefix $u \in L$ of $w \in X$. These remarks lead us to

9.2.6 THEOREM *If $L, M \in \mathcal{L}(A)$ are languages over an alphabet A and $\varepsilon \notin L$, then $X = L^*M \in \mathcal{L}(A)$ is the unique language so that $X = LX \cup M$.*

Proof In our discussion above, we have already proved that $L^*M \subseteq X$, and we prove here that $X \subseteq L^*M$. Choose some $w \in X$ and proceed by induction on the length of w. For the basis step, if $||w|| = 0$ then $w = \varepsilon$, so $w = \varepsilon \in X = LX \cup M$. We must therefore conclude that $\varepsilon \in M$ (since we assume that $\varepsilon \notin L$ and hence $\varepsilon \notin LX$). Thus, $\varepsilon \in M \subseteq L^*M$ by Theorem 9.2.5b.

For the induction step of our strong induction, consider a word $w \in X$ and suppose that for each word $v \in X$ satisfying $||v|| < ||w||$ we do indeed have $v \in L^*M$. We must show that also $w \in L^*M$. To this end, since $w \in X = LX \cup M$, we must have either $w \in M$ or $w \in LX$. In particular, if $w \in M$, then $w \in L^*M$ by Theorem 9.2.5b. If $w \in LX$, then $w = uv$ where $u \in L$ is a prefix and $v \in X$. Since $\varepsilon \notin L$, $u \neq \varepsilon$ must be non-trivial, and we therefore have $||v|| < ||w||$ for $v \in X$. Thus, $v \in L^*M$ by the induction hypothesis, and it follows, therefore, that

$$uv \in LL^*M = \cup_{i\geq 1} L^i M \subseteq \cup_{i \geq 0} L^i M = \subseteq L^*M,$$

where we have freely used Theorem 9.2.4 in writing the equalities. *q.e.d.*

EXERCISES

9.2 LANGUAGES

1. Prove that $(L^*M^*)^* = (L^* \cup M^*)^* = (L \cup M)^* = (M^*L^*)^*$.

2. Define singletons $A = \{a\}$ and $B = \{b\}$, and solve the following simultaneous equations for languages L, M over $\{a, b\}$.

   (a) $L = AL \cup BM$ and $M = (A \cup B)L \cup BM \cup \{\varepsilon\}$

   (b) $L = AL$ and $M = AB(L \cup \{\varepsilon\})$

3. A language L is said to have the "finite power property" or "satisfy FPP" if the set $\{L^i : i \geq 0\}$ is finite.

(a) Prove that the following three conditions are equivalent: (i) L has FPP; (ii) there is some $k \geq 0$ so that $L^k = L^{k+1}$; (iii) there is some $k \geq 0$ so that $L^k = L^*$.

(b) Show that no finite language containing a nonempty word satisfies FPP.

9.3 REGULAR EXPRESSIONS

We define and give examples of "regular expressions" here, which allow for the convenient specification of a class of languages, called "regular languages", over a given alphabet. We shall show in §9.5 that this class of regular languages can equivalently be described as the languages determined (in a sense to be made precise in the next section) by finite state automata.

A "regular expression" over an alphabet A is a special type of string to be specified below over the new alphabet $A \cup \{\varepsilon, *, \vee, (,)\}$, where we require that $\varepsilon, *, \vee, (,) \notin A$. We read "$\vee$" as the English word "vee" or "or", we read "$*$" as the English word "star", we shall use parentheses "(" and ")" as delimiters, and we intentionally use the same symbol ε here as for the empty word over A.

A single regular expression r (which has yet to be defined) over an alphabet A uniquely determines a language $L(r) \in \mathcal{L}(A)$ over A as will be described presently. We give an inductive definition of a *regular expression* r (*over* A) together with an inductive definition of its corresponding language $L(r) \in \mathcal{L}(A)$ as follows:

Basis Step: There are no regular expressions of length zero. The regular expressions r of length one are either $r = \varepsilon$ or $r = a$, where $a \in A$. To $a \in A$, we associate the singleton $L(a) = \{a\} \in \mathcal{L}(A)$, and to ε, we associate the singleton $L(\varepsilon) = \{\varepsilon\} \in \mathcal{L}(A)$. There is a unique regular expression of length two, namely $r = ()$, and the language it defines is the null set $L(()) = \emptyset \in \mathcal{L}(A)$.

Induction Step: If r, s are regular expressions, then so too are:

$$\begin{aligned}
(r) &\text{ and } L((r)) = L(r) \\
(r^*) &\text{ and } L((r^*)) = [L(r)]^* \\
(r \vee s) &\text{ and } L(r \vee s) = L(r) \cup L(s) \\
(rs) &\text{ and } L((rs)) = [L(r)][L(s)].
\end{aligned}$$

This simultaneously defines a regular expression r over A as well as its corresponding language $L(r) \in \mathcal{L}(A)$. Formally, a regular expression is thus a certain kind of string over the alphabet $A \cup \{\varepsilon, *, \vee, (,)\}$, and it is evidently convenient to imagine $*$ as a unary operation and $\vee$ and concatenation as binary operations on the collection of all regular expressions. According to the definition, concatenation of regular expressions corresponds to set product of their languages, the star operation $*$ corresponds to the Kleene closure, the $\vee$ operation corresponds to union, and the parentheses are included for definiteness.

As usual, we often omit parentheses in practice since for instance concatenation is associative and $\vee$ is commutative and associative. In order to omit further parentheses, we make the conventions that: $*$ is applied first, concatenation second, and $\vee$ third. For instance, if $A = \{a, b\}$, then $a \vee ab$ means $a \vee (ab)$, ab^* means $a(b^*)$, $a \vee b^*$ means $a \vee (b^*)$, and $a^*b \vee a^*$ means $((a^*)b)) \vee (a^*)$. Using some standard terminology from computer science, one could say here that $*$ "binds more tightly" than concatenation and that concatenation "binds more tightly" than $\vee$. As usual, we shall feel free to include parentheses for clarity when convenient.

■ **Example 9.3.1** Take the alphabet $A = \{0, 1\}$ and refer to the table below for several examples of regular expressions and their corresponding languages.

Regular Expression r	Corresponding Language $L(r)$
0	$\{0\}$
1	$\{1\}$
$0 \vee 1$	$\{0, 1\}$
01	$\{01\}$
0^*	$\{0\}^* = \{\varepsilon, 0, 00, 000, \ldots\}$
$(01)^*$	$\{01\}^* = \{\varepsilon, 01, 0101, 010101, \ldots\}$
0^*1	$\{0\}^*\{1\} = \{1, 01, 001, 0001, \ldots\}$
01^*	$\{0\}\{1\}^* = \{0, 01, 011, 0111, \ldots\}$
$0 \vee 1^*$	$\{0\} \cup \{1\}^* = \{0, \varepsilon, 1, 11, 111, \ldots\}$
$0^* \vee 1^*$	$\{0\}^* \cup \{1\}^* = \{\varepsilon, 0, 00, \ldots, 1, 11, \ldots\}$
ε^*	$\{\varepsilon\}^* = \{\varepsilon\}$
$(0 \vee 1)\varepsilon^*1^*$	$\{0, 1\}\{\varepsilon\}\{1\}^* = \{0, 01, 011, \ldots, 1, 11, 111, \ldots\}$

To recapitulate, we have given an inductive definition of a regular expression r over an alphabet A as a certain kind of string over $A \cup \{\varepsilon, *, \vee, (,)\}$ as well as the definition of its corresponding language $L(r)$ over A.

We say that a language $L \in \mathcal{L}(A)$ is *regular* if it arises in this way from some regular expression, that is, $L \in \mathcal{L}(A)$ is a regular language over A if $L = L(r)$ for some regular expression r over A. Not every language is regular, but those that are form an important class of languages for applications and are our main interest here.

We wish to emphasize that many different regular expressions can give rise to the same language. For instance, it follows from Theorem 9.2.5d that for any regular expressions r and s over any language A, we have $L((r^*s^*)^*) = L((r \vee s)^*)$. There is evidently a whole "calculus" here for determining when two regular expressions give rise to the same language in the same way that the predicate calculus describes when two predicate forms are logically equivalent; see Exercise 9.3.1 below.

■ **Example 9.3.2** We prove that for any two regular expressions r, s over a common alphabet A, we have $L((r^* \vee s^*)(r^* \vee s^*)) = L(r^*(\varepsilon \vee ss^*) \vee s^*(\varepsilon \vee sr^*))$. To prove this, we set $L = L(r), M = L(s) \in \mathcal{L}(A)$ and compute

$$\begin{aligned}(L^* \cup M^*)(L^* \cup M^*) &= L^*L^* \cup L^*M^* \cup M^*L^* \cup M^*M^* \\ &= L^* \cup M^* \cup (L^* \cup L^*MM^*) \cup (L^* \cup MM^*L^*) \\ &= L^* \cup M^* \cup L^*MM^* \cup MM^*L^* \\ &= L^* \cup M^* \cup L^*MM^* \cup M^*ML^* \\ &= L^*(\{\varepsilon\} \cup MM^*) \;\cup\; M^*(\{\varepsilon\} \cup ML^*),\end{aligned}$$

where we have freely used commutativity and associativity of $\cup$, the first and last equalities follow from distributivity of set product over unions Theorem 9.2.4, the second follows from part Theorem 9.2.5c and Example 9.2.5, and the fourth follows as in Example 9.2.4.

EXERCISES

9.3 REGULAR EXPRESSIONS

1. Here are a collection of valid identities of languages associated with arbitrary regular expressions r, s, t:

$$\begin{aligned} L(r \vee s) &= L(s \vee r), & L(r \vee (s \vee t)) &= L((r \vee s) \vee t) \\ L(r(s \vee t)) &= L(rs \vee rt), & L((r \vee s)t) &= L(rt \vee st) \\ L(r^*) &= L(r^*r \vee \varepsilon^*), & L(r^*) &= L((r \vee \varepsilon^*)^*), \\ L(\varepsilon^* r) &= L(r) = L(\varepsilon r), & L(r(st)) &= L((rs)t). \end{aligned}$$

There is furthermore the following rule of inference: If $L(r) = L(rs \vee t)$, then also $L(r) = L(ts^*)$ provided $L(s)$ does not contain the empty word.

(a) Prove the validity of each of the identities and inferences above.

(b) Show that all relations among languages associated with regular expressions can be derived from the identities and inferences above.

2. Define the *star height* $sh(r)$ of a regular expression r over the alphabet A recursively as follows: (i) the letters of A and ε have star height zero; (ii) if $sh(r) = i$ and $sh(s) = j$, then $sh(r \vee s) = sh(rs)$ is the maximum of i and j; (iii) if $sh(r) = i$, then $sh(r^*) = i + 1$. Finally, the *star height* of a regular language L is the least star height of any regular expression r so that $L(r) = L$.

(a) Prove the following two identities

$$\begin{aligned} L((rs^*t)) &= L(\varepsilon^* \vee r(s \vee tr)^*t), \\ L((sr \vee r^*rs)^*r^*) &= L((r \vee rs \vee sr)^*), \end{aligned}$$

which exhibit non-trivial reductions in the star height of regular expressions representing a given language.

(b) Show that the star height of a language is zero if and only if the language is finite.

9.4 FINITE STATE AUTOMATA

A "finite state automaton" is a mathematical model of how a certain electronic machine functions while running a program. At the same time, a finite state automaton together with an alphabet A determines a language $L \in \mathcal{L}(A)$ called the language "accepted" by the automaton which is roughly the set of all legitimate strings of input to the electronic machine. There is a valuable description of a finite state automaton in terms of a slight generalization of directed graphs in the sense of §5.2, and we use this description to prove that a language accepted by a finite state automaton is necessarily regular.

Before giving the precise definition of a "finite state automaton", we shall imagine a mythical electronic machine which reads a letter from a strip of magnetic tape and then alters its state depending upon the letter read.

Suppose that A is an alphabet and consider the corresponding collection A^* of all words with letters in A. An element $w \in A^*$ is a concatenation $w = a_1 a_2 \cdots a_m$ of letters, and we imagine a strip of magnetic tape on which we have recorded the letters $a_1, a_2, \ldots, a_m$, in this order.

Fix another finite set S, called the "state set" and imagine a machine which can be in exactly one of the states in S at any fixed moment. There is a subset $Y \subseteq S$ called the set of "accept states", and the machine has successfully terminated its task if it is one of these states in $Y \subseteq S$, where "Y" stands for "Yes". The state set moreover comes equipped with a distinguished element $s_0 \in S$, called the "initial state" or "start state" of the machine.

Imagine that we begin with our machine in the state s_0 and read the magnetic tape containing the string $w = a_1 a_2 \ldots a_m$ one letter at a time. Each letter a in w is meant to be a specific command to the machine to change its state in some specified way. Our machine serially reads the string $a_1 a_2 \cdots a_m$ in order from right to left (advancing the tape head) starting with a_m and ending with a_1 changing states as specified, and after exhausting the string w (so after m reads from the tape), we find the machine in some state $s \in S$. If $s \in Y$ so s is an "accept state", then our machine has successfully read the string w, and the machine "accepts" the string w.

Notice that while there are possibly many accept states, there must be a unique start state $s_0 \in S$ (though we shall relax this condition in the next section). Furthermore, the start state s_0 can be an accept state, i.e., $s_0 \in Y$, and in this case our machine accepts the empty string $\varepsilon \in A^*$.

To finish describing our machine, we must specify how it changes state for each letter $a \in A$, that is, for each $a \in A$ we must specify a corresponding function $d_a : S \to S$ to describe how the state of the machine should change upon reading the letter "a" for each $a \in A$. In other words, we must describe a function

$$\begin{aligned} A &\to S^S \\ a &\mapsto (d_a : S \to S). \end{aligned}$$

For another point of view on this function, define the associated *transition function*

$$\begin{aligned} d : A \times S &\to S \\ (a, s) &\mapsto d_a(s). \end{aligned}$$

Of course, the transition function itself determines the function $A \to S^S$ according to the rule $d_a(s) = d(a, s)$, so the two points of view are entirely equivalent. It is customary to consider the transition function $d : A \times S \to S$ rather than the associated function $A \to S^S$ as a description of the rules for the evolution of states of our machine.

Turning from this motivational image of a machine to more precise statements, we define a *finite state automaton* or simply an *automaton* to be a quintuple (S, A, d, Y, s_0), where

- S is a finite set called the *state set.*
- A is an alphabet as before.
- $d : A \times S \to S$ is a function called the *transition function.*
- Y is a (possibly empty) subset of S called the set of *accept states, success states*, or *final states.*
- s_0 is a distinguished element of S called the *initial state* or *start state.*

There is actually a generalization of this definition which we shall give in the next section, and we shall later refer to a finite state automaton (S, A, d, Y, s_0) as above as a *deterministic finite state automaton* to distinguish it from the generalization, which is called a "non-deterministic finite state automaton".

Before giving examples, we define a "digraph" G, called the *state diagram* of an automaton (S, A, d, Y, s_0), as follows: The vertices of G are labeled by the states S, so there is one vertex v_s of G for each $s \in S$; for each $s, t \in S$, we then adjoin an oriented edge from v_s to v_t with the label $a \in A$ if $d(a, s) = t$. We have used quotation marks above around the appellate digraph because the object G just described often fails to be a digraph in the sense of §5.2: There may be many edges with given initial and terminal point in the object G just constructed whereas in an honest digraph, there can be only one edge with this given data. For instance, if $d(a, s) = t = d(b, s)$ in the notation above, where $a, b \in A$, then there are two "edges" in G from v_s to v_t, namely, one corresponding to each of a and b.

This is the only sense in which our "digraph" G is not an honest digraph, and an object such as G is often called a "multi-digraph". Rather than go through the formalities of introducing multi-digraphs here, we accept this rough description as suitable for this chapter and adopt all the usual terminology of edges, edge-paths, cycles, and so on, in our current setting. This abuse should cause no confusion here, and we shall give a more careful discussion of multi-digraphs in Chapter 10.

Not only is the state diagram of a finite state automaton a multi-digraph, but each vertex is labeled by some element of S and each edge is labeled by some element of A. Notice that each vertex of G has exactly one oriented edge with label a beginning at it for each element $a \in A$ since d is a function.

It is customary to use small circles to denote the vertices in the state diagram of an automaton in order to put the labels of the vertices inside these circles. Furthermore, the initial state s_0 is often indicated by drawing an unlabeled arrow "$\rightarrow$" pointing at the corresponding vertex, and a vertex corresponding to an accept state is indicated by drawing two small concentric circles rather than a single small circle.

■ **Example 9.4.1** Let $S = \{s_0, s_1, s_2, s_3\}$, define the accept states $Y = \{s_2, s_3\}$, and take the alphabet $A = \{a, b\}$. In order to define the transition function d, we exhibit a table of elements of S, where the columns are indexed by elements $a \in A$, the rows are indexed by elements $s \in S$, and the corresponding entry in the table is $d(a, s)$.

For example, the table

	a	b
s_0	s_2	s_2
s_1	s_2	s_1
s_2	s_3	s_1
s_3	s_0	s_3

defines a transition function $d : A \times S \rightarrow S$ with $d(a, s_3) = s_0$, $d(b, s_1) = d(b, s_2) = s_1$, $d(a, s_0) = d(b, s_0) = d(a, s_1) = s_2$, and $d(a, s_2) = d(b, s_3) = s_3$. In other words, the column corresponding to $a \in A$ is simply the tuple of values of $d_a(s) = d(a, s)$ as s varies over S. This uniquely determines the finite state automaton (S, A, d, Y, s_0), and the corresponding state diagram is

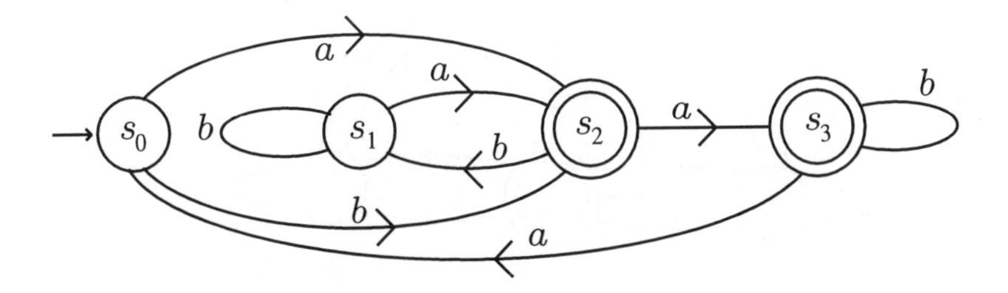

The table of values $d(a, s)$ indexed by $A \times S$ as in the previous example is

called the *state table* of the automaton, and this is the customary way of describing the transition function d of an automaton: exhibit the table of values $d(a, s)$ for $a \in A$ and $s \in S$. The fact that the transition function d can be specified by the state table has the practical consequence that actual computers can quickly access the transition function as a table lookup.

■ **Example 9.4.2** We construct here the finite state automaton (S, A, d, Y, s_0) determined by the state diagram

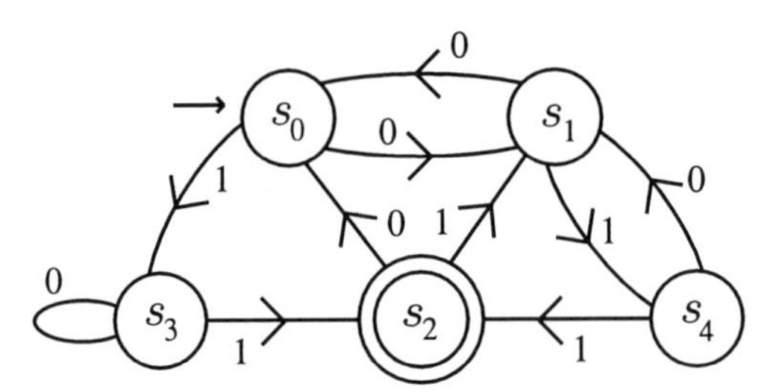

The state set is $S = \{s_0, s_1, s_2, s_3, s_4\}$, the alphabet is $A = \{0, 1\}$, the initial state is s_0, the accept states are s_2, s_4, and the transition function d is determined by the state table

	0	1
s_0	s_1	s_3
s_1	s_0	s_4
s_2	s_0	s_1
s_3	s_3	s_2
s_4	s_1	s_2

■ **Example 9.4.3** Consider the state diagram

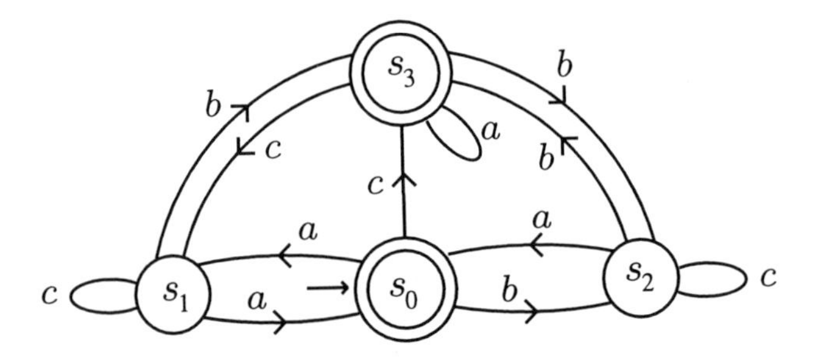

with corresponding alphabet $A = \{a, b, c\}$, state set $S = \{s_0, s_1, s_2, s_3\}$, initial state $s_0 \in S$ and final states $Y = \{s_0, s_3\}$. The transition function is given by the state table

	a	b	c
s_0	s_1	s_2	s_3
s_1	s_0	s_3	s_1
s_2	s_0	s_3	s_2
s_3	s_3	s_2	s_1

Let (S, A, d, Y, s_0) be a finite state automaton and reconsider our imaginary machine above. Suppose that $w = a_1 a_2 \cdots a_m \in A^*$ is a concatenation of $m \geq 0$ letters, and let our machine begin in the start state s_0. We read a_m from the magnetic tape provided $m \geq 1$, change states from s_0 to $d_{a_m}(s_0)$, and advance the tape head. We then read a_{m-1} from the magnetic tape provided $m \geq 2$, change states from $d_{a_m}(s_0)$ to $d_{a_{m-1}} \circ d_{a_m}(s_0) = d_{a_{m-1}}(d_{a_m}(s_0))$, and advance the tape head. When we have in this way read all of the letters of w after m reads from the magnetic tape, the start state has evolved to

$$d_{a_1} \circ d_{a_2} \circ \cdots \circ d_{a_m}(s_0),$$

so we are naturally led to consider this composition of the functions d_{a_1}, d_{a_2}, $\ldots, d_{a_m} \in S^S$.

Define the *w-transition*

$$d_w = d_{a_1} \circ d_{a_2} \circ \cdots \circ d_{a_m} : S \to S$$

for any word $w = a_1 a_2 \cdots a_m \in A^*$, or more formally, make the following inductive definition of $d_w \in S^S$ for each $w \in A^*$:

Basis Step: d_ε is the identity map $1_S : S \to S$.

Induction Step: if $w = ua$ for $a \in A$, then $d_w = d_{ua} = d_u \circ d_a$ for any $a \in A$.

This assignment

$$A^* \to S^S$$
$$w \mapsto d_w,$$

allows us to determine the state $d_w(s_0)$ to which the start state s_0 of our imaginary machine evolves upon reading the string w.

In keeping with the discussion of our imaginary machine, we say that a finite state automaton (S, A, d, Y, s_0) *accepts* or *recognizes* the string $w \in A^*$ if $d_w(s_0) \in Y$, i.e., if the initial state maps to an accept state under the w-transition. The state diagram of the automaton (S, A, d, Y, s_0) is a useful tool for "seeing" whether a given word $w \in A^*$ is accepted by it, as we illustrate in

■ **Example 9.4.4** Recall the finite state automaton (S, A, d, Y, s_0) defined in Example 9.4.2 over the alphabet $A = \{0, 1\}$. Consider the string 0010 over A.

Inspecting the state diagram above, we start in the initial state s_0 and serially read the letters: 0 (to produce s_1), then 1 (to produce s_4), then 0 (to produce s_1), then 0 (to finally produce s_0). Since $s_0 \notin Y = \{s_2, s_4\}$, the word 0010 is not accepted by (S, A, d, Y, s_0). For a positive example with the same automaton, consider the word 11101. We start in the initial state s_0 and serially read the letters: 1 (to produce s_3), then 0 (to produce s_3 again), then 1 (to produce s_2), then 1 (to produce s_1), then 1 (to finally produce s_4). Since $s_4 \in Y = \{s_2, s_4\}$, we conclude that 11101 is accepted by (S, A, d, Y, s_0).

Given $w = a_1 a_2 \cdots a_m \in A^*$ and a finite state automaton (S, A, d, Y, s_0), we are led to consider the oriented edge-path

$$s_0, d_{a_m}(s_0), d_{a_{m-1}a_m}(s_0),, \ldots, d_{a_1 a_2 \cdots a_m}(s_0) = d_w(s_0)$$

on the state diagram. This oriented edge-path has the start state s_0 as initial point and the state $d_w(s_0)$ as terminal point. The word w is evidently accepted by the automaton (S, A, d, Y, s_0) if and only if the corresponding oriented edge-path terminates at a vertex in Y. Thus, oriented edge-paths on the state diagram beginning at s_0 and ending in Y correspond exactly to words accepted by the automaton.

The state diagram itself therefore provides a convenient graphical description of an algorithm for determining when a given word $w \in A^*$ is accepted by the automaton as we next illustrate.

■ **Example 9.4.5** Continuing with the finite state automaton (S, A, d, Y, s_0) defined in Example 9.4.2, consider the word 110110. The corresponding oriented edge-path is

$$s_0, s_1, s_4, s_2, s_0, s_3, s_2$$

as the reader may verify by inspecting the state diagram above. Since this oriented edge-path terminates at the accept state $s_2 \in Y = \{s_2, s_4\}$, we conclude that the word 01101 is accepted by the automaton.

■ **Example 9.4.6** Consider the finite state automaton in Example 9.4.3, where we illustrated the state diagram. Since s_0 is an accept state, the automaton accepts the empty word ε. Furthermore, the oriented edge-path s_0, s_3, s_2, s_0 is a closed loop on the state diagram corresponding to the word abc which begins and ends at $s_0 \in Y$. It follows that any word $w = (abc)^n$ is accepted by the automaton for any $n \geq 0$ (as the reader should prove by induction). In the same way, for any $m, n \geq 0$, the word $w = a^m c(abc)^n$ is also accepted by the automaton.

A finite state automaton (S, A, d, Y, s_0) thus determines a corresponding language

$$L(S, A, d, Y, s_0) = \{w \in A^* : w \text{ is accepted by } (S, A, d, Y, s_0)\} \in \mathcal{L}(A)$$

called the language *accepted* or *recognized* by the automaton. The state diagram itself describes an algorithm for deciding whether a given string w is a word in $L(S, A, d, Y, s_0)$, namely, read off (from right to left) the oriented edge-path corresponding to w starting at s_0, and conclude that $w \in L(S, A, d, Y, s_0)$ if and only if the endpoint of this edge-path lies in Y.

Using this identification between oriented edge-paths on the state diagram and words accepted by the automaton, we next prove that a language accepted by an automaton must be regular. This assertion together with its converse (which is proved in the next section) is called "Kleene's Theorem".

9.4.1 LEMMA *Let (S, A, d, Y, s_0) be a finite state automaton over the language A accepting the language $L = L(S, A, d, Y, s_0)$ over A. Then there is a regular expression r over A so that $L = L(r)$ is defined by r.*

Proof Given an oriented edge-path $s_1, s_2, \ldots, s_m$ in the state diagram corresponding to (S, A, d, Y, s_0), we say that the path *visits* the states $s_2, \ldots, s_{m-1}$, so an oriented edge-path may or may not visit its own initial point s_1 and terminal point s_m. Given any ordered pair $(s_1, s_2) \in S^2$ of states, we define the language $L(s_1, s_2, X)$ to be the collection of all $w \in A^*$ whose corresponding oriented edge-path has initial point s_1, terminal point s_2, and visits only states in X. In particular, notice that

$$L(S, A, d, Y, s_0) = \cup_{y \in Y} L(s_0, y, S)$$

by definition. We shall prove that each $L(s_1, s_2, X)$ is actually a regular language by induction on the size of X.

For the basis step $X = \emptyset$, there are a finite number of oriented edges with initial point s_1 and terminal point s_2, and we let their labels be $x_1, x_2, \ldots, x_m \in A$. In this case, we have

$$L(s_1, s_2, X) = \begin{cases} L(x_1 \vee x_2 \vee \cdots \vee x_m), & \text{if } m \neq 0; \\ \emptyset, & \text{if } m = 0 \text{ and } s_1 \neq s_2; \\ \{\varepsilon\}, & \text{if } m = 0 \text{ and } s_1 = s_2, \end{cases}$$

so if $X = \emptyset$, then $L(s_1, s_2, X)$ is indeed defined by a regular expression over A.

If $X \neq \emptyset$, then let us choose $s \in S$ and define the languages $L_1 = L(s_1, s_2, X - \{s\})$, $L_2 = L(s_1, s, X - \{s\})$, $L_3 = L(s, s, X - \{s\})$, $L_4 = L(s, s_2, X - \{s\})$. The identity

$$L(s_1, s_2, X) = L_1 \cup (L_2 L_3^* L_4)$$

simply expresses the fact that any edge-path from s_1 to s_2 either does not visit s or if it does, then it may be written as the concatenation of a edge-path from s_1 to s then a loop based at s and finally a path from s to s_2. We invite the reader to give a formal proof here by induction on the number of appearances of s in the edge-path.

By the induction hypothesis, each of the languages L_1, L_2, L_3, L_4 is defined by a respective regular expression r_1, r_2, r_3, r_4, and hence $L(s_1, s_2, X)$ is itself defined by the regular expression $r_1 \vee r_2 r_3^* r_4$.

In particular, for each $y \in Y$, we conclude that $L(s_0, y, S)$ is defined by some regular expression r_y, and so $L(S, A, d, Y, s_0)$ itself is determined by the regular expression $\vee_{y \in Y} r_y$ over A. *q.e.d.*

It is worth emphasizing that the proof above actually gives an effective algorithm for determining a regular expression defining the language accepted by a given finite state automaton. We give a simple example of applying this algorithm in

■ **Example 9.4.7** Consider the finite state automaton (S, A, d, Y, s_0) defined by the state diagram

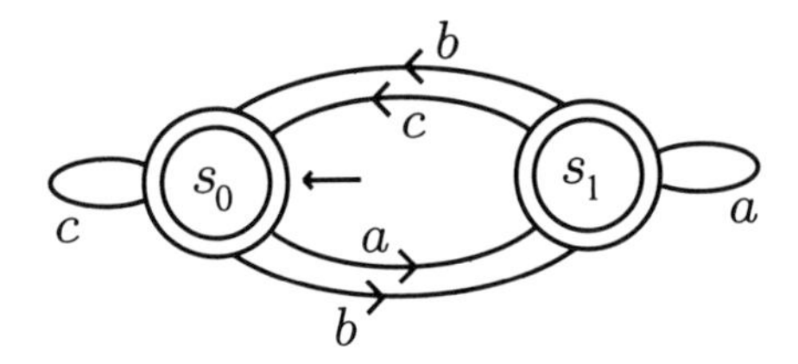

so the state set is $S = \{s_0, s_1\}$. We first compute the regular expressions for $L(s, t, \emptyset)$ as s, t vary over S^2 to be

$$L(s_0, s_0, \emptyset) = L(c), \qquad L(s_0, s_1, \emptyset) = L(a \vee b),$$
$$L(s_1, s_0, \emptyset) = L(b \vee c), \qquad L(s_1, s_1, \emptyset) = L(a).$$

Using the proof of the previous lemma, we may then compute

$$\begin{aligned} L(s_0, s_1, \{s_0\}) &= L(s_0, s_1, \emptyset) \cup L(s_0, s_0, \emptyset)\ \big[L(s_0, s_0, \emptyset)\big]^*\ L(s_0, s_1, \emptyset) \\ &= (a \vee b) \vee cc^*(a \vee b) \end{aligned}$$

and

$$\begin{aligned} L(s_1, s_1, \{s_0\}) &= L(s_1, s_1, \emptyset) \cup L(s_1, s_0, \emptyset)\ \big[L(s_0, s_0, \emptyset)\big]^*\ L(s_0, s_1, \emptyset) \\ &= a \vee (b \vee c)c^*(a \vee b). \end{aligned}$$

Again applying the proof of the previous lemma, we find

$$\begin{aligned} L(S, A, d, Y, s_0) &= L(s_0, s_1, \{s_0, s_1\}) \\ &= L(s_0, s_1, \{s_0\}) \cup L(s_0, s_1, \{s_0\})\big[L(s_1, s_1, \{s_0\})\big]^* L(s_1, s_1\{s_0\}), \end{aligned}$$

which is defined by the regular expression

$$r = [a \vee b \vee cc^*(a \vee b)] \ \vee \ [a \vee b \vee cc^*(a \vee b)]\,[a \vee (b \vee c)c^*(a \vee b)]^*\,[a \vee (b \vee c)c^*(a \vee b)],$$

and so we find $L(S, A, d, Y, s_0) = L(r)$.

Evidently, each linear ordering on S describes a different algorithm for determining a regular expression defining $L(S, A, d, Y, s_0)$, and the various regular expressions so described all determine the common language $L(S, A, d, Y, s_0)$.

EXERCISES

9.4 FINITE STATE AUTOMATA

1. Show that any finite language is accepted by some finite state automaton.

2. (a) How many different finite state automata are there with s states over an alphabet with n letters?

(b) We say that a finite state automaton is *simply minimal* if for all states s_i and all $s_j \neq s_i$, there is some letter $a = a_{ij} \in A$ so that $d(a, s_i) \neq d(a, s_j)$. How many simply minimal finite state automata are there with s states over an alphabet with n letters?

3. Show that a regular language L recognized by a finite state automaton with n vertices is infinite if and only if it contains a word w so that $n \leq ||w|| < 2n$. Conclude the following "pumping lemma": Any sufficiently long word w in a regular language L can be written in the form $w = \lambda\mu\nu$, where $\mu \neq \varepsilon$ and $\lambda\mu^i\nu \in L$ for any $i \geq 0$.

4. We say that a partition π on the set S of states of a finite state automaton satisfies the "substitution property" or "has SP" if for any states s, t if $s \sim_\pi t$, then also $d(a, s) \sim_\pi d(a, t)$ for any letter $a \in A$.

(a) Show that SP is preserved under meet and join of partitions.

(b) Show that the set of all SP partitions is a lattice under the usual ordering of refinement.

(c) Show that the lattice of SP partitions is not in general distributive.

5. Explain the sense in which a finite state automaton may be regarded as a dynamical system in the sense of §6.7.

(†) 9.5 KLEENE'S THEOREM

Generalizing the deterministic finite state automata discussed before, we introduce here "non-deterministic finite state automata". Not only is this generalization intrinsically interesting, but it is also a basic tool in our proof of Kleene's Theorem that a language is regular if and only if it is accepted by some deterministic finite state automaton. A non-deterministic finite state automaton again "accepts" a language, and in fact a language is accepted by a deterministic finite state automaton if and only if it is accepted by a non-deterministic one. Thus, we actually have three different characterizations of regular languages here in terms of regular expressions, deterministic and non-deterministic finite state automata. Our proof of Kleene's Theorem is constructive in the sense that it provides effective algorithms for determining a regular expression from a finite state automaton and a finite state automaton from a regular expression.

A *non-deterministic finite state automaton* is a quintuple (S, A, D, Y, S_0) where as before S is a finite set called the *state set*, A is an alphabet, and $Y \subseteq S$ is a collection of *accept* or *success* states. This time, we have a *subset* $S_0 \subseteq S$ of *start states* (rather than just a single start state $s_0 \in S$ for deterministic finite state automata).

It remains only to define D (the analogue of the transition function of a deterministic finite state automaton), and we require some terminology as follows. An *arrow* is a triple (s, x, t), where s, t are elements of S and $x \in A \cup \{\varepsilon\}$. The *source* of the arrow (s, x, t) is $s \in S$, $t \in S$ is its *target*, and $x \in A \cup \{\varepsilon\}$ is its *label*. To complete the definition of a non-deterministic finite state automaton (S, A, D, Y, S_0), we finally require simply that $D \subseteq S \times (A \cup \{\varepsilon\}) \times S$ is some collection of arrows.

In particular, the transition function $d : A \times S \to S$ of a deterministic finite state automaton (S, A, d, Y, s_0) gives rise to the set

$$D_d = \{(s_1, a, s_2) \in S \times A \times S : d(a, s_1) = s_2\}$$

of arrows, so a deterministic finite state automaton (S, A, d, Y, s_0) automatically gives rise to a corresponding non-deterministic one $(S, A, D_d, Y, \{s_0\})$ where

- there are no arrows labeled ε,
- $S_0 = \{s_0\}$ has exactly one element,
- for each state $s \in S$ and each letter $a \in A$, there is a unique arrow in D labeled a with source s.

Conversely, it is clear from the definitions that these three conditions on a non-deterministic finite state automaton guarantee that it arises in this way from a deterministic one.

There is a sensible *state diagram* G (which is again a multi-digraph with labels on its edges and vertices) associated to a non-deterministic finite state automaton

(S, A, D, Y, S_0), where the vertices of G are again labeled by S, the vertices in S_0 are indicated with a "$\rightarrow$", the vertices in Y are indicated with concentric circles, and there is one oriented edge in G labeled x with initial point s_1 and terminal point s_2 for each arrow $(s_1, x, s_2) \in D$.

■ **Example 9.5.1** Let (S, A, D, Y, S_0) be a non-deterministic finite state automaton over the alphabet $A = \{a, b, c\}$ with state set $S = \{s_0, s_1, s_2, s_3\}$, start set $S_0 = \{s_0, s_1\}$, and accept set $Y = \{s_0, s_3\}$ with the set of arrows

$$D = \{(s_0, \varepsilon, s_1), (s_0, a, s_3), (s_1, a, s_2), (s_1, c, s_0),$$
$$(s_2, \varepsilon, s_0), (s_2, b, s_1), (s_2, b, s_3), (s_3, \varepsilon, s_3)\}.$$

The corresponding state diagram is

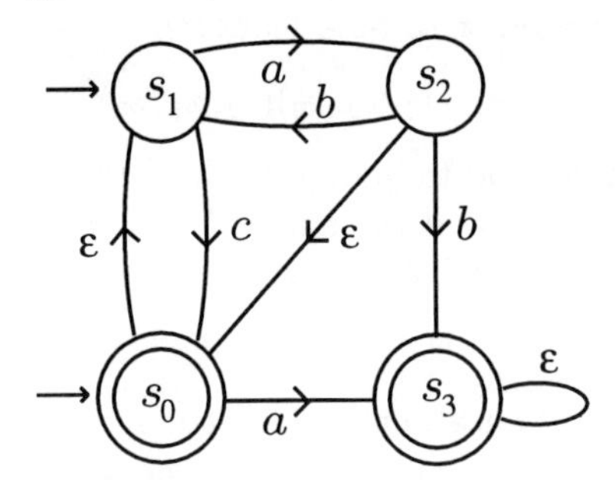

We might also take the alphabet $\{a, b, c, d\}$ even though there are no labels "d" on the edges in the state diagram to regard the previous state diagram as that of a different non-deterministic finite state automaton $(S, A \cup \{d\}, D, Y, S_0)$. Thus, unlike the situation for deterministic automata, we must here be careful to specify the alphabet as well as the state diagram to define a non-deterministic automaton.

■ **Example 9.5.2** Take a three-letter alphabet $A = \{a, b, c\}$ and define a non-deterministic finite state automaton (S, A, D, Y, S_0), with state diagram

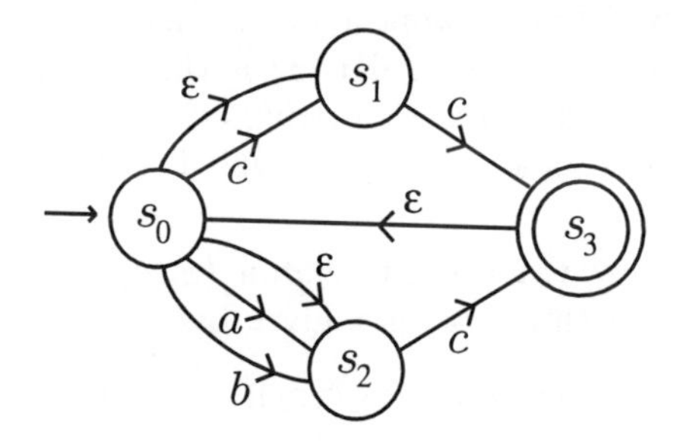

so the state space is $S = \{s_0, s_1, s_2, s_3\}$, there is a unique start state s_0, a unique accept state s_3, and the set of arrows is

$$D = \{(s_0, \varepsilon, s_1), (s_0, c, s_1), (s_0, \varepsilon, s_2), (s_0, a, s_2), (s_0, b, s_2)\}$$
$$\cup \ \{(s_1, c, s_3), (s_2, c, s_3), (s_3, \varepsilon, s_0)\}.$$

To describe the sense in which a non-deterministic automaton (S, A, D, Y, S_0) corresponds to an electronic machine as before, one should again imagine a machine whose state evolves by reading a string $w \in A^*$ over A. Upon reading a letter $a \in A$, a state $s \in S$ may evolve in one of several different ways corresponding to the several possible arrows with source s and label a. Furthermore, at any instant, a state s may evolve to another state t provided $(s, \varepsilon, t) \in D$. The machine is therefore "non-deterministic" in these senses: there may be several different evolutions of a given state upon reading a given letter, and at any time, the state of the machine might evolve using the arrows labeled ε. Thus, we cannot determine exactly what state the machine is in given the information of a start state and a string in A^*, so the machine is "non-deterministic".

In order to make these remarks about the non-deterministic machine precise, we define a *path of arrows* π to be a sequence

$$\pi = (s_0, x_1, s_1)(s_1, x_2, s_2)(s_2, x_3, s_3)\dots(s_{m-1}, x_m, s_m)$$

of arrows whose sources and targets coincide in this way. The state $\sigma(\pi) = s_0 \in S$ is called the *source* of π, and the state $\tau(\pi) = s_m \in S$ is called the *target* of π. Furthermore, the oriented edge-path π gives rise to its *label*

$$\ell(\pi) = x_m x_{m-1} \cdots x_2 x_1,$$

which is regarded as a concatenation of letters in the "alphabet" $A \cup \{\varepsilon\}$, where the quotation marks here are explained by the fact that we have required before that ε not be a letter in an alphabet.

Just as for deterministic finite state automata, a path of arrows on the automaton can be identified with an oriented edge-path on the corresponding state diagram, and the reader might construct her or his own examples of paths of arrows and their concatenations using the state diagrams in the examples above.

Let π be a path of arrows with label $\ell(\pi) = x_m \cdots x_2 x_1$, so each $x_i \in A \cup \{\varepsilon\}$. We may simply "erase" all of the letters ε from $\ell(\pi)$ to define a new string $w(\pi) \in A^*$ called the *reduced word* of π in the natural way. For instance, if $\ell(\pi) = \varepsilon a_1 \varepsilon^2 a_1 a_2 \varepsilon a_3$, where $a_1, a_2, a_3 \in A$, then the reduced word is $w = a_1^2 a_2 a_3$. In the extreme case that $\ell(\pi)$ is a word in just the letter ε or if π is the empty word, then the corresponding reduced word $w(\pi)$ is the empty word $w(\pi) = \varepsilon \in A^*$. Thus, a path of arrows π gives rise to a string $w(\pi) \in A^*$ called the reduced word which arises by erasing ε as above from the label $\ell(\pi)$ of π.

■ **Example 9.5.3** On the automaton (S, A, D, Y, S_0) defined in Example 9.5.1, consider the path of arrows π given by

$$(s_1, c, s_0)(s_0, \varepsilon, s_1)(s_1, a, s_2),$$

so the source is $\sigma(\pi) = s_1 \in S_0$, the target is $\tau(\pi) = s_2 \in S$, the label is $\ell(\pi) = a\varepsilon c$, and the reduced word is $w(\pi) = ac$. For another family of examples, consider the paths of arrows

$$\pi_n = \underbrace{(s_0, \varepsilon, s_1)(s_1, c, s_0)(s_0, \varepsilon, s_1)(s_1, c, s_0) \cdots (s_0, \varepsilon, s_1)(s_1, c, s_0)}_{2n \text{ arrows}},$$

for each $n \geq 1$. The common source and target of π_n is $\sigma(\pi_n) = s_0 = \tau(\pi_n) \in S_0 \cap Y$, the label is $\ell(\pi_n) = (c\varepsilon)^n$, and the reduced word is $w(\pi_n) = c^n$ for each $n \geq 1$.

We say that a non-deterministic finite state automaton (S, A, D, Y, S_0) *accepts* or *recognizes* a word $w \in A^*$ if there is some path of arrows π so that $w = w(\pi)$, where the source $\sigma(\pi) \in S_0$ is a start state and the target $\tau(\pi) \in Y$ is an accept state.

■ **Example 9.5.4** The automaton (S, A, D, Y, S_0) defined in Example 9.5.1 accepts the words c^n for each $n \geq 1$ as before. The path of arrows

$$(s_1, c, s_0)(s_0, \varepsilon, s_1)(s_1, a, s_2)$$

considered before does *not* show that ac is accepted by the automaton since its target is not an accept state. On the other hand, the path of arrows

$$(s_1, c, s_0)(s_0, a, s_3)$$

guarantees that the string ac is actually accepted by the automaton.

A non-deterministic finite state automaton (S, A, D, Y, S_0) determines a language

$$L(S, A, D, Y, S_0) = \{w \in A^* : (S, A, D, Y, S_0) \text{ accepts } w\} \in \mathcal{L}(A)$$

called the language *accepted* or *recognized* by the automaton.

A deterministic finite state automaton (S, A, d, Y, s_0) accepts the language $L(S, A, d, Y, s_0)$ as described in the previous section. On the other hand, it also gives rise to the non-deterministic automaton $(S, A, D_d, Y, \{s_0\})$ as above, which accepts the language $L(S, A, D_d, Y, \{s_0\})$. It follows from the definitions (as the reader should verify by tracing through them) that these two languages coincide, that is, we have

$$L(S, A, d, Y, s_0) = L(S, A, D_d, Y, \{s_0\})$$

for each deterministic finite state automaton (S, A, d, Y, s_0).

9.5.1 LEMMA *A language $L \in \mathcal{L}(A)$ is recognized by a non-deterministic finite state automaton over A if and only if it is recognized by some deterministic one.*

Proof We have already mentioned above that a deterministic finite state automaton and its corresponding non-deterministic version accept the same languages, so one implication has already been proved.

Conversely, suppose that $L \in \mathcal{L}(A)$ is a language accepted by the non-deterministic finite state automaton (S, A, D, Y, S_0), and let π be a path of arrows, say with source $s \in S$ and target $t \in S$. We say that π is an *ε-transition from s to*

t if the label of each arrow in π (if any) is ε. In other words, π is an ε-transition if its label $\ell(\pi)$ lies in $(\{\varepsilon\})^*$, so its corresponding reduced word is $\varepsilon \in A^*$.

If $X \subseteq S$ is a collection of states, then we define the *ε-closure* of X to be the set

$$\overline{X} = \{t \in S : \text{ there is an } \varepsilon - \text{transition } \pi \text{ so that } \sigma(\pi) \in X, \tau(\pi) = t\}.$$

In other words, the ε-closure of a set X is the union of X with the collection of targets of all paths of arrows each labeled ε with source in X. Notice that by definition of ε-closure, we have

$$\overline{(X_1 \cup X_2)} = \overline{X}_1 \cup \overline{X}_2, \text{ for } X_1, X_2 \subseteq S,$$

so ε-closure distributes over unions. In particular, we have

$$\overline{X} = \cup_{x \in X} \overline{\{x\}},$$

so the ε-closure of a set is simply the ε-closure of the singletons contained in the set. Furthermore, we have

$$\overline{(\overline{X})} = \overline{X}, \text{ for } X \subseteq S,$$

so the ε-closure of the ε-closure is simply the ε-closure of a set.

■ **Example 9.5.5** For the non-deterministic automaton in Example 9.5.1 above, we compute the ε-closures

$$\overline{\{s_0\}} = \{s_0, s_1\}, \ \overline{\{s_1\}} = \{s_1\}, \ \overline{\{s_2\}} = \{s_0, s_1, s_2\}, \ \overline{\{s_3\}} = \{s_3\}$$

of the singletons. It follows that for instance

$$\begin{aligned} \overline{\{s_0, s_1, s_3\}} &= \overline{\{s_0\}} \cup \overline{\{s_1\}} \cup \overline{\{s_3\}} = \{s_0, s_1, s_3\}, \\ \overline{\{s_2, s_3\}} &= \overline{\{s_2\}} \cup \overline{\{s_3\}} = \{s_0, s_1, s_2, s_3\}. \end{aligned}$$

A subset $X \subseteq S$ is said to be *ε-closed* if $\overline{X} = X$, so in particular, both $\emptyset$ and S are ε-closed. Furthermore, any ε-closure $\overline{X}$ for $X \subseteq S$ is itself ε-closed as was discussed above. By distributivity of ε-closure over unions, the elements of $X \subseteq S$ determine its closure by $\overline{X} = \cup_{x \in X} \overline{\{x\}}$.

Define a finite set S' (which will be our new state set) to be the collection of all ε-closed subsets of S. Notice that elements of S' are subsets of S by definition so $S' \subseteq \mathcal{P}(S)$, and furthermore $\emptyset, S \in S'$. The new start state $s_0' \in S'$ is the ε-closure $\overline{S}_0$ of the start set $S_0 \subseteq S$, and the new accept states

$$Y' = \{\overline{X} \in S' : \ y \in \overline{X} \text{ for some } y \in Y\}$$

are those ε-closures $\overline{X}$ which contain an accept state $y \in Y$. In particular, $S \in S'$ is always an accept state.

■ **Example 9.5.6** Continuing with the analysis of the automaton in Example 9.5.1, there are exactly eight ε-closed subsets of S, namely

$$S' = \{S, \emptyset, \{s_0, s_1\},\ \{s_1\},\ \{s_0, s_1, s_2\},\ \{s_3\},\ \{s_0, s_1, s_3\},\ \{s_1, s_3\}\}.$$

The start state is $\overline{\{s_0, s_1\}} = \{s_0, s_1\}$, and every new state *except* $\{s_1\}$ contains either s_0 or s_3 and is therefore an accept state.

Taking the same alphabet A, it remains to define a transition function $d' : A \times S' \to S'$, and given $a \in A$ and $\overline{X} \in S'$, we define $d'(a, \overline{X})$ to be the ε-closure of the set

$$\{t \in S : \tau(\pi) = t \text{ and } \sigma(\pi) \in \overline{X} \text{ for some path of arrows } \pi\}.$$

■ **Example 9.5.7** Finally finishing up our ongoing example, the deterministic finite state automaton corresponding to the non-deterministic one in Example 9.5.1 is described by the state diagram

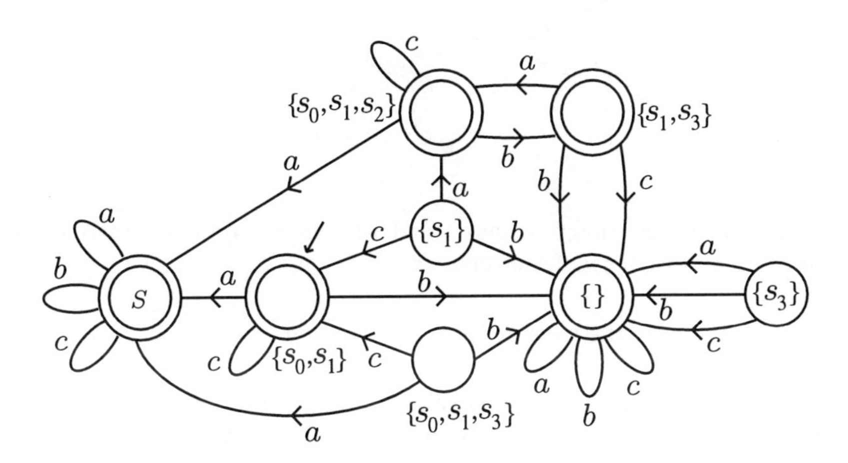

For instance, there is an arrow labeled c from $\{s_1\}$ to $\{s_0, s_1\}$ since only s_0 and s_1 can be the targets of paths of arrows with source s_1 and label some concatenation of letters c and ε. The automation accepts any non-trivial string over A.

At this point, we have defined the construction of a deterministic finite state automaton (S', A, d', Y', s_0') from the non-deterministic one (S, A, D, Y, S_0), and it remains only to check that these two automata accept the same language. To this end, a word $w = a_1a_2 \cdots a_m \in A^*$ is accepted by (S', A, d', Y', s_0') if and only if there is a sequence $\overline{X}_0, \overline{X}_1, \ldots, \overline{X}_m$ of ε-closed subsets of S for some $m \geq 1$ so that $\overline{X}_0 = \overline{S}_0$, $\overline{X}_m \cap Y \neq \emptyset$, and $d'(a_i, \overline{X}_{i-1}) = \overline{X}_i$ for each $i = 1, 2, \ldots, m$. By definition of the ε-closure and the transition function d', this is in turn equivalent to the condition that there is a sequence $x_0, x_0', x_1, x_1', \ldots x_{m-1}, x_{m-1}', x_m, x_m'$ so that $x_0 \in S_0$, $x_m' \in Y$, there is a ε-transition with source x_i and target x_i', for each $i = 0, 1, \ldots, m$, and there is an arrow $(x_{i-1}', a_i, x_i) \in D$, for each $i = 1, 2, \ldots, m$. Finally, using the definition of words accepted by a non-deterministic finite state automaton, this is in turn equivalent to the condition that $w = a_1a_2 \ldots a_m$ is accepted by (S, A, D, Y, S_0). *q.e.d.*

■ **Example 9.5.8** Analyzing the non-deterministic automaton (S, A, D, Y, S_0) in Example 9.5.2 above, we find the deterministic state diagram

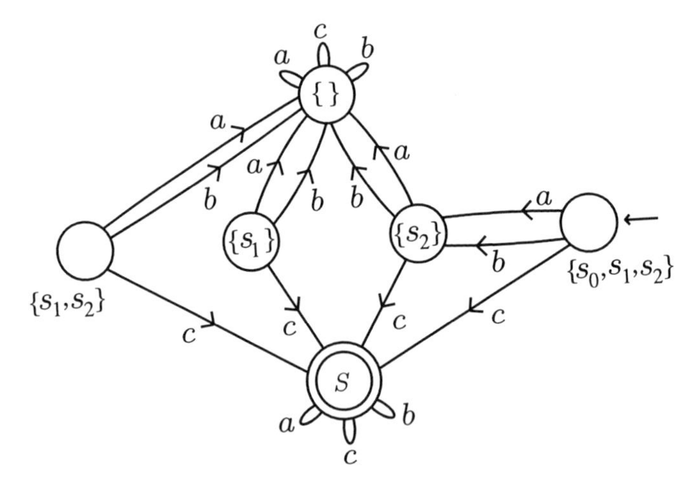

Inspection of this state diagram shows that the automaton only accepts strings which contain c in their two-letter prefixes.

We turn finally to the main result of this section

9.5.2 THEOREM [Kleene, Rabin, Scott] *Let A be any alphabet. The following three conditions on a language $L \in \mathcal{L}(A)$ over A are equivalent.*

(1) *L is accepted by a deterministic finite state automaton over A.*

(2) *L is accepted by a non-deterministic finite state automaton over A.*

(3) *$L = L(r)$ is defined by some regular expression r over A.*

Proof The previous lemma shows that conditions (1) and (2) are equivalent. Furthermore, Lemma 9.4.1 proves that (1) implies (3). To finish the proof, then, it remains to show that (3) implies (2).

To this end, given a regular expression r over A, we shall construct a corresponding non-deterministic finite state automaton which accepts the language defined by r. In each case, the automaton we construct will have only one start state and one accept state. If the regular expression r consists of the single letter $r = a \in A$, then there are exactly two states, namely the start state s and the unique accept state t, and there is only one arrow (s, a, t). If the regular expression r consists of the single letter $r = \varepsilon$ or if r is the expression "()", then there are no arrows and only one state which is both a start state and an accept state.

Inductively suppose now that r_1, r_2 is a regular expression over A defining a language which is accepted by a non-deterministic finite state automaton with one start state and one accept state, and consider the corresponding state diagrams G_1, G_2.

- We can construct another automaton whose state diagram arises from G_1 by adding one arrow labeled ε with source the accept state and target the start state and another arrow going the other way labeled ε with source the start state and target the accept state. The corresponding automaton evidently accepts the language defined by r^*.

- We can construct an automaton from G_1 and G_2 by connecting them by an arrow labeled ε with source the accept state of G_1 and target the start state of G_2, and the corresponding automaton evidently accepts the language defined by $r_1 r_2$.

- We can construct an automaton from G_1 and G_2 by creating a new start state s' and a new accept state t' where we add to D four arrows: one from s' to each of the start states of G_1 and G_2 and one from each of the accept states of G_1 and G_2 to t'. This automaton evidently accepts the language defined by $r_1 \vee r_2$.

By performing these operations a finite number of times, that is, by induction, we may therefore effectively construct a non-deterministic finite state automaton which accepts the language determined by a given regular expression, and this completes the proof. *q.e.d.*

■ **Example 9.5.9** Let the following state diagrams

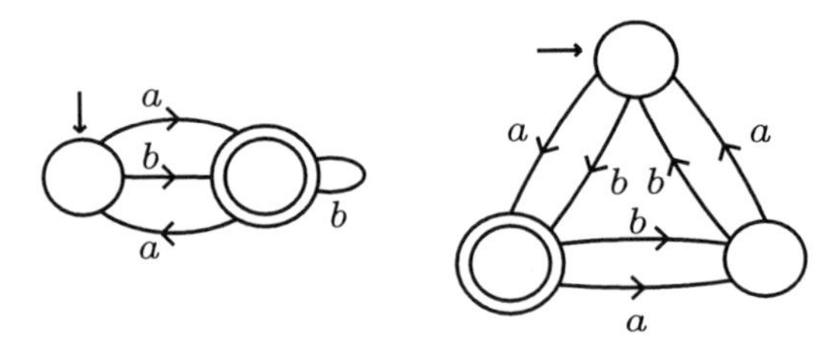

determine finite state automata accepting the respective languages L_1 and L_2. Then the state diagram

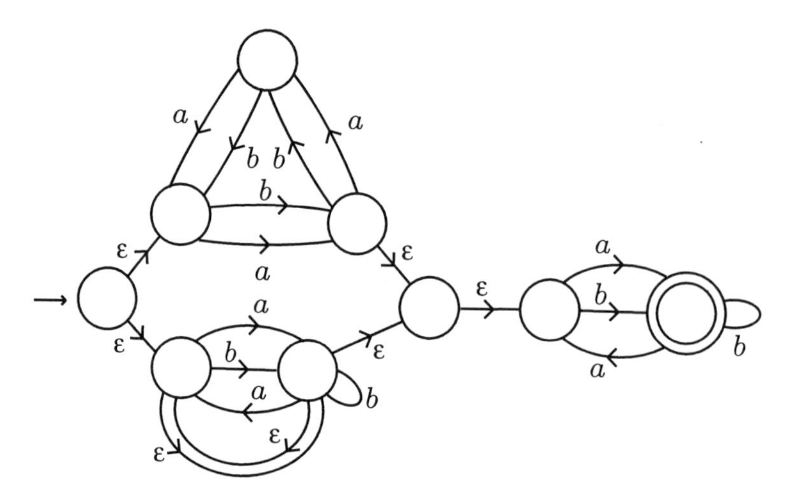

describes a non-deterministic automaton which accepts the language $(L_1^* \cup L_2)L_1$.

It is worth emphasizing that our arguments produce explicit algorithms to:

- Construct a regular expression for a language from a deterministic finite state automaton accepting that language in Lemma 9.4.1.
- Construct a deterministic finite state automaton accepting the same language as a non-deterministic one in Lemma 9.5.1.
- Construct a non-deterministic finite state automaton accepting the language given by a regular expression in Theorem 9.5.2.

Thus, one may algorithmically pass between the three equivalent descriptions of a regular language, and we leave the pursuit of such sample calculations to the untiring reader.

EXERCISES

9.5 KLEENE'S THEOREM

1. If G is the digraph of a finite deterministic automaton, then define the *iterate* G^* of G to be the following (typically non-deterministic) automaton. The vertex set of G^* consists of the vertex set of G together with another vertex labeled I. The initial and accept states as well as the edges and their labeling in G are maintained in G^*, and we add edges to obtain G^* as follows: whenever there is an edge in G from a vertex v to an accept state vertex, labeled by a, then there is an edge in G^* labeled by a from v to I; whenever there is an edge in G, labeled by a, from the initial state vertex to a vertex v, then there is an edge in G^* labeled by a from I to v; if, moreover, v is an accept state, then we add an edge labeled a from I to itself. Prove that for any finite deterministic automaton, we have

$$(L(G))^* = L(G^*) \cup \{\varepsilon\}.$$

2. The *iteration number* $in(w)$ of a word corresponding to an edge-path on the iterate G^* is defined to be the smallest integer n so that there is some path in G^* labeled by w from the initial state vertex to an accept state vertex which passes n times through the vertex I. Prove that if w is a non-empty word in $L(G^*)$ and $in(w) = n$, then w is in $(L(G))^{n+1}$ but not in any $(L(G))^k$ where $k < n + 1$.

(†) 9.6 SEMIGROUPS AND MONOIDS

"Semigroups" and "monoids" are basic mathematical structures which we define here. We have actually already encountered numerous examples of these structures in the previous parts of this chapter. Elementary mathematical aspects of monoids and semigroups are developed here with an emphasis on languages and automata.

A *semigroup* is an ordered pair $(X, \otimes)$, where X is some set and $\otimes : X \otimes X \to X$ is some *associative* binary operation on X, so $x \otimes (y \otimes z) = (x \otimes y) \otimes z$ for all $x, y, z \in X$. One sometimes refers to X itself as a semigroup when the binary operation $\otimes$ is fixed or understood. An element $e \in X$ is called an *identity element* of X if

$$e \otimes x = x \otimes e = x \quad \text{for all } x \in X.$$

A *monoid* is simply a semigroup which contains an identity element. A semigroup or monoid is said to be *commutative* if its corresponding operation $\otimes$ is commutative, that is, $x \otimes y = y \otimes x$ for all $x, y \in X$.

Before giving examples of semigroups and monoids (and we shall give many examples), observe that a semigroup can contain at most one identity element, or more precisely:

9.6.1 THEOREM *If $e, e' \in X$ are identity elements in the semigroup $(X, \otimes)$, then $e = e'$.*

Proof Since e is an identity element, $e \otimes e' = e'$, and since e' is an identity element $e \otimes e' = e$. Equating these two expressions for $e \otimes e'$ gives the desired equality. *q.e.d.*

■ **Example 9.6.1** $(\mathbb{N}, +)$ is a commutative semigroup but not a monoid, whereas $(\mathbb{N} \cup \{0\}, +)$ is a commutative monoid with identity $e = 0$. On the other hand, $(\mathbb{Z}, -)$ is not a semigroup since the binary operation $-$ of subtraction is not associative.

■ **Example 9.6.2** Let Z be any set and take the power set $X = \mathcal{P}(Z)$. Set-theoretic union $\cup$ is an associative and commutative binary operation on $\mathcal{P}(Z)$, and $(\mathcal{P}(Z), \cup)$ is a commutative monoid with identity $e = \emptyset$. Intersection $\cap$ is also associative and commutative, so $(\mathcal{P}(Z), \cap)$ is another commutative monoid with identity element Z.

The previous example shows that there can be multiple semigroup or monoid structures on a fixed set.

■ **Example 9.6.3** Define the *right zero semigroup* on the set $X = \{a, b\}$ by defining a binary operation $\otimes$ on X, where $a \otimes a = b \otimes a = a$ and $a \otimes b = b \otimes b = b$. It is easy (and an exercise) to verify that $\otimes$ is associative but not commutative, and $(X, \otimes)$ is a semigroup but not a monoid. We can construct a binary operation $\oplus$ on $X \cup \{e\}$ for $e \notin X$ where $x \oplus y = x \otimes y$ if $x, y \in X$, and we "formally adjoin an identity e" by defining $e \oplus x = x = x \oplus e$.

Of course, one may always construct a monoid from a semigroup by formally adjoining an identity element as in the previous example. Notice that if X is a monoid with identity e and we formally adjoin a new identity $f \notin X$, then f (not e) is the new identity in $X \cup \{f\}$.

■ **Example 9.6.4** Define the binary operation $\oplus$ on the set $\{0, 1\}$ according to the rules $0 \oplus 1 = 1 = 1 \oplus 0$ and $0 \oplus 0 = 0 = 1 \oplus 1$. We leave it for the reader to check that $(\{0, 1\}, \oplus)$ is a commutative monoid with identity 0.

Turning to examples of monoids and semigroups which have already arisen in this chapter, we have

■ **Example 9.6.5** Let A be any alphabet, and take $X = A^*$ which supports the associative operation of concatenation. The null word ε acts as the identity element, so the collection A^* of all words over A is a monoid. Furthermore, $A^* - \{\varepsilon\}$ is a semigroup again using the operation of concatenation, and the monoid A^* arises from the semigroup $A^* - \{\varepsilon\}$ by formally adjoining the identity ε.

If B is any set (not necessarily finite), then we may consider the collection

$$B^* = \{\varepsilon\} \cup \{b_1 b_2 \cdots b_n : b_i \in B \text{ for } n \geq 1\}$$

of all finite (and possibly empty) strings over B. The set B^* is a monoid under concatenation of strings with identity ε as before and is called the *free monoid generated by* the set B. The set $B^* - \{\varepsilon\}$ with the operation of concatenation is called the *free semigroup generated by* B. In particular, if $B = A$ is an alphabet, then the monoid A^* in the previous example is simply the free monoid generated by A.

■ **Example 9.6.6** The free monoid $\mathbb{N}^*$ generated by $\mathbb{N}$ consists of all (possibly empty) finite strings $n_1 n_2 \cdots n_\ell$, where $\ell \in \mathbb{N} \cup \{0\}$ and $n_i \in \mathbb{N}$, for each $i = 1, \ldots \ell$, with the operation of concatenation of strings.

■ **Example 9.6.7** Fix an alphabet A and let L be any language over A. Concatenation of words in L is an associative binary operation on the Kleene closure L^*, and L^* together with this binary operation is a monoid again with identity ε. Indeed, this monoid L^* is exactly the free monoid generated by the possibly infinite set $L \subseteq A^*$.

Thus, the set A^* of all words over an alphabet A and the Kleene closure L^* of a language L under concatenation are examples of free monoids. For more interesting examples of monoids, we have

■ **Example 9.6.8** Let S be any set, and consider the collection S^S of all functions from S to S. Composition $\circ$ of functions is associative and the identity map $1_S : S \to S$ acts as an identity for composition, so $(S^S, \circ)$ is yet another example of a monoid for any set S. For instance, if $f, g : S \to S$ are bijections which commute, i.e., $f \circ g = g \circ f$, then $f^{-1} \circ g^{-1} \circ f \circ g = 1_S$. In a monoid such as S^S, there can thus be non-trivial compositions which yield the identity, and this is in contrast to the free monoids described above.

The next two examples are evidently closely related.

■ **Example 9.6.9** Fix an alphabet A and consider the set $\mathcal{L}(A)$ of all languages over A. Set product of languages over A is an associative binary operation on $\mathcal{L}(A)$, and with this operation $\mathcal{L}(A)$ is a monoid with identity $\{\varepsilon\} \in A^*$.

■ **Example 9.6.10** Fix an alphabet A and consider the set $\mathcal{R}(A)$ of all regular expressions over A. Concatenation of regular expressions is an associative binary operation on $\mathcal{R}(A)$, and with this operation $\mathcal{R}(A)$ is a monoid with identity $\varepsilon \in \mathcal{R}(A)$.

Suppose that $(X, \otimes)$ is a semigroup or monoid and $F \subseteq X$ is any set. We say that F *is closed under* $\otimes$ if $f_1 \otimes f_2 \in F$ for any $f_1, f_2 \in F$.

■ **Example 9.6.11** The collection of all bijections in S^S is closed under composition $\circ$ of functions.

■ **Example 9.6.12** Suppose that A is an alphabet and $L \subseteq A^*$ is closed under concatenation, where $L - \{\varepsilon\} \neq \emptyset$. Choose $w \in L - \{\varepsilon\}$ and consider the concatenations $w^n \in L^*$, for $n \geq 1$. Since L is closed under concatenation, it follows by induction that each w^n is also in L for $n \geq 1$. Furthermore, each concatenation w^n is a distinct element of L^* for $n \geq 1$ (for instance, because each of these elements has a different length), so L must be infinite. We have shown that there can be no finite subset $L \subseteq A^*$ which is closed under concatenation other than $\emptyset$ and $\{\varepsilon\}$.

Continuing to let $(X, \otimes)$ denote a semigroup and $F \subseteq X$ an arbitrary subset, we define

$$< F >= \{x_1 \otimes x_2 \otimes \cdots \otimes x_m \in X : m \geq 1 \text{ and } x_i \in F \text{ for each } i = 1, \ldots, m\}.$$

It is not difficult (and left to the reader) to check that $< F > \subseteq X$ is closed under $\otimes$, so $\otimes$ induces an associative binary operation, also to be denoted $\otimes$, on $< F >$ itself. The pair $(< F >, \otimes)$ is called the *semigroup generated by F in X*. In case X happens to be a monoid with identity $e \in X$ and $e \in F \subseteq X$, then $e \in < F >$ is an identity for $\otimes$ on $< F >$, and $(< F >, \otimes)$ is actually a monoid called the *monoid generated by F in X*.

Let $(X, \otimes)$ be a semigroup. A non-empty subset $Y \subseteq X$ which is closed under $\otimes$ is called a *sub-semigroup*, where the binary operation on Y is simply the restriction of $\otimes$ to $Y \times Y \subseteq X \times X$. If $(X, \otimes)$ happens to be a monoid with identity $e \in X$, then a subset $Y \subseteq X$ containing e which is closed under $\otimes$ is called a *sub-monoid*, where again the binary operation of Y is simply the restriction of $\otimes$.

■ **Example 9.6.13** $(\mathbb{N}, +)$ is a sub-semigroup of the monoid $(\mathbb{Z}, +)$.

■ **Example 9.6.14** If $X \subseteq Y$, then $(\mathcal{P}(X), \cup)$ is a sub-monoid of $(\mathcal{P}(Y), \cup)$.

■ **Example 9.6.15** If L and M are languages over an alphabet A and $L \subseteq M$, then L^* is a submonoid of M^*, where each monoid is given the operation of concatenation. Furthermore, each of L^* and M^* are sub-monoids of A^*

■ **Example 9.6.16** Suppose that $(X, \otimes)$ is a monoid with identity $e \in X$, $F \subseteq X$ with $e \in F$, and let $< F >$ denote the monoid generated by F. Then $< F >$ is a sub-monoid of X.

Given semigroups $(X, \otimes)$ and $(Y, \oplus)$, a function $\phi : X \to Y$ is called a *homomorphism* if $\phi(x_1 \otimes x_2) = \phi(x_1) \oplus \phi(x_2)$ for any $x_1, x_2 \in X$. In other words, a function $f : X \to Y$ is a homomorphism if it "respects" the multiplications on X and Y. These are the basic functions between semigroups and between monoids.

■ **Example 9.6.17** Consider the alphabet $A = \{a, b\}$ and the free monoid A^* generated by it under concatenation, and recall the monoid $(\{0, 1\}, \oplus)$ defined in Example 9.6.4 above. Define the function

$$\begin{aligned} \phi : A^* &\to \{0, 1\} \\ w &\mapsto \begin{cases} 0, & \text{if } w \text{ contains an even number of letters } a; \\ 1, & \text{if } w \text{ contains an odd number of letters } a, \end{cases} \end{aligned}$$

so for instance $f(aabaab) = 0$ and $f(abaab) = 1$.

To see that ϕ is a homomorphism, consider the concatenation $w_1 w_2$ of words $w_1, w_2 \in A^*$. Since the number of letters a in $w_1 w_2$ is simply the number of letters a in w_1 plus the number of letter a in w_2, we have $\phi(w_1 w_2) = \phi(w_1) \oplus \phi(w_2)$ (as the reader should carefully check).

■ **Example 9.6.18** Suppose that $(X, \otimes)$ is a semigroup, and let $Y \subseteq X$ be any non-empty subset which is closed under $\otimes$. The restriction of $\otimes$ to Y produces a semigroup $(Y, \otimes)$, and the inclusion mapping $Y \to X$ is evidently a homomorphism.

■ **Example 9.6.19** In the construction of the language accepted by a non-deterministic finite state automaton over an alphabet A, we were led to consider the assignment of a "reduced word" $w \in A^*$ to a string $\ell \in (A \cup \{\varepsilon\})^*$, where w arises from ℓ by simply erasing all of the entries ε. The assignment $\ell \mapsto w$ is a homomorphism from the free monoid $(A \cup \{\varepsilon\})^*$ to the free monoid A^* as the reader may check.

■ **Example 9.6.20** Fix an alphabet A and recall from Examples 9.6.9 and 9.6.10 the respective monoids $\mathcal{L}(A)$ under the operation of set product and $\mathcal{R}(A)$ under the operation of concatenation. The function

$$\begin{aligned} \mathcal{R}(A) &\to \mathcal{L}(A) \\ r &\mapsto L(r) \end{aligned}$$

which assigns to a regular expression r its corresponding language is a homomorphism as the reader may check.

■ **Example 9.6.21** Recall that if (S, A, d, Y, s_0) is a deterministic finite state automaton, then the transition function $d : A \times S \to S$ determines a family of functions $d_a(s) = d(a, s)$, one function $d_a : S \to S$ for each $a \in A$. Define the finite set

$$F = \{1_S\} \cup \{d_a \in S^S : a \in A\} \subseteq S^S,$$

and take the closure $< F >$ of F in S^S. According to the discussion above, $(< F >, \circ)$ is a monoid, which is called the *monoid of the automaton* (S, A, d, Y, s_0). Notice that the state table of the automaton alone determines the set F and hence the monoid of the automaton.

For another point of view, recall the "w-transition"

$$d_w = d_{a_1} \circ d_{a_2} \circ \cdots \circ d_{a_m} \in S^S$$

of a string $w = a_1 a_2 \cdots a_m \in A^*$ associated with the automaton. The assignment

$$\begin{aligned} \phi : A^* &\to S^S \\ w &\mapsto d_w \end{aligned}$$

is a homomorphism as the reader may check, and the image $\phi(A^*) \subseteq S^S$ is the sub-monoid $< F >$ of S^S defined above. The monoid of the automaton is therefore the image of A^* under the homomorphism ϕ, and while this image $< F >= \phi(A^*) \subseteq S^S$ is a homomorphic image of a free monoid A^*, it is typically not itself a free monoid.

Suppose that $\phi : X \to Y$ is a homomorphism from the semigroup $(X, \otimes)$ to the semigroup $(Y, \oplus)$. We say that ϕ is an *isomorphism* if ϕ is moreover a bijection whose inverse ϕ^{-1} is also a homomorphism, and we furthermore say that $(X, \otimes)$ and $(Y, \oplus)$ are *isomorphic* if there is an isomorphism between them. It can be convenient to regard isomorphic semigroups or monoids as identical in certain contexts.

■ **Example 9.6.22** Suppose that A and B are alphabets and there is a bijection $\psi : A \to B$. The map ψ induces a corresponding map

$$\begin{aligned} \phi : A^* &\to B^* \\ a_1 a_2 \cdots a_n &\mapsto \psi(a_1)\psi(a_2) \cdots \psi(a_n). \end{aligned}$$

For instance, if $A = \{a, b, c\}$, $B = \{\alpha, \beta, \gamma\}$, and $\psi(a) = \alpha$, $\psi(b) = \gamma$, and $\psi(c) = \beta$, then $\phi(aabcc) = \alpha\alpha\beta\gamma\gamma$.

The induced mapping $\phi : A^* \to B^*$ is evidently a bijection (as we leave it for the reader to verify by constructing a two-sided inverse). Furthermore, ϕ evidently respects concatenation and is therefore a homomorphism (as the reader should verify directly), hence an isomorphism.

Of course our treatment of this topic of monoids from abstract algebra, a field of math, is little more than cursory. Though the intuition of concatenation is a sensible and appealing first view of algebraic structures, there are other related algebraic objects which are perhaps even more fundamental, and we at least define a few of these related and basic objects in the exercises.

EXERCISES

9.6 SEMIGROUPS AND MONOIDS

1. A *group* G is a "monoid with inverses", or in other words, a set G endowed with an associative binary operation $\cdot$ where: (i) there is an identity element $e \in G$ so that $e \cdot g = g = g \cdot e$ for each $g \in G$; and (ii) for each element $g \in G$, there is an element $g^{-1} \in G$ so that $g \cdot g^{-1} = e = g^{-1} \cdot g$. For instance, $\mathbb{Z}$ is a group under addition while $\mathbb{N}$ is not. A *homomorphism* of groups is a function $f : G_1 \to G_2$ between groups so that $f(g \cdot h) = f(g) \cdot f(h)$ for any $g, h \in G_1$, an *isomorphism* further has an inverse $G_2 \to G_1$ which is itself a group homomorphism, and an *automorphism* is an isomorphism $f : G \to G$ from a group to itself.

(a) Prove that composition of functions is a group operation on the set $A(G)$ of all automorphisms of G. Given an element $g \in G$, there is a corresponding *inner automorphism* $i_g : G \to G$ defined by $x \mapsto gxg^{-1}$; show that the set of all inner automorphisms is a subgroup $I(G)$ of $A(G)$.

(b) We say that a subgroup H of G is *normal* if $i_g(H) = H$ for every $g \in G$. Prove that $I(G)$ is a normal subgroup of $A(G)$.

(c) Fix a normal subgroup H of G, and define an equivalence relation on G, where $x \sim y$ if $xy^{-1} \in H$; equivalence classes are called "right cosets" of H and written gH for $g \in G$, and the quotient of G by this equivalence relation (i.e., the set of equivalence classes) is written G/H. Prove that $(xH)(yH) = (xyH)$ is a well-defined group operation on G/H.

(c) Define the *kernel* $ker(f)$ of a homomorphism $f : G_1 \to G_2$ to be $\{g \in G_1 : f(g) = e \in G_2\}$. Prove that $ker(f)$ is a normal subgroup of G_1. Prove that the image $Im(f)$ of f is a subgroup of G_2. Prove that the quotient $G_1/ker(f)$ is isomorphic as a group to $Im(f)$.

2. Define a *field* to be a set F together with two binary operations $\times$ and $+$ so that: (i) $(S, +)$ is a commutative group with identity e_+; (ii) $F - \{e_+\}$ is a commutative group with identity; and (iii) $\times$ and $+$ distribute across one another. Prove that $\mathbb{Q}$ is a field while $\mathbb{Z}$ is not. Show, however, that the usual multiplication and addition of integers induce corresponding binary operations on $\mathbb{Z}/n$ for any $n \in \mathbb{N}$, and that $\mathbb{Z}/n$ is actually a field with these operations if and only if n is prime.

Chapter 10

Graphs

A "graph" is roughly a set of points joined by a family of line segments, and there is a rich and venerable combinatorial theory of graphs with myriad applications to the sciences. On the other hand, the mathematician J.H.C. Whitehead referred to graph theory as the "gutter of topology" presumably because graphs give only trivial examples of the great theorems in the branch of mathematics called "topology".

We begin with the basic definitions and discuss an especially simple but important class of graphs called "trees". We study "Hamiltonian and Eulerian cycles" on graphs as well as the basics of "planar" graphs including a simple version of a deep and important theorem in topology called "Poincaré duality" in this setting. Various combinatorial problems are solved for graphs in the last section, where we describe famous theorems of Menger, Hall, and Ramsey, as well as describe the "chromatic polynomial" of a graph.

Our treatment of Poincaré duality in a sense belies Whitehead's pejorative remark above, for even this elementary version of Poincaré duality is beautiful and non-trivial. Furthermore, recent developments in geometry, topology, and physics express deep new "invariants" in terms of graphical enumeration problems, and we would therefore argue that the jury is in any case still out on whether it is to be gutter or glitter for graphs.

10.1 DEFINITIONS AND EXAMPLES

A *graph* G is an ordered pair (V, E) of sets, where E is a collection of unordered pairs $\{u, v\}$ of elements $u, v \in V$ so that $u \neq v$ (or in other words, E is a set of doubletons of V). We shall typically assume that V and hence E are finite sets and thus consider only *finite graphs* though there is an analogous theory for general graphs. If G is a graph, then the elements of the corresponding $V = V(G)$ are called *vertices* or *nodes* while elements of $E = E(G)$ are called *edges*; we shall let $e(G)$ denote the number of edges of the graph G and $v(G)$ denote the number of vertices of G. An edge $\{u, v\} \in E$ will be denoted simply $uv = vu$, and it is said to *join* its *endpoints* $u, v \in V$. Two vertices u, v are said to be *adjacent* if there is an edge joining them, and we say the edge uv is *incident* on its endpoints u, v. We shall also say two edges are *adjacent* if they share a common vertex.

Of course, one thinks of a graph not formally as a pair of sets in the sense of this definition but rather as a collection of vertices in the plane joined by certain unoriented edges.

■ **Example 10.1.1** Taking

$$V = \{0, \ldots, 10\},$$
$$E = \{01, 02, 03, 04, 05, 06, 56, 34, 13, 25, 78, 89, 97\}$$

determines the labeled graph

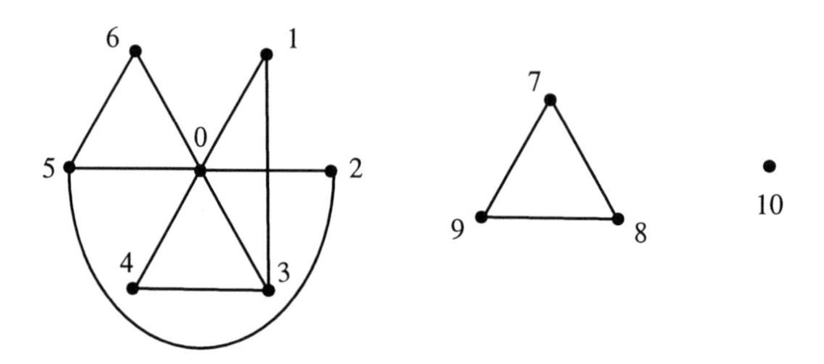

which illustrates typical phenomena, for instance, that a vertex may not be the endpoint of any edge and it is not required (and it may not even be possible as we shall discover in §10.4 below) to draw the edges simultaneously in the plane without having them cross.

■ **Example 10.1.2** For any natural number $n \geq 1$, take $V = \{1, 2, \ldots, n\}$, and define the *complete graph* on n vertices to be the graph K_n with the maximal set $E = \{uv : u, v \in V \text{ and } u \neq v\}$ of edges. At the other extreme, the *trivial graph* E_n on $n \geq 1$ vertices is the graph with no edges.

■ **Example 10.1.3** Given natural numbers m and n where $2m \leq n$, the *Kneser graph* K_n^m has one vertex for each subset of $\{1, \ldots, n\}$ with m elements, where two vertices are joined by an edge if and only if they correspond to disjoint subsets. In particular, K_5^2 is called the *Petersen graph*, and K_6^2 is called the *Grötzsch graph* as illustrated below.

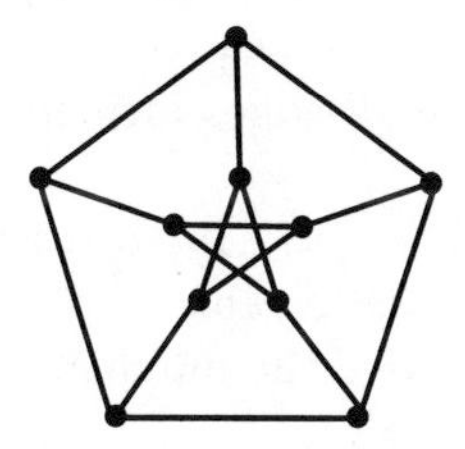

Petersen graph

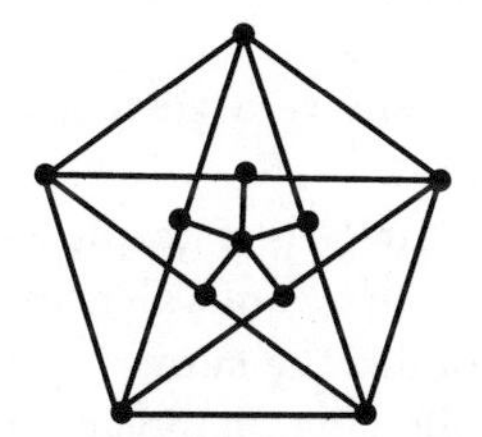

Grötzsch graph

We say $G' = (V', E')$ is a *subgraph* of $G = (V, E)$ and that G is a *supergraph* of G' provided $V' \subseteq V$ and $E' \subseteq E$. A subgraph G' of G is said to be *full* if $E' = E \cap (V')^2$. For instance, any K_n is a full subgraph of K_m provided $n \leq m$, and any graph G other than a complete graph is a non-full subgraph of $K_{v(G)}$.

A set (or a sequence) S of vertices is said to *span* a graph G if $S = V(G)$ (or $\cup S = V(G)$), and likewise a subgraph G' of G *spans* G if $V(G') = V(G)$.

The *(disjoint) union* of graphs G, H is defined to be the common supergraph $G \cup H$ whose vertex set consists of the disjoint union of $V(G)$ and $V(H)$, and the edges $E(G)$ (and $E(H)$ respectively) are taken as edges of $G \cup H$ joining only vertices of the former (and latter) set. For instance, the previous figure illustrates the disjoint union of the Petersen and Grötzsch graphs.

Two graphs are said to be *isomorphic* if there is a bijection between their vertex sets that preserves adjacency, so $G = (V, E)$ is isomorphic to $G' = (V', E')$ if there is some bijection $f : V \to V'$ so that $u, v \in V$ if and only if $f(u)f(v) \in E'$. We shall typically not distinguish between isomorphic graphs in the sequel unless we are considering graphs with some sort of labeling.

The set of vertices adjacent to a given vertex u is called the *link* of u in G, and the *valence* or *degree* of u is the number of elements in its link to be denoted $\nu(u) \geq 0$. A graph is said to be r*-regular* if every vertex has a common valence r, it is said to be simply *regular* if it is r-regular for some r, and in particular, a 3-regular graph is called *cubic*. For instance, the Petersen graph is cubic, and the Grötzsch graph is not regular.

We say that a graph $G' = (V', E')$ arises from $G = (V, E)$ by a *series insertion* if some edge $uv \in E$ is replaced by a pair $uw, wv \in E'$ of edges, where w is a new bivalent vertex, i.e., $\nu(w) = 2$. If a graph G' arises from a graph G by a finite sequence of series insertions, then we say that G' is a *subdivision* of G. Also, define a *series deletion* to be the opposite of a series insertion. Two graphs G_1 and G_2 are said to be *homeomorphic* if there is some sequence of series insertions and/or deletions on G_1 which produce a graph isomorphic to G_2.

The *star* of a vertex is the set of edges incident on it, so the star of a vertex

u consists of the edges joining u to a vertex in the link of u. It will be necessary sometimes to remove edges and vertices from a graph, and we establish the notation that if $E' \subseteq E$, then $G - E'$ denotes the graph $(V, E - E')$ whereas if $V' \subseteq V$, then $G - V'$ denotes the graph

$$\big(V - V', E - \{uv : u \in V' \text{ and } v \text{ is in the link of } u\}\big);$$

thus, when we remove a vertex, we also remove all edges incident on it (so as to produce a graph).

Suppose that G is a graph with vertices $V = \{u_1, \ldots, u_n\}$, and consider the tuple $\nu(u_1), \ldots, \nu(u_n)$ of valences, called the *valence tuple* of G. Since each edge has two endpoints, the sum of the valences is twice the number of edges, that is, we have the following "handshaking lemma".

10.1.1 LEMMA *For any graph G, we have $\sum_{i=1}^{n} \nu(u_i) = 2e(G)$.*

In particular, the sum of degrees is even as must also therefore be the number of vertices of odd valence.

An *(oriented) path* P in the graph $G = (V, E)$ is a sequence $u_0, u_1, \ldots, u_n$ of vertices of G where we demand that $u_i u_{i+1} \in E$ for each $i = 0, 1 \ldots, n-1$. P is said to have *length* $n \geq 1$, u_0 is the *initial point*, u_n the *final point* of P, and we say P runs *from* u_0 to u_n or *joins* u_0 to u_n. P is said to *visit* the vertices $u_0, \ldots, u_n$ and to *traverse* the edges $u_i u_{i+1}$ for $i = 0, \ldots, n-1$. P is said to be *closed* if $u_0 = u_n$, and a closed path is called simply a *cycle*. The path P is said to be *simple* if the vertices $u_0, u_1, \ldots, u_{n-1}$ are all distinct and the vertices $u_1, u_2, \ldots, u_n$ are all distinct (so we allow $u_0 = u_n$ but no other duplications).

In order to point out the vagaries of these definitions, observe that a path P of length zero must be of the form $P = u$ for some vertex u, hence P is a simple cycle (of an especially trivial sort); a path $P = u, v$ of length one describes a oriented edge from u to v in G and cannot be a cycle (since there are no edges of the form uu allowed); a path P of length two either is not closed and corresponds to two adjacent edges of G, or it is closed and reads $P = u, v, u$ for some edge uv of G. Thus, a cycle of length two just traverses a single edge, once in one direction and then in the reverse direction. In particular, cycles of length less than three are called *trivial*, and when we discuss cycles in the sequel, we shall usually assume that they are non-trivial, i.e., have length at least three.

A pair X, Y of paths, cycles, or subgraphs is said to be *edge disjoint* (or *vertex disjoint* respectively) if X and Y traverse or contain no common edge (or visit no common vertex). A set of paths is said to be *independent* if its members are pairwise vertex disjoint except perhaps for their endpoints.

■ **Example 10.1.4** Define the *girth* of a graph G to be the length of a shortest non-trivial simple cycle in G. Define an *n-cage* to be a cubic graph of girth n with the minimal number of vertices. The Petersen graph K_5^2 in Example 10.1.3 is in fact the unique 5-cage, and the unique 6-cage is the following *Heawood graph.*

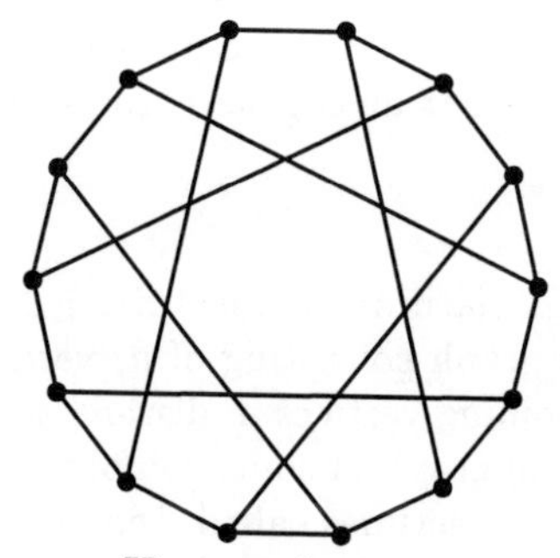

Heawood graph

A graph G is *connected* if for each pair u, v of distinct vertices there is a path in G from u to v. In particular, a connected graph with at least two vertices cannot have any vertices of valence zero. A maximal connected subgraph of G is called a *component* of G, and an edge $x = e$ or a vertex $x = v$ of G is said to be *separating* if $G - \{x\}$ has more components than does G and *non-separating* otherwise. (A separating vertex is often called a *cutvertex*, and a separating edge is often called a *bridge*, but we shall not use this terminology.)

We have the following useful characterization of components.

10.1.2 LEMMA *If u is a vertex of a graph G, then let C denote the set of vertices of G that lie in the component of G containing u. Then there are the following characterizations of C.*

- $C = \{v \in V(G) : \text{there is a path from } u \text{ to } v\}$
- $C = \{v \in V(G) : \text{there is a simple path from } u \text{ to } v\}$
- *Define a binary relation R on $V(G)$, where uRv if and only if $uv \in E(G)$. Let $\overline{R}$ denote the smallest equivalence relation on V containing R. Then C is the $\overline{R}$-equivalence class of u.*

Proof If $P = u_0, u_1, \ldots, u_n$ is a path in G which is not simple, then there must exist $u_j = u_k$ with $0 \le j < k \le n$. We may thus construct a path $u_0, \ldots, u_{j-1}, u_j, u_{k+1}, \ldots u_n$ with the same endpoints which is shorter, and the existence of a simple path thus follows by induction. This proves the equivalence of the first two characterizations, and we leave consideration of the third characterization (which we shall not require below) to the untiring reader. *q.e.d.*

It follows that each vertex lies in a unique component, so every graph is the vertex disjoint union of its components. Furthermore, an edge is separating if and only if it is not contained in a non-trivial simple cycle. Finally, a *forest* is a graph with no non-trivial simple cycles, and a connected forest is a *tree*; we shall study trees in the next section.

A graph G is said to be *bipartite* if there are sets V_1, V_2 so that $V(G) = V_1 \cup V_2$, $V_1 \cap V_2 = \emptyset$, and every edge of G joins a vertex of V_1 to a vertex of V_2. More generally, we say that G is *r-partite* if there are pairwise disjoint sets $V_1, \ldots, V_r$ of vertices whose union is $V(G)$ so that no edge of G joins two vertices in a common set V_j, for any $j = 1, \ldots, r$.

■ **Example 10.1.5** Given natural numbers $n_1, n_2, \ldots, n_r$, we let $K_{n_1,\ldots,n_r}$ denote the *complete r-partite* graph consisting of n_i vertices in the ith set V_i for each i and all possible edges joining vertices in distinct sets. A graph is evidently r-partite if and only if it is a subgraph of some complete r-partite graph. Just as a point of terminology, $K_{3,3}$ is sometimes called *Thomsen's graph* while $K_{3,3}$ and the complete graph K_5 together are called the *Kuratowski graphs.*

Of course, in the definition of graph we have prohibited a *loop*, namely, an edge uu joining a vertex u to itself and have furthermore outlawed the possibility of *multiple edges*, namely, several distinct edges with the same endpoints. Allowing multiple loops and multiple edges in graphs produces an extension of graphs called *multigraphs*, whose analogous theory we shall discuss further at various points below.

Instead of taking unordered pairs of vertices to be edges as for graphs, we might instead take *ordered* pairs to get the notions of *directed graph* (as discussed in Chapter 5) and *directed multigraph* (as discussed in Chapter 9). An ordered pair (u, v) is said to *run from u to v* and is denoted $\vec{uv}$. The constructions and formalisms discussed above for graphs extend in the natural manner to multigraphs, directed graphs, and directed multigraphs.

Notice that if u is a vertex of a directed graph or multigraph G, then there are two different kinds of edges of G incident on u, namely, those that begin at u and those that terminate at u; the number of edges of the former type is called the *outdegree* and of the latter type the *indegree* of u. A directed graph or multigraph is said to be *connected* if the underlying graph or multigraph is connected.

An *orientation* on a graph G is the specification of an orientation $\vec{uv}$ or $\vec{vu}$ for each edge $\{u, v\}$ of G. Thus, an oriented graph is a directed graph in which exactly one of $\vec{uv}$ or $\vec{vu}$ occurs for any $uv \in E$.

■ **Example 10.1.6** There is another natural extension of the notion of graph which leads to a construction of bipartite graphs as follows. A *hypergraph* is a pair (V, E) so that $V \cap E = \emptyset$ as before, but now E is any subset of $\mathcal{P}(V)$ (rather than just a collection of doubletons). Indeed, there is a bijective correspondence between the collection of all hypergraphs and certain bipartite graphs as follows: Given the hypergraph (V, E), construct a bipartite graph with vertices $V \cup E$ and add an edge joining a vertex $u \in V$ to a "hyperedge" $S \in \mathcal{P}(V)$ if and only if $u \in S$. We shall not actually consider hypergraphs again in the sequel.

EXERCISES

10.1 DEFINITIONS AND EXAMPLES

1. (a) Prove that in a graph G there is a set of cycles so that each edge of G belongs to exactly one of the cycles if and only if every vertex has even degree.

(b) Prove that a graph is bipartite if and only if it does not contain a cycle of odd length.

2. (a) Show that $\nu_1 \leq \cdots \leq \nu_n$ is the valence tuple of a tree if and only if $\nu_1 \geq 1$ and $\sum_{i=1}^n \nu_i = 2n - 2$.

(b) Prove that every natural sequence $1 \leq \nu_1 \leq \cdots \leq \nu_n$ satisfying $\sum_{i=1}^n \nu_i = 2(n-k)$ is the valence tuple of a forest with k components.

(c) Characterize the valence tuples of forests.

3. Given a graph G, define an *automorphism* to be an isomorphism as above of G with itself. Prove that isomorphic graphs have identical collections of automorphisms.

4. A particular subdivision $G' = (V', E')$ of a graph or multigraph G is its *barycentric subdivision*, where we perform one series insertion for each edge of G to produce G'; thus, there is exactly one vertex in V' for each vertex of G as well as exactly one vertex in V' for each edge of G. (Compare with Example 10.1.6.)

(a) Prove that G' admits a canonical orientation.

(b) Prove that G' is a graph if G is a multigraph.

5. This problem requires some knowledge of elementary linear algebra. List the vertices of G in some order $u_1, \ldots, u_n$, where $n = v(G)$, to define the *adjacency matrix* $A = A(G) = (a_{ij})$, for $i, j = 1, \ldots, n$ by

$$a_{ij} = \begin{cases} 0, & \text{if } u_i u_j \notin E(G); \\ 1, & \text{if } u_i u_j \in E(G). \end{cases}$$

To define another related matrix associated to a graph G, we must enumerate the edges of G in some order $e_1, \ldots, e_m$, where $m = e(G)$, and furthermore choose an orientation on G to define the *incidence matrix* $B = B(G) = (b_{ij})$ by setting

$$b_{ij} = \begin{cases} 1, & \text{if } u_i \text{ is the final vertex of edge } e_j; \\ 0, & \text{if } u_i \text{ is not an endpoint of } e_j; \\ -1, & \text{if } u_i \text{ is the initial vertex of edge } e_j. \end{cases}$$

Prove that these matrices are related by $BB^t = \Delta - A$, where Δ is the diagonal matrix with i^{th} entry given by the valence $\nu(u_i)$, for $i = 1, \ldots, n$, and B^t denotes the transpose of B.

10.2 TREES

Recall that a cycle is said to be *non-trivial* if it has length at least three, a *forest* is a graph with no non-trivial simple cycles, and a *tree* is a connected forest. Here is an example of a forest with four component trees

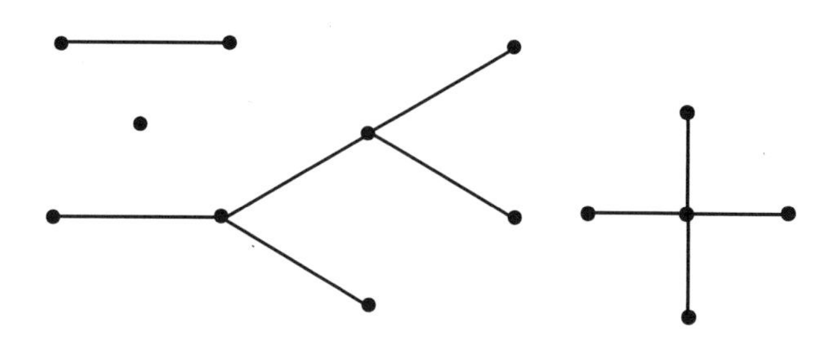

10.2.1 THEOREM *A graph G is a forest if and only if for any pair u, v of distinct vertices of G, there is at most one path from u to v .*

Proof If $u_1, u_2, \ldots u_L, u_{L+1}$ is a cycle in G with $L \geq 2$, then $u_1, u_2, \ldots u_L$ is a path in G from u_1 to u_L, whereas since u_L, u_{L+1} occurs in the cycle, there is also an edge joining u_L and $u_{L+1} = u_1$ giving another path from u_1 to u_L.

For the converse, suppose that $P_1 = u_0, u_1 \ldots, u_L$ and $P_1 = u_0, v_1, \ldots, v_K, u_L$ are two distinct paths in G from u_0 to u_L. Let i denote the least index for which $u_{i+1} \neq v_{i+1}$, and let j denote the least index for which $j \geq i$ and v_{j+1} is a vertex occurring in P_1, say $v_{j+1} = u_k$. Then $u_i, u_{i+1}, \ldots, u_k, v_j, v_{j-1}, \ldots, v_{i+1}$ is a non-trivial simple cycle in G. *q.e.d.*

There are many equivalent characterizations of trees, some of which are discussed in the exercises, and several of which we next describe.

10.2.2 THEOREM *The following properties of a graph G are equivalent.*

(a) *G is a tree.*

(b) *G is a minimal connected graph in the sense that G is connected and for any edge $uv \in E(G)$, $G - \{uv\}$ is disconnected.*

(c) *G is a maximal graph without non-trivial simple cycles in the sense that if u, v are vertices of $G = (V, E)$ which are not adjacent, then $(V, E \cup \{uv\})$ contains a non-trivial simple cycle.*

Proof First, suppose that G is a tree, let uv be an edge of G, and notice that $G - \{uv\}$ cannot contain a simple path from u to v for otherwise G would contain a non-trivial simple cycle. We conclude that $G - \{uv\}$ must be disconnected, so

(a) implies (b). In the same way, if u, v are vertices of the tree G which are not adjacent, then the simple path in G from u to v gives a cycle in $(V, E \cup \{uv\})$, so (a) likewise implies (c).

Next, suppose that G satisfies (b) and yet contains a simple cycle u, w_1, w_2, $\ldots w_k$, v, u. In any path in G, the edge uv can be replaced by the path u, w_1, w_2, $\ldots$, w_k, v, so $G - \{uv\}$ is connected. This contradicts the minimality of G, so (b) implies (a).

Finally, suppose that G satisfies (c). Notice that if vertices u and v lie in distinct components of G, then adding the edge uv to G cannot create a non-trivial simple cycle. We conclude that G must be connected and contain no non-trivial simple cycles, that is, G is a tree, whence (c) implies (a). *q.e.d.*

Recall that we say that a subgraph G' *spans* a graph G if $V(G') = V(G)$, i.e., G' contains every vertex of G. Taking a minimal connected spanning subgraph of a given graph, we conclude

10.2.3 COROLLARY *Every connected graph contains a spanning tree.*

There are many algorithms for constructing spanning trees, and indeed, a given graph typically has many different spanning trees as discussed further in the exercises. Let us give just one such algorithm here. Pick any vertex u of G and let T_1 denote the tree consisting of this single vertex u. Suppose we have inductively constructed trees $T_1 \subseteq T_2 \subseteq \cdots \subseteq T_k \subseteq G$, where tree T_i has i vertices. If $k < n = v(G)$, then by connectedness of G, there is some vertex $v \in V(G) - V(T_k)$ that is adjacent in G to a vertex $w \in T_k$, and we define T_{k+1} by adjoining to T_k the vertex v and the edge vw. Thus, T_{k+1} is connected and since v, w cannot occur in a cycle of T_{k+1}, it furthermore contains no non-trivial simple cycles and is thus a tree. Continuing this process therefore to $T_0 \subseteq \cdots \subseteq T_n$ produces the desired spanning tree T_n.

10.2.4 COROLLARY *A tree with n vertices has $n-1$ edges, and a forest with n vertices and k components has $n-k$ edges.*

Proof First, observe that the construction above produces a spanning tree with $n-1$ vertices since $e(T_0) = 0$ and $e(T_k + 1) = e(T_k) + 1$.

Theorem 10.2.2 shows that every tree has a *unique* spanning tree (namely, itself!) which proves the assertion about trees, from which follows the assertion about forests. *q.e.d.*

10.2.5 COROLLARY *A tree with at least one edge has at least two vertices of valence one.*

Proof Let $\nu_1 \leq \nu_2 \leq \cdots \leq \nu_n$ be the valence tuple of a tree T with at least one edge. Since T is connected, we must have $\nu_1 \geq 1$, so if T had at most one vertex of valence one, then by the handshaking lemma 10.1.1 and the previous result, we find

$$2n - 2 = 2e(T) = \sum_{i=1}^{n} \nu_i \geq 1 + 2(n-1),$$

which is absurd. *q.e.d.*

A vertex of valence one is called an *external node* of the graph G while a vertex of valence greater than one is called an *internal node* of G, so the previous result simply asserts the existence of at least two external nodes in any tree.

In practice, one often chooses, once and for all, a distinguished vertex u_* of a tree T to be called the *root*, and a tree together with a choice of root is a *rooted tree* (T, u_*). In fact, one often also chooses the root from among the external nodes and refers to it as an *external root.* Of course, by definition if T is a tree with root u_* and $u \neq u_*$ is any other vertex of T, then there is a unique simple path $P(u)$ in T from u_* to u. Furthermore, if uv is an edge of T, then $P(u)$ must be a subpath or superpath of $P(v)$ in order to guarantee that T have no non-trivial simple cycles. It follows that the edge uv has a well-defined orientation, that is, a canonical specification of $\vec{uv}$ or $\vec{vu}$ (inherited from its traversal in the longer path among $P(u)$ and $P(v)$), and we have thus proved

10.2.6 THEOREM *There is a canonical orientation on any rooted tree.*

Thus, a rooted tree T determines uniquely the structure of a corresponding directed graph, i.e., a suitable binary relation on $V(T)$.

In a distinctly unbotanical manner, we usually imagine the root of a rooted tree at the top and the specified orientation is given top-to-bottom in a figure drawn in the plane such as

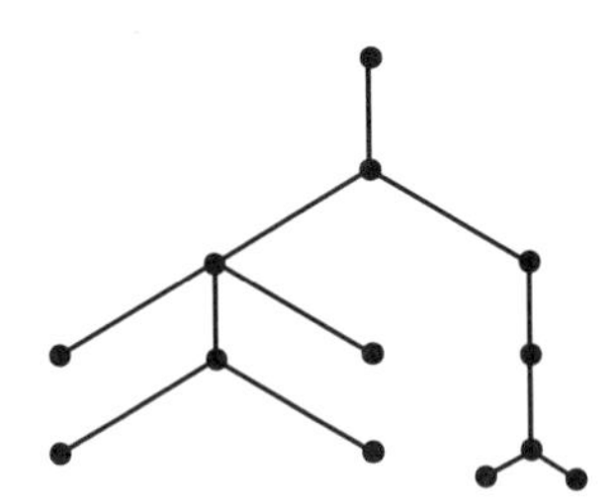

One should imagine a family tree. Indeed, one defines the *generation* of a vertex u to be the length of the simple path $P(u)$ from the root to u; the *offspring* of u (if any) are the vertices v of one larger generation than u so that $\vec{uv}$ is the correctly oriented edge in T.

Upon drawing such a figure in the plane corresponding to the tree T with root u_*, there is linear ordering (from left to right) of points in a given generation, for instance, if a given node has two offspring, then one lies to the right of the other in a figure in the plane. Just to emphasize the point: This distinction between right and left is not part of the structure of a rooted tree, rather it is an artifice of the drawing in the plane.

■ **Example 10.2.1** In a *dyadic* or *binary* rooted tree T there is an internal root, every node either gives rise to no offspring (and so is external) or it gives rise to two offspring, one called "right" and one called "left" (and we make this determination arbitrarily as was stressed above). In the *complete dyadic tree* we require every node to have exactly two offspring. Thus, the complete dyadic rooted tree with four generations is

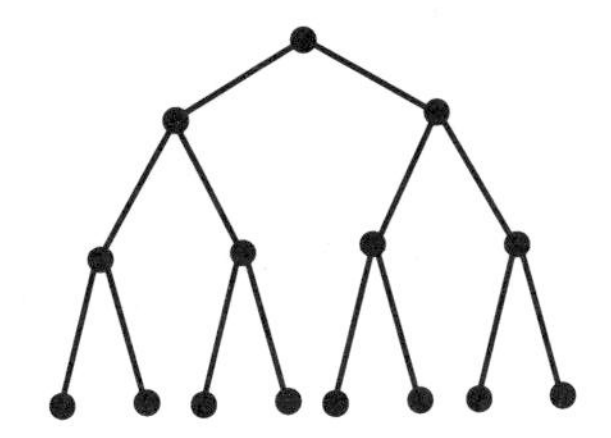

For any node u of T, define the "right" and "left" subtrees (if they exist), respectively, to be the full subtrees of T whose vertices are descendents (i.e., either offspring or descendents of offspring) of the right and left offspring of u. Of course, each right and left subtree is again a dyadic tree in its own right and has its own canonical root.

In fact, there are three natural linear orderings on the vertices in a rooted dyadic tree (depending on whether the root comes first, second, or third) which are useful in practice and are defined recursively as follows.

- *Preorder*
  1. Take the root of T.
  2. If the right subtree exists, take its vertices in preorder.
  3. If the left subtree exists, take its vertices in preorder.
- *Inorder*
  1. If the right subtree exists, take its vertices in inorder.
  2. Take the root of T.
  3. If the left subtree exists, take its vertices in inorder.
- *Postorder*
  1. If the right subtree exists, take its vertices in postorder.
  2. If the left subtree exists, take its vertices in postorder.
  3. Take the root of T.

For example, the nodes of the following binary trees are labeled to indicate the corresponding linear orderings

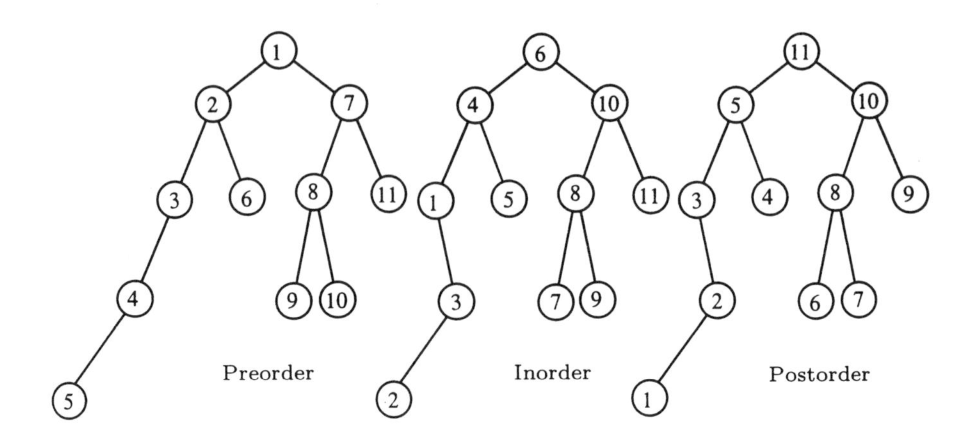

One can likewise define an *n-adic tree* (where every node has either n or zero offspring) as well as the *complete n-adic tree* (where every node has exactly n offspring) for any $n \geq 2$ and devise analogous linear orderings on its vertices.

■ **Example 10.2.2** The *Fibonacci trees* F_n, for $n \geq 0$, are defined recursively by setting $F_0 = F_1$ to be the tree with one vertex, and for $k \geq 2$, the left subtree is the Fibonacci tree F_{k-1}, and the right subtree is the Fibonacci tree F_{k-2}. The Fibonacci tree F_k has $f_{k+1} - 1$ internal and f_{k+1} external nodes, where f_k denotes the kth Fibonacci number, and there is furthermore a useful linear ordering on the internal and external vertices of the Fibonacci tree defined as follows: The labels on the left subtree are those of F_{k-1} while the labels on the right subtree are the labels of the Fibonacci tree F_{k-2} increased by adding f_k. For instance, the Fibonacci tree F_6 and its separate orderings on the internal and external nodes is illustrated below.

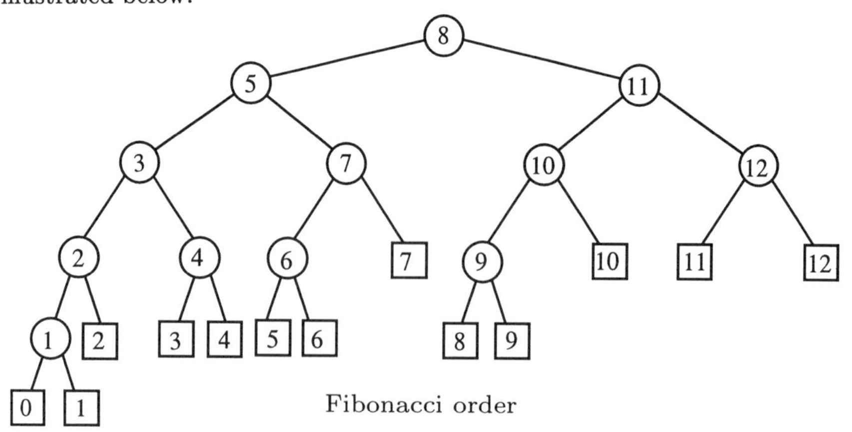

EXERCISES

10.2 TREES

1. (a) Let T_n denote the number of distinct labeled trees on n vertices. Prove that $T_n = \sum_{k=1}^{n-1} k\binom{n-2}{k-1} T_k T_{n-k}$ for each $n \geq 2$.

(b) Derive *Cayley's formula* that $T_n = n^{n-2}$.

2. Let G be a graph with $n = e(G)$ vertices. Show that the following properties of G are equivalent.

(a) G is a tree.

(b) G has $n - 1$ edges and is connected.

(c) G has $n - 1$ edges and no non-trivial simple cycles.

(d) $G = K_n$ if $n = 1, 2$ and otherwise the addition of an edge to G produces exactly one new cycle.

3. A well known problem in optimization theory is as follows: Given a graph $G = (V, E)$ and a nonnegative "cost function" $\phi : E \to \mathbb{R}$, find a connected spanning subgraph $T = (V', E') \subseteq G$ for which the "total cost" $\phi(T) = \sum_{uv \in E'} \phi(uv)$ is minimal. First observe that T must be a tree since otherwise we might remove an edge and still retain a spanning set without increasing the cost. In fact, such a minimal spanning subtree may not be unique, but it always exists; we next present two algorithms for finding it.

• First choose one of the cheapest edges of G and recursively proceed to choose among the cheapest remaining edges subject to the constraint that we may not select all the edges in any non-trivial simple cycle. By Theorem 10.2.2(c), the procedure terminates with a spanning tree.

• Start with the graph G and recursively proceed to delete from it one of the costliest edges which is not separating. By Theorem 10.2.2(b), the procedure terminates with a spanning tree.

(a) Prove that each of the algorithms above indeed produces a minimal spanning subtree.

(b) Prove that if no two edges have the same cost, then there is a unique minimal spanning subtree.

4. This problem requires some knowledge of elementary linear algebra. As was observed in the text, a tree has a unique spanning tree, namely, itself. More generally, the number of spanning trees in a graph G can be computed from the incidence matrix, as we next describe. Choose any spanning tree T in G (which we suppose is not itself a tree), and delete from the incidence matrix $B = B(G)$ (defined in Exercise 10.1.5) a row corresponding to an edge in G which is not in T to get a modified matrix β.

(a) Prove that the modulus of the determinant of an $(n-1) \times (n-1)$ submatrix of β is unity if the edges corresponding to the columns form a tree and zero otherwise.

(b) Prove that the number of spanning trees of G is the determinant $det[\beta\beta^t]$ using the following Cauchy-Binet formula: if K is a $p \times q$ matrix and L is a $q \times p$ matrix, where $p \leq q$, then $detKL = \sum_S detK_S \; detL_S$, where the summation is over all subsets $S \subseteq \{1, \ldots, q\}$ with p elements, K_S is the submatrix of K formed by the columns of K indexed by S, and L_S is the submatrix of L formed by the rows indexed by elements of S.

(†) 10.3 HAMILTONIAN AND EULERIAN PATHS

Suppose that $P = u_0, u_1, \ldots, u_n$ is a path on the graph G, and consider the corresponding sequence $e_1, e_2, \ldots, e_n$ of edges of G, where $e_j = u_{j-1}u_j$, for $j = 1, \ldots, n$.

We say that P is an *Eulerian path* for G if P traverses each edge of G exactly once, i.e., $\{e_1, \cdots, e_n\} = E$ and $n = e(G)$; P is likewise called an *Eulerian cycle* if it is an Eulerian path which is also a cycle.

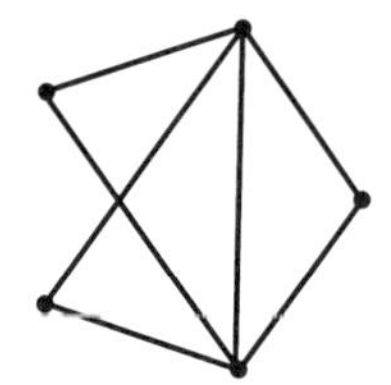

admits Eulerian cycle

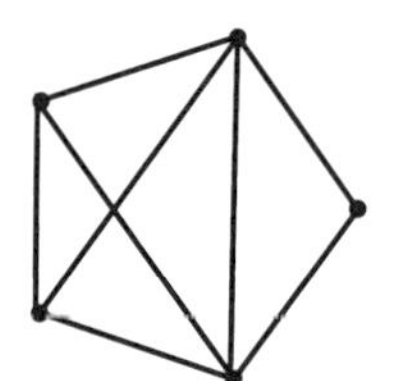

admits Eulerian path

Analogously, we say that P is a *Hamiltonian path* for G if P visits each vertex exactly once, i.e., $\{u_0, \cdots u_n\} = V$ and $n+1 = v(G)$; P is likewise called a *Hamiltonian cycle* if $\{u_1, \cdots u_n\} = V$ and $n = v(G)$. Here are some sample Hamiltonian paths on the complete graph K_6.

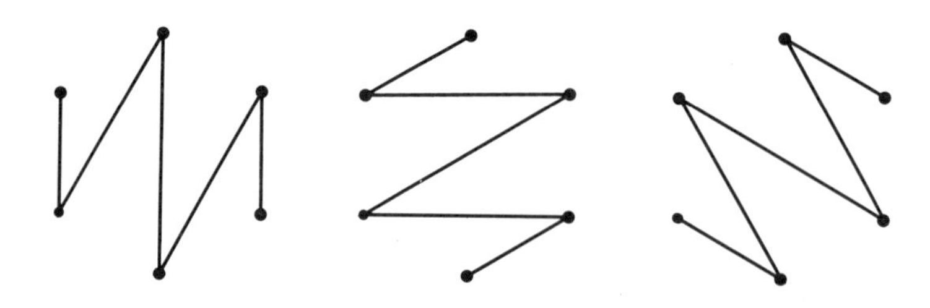

The seeming similarity between these two conditions is illusory, and the characterizations of graphs containing Eulerian paths/cycles as well as algorithms for their construction and other results are far simpler than for Hamiltonian paths/cycles. Indeed, we begin with a complete characterization of graphs containing Eulerian cycles (whereas there is no such simple characterization for Hamiltonian cycles).

10.3.1 THEOREM *A connected graph with at least one edge has an Eulerian cycle if and only if each vertex has even valence. A connected graph admits an Eulerian path from vertex u to vertex $v \neq u$ if and only if u and v are the unique vertices of G of odd valence.*

Proof Suppose first that $P = u_0, u_1, \ldots, u_n$ is an Eulerian path on the graph G and consider the corresponding edges $e_j = u_{j-1}u_j$, for $j = 1, \ldots, n$. Each time P visits a vertex u_j of G, it traverses *two* edges e_j and e_{j+1} except for its initial and final vertices. It follows that the condition above for the existence of an Eulerian path is necessary, and one likewise sees that the condition above for the existence of an Eulerian cycle is also necessary.

As to sufficiency of the first condition, we proceed by induction on the number of edges, and the basis steps of zero or one edges hold trivially. For the induction, suppose that G is a connected graph with $e(G) > 1$ edges in which each vertex has even valence. It follows that the minimum valence of vertices of G is at least two, so G must contain a non-trivial simple cycle by Corollary 10.2.5, and we may therefore take a non-trivial cycle C in G with the maximal number of edges.

We claim that C must be Eulerian, and suppose to the contrary that C contains a vertex u which lies in some component H of

$$G - \{\text{edges of } G \text{ traversed by } C\},$$

where H has at least one edge. Every vertex of H has even valence in H, so by the induction hypothesis, H must contain an Eulerian cycle Z. The cycles C and Z can have no common edges yet must have a common vertex and hence can be combined to produce a longer cycle; this contradicts the maximality of C, proving that C is indeed Eulerian and establishing the first part of the theorem.

Suppose finally that $G = (V, E)$ is connected and has exactly two vertices u and v with odd valence. Build a graph

$$\big(V \cup \{w\}, E \cup \{wu, wv\}\big),$$

which has only vertices of even valence and thus has an Eulerian cycle which gives rise to the desired Eulerian path on G itself. *q.e.d.*

As to Hamiltonian paths/cycles, we begin by showing among other things that the complete graphs always admit Hamiltonian paths. Recall that we say that a collection of paths or cycles on a graph G are *edge disjoint* if the paths or cycles traverse no common edges.

10.3.2 THEOREM *The complete graph K_n for $n \geq 3$ is can be decomposed into edge disjoint Hamiltonian cycles if and only if n is odd. The* *complete graph K_n for $n \geq 2$ can be decomposed into edge disjoint Hamiltonian paths if and only if n is even.*

Proof For the first part, since K_n is $(n-1)$-regular and a Hamiltonian cycle is 2-regular, $n-1$ must be even, i.e., n must be odd, for the existence of a decomposition into cycles.

Assuming then that $n \geq 3$ is odd, we may delete a vertex of K_n in order to see that K_{n-1} is the union of $N = (n-1)/2$ Hamiltonian paths if K_n is the union of N Hamiltonian cycles. As in the example above of Hamiltonian paths on K_6, for odd values of n, K_{n-1} indeed admits the required decomposition. In this decomposition of K_{n-1} into N Hamiltonian paths, each vertex is the endpoint of exactly one such path. Indeed, this holds for each such decomposition since each vertex u of K_{n-1} has odd degree, so at least one such path must terminate at u.

It follows that if we add a vertex u to K_{n-1} to get K_n and complete each Hamiltonian path on it to a Hamiltonian cycle through u, then we obtain the required decomposition of K_n. *q.e.d.*

There is finally the following sufficient condition for a graph to possess a Hamiltonian cycle.

10.3.3 THEOREM *Suppose that G is a graph with $v(G) \geq 3$ so that for every pair u, v of nonadjacent vertices, we have $\nu(u) + \nu(v) \geq v(G)$. Then there is a Hamiltonian cycle on G.*

Proof The proof is by contradiction, and we suppose that G is a graph satisfying the conditions of the theorem with no Hamiltonian cycle, where $v(G) = n$ and $e(G)$ is greatest among all such graphs. If u, v are vertices of G which are not adjacent, then $G \cup \{uv\}$ must itself admit a Hamiltonian cycle, so there must be a Hamiltonian path $P = u_1, u_2, \ldots, u_n$ from $u = u_1$ to $v = u_n$ in G. If u is adjacent to vertex u_j, then v cannot be adjacent to u_{j-1}, for otherwise, the sequence $u_1, u_2, \ldots, u_{j-1}, v, u_{n-1}, \ldots, u_j, u_1$ would determine a forbidden Hamiltonian cycle on G. It follows that for every vertex in G that is adjacent to u, there must be a vertex in G which is not adjacent to v. Thus, we find that $\nu(v)$ is at most $n - 1 - \nu(u)$, which is contrary to hypothesis. *q.e.d.*

EXERCISES

10.3 HAMILTONIAN AND EULERIAN PATHS

1. Define a *tournament* to be a complete *oriented* graph, that is, a directed graph so that for any distinct vertices u and v either there is an edge from u to v or from v to u, but not both. Prove that every tournament contains a directed Hamiltonian path.

2. Prove that a connected digraph contains a directed Eulerian cycle if and only if each vertex has identical indegree and outdegree.

3. Below you will find an illustration of the Pregel River in the ancient Prussian city of Königsberg (now Kaliningrad, Russia) and of its seven bridges.

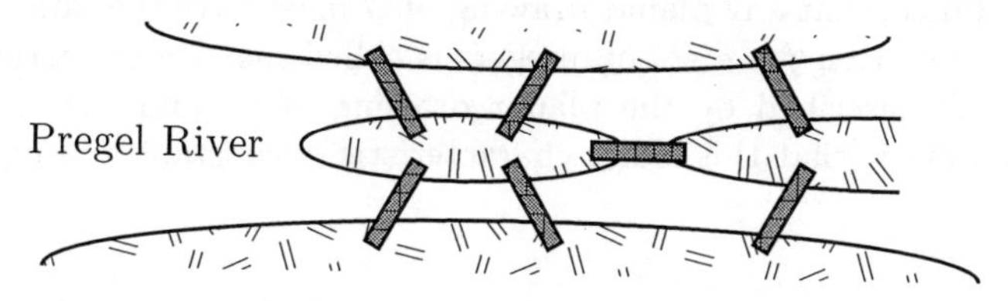

Euler characterized graphs containing what we now call Eulerian cycles and was led to this by the puzzle of planning a walk in Königsberg crossing each bridge once and only once. Solve the puzzle or prove there is no solution.

4. *Fleury's algorithm* for producing an Eulerian path in a graph G repeatedly extends a "partial Eulerian path" T from its terminal vertex v by appending to T any edge incident on v which is not separating in $G - T$ and whose removal would not isolate the initial vertex of T before all the edges of G are traversed. Prove that Fleury's algorithm succeeds in producing an Eulerian path provided that G contains an Eulerian path.

5. Given a graph G, define its *closure* to be the graph obtained from G by adding to its edges additional edges between every pair of nonadjacent vertices u, v where $\nu(u) + \nu(v) \geq v(G)$ (or which become so recursively as the result of adding edges). Prove that a graph admits a Hamiltonian cycle if and only if its closure admits a Hamiltonian cycle.

6. A notoriously difficult problem in optimization theory is the following *traveling salesman problem*: Given a graph G, suppose that $c : E(G) \to \mathbb{R}$ is a "cost function" taking positive values. Consider a path or cycle P visiting each node of G traversing edges $e_1, e_2, \ldots, e_n$, and define the "cost" of P itself to be $c(P) = \sum_{i=1}^{n} c(e_i)$. The general problem is to find a cycle on G which visits each node and minimizes the cost. We say that the function c above is "Euclidean" provided $c(u, v) + c(v, w) \geq c(u, w)$ for any three $u, v, w \in V(G)$. Prove that if the cost function is Euclidean, then a Hamiltonian cycle on G of least cost is automatically a solution to the traveling salesman problem.

(†) 10.4 PLANARITY

A graph G is said to be *planar* if it can be drawn in the plane $G \subseteq \mathbb{R}^2$ in such a way that its edges meet only at vertices. As before, we say two points $x, y \in \mathbb{R}^2 - G$ lie in a common *component* if and only if there is a path in the plane from x to y which is disjoint from G. Each such component of $\mathbb{R}^2 - G$ is called a *face* or *plaquette* of the graph G in the plane.

10.4.1 THEOREM *Let G be a connected planar graph with $v = v(G)$ vertices and $e = e(G)$ edges and suppose that there are f faces in some planar drawing of G. Then $f - e + v = 2$.*

It follows, for instance, that any planar drawing of G must have the same number of faces. This combination $f - e + v$ of integers is called the "Euler characteristic of the polyhedron" described by the planar drawing of G, and the content of the previous theorem is that this Euler characteristic is constant and equal to 2 independent of G.

Proof We proceed by induction on the number f of faces of the drawing of G in the plane, and for the basis step $f = 1$, there can be no cycle on G. Thus, G is a tree, and the result holds by Corollary 10.2.4.

For the induction, suppose that $f > 1$, and let uv be an edge in G contained in some cycle. Since a non-trivial simple cycle separates the plane into an "inside" and "outside", uv must be in the boundary of two different faces, say the faces are F_1, F_2. Removing uv from G produces a planar drawing of a new graph G' with one fewer face than for G, where $v(G') = v(G)$ and $e(G') = e(G) - 1$. The result finally follows from the inductive hypothesis. *q.e.d.*

The reader is invited to prove the result used above (that a non-trivial simple cycle in a planar graph separates the plane into an "inside" and an "outside" of which it is the common boundary) by induction on the length of the cycle under the assumption that the edges are straight line segments (compare Exercise 10.4.1 below). Under progressively weaker assumptions on the edges, one still proves this separation property of non-trivial simple cycles in the plane. The most general version of this result (finally just assuming that the edges are continuous injective images of straight line segments) is a famous result which resisted proof for several decades and is called the "Jordan curve theorem". It was first proved by Veblen in the early part of the 1900's. We shall only require in the sequel simpler versions of this major theorem such as the one above, which the untiring reader may prove directly.

10.4.2 COROLLARY *If G is a planar graph, then $e(G) \leq 3v(G) - 6$.*

Proof Let ϕ_i denote the number of edges of G bounding the ith face of some fixed planar drawing of G with f faces. Each face necessarily contains at least three bounding edges, so

$$\sum_{j=1}^{f} \phi_j \geq 3f.$$

Since each edge lies in the border of exactly two faces, this sum counts each edge twice, so

$$\sum_{j=1}^{f} \phi_j = 2e.$$

The desired inequality finally follows from these inequalities and Euler's formula $f - e + v = 2$. *q.e.d.*

We may apply these results to give examples of nonplanar graphs, as follows.

10.4.3 COROLLARY *The Kuratowski graphs K_5 and $K_{3,3}$ are nonplanar.*

In fact, these graphs are so named because of Kuratowski's remarkable results (whose proofs are elementary but too lengthy to be included here) that a graph is planar if and only if it contains no subgraph homeomorphic to $K_{3,3}$ or K_5.

Proof Nonplanarity of K_5 follows from the previous result since the inequality $10 = e \leq 3v - 6 = 9$ fails.

For $K_{3,3}$ to be planar, there must be $f = 2 + e - v = 5$ faces in a planar drawing by Euler's formula. On the other hand, every cycle in $K_{3,3}$ has length at least four, and we may refine the proof of Corollary 10.4.2 in this case to prove that in fact $f \leq 4$, which is a contradiction. To explain this refinement, notice that if $K_{3,3}$ were planar, then $\sum_{j=1}^{f} \phi_j \geq 4f$ in the notation of Corollary 10.4.2, so $9 = e \geq 2f$, so indeed $f \leq 4$. *q.e.d.*

10.4.4 COROLLARY *In a planar drawing of a graph G, there is a single face if and only if G is a tree.*

Proof Euler's formula $2 = f - e + v$ holds with $f = 1$ if and only if $e(G) = v(G) - 1$, and this is equivalent to G being a tree by Corollary 10.2.4.

q.e.d

Given a planar drawing of a multigraph $G \subseteq \mathbb{R}^2$, define the *(Poincaré) dual* G^* to be the multigraph in the plane whose vertex set is given by the faces G (and we say that a vertex of G^* is "dual" to the corresponding face of G) and where there is one edge of G^* for each edge of G, namely, if the edge $e \in E(G)$ separates components C and D of $\mathbb{R} - G$, then there is a corresponding "dual" edge e^* joining the vertices of G^* indexed by C and D. Similarly, we say that a vertex of G is "dual" to the corresponding face of G^*. For instance, here is an example of a multigraph G^* dual to a planar multigraph $G \subseteq \mathbb{R}^2$

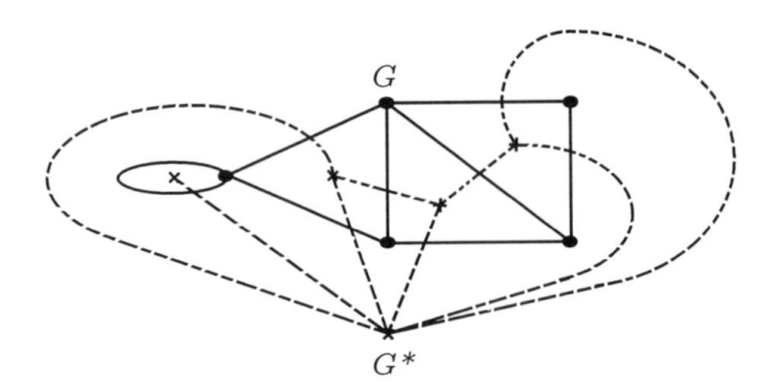

It is worth emphasizing that a graph begets a dual *multi*graph *not* a dual *graph*, and this is the rationale for pursuing this discussion in the context of multigraphs (rather than just graphs) from the outset.

Here is the promised elementary formulation of the deep theorem from topology called Poincaré duality.

10.4.5 THEOREM *Fix any planar graph G and consider its dual G^*. If $T \subseteq G$ is a spanning tree, then let $T' \subseteq G^*$ denote the collection of edges of G^* which are not dual to edges of T^*. Then the assignment $T \mapsto T'$ establishes a bijection between the set of all spanning trees of G and the set of all spanning trees of G^*. In particular, G and G^* have the same number of spanning trees.*

Proof If $T \subseteq G$ and $T' \subseteq G^*$ are as above, then we first claim that T' can contain no non-trivial simple cycle. Indeed, if there were such a cycle, then it must separate the plane into two components by (our simple version of) the Jordan curve theorem; vertices of G inside this cycle could not be joined to vertices outside by a path in T, and this contradicts that T is a spanning tree of G.

We next claim that T' must furthermore be connected (and hence be a tree), for if not, say $T' = T_1 \cup T_2$ with $T_1 \cap T_2 = \emptyset$. Let F denote the union of the faces of G whose dual vertex lies in T_1, so V cannot be the entire plane (since no vertex of T_2 can belong to V). V therefore has a non-empty boundary B consisting of certain edges of T, each having a face in V on one side and a face not in V on the other. There can therefore be no vertex of valence one in B, so B cannot be a tree by Corollary 10.2.5 and thus must contain a simple cycle in contradiction to the previous paragraph.

It follows that $T' \subseteq G^*$ is indeed a tree, and arguing as above, one sees that T' is in fact a *spanning* tree of G^*. The assignment $T \mapsto T'$ is obviously an injection, and it remains to prove only that it maps onto the set of spanning trees of G^*. To this end, if $\tau \subseteq G^*$ is a spanning tree, then let $T \subseteq G$ denote the set of edges of G which are not dual to edges of τ *and* which are not loops. Arguing as above, one finds that T is a spanning tree of G and $T \mapsto \tau$ (i.e., $\tau = T'$), completing the proof. *q.e.d*

EXERCISES

10.4 PLANARITY

1. Prove that every planar graph has a drawing in the plane in which every edge is a straight line segment. [HINT: Argue by induction on the number of vertices of a maximal planar graph by removing a suitable vertex.]

2. Prove that a planar graph with n vertices and girth at least $g \geq 3$ has at least $\max\{n-1, g(n-2)/(g-2)\}$ edges.

3. Prove that there are only five regular planar graphs in which each face has the same number of bounding edges.

4. Suppose that G is a planar graph with no separating vertices. Prove that G is bipartite if and only if the dual G^* of G is Eulerian.

5. We say that multigraph G' arises from a multigraph G by *contracting* the edge uv of G if we delete uv as an edge and identify its endpoints u and v to form a new single vertex w as in

where we furthermore require that any edge of G other than uv which is incident on either u or v is incident on w in G'. The reverse operation from G' to G is called an *expansion*. We say that G_1 and G_2 are *related (by contractions/expansions)* if there is a sequence of contractions and expansions carrying G_1 to G_2.

Define the *Euler characteristic* of a connected graph G to be $\chi(G) = v(G) - e(G)$. It is the purpose of this exercise to sketch a proof that $\chi(G) = \chi(G')$ for two graphs G, G' if and only if G and G' are related by contractions/expansions.

(a) Show that we may assume without loss that G and G' are cubic.

(b) Define a *rose* R to be a graph with only one vertex. Show that any connected graph is related by contractions/expansions to a rose. Conclude that we may assume without loss that G and G' are planar.

(c) Define a *triangulation* of a polygon P to be a decomposition of P into triangles any two of which meet (if at all) either at a common vertex or along a common edge. Consider a triangulation of a polygon in the plane as a planar graph G, so its dual is a cubic tree T. Each edge of G separates two triangles which together form a quadrilateral; show that a contraction/expansion of an edge of T corresponds to removing an edge of G separating two triangles and replacing it by the other diagonal of the corresponding quadrilateral as in

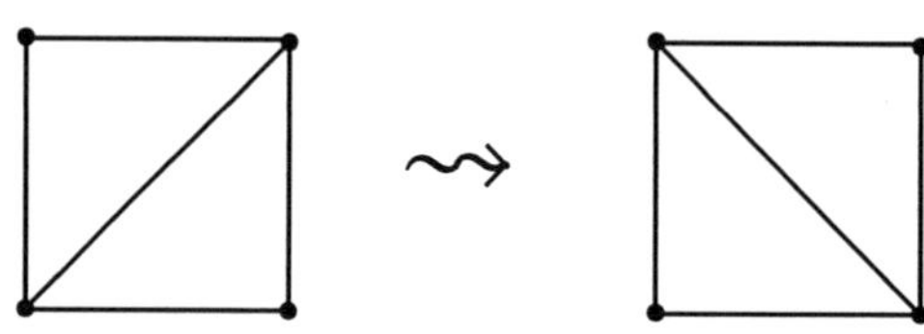

Such a modification of G is called a *Whitehead move.*

(d) Prove that given any two triangulations of a fixed polygon P, there is a sequence of Whitehead moves carrying one triangulation to the other. [HINT: Argue by double induction, where we first induct on the number of sides of the polygon, and if two triangulations τ_1, τ_2 of P do not share a common edge, then we further induct on the minimum over all edges e of τ_1 of the number of points of $e \cap \tau_2$.]

(e) Derive the asserted result that two graphs have the same Euler characteristic if and only if they are related by contractions/expansions.

(†) *10.5 CONNECTIVITY, MATCHING, AND COLORING*

This section is dedicated to a survey of various combinatorial results in graph theory. Owing to the enormity of this subject, our discussion here can only scratch the surface, and our aim is simply to survey the central ideas and basic results in several directions.

If G is a connected graph and S is a set of edges or vertices, then we say that S *separates* G if $G - S$ is disconnected. We furthermore say that S separates two vertices u, v of G if u and v belong to different components of $G - S$. G is said to be k*-connected* for some $k \geq 2$ if either G is the complete graph K_{k+1} on $k+1$ vertices or else it has more than $k+1$ vertices and no set of $k-1$ vertices separates it. Likewise, G is k*-edge-connected* if it has at least two vertices and no set of at most $k-1$ edges separates it. The maximal value of k for which a connected graph is k-connected is called its *connectivity* and is denoted $k(G)$, where we set $k(G) = 0$ for a disconnected graph G by convention. We likewise define the *edge-connectivity* $\ell(G)$ of a graph G.

Notice that a graph is 2-connected if and only if it is connected, contains no separating vertex, and has at least three vertices. Analogously, a graph is 2-edge-connected if and only if it is connected, contains no separating edge, and has at least two vertices. For some further examples, we have $k(K_n) = \ell(K_n) = n - 1$ for complete graphs and $k(K_{m,n}) = \ell(K_{m,n}) = m$ for complete bipartite graphs.

On the other hand, if G is obtained from two disjoint copies G_1, G_2 of K_m by adding one new vertex v together with an edge from v to each vertex of G_1 or G_2, then $k(G) = 1$ (since v is a separating vertex) while $\ell(G) = m$. Furthermore, we have by definition that

$$k(G) - 1 \leq k(G - \{v\}), \quad \text{and } \ell(G) - 1 \leq \ell(G - \{uv\}) \leq \ell(G),$$

for any vertex v and edge uv of G.

Recall that two paths from u to v are said to be *independent* if they have only the vertices u and v in common. We have the following celebrated theorem of Menger from the 1920's.

10.5.1 THEOREM [Menger] *Let u and v be distinct vertices of a graph.*

(a) *If u and v are not adjacent, then the maximal number of independent paths from u to v equals the minimal number of vertices separating u from v.*

(b) *The maximal number of edge disjoint paths from u to v equals the minimal number of edges separating u from v.*

Proof Let $\kappa = \kappa(u, v)$ denote the minimal number of vertices which separate u from v. It is clear that there are then at most κ independent paths from u to v, and for $k \leq 1$, there are indeed k independent paths from u to v.

Suppose to get a contradiction that the theorem fails, let $\kappa \geq 2$ be a least value for which there is a counterexample to the theorem, and let G be a counterexample (for this minimal value of κ) with the least number of edges. There can be at most $\kappa - 1$ independent paths from u to v, and no vertex w of G can be joined to both u and v, for otherwise, $G - \{w\}$ would be a counterexample with lesser $\kappa(u, v) = \kappa - 1$.

Let S be any set of κ vertices which separate u from v and suppose that neither u nor v is adjacent to each vertex in S. Let G' be obtained from G by replacing the component of $G - S$ containing u by a single vertex u' which is then joined to each vertex in S. There are still κ vertices required to separate u' from v in G', and G' has fewer edges than G since the component of $G - S$ we collapsed to form u' had at least one edge. Since G is by assumption a counterexample with a minimal number of edges, there must be κ independent paths from u' to v in G'. Any two subpaths of these κ paths from v to S must have only the vertex v in common, and thus, for every $w \in S$, exactly one of these subpaths joins v to w.

We may perform the analogous construction on G using the vertex v instead of u to again find κ paths from u to S and combine these two collections of paths in the natural way to give κ independent paths from u to v, in contradiction to our hypotheses. It follows that for any set S of κ vertices separating u from v, either u or v must be adjacent to each vertex of S.

Finally, let $u, w_1, \ldots, w_\ell, v$ be a shortest path from u to v in G, so we must have $\ell \geq 2$. By minimality of G, there is a set S_0 of $\kappa - 1$ vertices in $G - \{w_1 w_2\}$

separating u from v. Thus, both $S_1 = S_0 \cup \{w_1\}$ and $S_2 = S_0 \cup \{w_2\}$ are sets with κ elements separating u from v in G. Since v is not joined to w_1, u must be joined to every vertex in S_2, and this contradicts that there is a vertex which is adjacent to both u and v, for every vertex in S_0 is adjacent to both u and v, and S_0 contains $\kappa - 1 \geq 1$ vertices. *q.e.d.*

There is the following immediate corollary giving interesting necessary and sufficient conditions for k-connectivity and k-edge-connectivity of a graph.

10.5.2 COROLLARY *A graph G is k-connected for $k \geq 2$ if and only if $v(G) \geq 2$ and any two vertices can be joined by k independent paths. A graph G is k-edge-connected for $k \geq 2$ if and only if $v(G) \geq 2$ and any two vertices can be joined by k edge disjoint paths.*

Given a family $\mathcal{S} = \{S_1, S_2, \ldots, S_n\}$ of subsets of some set X, we next turn to the question of whether it is possible to choose n *distinct* elements of X, one from each set S_i. A set $\{x_1, \ldots, x_n\}$ with these properties (namely, $x_i \in S_i$ for each i and $x_i \neq x_j$ for all $i \neq j$) is called a *set of distinct representatives* of $\mathcal{S}$. The collection $\mathcal{S}$ naturally gives rise to a bipartite graph G with two vertex classes $V_1 = \mathcal{S}$ and $V_2 = X$ where $S_i \in \mathcal{S}$ is joined to every $x \in X$ contained in S_i. A system of distinct representatives is thus given by a set of m independent edges (so each vertex in V_1 has one incident such edge). In this case, we also say that there is a *complete matching* from V_1 to V_2.

A traditional formulation of this is called the *marriage problem*: Given n boys and m girls, when is it possible to marry all the girls provided we demand that a girl must be married to a boy that she knows? There is the obvious necessary condition that if there are k girls who know among them only at most $k-1$ boys, then there is no solution. In other words, if there is a complete matching from V_1 to V_2, then for every subset $T \subseteq V_1$ with t elements, there must be at least t vertices of V_2 adjacent to a vertex in T.

In fact, the following result from the 1930's, called "Hall's Marriage Theorem", asserts that this obvious necessary condition is in fact sufficient.

10.5.3 THEOREM [Hall] *Suppose G is a bipartite graph with vertex sets V_1, V_2. Then G admits a complete matching from V_1 to V_2 if and only if for every subset $T \subseteq V_1$ with t elements, there must be at least t vertices of V_2 adjacent to any vertex in T.*

Proof The proof is by induction on the number m of girls, and for $m = 1$, the condition is certainly sufficient.

Assuming now by induction that $m \geq 2$, if any k girls for $1 \leq k \leq m$ know at least $k+1$ boys, then we can marry off one boy/girl couple arbitrarily so that the remaining unmarried boys and girls still satisfy the required condition, and the remaining $m-1$ girls can then be married by induction.

We therefore suppose that for some $k < m$, there are k girls who taken together know exactly k boys. These girls can evidently be married by induction, and the remaining girls can also be married by induction provided they satisfy the required condition (where we must ignore the boys who have already been married). Finally, the required condition must indeed be satisfied since otherwise if some ℓ girls to be married know fewer than ℓ remaining boys, then these girls along with the first k girls would then know fewer than $k + \ell$ boys. *q.e.d*

The following result is the direct reformulation of Hall's Marriage Theorem in terms of distinct representatives.

10.5.4 THEOREM *A collection $\mathcal{S} = \{S_1, \ldots, S_m\}$ of sets admits a set of distinct representatives if and only if for every $I \subseteq \{1, \ldots, m\}$ with t elements, $\cup_{i \in I} S_i$ also has at least t elements.*

We turn finally to colorings of graphs. Our first topic here touches upon a collection of related theorems in various branches of mathematics which go under the rubric of "Ramsey Theorems". Ramsey worked also in the 1930's, and he died in his late twenties as an already accomplished mathematician and philosopher.

We consider partitions of the set of *edges* $E(G)$ of a graph G and refer to a partition with $r \geq 1$ classes as an *r-edge-coloring* of G. As is traditional for $r = 2$, we think of the two blocks of the partition as colors "red" and "blue" on the edges of the graph, and a subgraph is said to be simply "red" (or "blue" respectively) if all of its edges are red (or blue).

Define the *Ramsey number* $R(s,t)$ to be the minimum n for which every red/blue coloring of K_n contains a red K_s or blue K_t, where we assume that $s,t \geq 2$, and every K_1 is both blue and red (since it has no edges).

It is not immediately clear even that $R(s,t)$ is finite for every s,t, but at least we can conclude

$$R(s,t) = R(t,s) \text{ and } R(s,2) = R(2,s) = s, \text{ for every } s,t \geq 2,$$

since in a red/blue coloring of K_s either every edge is red or there is a blue edge.

We next prove that the Ramsey numbers are indeed finite by giving upper bounds on them.

10.5.5 THEOREM [Ramsey] *Provided $s,t > 2$, we have $R(s,t) \leq R(s-1,t) + R(s,t-1)$, and thus $R(s,t) \leq \binom{s+t-2}{s-1}$.*

Proof For the first part, we may assume by induction that $R(s-1,t)$ and $R(s,t-1)$ are finite and let $n = R(s-1,t)+R(s,t-1)$. Consider a red/blue coloring of K_n and let u be a vertex of K_n. Since $\nu(u) = n-1 = R(s-1,t)+R(s,t-1)-1$, either there are at least $n_1 = R(s-1,t)$ red edges incident on u or there are at least $n_2 = R(s,t-1)$ blue edges incident on u. By symmetry of the two cases, we may assume without loss of generality that the first case holds, and consider a full subgraph K_{n_1} of K_n which is spanned by n_1 vertices joined to u by red edges. If K_{n_1} has a blue K_t, then the argument is complete, and otherwise, K_{n_1} contains a red K_{s-1} which together with u forms a red K_s proving the first part.

The second part holds if $t = 2$ or $s = 2$, and indeed, we find equality in this case since $R(s, 2) = R(2, s) = s$. Assume that $s, t > 2$ and inductively that the result holds for every s_1, t_1 with $s_1, t_1 \geq 2$ and $s_1 + t_1 < s + t$. According to the first part, we have

$$\begin{aligned} R(s,t) &\leq R(s-1,t) + R(s,t-1) \\ &\leq \binom{s+t-3}{s-2} + \binom{s+t-3}{s-1} \\ &= \binom{s+t-2}{s-1}. \end{aligned}$$

q.e.d.

There is the following extension of the previous result to r-edge-colorings for $r > 2$: Given r and natural numbers $s_1, \ldots, s_r$, if n is sufficiently large, then for every r-edge-coloring of the complete graph K_n, there must be some K_{s_i} with color i. We prove this by induction on r by replacing the first two colors by a new color. For n sufficiently large, there must be a K_{s_i} with color i for some $i = 3, \ldots, k$, or there must be a K_m, where $m = R(s_1, s_2)$ with the new color (i.e., colored with the first two original colors). The theorem follows in the former case, while in the latter, for $i = 1, 2$, there must be a K_{s_i} in K_m with color i.

In fact, very few of the Ramsey numbers are actually known explicitly (and even estimates of them are hard to come by) even for $r = 2$. One can check easily that $R(3, 3) = 6$ and can verify with further elementary enumerative work that $R(3, 4) = 9$, $R(3, 5) = 14$, $R(3, 6) = 18$, and $R(4, 4) = 18$ for instance.

Turning from this glimpse of Ramsey theory to our final discussion of what are usually referred to as coloring problems on graphs, we now wish to partition the *vertices* of a graph G by coloring them as before but now also subject to the constraint that if two vertices are adjacent, then they are required to have different colors. Such a partition on $V(G)$ is called a *(vertex) coloring* of G, it is an *r-coloring* if there are r colors, and the least r required for G to admit an r-coloring is called the *chromatic number* $\kappa(G)$ of G. For instance, $\kappa(G) \geq 2$ if and only if G contains an edge. Furthermore, we have seen (in Exercise 10.1.1b) that a graph is bipartite if and only if it does not admit an odd simple cycle, so $\kappa(G) \geq 3$ if and only if G contains an odd cycle. Similarly simple characterizations of graphs with a fixed chromatic number $\kappa \geq 4$ are not known, but there are, for instance, the following trivial estimates on chromatic numbers: If there are not $h + 1$ independent vertices in G, then $\kappa(G) \geq v(G)/h$, and if G contains K_n as a subgraph, then $\kappa(G) \geq n$.

(We should mention that the traditional edge coloring problem for graphs is not the Ramsey-type coloring described before but rather as follows: An "edge coloring" is a partition of the edges so that no two adjacent edges have the same color. We shall not further discuss these edge colorings here and stick to vertex colorings in the sequel.)

Here is the so-called *greedy algorithm* for coloring a graph G, which depends upon a linear ordering $u_1, \ldots, u_n$ on $V(G)$: Color u_1 with color 1, then color u_2 with color 1 *unless* $u_1 u_2 \in E(G)$, in which case we color u_2 with color 2; continue in this way using the least color possible at each stage. With a bad initial linear ordering on $V(G)$, the greedy algorithm can be very inefficient (i.e., use far more than $\kappa(G)$ colors); on the other hand, it is easy to see that there is a linear ordering on $V(G)$ so that its greedy algorithm uses precisely $\kappa(G)$ colors.

10.5.6 THEOREM *Define $L = \max_H \min_{u \in V(H)} \nu_H(u)$, where the max is over all full subgraphs H of G and $\nu_H(u)$ denotes the valence of u in H. Then, $\kappa(G) \leq L + 1$.*

Proof Let u_n denote a vertex of degree at most k and set $H_{n-1} = G - \{u_n\}$. By hypothesis, H_{n-1} has a vertex of valence at most k, so we may choose such a vertex u_{n-1} and set $H_{n-2} = H_{n-1} - \{u_{n-1}\}$. Continuing in this manner, we enumerate all the vertices. By construction of the ordering $u_1, \ldots, u_n$, each u_j is adjacent to at most L vertices before it, so the greedy algorithm requires at most $L + 1$ colors. *q.e.d*

Another algorithm for coloring arises from the following "reduction" of a given graph G into graphs G', G''. Suppose that $u, v \in V(G)$ are not adjacent; let G' denote the graph obtained from G by adding the edge uv, and let G'' denote the graph obtained from G by identifying the vertices u and v. Thus, there is a new vertex "uv" in G'' (arising from the coalescence of u and v), and uv is adjacent to the vertices which are adjacent to u or v in G, as in

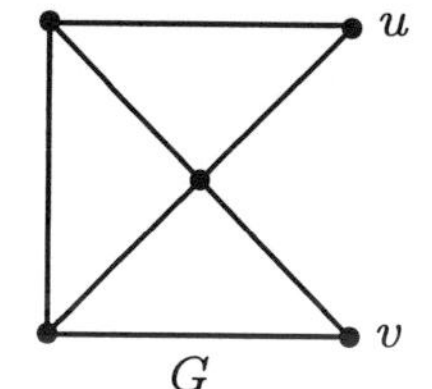

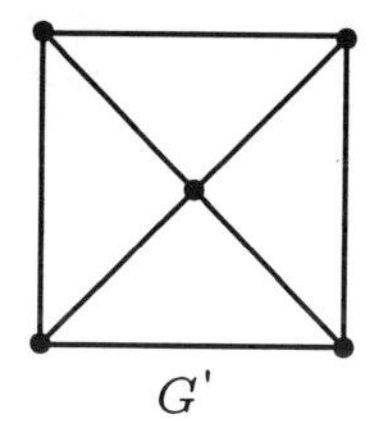

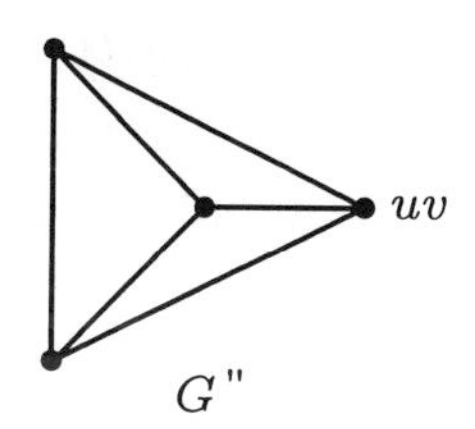

It is clear that the collection of colorings of G in which u and v get different colors is in natural bijection with the set of colorings of G' while the collection of colorings of G in which u and v get the same color is in natural bijection with the set of colorings of G''.

Given a natural number n and a graph H, define $\chi_H(n)$ to be the number of colorings of H with n colors. It follows from the remarks above that

$$\chi_G(n) = \chi_{G'}(n) + \chi_{G''}(n),$$

and since $\kappa(G)$ is the least n for which $\chi_G(n) \geq 1$, we conclude

$$\kappa(G) = \min \{\kappa(G'), \kappa(G'')\}.$$

The coloring algorithm based on the reduction of G to G', G'' depends upon constructing a sequence $G = G_0, G_1, \ldots$ as follows. Given G_i, if it is complete, then terminate the sequence with it, and otherwise take G_{i+1} to be either G'_i or G''_i. The sequence must terminate in a complete graph G_n say with m vertices. An m coloring of G_n naturally gives rise to an m coloring of G, so $\kappa(G) \leq m$. Indeed, since $\kappa(G) = \min \{\kappa(G'), \kappa(G'')\}$, we conclude that $\kappa(G)$ is the minimum number of vertices of a complete graph in which such a sequence $G_0, G_1, \ldots$ can terminate.

10.5.7 THEOREM *Suppose that H is a graph with $n \geq 1$ vertices, m edges, and k components. Then*

$$\chi_H(t) = \sum_{i=0}^{n-k} (-1)^i c_i t^{n-i},$$

where $c_0 = 1$, $c_1 = m$, and $c_i > 0$ for every $i \leq n-k$.

Proof We proceed by induction on $m+n$, and the basis step $m+n=1$ is trivial. Proceeding to the induction step, if $m=0$, then again the theorem follows since $k=n$ and every function from $V(H)$ to $\{1,2,\ldots,t\}$ is a coloring of H; we conclude that $\chi_H(t) = t^n$. For $m > 0$, pick two adjacent vertices u, v of H and set $G = H - \{uv\}$, so $H = G'$. Since $e(G) = m-1$ and $v(G'') + e(G'') \leq n-1+m$, the induction hypothesis assures us of the result for G' and G''.

Now, G'' has exactly k components while G has at least k components, hence

$$\chi_G(t) = t^n - (m-1)t^{n-1} + \sum_{i=2}^{n-k} (-1)^i b_i x^{n-i},$$

where $b_i \geq 0$ for all i, and furthermore

$$\chi_{G''}(t) = t^{n-1} - \sum_{i=2}^{n-k} (-1)^i a_i x^{n-i},$$

where $a_i > 0$ for each i. Since $\chi_G = \chi_{G'} + \chi_{G''}$, we find

$$\begin{aligned}\chi_H(t) = \chi_{G'}(t) &= \chi_G(t) - \chi_{G''}(t) \\ &= t^n - mt^{n-1} + \sum_{i=2}^{n-k} (-1)^i (a_i + b_i) t^{n-i} \\ &= t^n - mt^{n-1} + \sum_{i=2}^{n-k} (-1)^i c_i t^{n-i},\end{aligned}$$

where $c_i > 0$ for each i. *q.e.d.*

In light of the previous result, $\chi_H(t)$ is called the *chromatic polynomial* of the graph H, and it figures prominently in various enumerative problems which will not be taken up here. Actually, as a point of terminology, we say that G arises from G' by "deletion" and that G'' arises from G by "contraction". The elegant method of proof illustrated in the previous theorem is a paradigm for families of arguments in graph theory called the "deletion/contraction" method.

EXERCISES

10.5 CONNECTIVITY, MATCHING, AND COLORING

1. Let $(P, \leq)$ be a poset and recall that a *chain* in P is a linearly ordered subset of P and an *antichain* is a subset A of P so that for no $x, y \in A$ do we have $x < y$. Prove *Dilworth's Theorem* that if every antichain in a finite poset P has at most n elements, then P is the union of n chains.

2. Define an $r \times s$ *Latin rectangle on* $\overline{n} = \{1, \ldots, n\}$ to be an $r \times s$ array $M = (m_{ij})$, where each element of $\overline{n}$ occurs in each row and column of M at most once.

(a) Prove that a $r \times n$ Latin rectangle can be extended to an $n \times n$ Latin square. [HINT: Suppose that $r < n$, and extend M to an $(r+1) \times n$ Latin rectangle. Then define $M_j = \{k \in \overline{n} : k \neq m_{ij}\}$ and prove that $\{M_j : j \in \overline{n}\}$ has a set of distinct representatives.]

(b) Let M be an $r \times s$ Latin rectangle, and define $\#_M(i)$ to be the number of times i occurs in M. Show that M can be extended to an $n \times n$ Latin square if and only if $\#_M(i) \geq r + s - n$ for every $i \in \overline{n}$.

3. Prove the following two extensions of the marriage theorem (where we first ask how close we can come to successful matchmaking if the hypotheses of Hall's theorem fail, and we then consider a polygamous society where a boy can marry several girls).

(a) Suppose that the bipartite graph G with vertex sets V_1, V_2 has m edges and there are at least $s - d$ distinct vertices of G adjacent to a subset $S \subseteq V_1$ with s elements. Then G contains $m - d$ independent edges. [HINT: Add d new vertices to V_2 joining them to each vertex of V_1.]

(b) Suppose that G is a bipartite graph with vertex classes $V_1 = \{u_1, \ldots, u_m\}$ and $V_2 = \{v_1, \ldots, v_n\}$. Then G contains a subgraph H such that the valence $\nu_H(u_i)$ of u_i in H satisfies $\nu_H(u_i) = d_i$, for $i = 1, \ldots, m$, and $0 \leq \nu_H(v_j) \leq 1$, for $j = 1, \ldots, n$, if and only if for every subset $S \subseteq V_1$, there are at least $\sum_{u_i \in S} d_i$ distinct vertices in G adjacent to S. [HINT: Replace each vertex u_i by d_i vertices each of which is joined to each vertex in the link of u_i for each $i = 1, \ldots, m$.]

4. Prove that the first part of Theorem 10.5.7 holds also for hypergraphs, that is, for 2-colorings of the set $\{1, \ldots, n\}^r$ of r-tuples in $\{1, \ldots, n\}$. (This is a theorem originally proved by Ramsey.)

5. A famous recent exhaustive computer proof of a result with a rich history dating back to the 1850's is the *four-color theorem* that any planar graph is 4 colorable. *Tait's Conjecture* from the 1890's says that every 3-connected cubic plane graph has a Hamiltonian cycle.

(a) Show that Tait's conjecture implies the four-color theorem.

(b) Show that the following graph, which is called *Tutte's Counter-Example*, disproves Tait's Conjecture.

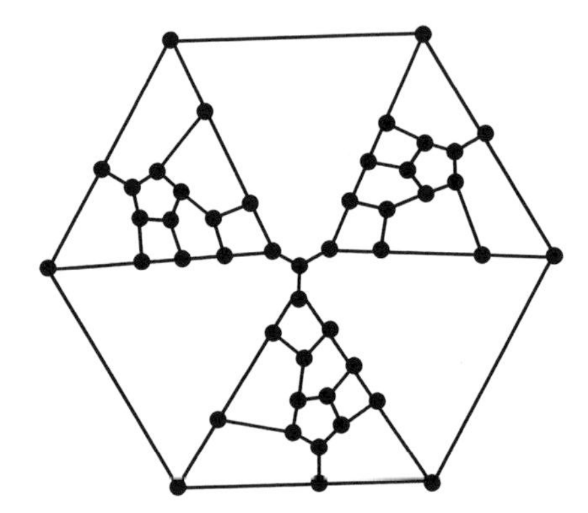

6. Prove that the Kneser graph K_n^m defined in Example 10.1.1 has chromatic number $\kappa(K_n^m) = n - 2m + 2$.

7. Let $\chi_G(t) = \sum_{i=0}^{n}(-1)^i c_i x^{n-i}$ be the chromatic polynomial of G. Show that $c_2 = \binom{e(G)}{2} - t(G)$, where $t(G)$ denotes the number of distinct simple cycles of length 3 in G.

Suggestions for Further Reading

Proofs

G. Polya, *How to Solve It*, Doubleday Anchor, 1957.

Set Theory and Cardinality

J. W. Dauben, *Georg Cantor*, Harvard University Press, 1979.

P. R. Halmos, *Naive Set Theory*, Springer-Verlag, 1974.

K. Kuratowski, *Introduction to Set Theory and Topology*, Pergamon Press, 1961 .

E. Nagel and J. R. Newman, *Gödel's Proof*, New York University Press, 1958.

G. Takeuti and W. M. Zaring, *Introduction to Axiomatic Set Theory*, Springer-Verlag, 1971.

Number Theory

G. H. Hardy and E. M. Wright, *An Introduction to the Theory of Numbers*, Oxford University Press, 1979.

I. Niven, H. S. Zuckerman, H. L. Montgomery, *An Introduction to the Theory of Numbers*, John Wiley and Sons, 1991.

Combinatorics

V. Bryant, *Aspects of Combinatorics*, Cambridge University Press, 1993.

D. E. Knuth, *The Art of Computer Programming*, Addison-Wesley, 1973.

L. Lovász, *Combinatorial Problems and Exercises*, North-Holland, 1993.

G. Szego, *Orthogonal Polynomials*, American Math Society, 1975.

Languages and Machines

T. L. Booth, *Sequential Machines and Automata Theory*, John Wiley and Sons, 1967.

J. Hartmanis and R. E. Stearns, *Algebraic Structure Theory of Sequential Machines*, Prentice-Hall, 1966.

A. Salomaa, *Jewels of Formal Language Theory*, Computer Science Press, 1981.

Graphs

N. Biggs, *Algebraic Graph Theory*, Cambridge University Press, 1993.

B. Bollobás, *Graph Theory*, Springer-Verlag, 1979.

J. A. McHugh, *Algorithmic Graph Theory*, Prentice-Hall, 1990.